Tumor Suppressor Genes in Human Cancer

CANCER DRUG DISCOVERY AND DEVELOPMENT
Beverly A. Teicher, Series Editor

TUMOR SUPPRESSOR GENES IN HUMAN CANCER

Edited by

DAVID E. FISHER, MD, PhD

Dana Farber Cancer Institute,
Harvard Medical School,
Boston, MA

HUMANA PRESS
TOTOWA, NEW JERSEY

Cover illustration: The cover shows the histologic appearance of a Wilm's tumor stained with hematoxylin and eosin (purple) or immunostained for the Wilm's tumor suppressor protein (black and white inset). Cover photos were generously provided by Scott Granter, MD, Department of Pathology, Brigham and Women's Hospital, Boston, MA.

Cover design by Patricia F. Cleary.

Due diligence has been taken by the publishers, editors, and authors of this book to ensure the accuracy of the information published and to describe generally accepted practices. The contributors herein have carefully checked to ensure that the drug selections and dosages set forth in this text are accurate in accord with the standards accepted at the time of publication. Notwithstanding, as new research, changes in government regulations, and knowledge from clinical experience relating to drug therapy and drug reactions constantly occurs, the reader is advised to check the product information provided by the manufacturer of each drug for any change in dosages or for additional warnings and contraindications. This is of utmost importance when the recommended drug herein is a new or infrequently used drug. It is the responsibility of the health care provider to ascertain the Food and Drug Administration status of each drug or device used in their clinical practice. The publisher, editors, and authors are not responsible for errors or omissions or for any consequences from the application of the information presented in this book and make no warranty, express or implied, with respect to the contents in this publication.
For additional copies, pricing for bulk purchases, and/or information about other Humana titles, contact Humana at the above address or at any of the following numbers: Tel.: 973-256-1699; Fax: 973-256-8341; E-mail: humana@humanapr.com or visit our Website: http://humanapress.com

This publication is printed on acid-free paper. ∞
ANSI Z39.48-1984 (American National Standards Institute) Permanence of Paper for Printed Library Materials.

Photocopy Authorization Policy:
Authorization to photocopy items for internal or personal use, or the internal or personal use of specific clients, is granted by Humana Press Inc., provided that the base fee of US $8.00 per copy, plus US $00.25 per page, is paid directly to the Copyright Clearance Center at 222 Rosewood Drive, Danvers, MA 01923. For those organizations that have been granted a photocopy license from the CCC, a separate system of payment has been arranged and is acceptable to Humana Press Inc. The fee code for users of the Transactional Reporting Service is: [0-89603-807-6/01 $8.00 + $00.25].

Printed in the United States of America. 10 9 8 7 6 5 4 3 2 1

Library of Congress Cataloging-in-Publication Data

Tumor suppressor genes in human cancer / edited by David E. Fisher.
    p.; cm.--(Cancer drug discovery and development)
  Includes bibliographical references and index.
  ISBN 0-89603-807-6 (alk. paper)
  1. Antioncogenes. I. Fisher, David E. (David Erich) II. Series.
  [DNLM: 1. Genes, Suppressor, Tumor. 2. neoplasms--genetics. QZ 202 T9248 2000]
  RC268.43 .T8634 2000
  61699'4042--dc21

                                                  00-027593
                                                  CIP

PREFACE

The war against cancer has seen enormous successes, but also painful frustration. While major conceptual breakthroughs have been made in our understanding of how cell proliferation is regulated, the translation of this information into effective treatment discoveries has lagged terribly behind. Modern molecular oncology has begun to inch closer to treatment-related questions because the pathways under study are now known to regulate events such as cell death, the precise goal of cancer treatment. Because tumor suppressor biology has not yet translated into a therapy-oriented discipline, the purpose of *Tumor Suppressor Genes in Human Cancer* is to present a view of the current field which simultaneously highlights the clinically relevant directions which have already emerged while stimulating the discovery of new ones.

Through the detailed presentation of tumor suppressor genes with a molecular biological and genetic perspective, two paradigms emerge: 1) a finite number of discrete pathways exist into which tumor suppressors and dominant oncogenes reside and 2) cancer biology rests heavily on both regulators of cell proliferation and cell death. In the current climate of informatics, genomics, and molecularly driven drug discovery, cancer research holds greater promise than ever. *Tumor Suppressor Genes in Human Cancer* first sets the stage by presenting the background of systems for the study of tumor suppressor genes as well as the fields of apoptotic cell death and cancer drug discovery. The second section of *Tumor Suppressor Genes in Human Cancer* proceeds to present detailed analyses of major tumor suppressors and, most important, the pathways into which they fit. The intended audience is the student of cancer biology, from those engaged in graduate or medical education to clinicians or drug development professionals seeking to understand the context of cancer cell biology and its promise for therapeutic gains in the coming years.

The concept that individual genes underlie the biology of malignant transformation hearkens back to the early 1900s with the discovery by Peyton Rous of avian sarcomas that were caused by infectious viruses. Many decades later, the identification of the *Src* oncogene placed into focus the notion of the dominant oncogene, a factor whose inappropriate activation confers cellular changes associated with malignant transformation. Alfred Knudsen predicted the existence of a second class of oncogenes whose contribution to cancer is recessively inherited. His hypothesis was based upon clinical observations of cancer risk in familial cancer inheritance patterns and the notion that disease predisposition may represent a multi-hit phenomenon with loss-of-function mutations contributing to the malignant phenotype. Thus the concept of tumor suppressor gene was born and has been abundantly validated by observations that span bench to bedside.

The most striking validation of the tumor suppressor concept comes from the discovery of inactivating mutations or deletions of candidate genes in cancer prone families. Originally discovered for retinoblastoma, the list has been dramatically extended to include p53, p16/Ink4a/ARF, the NF family, DNA mismatch repair genes, Wilms, von Hippel Lindau, Fanconi Anemia, and other genes. In these cases heterozygous germline disruption of a single allele is associated with cancer predisposition in affected family members. Loss of heterozygosity is frequently observed at the genetic locus within tumors that develop in affected individuals. Mechanisms for tumor suppressor inactivation are diverse and are still being discovered today. For example, in addition to traditional loss of function mutations or deletions, the more recently appreciated inactivating mechanisms include transcriptional silencing (e.g., p16/Ink4a), targeted protein degradation (e.g., p53), and functional disruption of tumor suppressing gene activities (e.g., bcl-2 or Mdm2). These diverse mechanisms of tumor suppressor inactivation highlight one of the most striking breakthroughs in cancer biology, the discovery of discrete pathways in which dominantly acting and tumor suppressing genes converge.

The functional convergence of dominant oncogenes and tumor suppressor genes in cellular growth or survival pathways represents such a powerful clue in the puzzle of carcinogenesis that the ability to fit into a known growth regulatory pathway has become a virtual requirement for a gene's acceptance as a true cancer modifier. Moreover much of the data defining these interactions stemmed from the convergence of clinically derived questions with more basic laboratory science. For example, the retinoblastoma tumor suppressor was found to be targeted by multiple dominant oncogenes discovered through the analysis of animal DNA tumor viruses. The existence of additional interactions between tumor suppressors and dominant oncogenes has cemented the notion that key cellular pathways produce the phenotypes associated with cancer, and the homeostatic regulators of these pathways are potent and common targets of carcinogenic disruption.

The pathways that tumor suppressor genes modulate in cancer have been found to cluster around regulation of the cell cycle, cell death, growth factor signaling, DNA damage responses, and other stress responses. Nearly all tumor suppressors are thought to act through modulation of one (or several) of these pathways. Cell cycle regulation has been the traditional pathway thought to be targeted in the etiology of cancer. More recent observations have added a dramatic new dimension to this view in suggesting that cell survival pathways exist as distinct, genetically selected entities and may profoundly influence behaviors we associate with malignancy. Either dysregulated growth or inefficient death (or both) are strongly associated with tumorigenesis. The current revolution in molecular oncology has been fueled largely by the ability to place individual cancer genes within such functional pathways of known importance. Perhaps more important, these functional classifications have in some cases led to investigations that relate more than ever before to cancer treatment.

The study of cancer cell death carries with it the hope of intervening in the very same process for therapeutic benefit. Rarely has a field of fundamental basic science become so mainstream in biologic inquiry while simultaneously focusing on questions of direct therapeutic importance. The interface between research on cell death and clinical ramifications of that work is well illustrated by the actions of tumor suppressor genes, many of which are now recognized to regulate cell survival.

Tumor Suppressor Genes in Human Cancer is not an attempt to fully synthesize cancer biology and treatment, since the field has not arrived at a stage where such a synthesis is yet possible. However the convergence of new technologies suggests that the coming years will begin to see treatment discoveries more directly interface with basic research. Genomics and systematic gene expression technologies will provide thorough catalogs of information whose discovery currently occupies substantial research effort. Linkage of these catalogs to clinical data including treatment responses (pharmacogenomics) promises to dramatically alter drug discovery and treatment design. The pillar of this revolution is the basic biology of disease, and tumor suppressor genes lie at the core of that pillar.

David E. Fisher

CONTENTS

CONTRIBUTORS

MARGARET ASHCROFT, PHD • *ABL Basic Research Program, NCI-FCRDC, Frederick, MD*

ANDRÉ BERNARDS, PHD • *Massachusetts General Hospital Cancer Center and Harvard Medical School, Charlestown, MA*

SRIKUMAR P. CHELLAPPAN, PHD • *Department of Pathology, College of Physicians and Surgeons, Columbia University, New York, NY*

ALAN D. D'ANDREA, MD • *Division of Pediatric Oncology, Dana Farber Cancer Institute, and Department of Pediatrics, Harvard Medical School, Boston, MA*

JINYAN DU • *Division of Pediatric Hematology and Oncology and Graduate Program in Biological and Biomedical Sciences, Harvard Medical School, Children's Hospital, and Dana Farber Cancer Institute, Boston, MA*

DAVID E. FISHER, MD, PHD • *Division of Pediatric Hematology and Oncology and Graduate Program in Biological and Biomedical Sciences, Harvard Medical School, Children's Hospital, and Dana Farber Cancer Institute, Boston, MA*

IRENE GARCIA-HIGUERA, PHD • *Division of Pediatric Oncology, Dana Farber Cancer Institute, and Department of Pediatrics, Harvard Medical School, Boston, MA*

LIYA GU, PHD • *Department of Pathology and Laboratory Medicine, University of Kentucky, Lexington, KY*

J. WADE HARPER, PHD • *Verna and Marrs McLean Department of Biochemistry, Baylor College of Medicine, Houston, TX*

MELANIE T. HARTSOUGH, PHD • *Women's Cancers Section, Laboratory of Pathology, Division of Clinical Sciences, National Cancer Institute, Bethesda, MD*

NICK D. HASTIE, PHD • *MRC Human Genetics Unit, Western General Hospital, Edinburgh, UK*

OTHON ILIOPOULOS, MD • *Dana Farber Cancer Institute and Harvard Medical School, Boston, MA*

WILLIAM G. KAELIN, JR., MD • *Howard Hughes Medical Institute, Boston, MA*

ALEXANDER KAMB, PHD • *Arcaris, Salt Lake City, UT*

JORDAN A. KREIDBERG, MD, PHD • *Department of Medicine, Children's Hospital, and Department of Pediatrics, Harvard Medical School, Boston , MA*

YANAN KUANG, PHD • *Division of Pediatric Oncology, Dana Farber Cancer Institute, and Department of Pediatrics, Harvard Medical School, Boston, MA*

GUO-MIN LI, PHD • *Department of Pathology and Laboratory Medicine, Markey Cancer Center, and the Graduate Center for Toxicology, and Multidisciplinary Ph.D. Program in Nutritional Sciences, University of Kentucky, Lexington, KY*

ALEX MATTER, MD • *Oncology Research, Novartis Pharma AG, Basel, Switzerland*

ANDREA I. MCCLATCHEY, PHD • *Massachusetts General Hospital Cancer Center and Harvard Medical School, Charlestown, MA*

KEN MCCORMACK, PHD • *Arcaris, Salt Lake City, UT*

SCOTT MCCULLOCH, PHD • *Graduate Center for Toxicology, University of Kentucky, Lexington, KY*

ASWIN L. MENKE, PHD • *MRC Human Genetics Unit, Western General Hospital, Edinburgh, UK*

THOMAS A. NATOLI, PHD • *Department of Medicine, Children's Hospital, and Department of Pediatrics, Harvard Medical School, Boston, MA*

ERIC NISBET-BROWN, MD, PHD • *Division of Pediatric Oncology, Dana Farber Cancer Institute, and Department of Pediatrics, Harvard Medical School, Boston, MA*

TAOUFIK OUATAS, PHD • *Women's Cancers Section, Laboratory of Pathology, Division of Clinical Sciences, National Cancer Institute, Bethesda, MD*

ANNA SAVOIA, MD • *Division of Pediatric Oncology, Dana Farber Cancer Institute, and Department of Pediatrics, Harvard Medical School, Boston, MA*

WILLIAM R. SELLERS, MD • *The Departments of Adult Oncology and Internal Medicine, Dana Farber Cancer Institute and Brigham and Women's Hospital, Harvard Medical School, Boston, MA*

PATRICIA S. STEEG, PHD • *Women's Cancers Section, Laboratory of Pathology, Division of Clinical Sciences, National Cancer Institute, Bethesda, MD*

LILIA STEPANOVA, PHD • *Verna and Marrs McLean Department of Biochemistry, Baylor College of Medicine, Houston, TX*

KAREN H. VOUSDEN, PHD • *ABL Basic Research Program, NCI-FCRDC, Frederick, MD*

JEN JEN YEH, MD • *The Departments of Adult Oncology and Internal Medicine, Dana Farber Cancer Institute and Brigham and Women's Hospital, Harvard Medical School, and the Department of Surgery, Boston University Medical Center, Boston, MA*

I

ANALYSIS AND CLINICAL IMPLICATIONS OF TUMOR SUPPRESSOR GENES

1

Animal Models for Tumor Suppressor Genes

Jordan A. Kreidberg, MD, PhD and
Thomas A. Natoli, PhD

CONTENTS

1. INTRODUCTION

The observation that bilateral retinoblastoma (RB) occurred with earlier onset than unilateral disease led Knudson to formulate the two-hit theory for tumorigenesis *(1)*. This theory provided the underpinning for the search for tumor suppressor genes (TSGs). In 1986–1987, the *Rb* gene was identified, appearing to fulfill the criteria of a TSG for RB, and therefore was among the early candidates for gene-targeting experiments in mice *(2–4)*. Since the initial development of gene-targeting technology, several hundred genes have been mutated, including many of the currently defined TSGs.

This chapter summarizes results obtained from animal models for TSGs, and discusses how these models have improved understanding of tumor development. Most of the results dealt with in this chapter are derived from mice carrying targeted deletions, and therefore a brief summary of this technology is appropriate. Readers desiring a

From: *Tumor Suppressor Genes in Human Cancer*
Edited by: D. E. Fisher © Humana Press Inc., Totowa, NJ

more comprehensive summary should consult one of several recent reviews of this area *(5–9)*.

The goal of gene targeting is to introduce specific mutations into the germ line of mice, so that heterozygous and/or homozygous mice can be derived. These experiments make use of embryonic stem (ES) cells, which are carried as cell lines that can be introduced into preimplantation mouse blastocysts, and contribute to embryonic tissue. ES cell lines were originally derived from the inner cell mass of the preimplantation mouse blastocyst; this is the group of cells that are totipotent and give rise to the entire embryo after implantation. As described below, it is possible to target mutations to specific genes in ES cells, then use mutated ES cells to derive mutant mice. The general approach to homologous recombination is shown and described in Fig. 1. Once the ES cells carrying the desired mutation are obtained, they are microinjected into mouse blastocysts, where they recombine with the inner cell mass, and contribute to the resultant mouse (termed a chimera, because of its dual origin). Chimeric mice whose germ cells are derived from ES cells are able to transmit the mutation to their offspring, which are then true heterozygotes for the mutation. Male and female heterozygotes can be mated, and homozygous mutants, if viable, can be obtained. Among the several hundred reported experiences with gene targeting, the full range of expected results has been observed, from phenotypes that are apparent in heterozygotes, to animals that are apparently normal as homozygous mutants. Mutations that result in observable phenotypes in homozygous animals range from those that cause lethality very early in embryogenesis, to those that result in no morphological impairment, but cause functional impairment after birth.

In this chapter, the emphasis is placed on those TSGs that have thus far received the most intense study. A comprehensive table (Table 1) of TSGs studied in mice or rats includes those not more fully discussed in the text. New TSGs are continually identified, and many of these have not been studied in animal models beyond their initial identification. Therefore, these genes will be given less attention, or are not discussed. Included are some emerging technologies aimed at identifying novel TSGs.

2. P53

p53 is the most commonly mutated TSG in human tumors, and therefore was an obvious candidate for which to obtain an animal model. As discussed in greater depth in Chapter of this volume, the p53 protein appears to function mostly as a transcription factor involved in the regulation of the progression through the cell cycle and entry into cell death pathways, particularly in response to DNA damage. One might expect that a protein involved in such basic activities important to all cell types would be essential for normal development, and that embryos unable to produce it might arrest early in development. In the event, the opposite was observed upon targeting the *p53* gene *(10–12)*. At least some portion of *p53–/–* mice are developmentally normal, although defects are observed, as is discussed below. Despite developing normally, all *p53* homozygous mutant mice succumb to tumors by about 9 mo of age *(10,12)*. It is generally assumed that most, if not all, genes that are found within the genome are present because of evolutionary selection against their loss. Therefore, when a mouse is found to survive and to reproduce without the function of a particular gene, it is usually suggested that the function of the gene product is redundant or overlapping with another

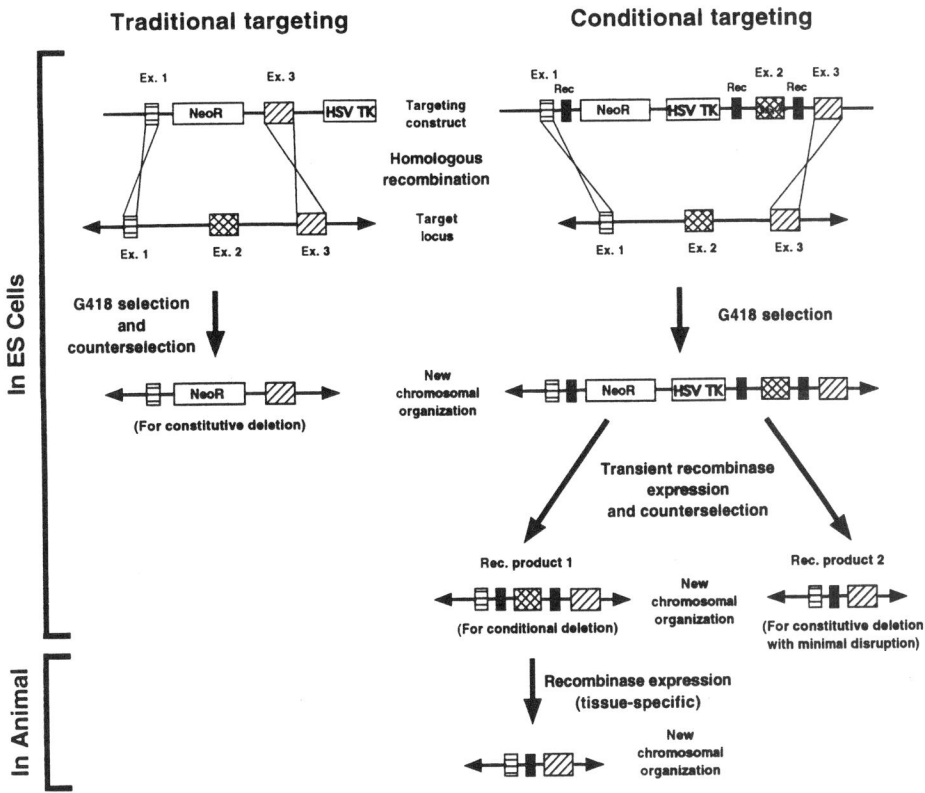

Fig. 1. Schematic representation of traditional and conditional gene-targeting approaches. Traditional targeting: Vectors for traditional gene targeting are designed with an antibiotic resistance marker (NeoR), introducing an interruption or deletion at the site to be mutated (in this figure, exon 2 is to be deleted). The *NeoR* gene is flanked upstream and downstream by DNA sequences homologous to those bordering to be mutated. The herpes virus thymidine kinase gene *(HSVTK)* is placed outside this region of homology. Upon transfection of ES cells with this vector, homologous recombination between the vector and genomic DNA inserts the *NeoR* gene, but not the *HSVTK* gene; random integration is more likely to insert both genes. ES cells are electroporated with the targeting construct, and selected with G418. G418 is used to select ES cells that have incorporated the *NeoR* gene, and gancyclovir is used to counterselect against those that have incorporated the *HSVTK* gene. Conditional targeting: Vectors for conditional targeting are designed with site-specific recombinase recognition sites *(lox* sites or Frt sites, indicated by black boxes labeled "Rec") flanking both the region to be deleted and the selective markers. After transfection of ES cells, G418-resistant colonies are screened for homologous recombination (generating the indicated chr organization). Cells from properly targeted colonies are transiently transfected with the recombinase gene *(cre* or *flp),* and gancyclovir is used to select cells that have lost the *neo* resistance and *HSVTK* genes, because site-specific recombination. There are two potential products of this reaction. Site-specific recombination between the left and center sites removes only the selective markers, leaving the exon to be deleted flanked by recombinase sites. Resultant mice from these ES cells are used for conditional mutations, in which expression of the recombinase under the control of a tissue-specific promoter allows cell-type-specific deletion of the exon. Recombination between the left and right sites deletes both the selectable markers and the exon. This type of deletion can be useful when it is desirable to eliminate a single exon, while otherwise maintaining expression of the gene. Because recombination sites are much smaller than the *Neo* gene, (e.g., 34 bp vs 2.5 kb), it is less likely that a single recombination site will interfere with gene expression.

Table 1

Phenotypes of Animals with Mutations in TSGs

Gene	Heterozygote phenotype	Homozygote phenotype	Ref(s).
APC	Intestinal adenomas	Embryonic lethal before E11	(74,76,139)
ATM	Normal	Normal development; lymphoma, radiation sensitivity	(67,68)
E-Cadherin	Normal	Embryonic lethal at E4.5	(140,141)
BRCA1	Normal	Embryonic lethal at E7.7–E13	(111,112)
BRCA2	Normal	Embryonic lethal at E8.5–9.5; lymphomas (truncation mutation)	(111,113,115,116)
E2F-1	Normal	Reproductive tract sarcomas lung adenocarcinomas, lymphomas	(49)
$G\alpha_{i2}$	Normal	Normal development; ulcerative colitis, colon adenocarcinoma	(142)
α-inhibin	Normal	Normal development; gonadal tumors	(143)
Mgmt	Normal	Normal development; lymphomas, adenomas induced with MNU	(107,144)
Mlh1	Normal	Defective spermatogenesis; DNA repair defects, intestinal adenomas, adenocarcinoma, lymphoma	(96,145)
Msh2	Normal	Lymphoma intestinal neoplasia in older mice	(93,95)
Msh6	Normal	Normal development; GI tumors, lymphoma	(100)
Nf1	Adrenal tumors in older mice	Embryonic lethal around E13.5 with heart defects	(146)
Nf2	Range of metastatic tumors in older mice	Embryonic lethal at E6.5–7	(147,148)
$p16^{Ink4a}$–$p19^{ARF}$	Normal	Normal development; sarcomas and lymphomas	(66)
$p21^{Cip1/Waf1}$	Normal	Normal development Cell cycle defects in vitro	(57)
$p27^{Kip1}$	Tumors after irradiation	Organ hyperplasia tumors after irradiation	(61–63)
p53	Sarcomas and lymphomas in older mice	Normal development or neural tube defects; lymphomas and sarcomas	(10,12,13)
$p57^{Kip}$	Normal or same as –/– (imprinted gene)	Neonatal lethal with areas of hyperplasia, renal dyspasia	(64)
p107	Normal	Normal; dysplastic retinas in p107–/–Rb+/–	(52)
p130	Normal	Normal; skeletal defects in p130/p107–/–	(51)
Pms2	Normal	Normal development; DNA repair defects, lymphomas and sarcomas	(96,97)
Rb1	Pituitary tumors	Embryonic lethal at E14–15 with defective hematopoeis and neurogenesis	(41,42)
Smad3	Normal	Small size; metastatic colorectal tumors	(92)
Smad4	Intestinal tumors in Apc+/– compound heterozygotes	Embryonic lethal at E7	(91)
Tsc2	(Eker rat) renal cell carcinoma	Embryonic lethal around E13	(149)
Vhl	Normal	Embryonic lethal at E9.5–10.5	(128)
WT1	Normal	Embryonic lethal E13–14 No kidneys or gonads	(65)
Xpa	Normal	Normal development; UV-induced skin tumors	(101,103)
Xpc	Normal	Normal development; UV-induced skin tumors	(102)

gene, or that some other gene is able to functionally compensate in its absence. However, in the case of the *p53* gene, there is probably not the need to invoke either of these explanations. Even though *p53–/–* mice do progress through embryogenesis, and in some cases are able to reproduce, the uniform development of tumors by several months of age implies that the TSG function of *p53* offered sufficient evolutionary advantage to maintain this gene.

Though a portion of *p53*-mutant mice were born without developmental defects, studies of large numbers of these homozygous mutants revealed that 23% of female *p53–/–* embryos failed to achieve normal closure of their neural tube, and developed anencephaly, leading to embryonic or neonatal death *(13)*. This phenotype was to some degree influenced by the genetic background of the mice. The reason for the female preponderance in the anencephaly phenotype is not yet understood, although a similar female predisposition to neural tube defects is seen in humans. Other defects observed in homozygous mutant embryos included dental and ocular anomalies.

A spectrum of tumor types is observed in *p53–/–* mice, although lymphomas account for nearly 75% of observed tumors *(10,12)*. Other tumors observed include hemangiosarcomas, rhabdomyosarcomas, fibrosarcomas, anaplastic sarcomas, and teratomas. Consistent with the original Knudson two-hit hypothesis for the development of tumors, mice heterozygous for the *p53* mutation develop tumors at later points, and, additionally, the spectrum is somewhat different *(10,12)*. Instead of the great predominance of lymphomas, the most frequently observed tumors are osteosarcomas, lymphomas, hemangiosarcomas, brain tumors, and rhabdomyosarcomas, as well as numerous other tumors at lower frequencies. Also consistent with the two-hit hypothesis, tumor cells from heterozygous mice have in many cases lost the remaining wild-type (wt) allele.

It is of obvious interest to compare *p53*-mutant mice to human individuals with the Li-Fraumeni syndrome, who are heterozygous for constitutional mutations of *p53* *(14)*. These individuals predominantly have sarcomas and tumors of the brain and breast. Therefore, there are some similarities, but not complete overlap, between *p53* mice and the Li-Fraumeni syndrome. It is not unexpected that humans and mice will differ in the spectrum of tumors that develop, even in response to similar mutations in identical oncogenes. Mice, in general, tend to develop a different spectrum of tumors than humans, with sarcomas being more frequent than carcinomas: The opposite is true in humans. This variability presumably results from differences in genetic background that exist between species, or even within noninbred individuals of a particular species. Examples of such differences may be slight but significant polymorphisms in protein structure, or differences in gene expression patterns.

An analysis of the *p53* mutation in different strains of mice reveals the effect of genetic background on tumor development *(15,16)*. It is well known that different strains of inbred mice are predisposed to distinct types of tumors or leukemias/lymphomas. This is probably related to their complements of endogenous retroviruses and oncogenes, and other genes that interact with them. Donehower et al. *(15)* compared survival rates and the spectrum of tumors in strains 129/Sv inbred and C57B1/6 X 129/Sv mixed-inbred mice, and observed some differences in these two groups. Survival was slightly poorer in 129-inbred mice, both in homozygous and heterozygous mutant mice. For example, they noted prominent difference in the frequency of malignant teratomas in the testicles of male mice. These tumors were frequent in 129-inbred males, infrequent in the mixed genetic background, and were not observed in wt 129 males with intact *p53* genes. This demonstrates that 129 males are predisposed to testicular tumors when *p53* mutations are present, and it will be of great interest to eventually determine the responsible gene(s) that are presumably polymorphic between strains 129 and C57, and responsible for this difference. Genetic background also affected the types of lymphomas that arose in *p53*-mutant mice. Lymphomas in *p53*–/– mice on a mixed 129Sv/C57BL/6 background are predominantly of CD4+CD8+ T-cell origin; pre-B-lymphomas made up about 13% of lymphomas observed in *p53*–/– C.B.17-C57BL/6 mice. The recent accomplishments of the mouse genome project *(17)*, in providing a detailed genetic map of the mouse, now make the identification of these modifier genes a reasonable goal.

  p53-mutant mice have become a valuable tool for experimentally approaching the role of this gene in tumorigenesis, as evidenced by numerous, recently published studies *(18–21)*. For example, *p53* homozygous mutant mice succumbed more quickly to chronic treatment with a carcinogen, dimethylnitrosamine, presumably because of a decreased ability of cells in treated animals to respond appropriately to carcinogen-induced DNA damage *(19)*. However, the absence of *p53* did not result in an increased accumulation of point mutations in a transgene target, either in untreated or mutagen-treated mice, suggesting that increased sensitivity to carcinogen in the abovementioned study may reflect DNA damage other than point mutations *(22)*. *p53*-mutant mice were also used to study the effect of the *p53* gene dosage in a model for chemically induced skin tumors. Topical treatment with dimenthylbenzanthracene (DMBA) resulted in initiation of similar numbers of papillomas in wt, heterozygous, or homozygous mutant mice *(23)*. In contrast, progression of these lesions to malignant carcinomas upon treatment with the tumor promoter, 12-O-tetradecanoyl-phorbol-13-acetate (TPA), occurred more rapidly in homozygous mutant mice than in the heterozygotes *(23)*. As in other studies, loss of heterozygosity occurred during the malignant transition, so that the heterozygotes lost the wt allele. Carcinomas that arose in *p53*-mutant mice appeared less differentiated than those from wt mice, which is again consistent with the general observation that absence of *p53* may accelerate the accumulation of genetic alterations in cells.

  Several studies have also examined susceptibility to radiation-induced mutation and tumors in *p53* transgenic and mutant mice, and cells derived from mutant mice *(24–27)*. These studies have demonstrated a decreased apoptotic response to irradiation, both in tissues and cells in culture, and an increase in the accumulation of mutations. In one study, transgenic mice expressing a dominant-negative mutant *p53* gene developed tumors, mostly sarcomas, in response to γ-irradiation *(27)*. Chromosomal damage induced by γ-irradiation in wt, *p53*-transgenic, or *p53*-null mice, was evaluated by examining the presence of micronuclei in red blood cells. Micronuclei were increased in both *p53*-null and transgenic mice in comparison with wt mice *(27)*. Thus, these results are consistent with the role of *p53* in surveillance and destruction of genetically damaged cells that are the likely precursors to tumors.

  The effect of *p53* mutations in mice has been studied in some detail in relation to mammary carcinomas, a tumor for which mouse models have existed for some time *(28–31)*. Mammary tumors in mice are often caused by the integration of mouse mammary tumor virus (MMTV), which carries a promoter/enhancer that confers mammary-specific expression. Mouse models for mammary carcinoma have included mice transgenic for the c-*myc,* v-Ha-*ras, neu, Wnt1,* or *int2* proto-oncogenes, under control of the MMTV promoter enhancer. Li et al. *(30)* expressed a mutant *p53* gene in transgenic mice, under control of the whey acidic protein promoter, to confer mammary-specific expression, and demonstrated that treatment with DMBA-reduced tumor latency. Nuclear morphology was more irregular in tumors in transgenic mice than in wt, again indicating a role for *p53* in the maintenance of genome integrity. Using mice carrying the targeted deletion of the *p53* gene, Donehower et al. *(28)* showed that a combination of a *Wnt1* transgene and the absence of *p53* resulted in acceleration of mammary tumor formation. Chromosomal aneuploidy, amplification, and deletion were more commonly observed in tumor cells arising in *p53*-nullizygous mice. Additionally, tumor cell proliferation was enhanced in the absence of *p53 (31).*

In contrast, another study examined mice carrying both an MMTV–c-*myc* transgene and homozygous *p53*-null alleles for the onset of tumors. These mice developed more aggressive lymphomas than mice carrying either transgene/mutation alone, but the presence of the *p53*-null alleles did not appear to accelerate the appearance of mammary carcinomas *(29).* Although these latter results correlate with earlier observations that mammary carcinomas are not frequently observed in *p53*-mutant mice, the formerly mentioned studies demonstrate that mutation or absence of *p53* can contribute to mammary carcinogenesis, in collaboration with a subset of oncogenes, or when *p53* mutant genes are deliberately expressed in mammary tissue.

The wt *p53* gene has also been used as an antitumor agent in a mammary tumor model. Adenovirus vectors, containing either *p53* wt or interleukin-2 cDNAs, were injected in combination into mammary tumors induced in transgenic mice expressing the polyoma middle T-antigen under control of the MMTV promoter/enhancer *(32).* Tumor regression was greatly enhanced by the combination therapy, compared to therapy with interleukin-2 alone, and only with the combination therapy were antitumor cytotoxic T-lymphocytes produced *(32).*

p53 interacts with many other proteins, and also serves as a transcription factor (TF) involved in the regulation of expression of target genes such as *p21*$^{WAF1/CIP1}$ *(33).* Therefore, it was of interest to study the effect of the *p53*-targeted-mutation in combination with other mutations, to determine the effect on viability and/or tumor formation (*p53/Rb* double mutants will be discussed in subheading 3 on the *Rb* mutant mouse). A striking effect was observed when mice carrying mutations in *p53* and *mdm2* were intercrossed *(34,35).* Amplification of *mdm2* is observed in human tumors, particularly frequently in sarcomas, defining it as an important human oncogene *(36).* Physical interaction of the *mdm2* gene product with p53 inactivates the transcriptional activation function of the p53 protein *(37).* Mouse embryos homozygous for a targeted deletion of the *mdm2* gene arrest around the time of implantation *(34,35),* which suggests that negatively regulating p53 function may be crucial during early embryogenesis, when cell division is rampant. This hypothesis was supported by the interesting observation that the *mdm2* mutant phenotype is rescued in the absence of *p53:* embryos carring homozygous mutations in both the *mdm2* and *p53* genes survive and grow to adults, demonstrating that a major, if not sole, function of *mdm2* is indeed to regulate p53 activity *(34,35).* A study of double-mutant mice demonstrated that the spectrum of tumors in *p53–/–mdm2–/–* double-mutant mice did not differ from that observed in *p53–/–* mice, further supporting the hypothesis that the only function of *mdm2* is to regulate p53 activity *(38).*

p53-mutant mice were also intercrossed with *scid* mice that carry a mutation in the gene encoding DNA protein kinase, which is involved both in the response to DNA damage, and in joining DNA ends that break during immunoglobulin gene rearrangement *(39)*. T-cell development was enhanced in double-mutant mice, and the development of lymphomas occurred earlier in double-mutant mice than in *p53–/–scid+/–* mice. Nacht et al. *(39)* suggested that p53 may be involved in eliminating cells with faulty immunoglobulin gene rearrangements that resulted in activation of oncogenes. Alternatively, the combined absence of the *scid* gene-encoded DNA protein kinase and *p53* may lead to a more severe defect in DNA repair than would result from the absence of either gene by itself.

The high frequency of tumor formation in *p53*-mutant mice allows them to be used as a model for chemoprevention and chemotherapy of tumors. In one such study, it was demonstrated that dyhydroepiandrosterone delayed the onset of tumors; several other compounds had no effect *(20)*.

3. RB AND ASSOCIATED GENES

The details of molecular pathways that involve the *Rb* gene are discussed in greater detail in Chapter 6 of this volume, and it suffices for this chapter to mention that it is involved in regulating transcription in a cell-cycle-dependent manner. The two salient features of this regulatory system are that the Rb protein is hyperphosphorylated by cyclin-dependent kinases during late G1 phase, and this hyperphosphorylated form is found mostly in the cytoplasm. The hypophosphorylated form is found in the nucleus during G0/early G1, where it complexes with TFs such as E2F-1, preventing E2F-1 from activating transcription of genes whose activity normally increases during S phase *(40)*. In this fashion, the *Rb* gene product functions as part of a molecular switch to control the proliferative state of cells. It is apparent that such a switch may figure prominently in determining entry of a cell into a tumorigenic state, but it is less obvious why cells of the retina may be most affected by loss of this ubiquitously expressed gene. Therefore, the *Rb* gene was mutated in mice, with the expectation of clarifying this issue, and contributing to the general understanding of tumor development.

Unlike viable *p53* homozygous mutant mice, *Rb–/–* mice do not survive past embryonic day 14 (E14), apparently because of a major defect in hepatic hematopoesis *(41,42)*. *Rb–/–* mice were also observed to have abnormal cell division and cell death in the nervous system *(41,42)*. No developmental abnormalities were observed in the eye itself, and the retina, did not display the cell death otherwise widely present in the nervous system. Although the embryonic demise of mutant mice did not allow observation of homozygous mutants for tumor development, this was possible with *Rb+/–* mice, and indeed these mice developed tumors between 2 and 11 mo of age *(42,43)*. Unlike humans, however, tumors in *Rb+/–* mice developed from melanotrophic cells in the intermediate lobe of the pituitary (PIT) gland *(42,43)*. As observed in tumors developing in *p53+/–* mice, these PIT tumors lost the remaining wt allele of the *Rb* gene, again confirming Knudson's original two-hit hypothesis for tumor development.

The *Rb* gene also demonstrated the use of conditional gene targeting to study TSGs. This emerging technology is employed to target mutations at specific points during or after embryonic development, in specific locations, so that the effect of that mutation can be observed in a specific tissue at a specific time, without affecting the entire embryo or animal *(5,6,44,45)*. This technology involves homologous recombination in ES cells to place target sites for site-specific recombination on either side of the portion of the chromosome (chr) to be deleted. A site-specific recombinase-encoding transgene is then expressed at a desired point, usually by placing it under the control of a tissue-specific promoter *(see* Fig. 1). In the case of the conditional targeting of the *Rb* gene, *Frt* sites, which are target sites for site-specific recombination catalyzed by the yeast enzyme, Flp, were placed flanking exon 19 of the *Rb* gene *(46)*. *Flp* recombinase was expressed under control of the rat pro-opiomelanocortin promoter, which primarily expresses in melanotrophic cells of the intermediate lobe. In mice carrying the *Flp* transgene and *Frt* sites in the *Rb* gene, PIT tumors were observed *(46)*. As observed with the original mutation, if only one *Rb* allele carried the *Frt* sites, tumors that developed subsequently lost the wt *Rb* allele. *Rb* loss appeared to be more widespread in the intermediate lobe than actual tumor development, and pretumor histological stages were observed, suggesting that additional genetic events, subsequent to losing *Rb* gene, may be involved in tumorigenesis *(46)*.

Human tumors have been identified in which both the *p53* and *Rb* genes have been inactivated *(see* refs. in ref. *47)*. In addition, both genes are involved in controlling progression through the cell cycle. These two important characteristics provided the motivation for intercrossing mice carrying *Rb* and *p53* mutations, to determine the phenotype resulting from the loss of these two prominent TSGs. Because *Rb–/–* mice are not viable, *p53–/–Rb+/–* mice were studied, as well as double heterozygotes. Similar to *Rb+/–* mice, PIT tumors were most frequently found in *p53–/–Rb+/–* mice, although earlier onset was found *(47,48)*. Medullary thyroid carcinoma was also frequently observed, as were lymphomas, sarcomas, and pancreatic islet cell carcinomas. Retinal dysplasia (but not hyperplasia) was also detected, but animals died of other tumors before retinal tumors might have arisen. Williams et al. *(47)* discuss three possible models for the cooperation observed between *p53* and *Rb* mutations: In the first, *p53* and *Rb* may each negatively regulate growth, so that double-mutant cells undergo relatively unrestricted proliferation; second, the absence of *p53* function leads to an increased rate of loss of the *Rb* gene, which relates to the general observation that accumulation of mutations is increased in the absence of *p53*; third, the absence of *p53* allows for the survival of *Rb–/–* cells, which takes into account the extreme cell death observed in *Rb–/–* embryos, and indicates that *p53* loss may allow these cells to transform instead *(47)*.

Because the *Rb* gene product exerts at least a portion of its regulatory effect by complexing with the TF E2F-1, it became of interest to target a mutation to the *E2F-1* gene, and observe potential similarities with *Rb*-mutant mice. In contrast to *Rb–/–* mice, mice homozygous for a mutation in the *E2F-1* gene survive and are fertile, although they demonstrate hypertrophy of exocrine glands, and males experience testicular atrophy *(49)*. This indicates that some vital functions of Rb do not rely on its interaction with E2F-1. *E2F-1* also appears to function as a tumor suppressor, and homozygous- and heterozygous-mutant mice develop reproductive tract sarcomas, lung adenocarcinomas, lymphomas, and other tumors, at lower frequencies *(49)*. Upon derivation of *E2F-1–/– Rb+/–* mice, the frequency of PIT and thyroid tumors was found to be reduced, and the life-span of animals lengthened, indicating that the tumorigenic phenotype of

Rb+/– mice is at least in part a consequence of the failure of Rb to adequately downregulate the expression of E2F-1-activated genes (50).

Two proteins with structural and functional similarities to Rb are known: p107 and p130. This raises the question of whether there is either overlap or the potential for compensation in their respective functions. As with p53, this issue is raised because Rb fulfills a basic role in the control of cell proliferation, and so it is not completely apparent why embryos progress normally more than halfway through embryogenesis in the absence of Rb function. The p107 and p130 genes were separately targeted, and in both cases homozygous mutant mice were viable, fertile, and without apparent major defects (51,52). In neither case did mutant mice develop tumors at increased frequency, even after long observation. p107–/–p130–/– double mutants were then obtained, and, although cartilage and skeletal abnormalities were observed, tumors were still not evident (51). However, when Rb+/–p107–/– mice were derived, pronounced growth retardation and early mortality were evident (52). The spectrum of tumors observed, however, did not differ from Rb+/– mice among those mice that survived to adulthood, nor did tumors occur with earlier onset. Rb+/–p107–/– mice developed retinal dysplasia, providing an indication of the importance of Rb function in this tissue. When PIT tumors from Rb+/–p107+/– were examined, it was found that the wt Rb allele was lost, but not the wt p107, indicating that Rb is the more important determinant of tumor formation, although both genes are important in growth control (52). Demonstrating that there was indeed functional overlap between Rb and p107, embryonic death occurred earlier in Rb–/–p107–/– embryos than in absence of Rb alone, with widespread apoptosis in the liver and nervous system (52).

The retinal dysplasia observed in Rb+/–p107–/– mice suggests that these two genes may serve as coordinately acting tumor suppressors in the retina, and raises the question of whether retinal tumors would be observed in double-homozygous mutants, if they were otherwise viable. This issue represents a general problem with gene-targeting technology: It is only possible to observe the earliest phenotype that results in arrest of development, and later potential phenotypes remain unseen. One way of approaching this problem is to use the previously discussed technology for conditional gene targeting.

An alternate approach that is perhaps technically less challenging, but has seen some important success, is to derive ES cells that have undergone targeting of both alleles of a gene, then use these ES cells to derive chimeric mice. Two approaches are generally available to target second alleles: a second targeting vector, carrying a different antibiotic-resistance gene than neo may be used, or it has been observed that in some instances, raising the G418 concentration in the ES cell culture media results in homozygosing of the original targeted allele. Once homozygously targeted ES cells are obtained and injected into wt blastocysts, they often contribute to various tissues in chimeric embryos or animals, so that cells missing both alleles are rescued beyond the point at which an entirely homozygous mutant tissue would be viable. When homozygous mutant cells are retained in particular tissues, they often represent a significant portion of the cells, so that an abnormal phenotype is observed, even though the whole animal remains viable.

This type of experiment was performed using ES cells that carried homozygous mutations in both the Rb and p107 genes. Chimeric mice derived with these doubly targeted ES cells developed RBs (53). Retinal cells in chimeras, derived using ES cells carrying only a double targeting of the *Rb* gene alone, underwent apoptosis, indicating that loss of *p107* was also required for retinal tumor development in the mouse. Furthermore, not all chimeras made with double-mutant ES cells developed RB, suggesting that additional mutation events may be required for development of the malignant phenotype *(53)*. These results demonstrate that *Rb* and *p107* are TSGs in the mouse, and further indicate that murine and human tissues are differentially sensitive to the loss of specific TSGs.

Adenovirus-mediated gene therapy has also been attempted using the *Rb* gene. Adenoviral vectors carrying the *Rb* gene were injected into PIT tumors arising in *Rb+/−* mice. Cells of injected tumors showed lower proliferative rates, and animals with injected tumors survived longer than controls, indicating a potential use for the *Rb* gene as a therapeutic agent *(54)*.

4. CELL CYCLE GENES

Mice carrying mutations in *p53* and *Rb,* two genes involved in the control of cell proliferation, have already been discussed. Several general observations suggested that targeting mutations to other genes, whose products are involved in control of the cell cycle, could also be informative about tumorigenesis. First, cancer is a disease of abnormal cell proliferation. Therefore, genes involved in the control of cell proliferation are obvious candidates to study in an animal model. Second, well-described TSGs, such as *p53*, *p16^{Ink4a}*, and *Rb,* were shown to be involved in pathways that control passage through the cell cycle. Finally, cell cycle regulatory proteins, such as cyclins, were found to be mutated in human cancers *(55)*. Two set of genes, both functioning as CDK inhibitors, exert negative regulatory effects on passage through the cell cycle. One set, the Kips, includes *p21^{Cip1/Waf1}*, *p27^{Kip1}*, and *p57^{Kip2}*. The other, the Inks, includes *p16^{Ink4a}*, *p15*, *p18,* and *p19^{Arf}*. Overall, the study of additional members of these sets of genes in mutant mice has been less yielding of actual models of tumorigenesis, but, nonetheless, has been informative about the importance of the cell proliferation regulation at the level of the whole organism. Readers wishing a more comprehensive review of these mutant mice should consult the recent review by Kiyokawa and Koff *(56)*.

As discussed in subheading 2, *p21^{Cip1/Waf1}* is a regulatory target of *p53,* which raises the question of whether the tumor suppressor activity of *p53* is mediated by *p21^{Cip1/Waf1}* and downstream targets. However, very few *p21^{Cip1/Waf1}* mutations have been found thus far in human tumors. When *p21^{Cip1/Waf1}* knockout mice were derived, homozygous mutants experienced normal development *(57)*. Unlike *p53* mutant mice, no tumors developed in *p21^{Cip1/Waf1}−/−* mice. In addition, the absence of *p21^{Cip1/Waf1}* did not rescue the lethality of the *mdm2* mutation in mice *(58),* further indicating that *p53* must act more broadly than simply regulating *p21^{Cip1/Waf1}* expression. Although *p21^{Cip1/Waf1}* mice appeared normal, study of *p21^{Cip1/Waf1}*-deficient cells in vitro did reveal abnormalities. When *p21^{Cip1/Waf1}−/−* cells were observed in tissue culture, defects in G1 checkpoint control were found, and cells entered G0 at higher densities than normal, a phenomenon also displayed by *p53−/−* cells *(57,59)*. In contrast to *p53−/−* thymocytes, which are refractory to apoptosis, apoptosis could be induced in *p21^{Cip1/Waf1}−/−* thymocytes identically to wt cells *(57)*. Another study examining the potential involvement of *p21^{Cip1/Waf1}* in tumorigenesis demonstrated that ras-transformed primary keratinocytes from $p21^{Cip1/Waf1}$–/– mice developed more aggressive tumors than did transformed keratinocytes from wt mice *(60)*.

Disruption of the $p27^{Kip1}$ gene in mice resulted in large mice with multiorgan hyperplasia, indicating a failure to properly regulate cell division *(61,62)*. Nodular hyperplasia and adenomas of the PIT gland intermediate lobe were found in $p27^{Kip1}$–/– mice, reminiscent of the phenotype of *Rb*+/– mice, suggesting a particular sensitivity of this tissue to defects in control of the cell cycle *(61,62)*. Although low levels of $p27^{Kip1}$ are found in human tumor cells, $p27^{Kip1}$ mutations have generally not been found in human tumors. Likewise, with the exception of the PIT adenomas, tumors do not develop in $p27^{Kip1}$–/– mice, despite increased cell proliferation. However, when $p27^{Kip1}$–/– or +/– mice are challenged with γ-irradiation or carcinogens, tumors in multiple tissues are obtained, with homozygous mutants developing tumors at approximately twice the frequency of heterozygous mice *(63)*. In contrast to observations with *p53*- or *Rb*-mutant mice, tumors in heterozygotes appear not to have undergone loss of heterozygosity (LOH) at the $p27^{Kip1}$ locus *(63)*. Thus, $p27^{Kip1}$ appears to be a TSG in mice, but, in contrast to most TSGs, a single copy of this gene appears to be insufficient to prevent tumors.

Mice carrying a targeted deletion of the $p57^{Kip2}$ gene also have a phenotype characteristic of a defect in the regulation of cellular proliferation, including omphalocele, cleft palate, incomplete differentiation of chrondrocytes, renal medullary dysplasia, lens cell hyperplasia, and adrenal cortical hyperplasia *(64)*. $p57^{Kip2}$ is an imprinted gene, meaning that it is only expressed from the chromosome inherited from one parent; in the case of $p57^{Kip2}$, expression of the maternally inherited chromosome is observed. Therefore, the full mutant phenotype can be found in heterozygotes that have inherited the mutant allele from their female parent. The $p57^{Kip2}$ gene maps to the chromosome locus that harbors deletions in individuals with the Beckwith-Wiedeman syndrome (BWS). Several aspects of the murine $p57^{Kip2}$ phenotype are also characteristic of the BWS in humans, including omphalocele, renal dysplasia, and the adrenal hyerplasia. A portion of BWS patients also develop Wilms' tumors, a tumor of the kidney, but this is not observed in mutant mice, suggesting that mice may not be a good model for this tumor. Mice carrying a mutation in the *WT1* gene, another tumor suppressor for Wilms' tumor, also fail to develop kidney tumors *(65)*. Nevertheless, the similarity of $p57^{Kip2}$-mutant mice to the premalignant condition of BWS individuals suggests that $p57^{Kip2}$ may be exerting a tumor-suppressive effect that is more vital in humans than in mice *(64)*.

In contrast to the previous CDK inhibitor mutant mice, mutant mice that carry a targeted mutation that eliminates both $p16^{Ink4a}$ and the overlapping gene, $p19^{Arf}$, provide clear evidence for the tumor-suppressive role of the $p16^{Ink4a}$ gene, even without treatment with carcinogens *(66)*. Homozygous mutant mice were viable and fertile, but developed tumors, mostly sarcomas and lymphomas, starting at 18 wk after birth. Treatment with DMBA and UVB irradiation accelerated tumor development. Cells from $p16^{Ink4a}$–/– mice also proliferated faster in culture. Furthermore, introduction of the Ha-ras gene into fibroblasts derived from –/– mice, but not from +/+ or +/– mice, resulted in focus formation in soft agar, again demonstrating the tumor-suppressive effects of $p16^{Ink4a}$ and $p19^{Arf}$ (66).

Mutations in this set of genes provide strong evidence for their role in regulating cellular proliferation in diverse tissues, and demonstrate that this is an important and perhaps central role of TSGs.

5. ATAXIA TELANGIECTASIA

Ataxia telangiectasia (AT) is an autosomal recessive disease with diverse clinical manifestations, including progressive neurodegeneration, immunodeficiency, and lymphoreticular malignancies, as well as several other components. The gene responsible for this disease, *ATM,* encodes a protein with kinase activity that appears to function upstream of *p53* in the cellular response to ionizing radiation. Upon targeting the *ATM* gene, homozygous mutant mice display many phenotypic features of the human disease, including the development of lymphomas *(67,68).* The relationship between *p53* and *ATM* has been studied in considerable detail, and the results and conclusions obtained thus far are complex, which is not surprising, given the tremendous variety of human cancer.

Tumor formation, consisting mostly of lymphomas, is accelerated in *ATM–/–p53–/–* mice, (but not in *ATM/p21$^{Cip1/Waf1}$* double mutants), compared to either single mutant *(69,70).* p53 protein levels are usually elevated in thymuses of mice exposed to γ-irradiation, but this failed to occur in *ATM–/–* mice *(71).* Lymphoma formation in *ATM* mice may therefore result from a failure to respond to DNA damage appropriately, especially in cells in which the damage is not lethal, but sufficient to cause malignant transformation.

Mouse embryo fibroblasts, cultured from *ATM–/–* mice, grew poorly, and underwent early senescence, ascribing a role to ATM in the maintenance of proliferation *(69,70).* This early senescence was rescued in both *ATM/p53* or *ATM/p21$^{Cip1/Waf1}$* double-mutant cells *(70,72).* The observation that p21$^{Cip1/Waf1}$ protein was elevated in *ATM–/–* cells, probably because of increased stabilization of the protein since mRNA levels were actually reduced in mutant cells, led Xu et al. *(70)* to suggest that elevated p21$^{Cip1/Waf1}$ was responsible for the early senescence. In addition, irradiated mouse embryo fibroblasts from *ATM–/–* mice continued to enter S phase, rather than arresting at the G1–S checkpoint, as is normally observed *(70,71).* This defect is more severe in both *ATM/p53* and *ATM/p21$^{Cip1/Waf1}$* double-mutant cells, indicative of a correlation between defective G1–S checkpoint control and early senescence. Together, these results suggest that *ATM* and *p53* cooperate in regulating passage through the cell cycle, and that failure to properly regulate this pathway leads to tumorigenesis.

Additional studies demonstrated that *ATM* is likely to regulate both *p53*-dependent and -independent pathways. γ-irradiation-induced thymic apoptosis, a *p53*-mediated event, occurred normally in *ATM–/–* mice in vivo, but was suppressed in *ATM/p53* double-mutant mice *(69,71).* In vitro, thymocytes from *ATM–/–* are also somewhat resistant to γ-irradiation-induced apoptosis, but thymocytes from *ATM/p53* double-mutant mice are completely resistant *(69,71).* These results confirm that the apoptotic response to irradiation is regulated through *p53*-dependent pathways, but some of these pathways may not involve *ATM.*

Although thymocytes in *ATM–/–* mice were relatively radiation-resistant, other tissues, the gastrointestinal (GI) tract and skin, in particular, were markedly more sensitive to γ-radiation treatment than wt *p53–/–* mice *(68).* Presumably, this response involves more than simple apoptosis, but also demonstrates that results in one tissue cannot be generalized to the whole organism. This extreme tissue radiation sensitivity was identical in *ATM/p53* double mutants *(73),* again contrasting with the protective effect observed with thymic apoptosis, suggesting that radiation-induced tissue damage is not mediated through *p53*. Thus, although it is clear that *ATM* and *p53* are both intimately involved in the determination of cell proliferation and survival, the interrelationships of the specific pathways are complex.

6. COLON CANCER

The *APC* gene first came to notice in mice under the guise of the *Min* gene, originally defined as a dominant mutation identified in a screen of ethylnitrosourea-treated mice, which predisposed them to multiple intestinal neoplasia *(74)*. *Min* mice develop multiple tumors throughout their intestinal tract that resemble various forms of adenomas, which in some cases become invasive, but not metastatic. This phenotype bears a strong resemblance to the human condition of adenomatous polyposis coli, in which afflicted individuals develop hundreds of polyps of the colon that can progress to carcinoma. *APC* was identified to be the gene harboring mutations in humans with this disease, and it was subsequently found that the *APC* gene was mutated in *Min* mice (now designated *APC^Min*) *(75)*. This provides an important demonstration that mouse models of TSGs genes can result from experiments other than standard gene targeting (although other mutations in the *APC* gene have subsequently been introduced using that methodology). Indeed, the introduction of more drastic mutations in the *APC* gene has revealed that homozygous mutants do not survive embryogenesis, which ascribes a developmental role to APC *(76)*. Further study of APC has demonstrated that it plays an important role in signaling pathways related to transcriptional activation of certain genes by the LEF-1 TF, and possibly to cell adhesion mediated by β-catenin and E-cadherin *(77)*.

Adenomas in *APC^Min+/–* mice have lost the remaining wt allele of *APC,* consistent with findings in humans with colon carcinoma *(78,79)*. This was also true of mice carrying targeted mutations in the *APC* gene *(76,80)*. In the latter case, adenomatous cells were found to have lost the wt allele at the microadenoma stage, suggesting, as is also suspected in humans, that the loss of *APC* signifies an early event in the development of colon carcinomas *(76,80)*. Another study, that involved a different targeted *APC* mutation, utilized a novel approach for conditional gene targeting by first using homologous recombination to place *lox* site-specific recombination sites (an alternative to the previously discussed *Frt*), flanking exon 14 of the *APC* gene *(81)*. Because the *lox* sites did not interfere with expression of *APC,* mice homozygous for the targeted allele were derived, and were normal. An adenoviral vector encoding the cre recombinase, an enzyme that deletes sequences between two lox sites, was then introduced into the intestine transrectally. Subsequently, infected cells that had deleted a portion of the *APC* gene gave rise to adenomas *(81)*. As with the previously discussed conditional targeting of the *Rb* gene, this conditional approach allowed the study of *APC–/–* cells, while avoiding the problem that *APC–/–* embryos do not survive embryogenesis.

It was also shown that the cyclooxygenase 2 gene is activated during polyp formation, and that a targeted mutation in the cyclooxygenase 2 gene suppresses polyp formation in *APC^Δ716* mice, possibly by affecting angiogenesis *(82)*. This observation may provide an insight regarding recent reports that aspirin, a cyclooxygenase inhibitor, may suppress colon polyp formation in humans *(83,84)*. *APC^Min* mice have also been crossed with *p53*-mutant mice. *APC^Min+/–p53–/–* mice developed lymphomas and sarcomas similar to *p53–/–* mice, but also developed pancreatic dysplasia *(85)*. In some cases, the pancreatic dysplasia could be classified as an adenocarcinoma, in which the wt *APC* allele was lost. Despite the increase in pancreatic abnormalities, no increase was observed in intestinal malignancy caused by absence of *p53*. This was surprising, given that *p53* is clearly mutated in tumors of the human colon, and that the *p53* mutation had also been shown to protect against irradiation-induced apoptosis in mice. These results may be indicative, as discussed previously, that *p53* may be involved in protecting against certain types of chromosome insults, but not the mutational events that finally lead to adenoma formation in *APC^Min* mice.

Study of *APC^Min* mice also provided a window through which to examine another important aspect of how animal models can contribute to the identification of novel TSGs. The degree of polyp formation in *Min* mice is heavily dependent on the genetic background: C57BL/6 is the strain on which tumors appear most rapidly *(74)*. This observation implies that other genes interact with *APC* to affect the tumor phenotype, which are polymorphic between different stains of mice. The mouse genome project has produced a fine genetic map of the mouse that allows the mapping and eventual cloning of so-called "modifier genes," based on differential phenotypes between strains *(17)*. One such gene has been identified and designated *MOM* (modifier of *Min*) *(86,87)*. A candidate gene found in the *MOM* locus is intestinal phospholipase A2, though it remains unexplained how this particular enzyme may effect rates of adenoma formation *(88,89)*. The pursuit of new TSGs and their genetic modifiers promises to develop into a major field of study. As discussed in a recent review by Balmain and Nagase *(90)*, it is now possible use the fine genetic map of the murine genome to begin to identify loci that control inbred mouse strain-dependent predisposition to the development of various tumors, with the expectation of identifying novel TSGs.

Recently, gene knockouts of two *SMAD* genes have provided what should prove to be very informative models for intestinal tumorigenesis *(91,92)*. SMADs are intracellular proteins that are involved in transducing signals from the transforming growth factor β (TGF-β) receptor. *SMAD4* (also known as *Dpc4*) maps to the 18q locus, a chr segment often deleted in human colon carcinoma. The *DCC* gene was originally cloned as a candidate tumor suppressor from this locus, although this now appears not to be the relevant TSG. Homozygous mutant *Smad4* mice do not survive embryogenesis, but heterozygotes appear normal *(91)*. When *Smad4* heterozygotes were crossed with *APC^Δ716* mice, however, the double heterozygotes developed more malignant tumors than were observed in simple *APC^Δ716* heterozygotes, including adenocarcinomas that invaded the submucosa *(91)*. As before, tumors had lost the wt alleles. In tumor cells that underwent LOH, it appeared that the chr containing mutant alleles of both *APC* and *Smad4* had subsequently been duplicated, restoring diploid levels of all other genes, *(91)*. Because murine homologs of the *APC* and *Smad4* genes are on the same chr, separated by approx 30 cM, this experiment first required mating single heterozygotes and screening for mice that had both mutant alleles on the same chr as a result of meiotic recombination. This indicated that *APC* and *Smad4* were probably the only TSGs involved in colon tumor formation on that chr.

A striking phenotype was obtained upon targeting of a second Smad gene, *Smad3*. *Smad3–/–* mice are viable and fertile, but develop metastatic colorectal adenocarcinomas between 1 and 6 mo of age, implicating TGF-β signaling in the development of tumors of the GI tract, and providing the most authentic model of human colon carcinoma in a genetically engineered mouse obtained to date *(92)*. These tumors extended into the submucosa, and metastasized to lymph nodes. APC was not lost in these tumors, and appeared to be expressed in tumor tissue. Along with the more malignant-appearing lesions, other abnormal tissue was observed, so that future analysis of these mice should reveal whether they progress through similar genetic changes that occur in human malignancy. Whether tumor formation in Smad-mutant mice relates to a growth-suppressive property of TGF-β, or an as-yet unknown function, remains to determined. The present mouse models should provide ample experimental material for these studies.

7. DNA REPAIR GENES

Genes involved in DNA repair comprise an important group of tumor suppressors, and mice with mutations in these genes have provided models for human syndromes such as xeroderma pigmentosum and hereditary nonpolyposis colon cancer (HNPCC). Three classes of genes have been studied, chiefly because of their involvement in human cancer-predisposition syndromes. The first class includes the DNA mismatch repair (MMR) genes, *Mlh1, Msh2, Msh6, Pms1,* and *Pms2;* the second class contains two genes involved in nucleotide excision repair, *Xpa* and *Xpc;* and the final class is comprised of the nucleotide repair gene that encodes o(6)-methylguanine-DNA methyltransferase.

HNPCC is a relatively common cancer predisposition syndrome in humans that results from mutations in DNA MMR genes, most often *mlh1* and *msh2,* and less commonly *pms1* and *pms2*. *Msh2–/–* mice are viable, and develop GI, skin, and lymphoid tumors *(93,94)*. In *Msh2–/–* mice, expression of the APC protein appeared to be lost in their adenomas, presumably contributing to the neoplastic progression *(94)*. Furthermore, accelerated intestinal tumorigenesis occurred in *Msh2–/–Apc^{Min}+/–* mice *(95)*. LOH at the *Apc* locus in adenomatous tissue always occurred in *Apc+/-* mice, but was sporadic in *Msh2–/– Apc^{Min}+/–* mice *(95)*. This observation suggests that the absence of *Msh2* activity leads to loss of *Apc* function, through structural mutations, rather than loss of the *Apc* chromosomal locus by large-scale deletions.

As might be expected in animals deficient in DNA repair genes, an increased DNA mutation rate can be detected in intestinal tissue of *Mlh1* and *Pms2* homozygous mutant mice, although not in *Pms1* mutants *(96)*. *Mlh1*-mutant mice develop GI tumors, but *Pms2* mutants only develop lymphomas and sarcomas, despite the increased mutation rate in the intestine *(96,97)*. Similar to *Msh2* mice, however, *Pms2–/– Apc^{Min}+/–* mice gave rise to 3–4× the number of intestinal adenomas as *Apc+/–* mice, further implicating MMR genes in intestinal tumorigenesis *(98)*. An additional MMR gene, *Msh6,* which has recently been shown to be involved in human colorectal tumors *(99)*, also appears to function as a colon TSG in mice. *Msh6–/–* mice developed both GI and lymphoid tumors, suggesting that it may be important to further examine the involvement of this gene in human cancer *(100)*. Hence, not all DNA MMR genes are of equal significance in tumorigenesis, and further study of these mutations may provide insight into their respective functions.

Xeroderma pigmentosum is a rare autosomal recessive skin disease characterized by increased sensitivity to ultraviolet radiation (UV), and frequent development of skin tumors. Enzymatic defects in nucleotide-excision repair also define this disorder, and eight complementation groups have been defined in humans. Two of the known XP genes, *Xpa* and *Xpc,* have been mutated in mice, and, in both cases, homozygous mutant mice were highly susceptible to UV-induced squamous skin carcinomas *(101–103).* With the exception of some solid organ adenomas in older *Xpa–/–* mice, spontaneous tumors were not observed in these mice *(101–103).* Prolonged treatment of *Xpa+/–* mice with UV did not result in an increased incidence of tumors, suggesting that LOH at this locus is a rare event *(104).* As observed with other mutations, *Xpc/p53* double-mutant mice experience an accelerated rate of skin tumor formation, compared with the *Xpc* mutation alone, implicating *p53* in pathways that protect against UV-induced DNA damage in the skin *(105).*

The O(6)-methylguanine-DNA methyltransferase gene *(Mgmt)* encodes an enzyme that repairs mutagenic alkylation of the O^6 position of guanine. Transgenic mice were generated that specifically overexpressed this enzyme in skin cells, and this overexpression was found to be protective against DMBA/TPA-induced skin tumors *(106).* *Mgmt–/–* mice have also been generated, and were found to have significantly lower $LD_{50}s$ upon treatment with the mutagen, methylnitrosourea (MNU) *(107).* Mutant mice developed thymic lymphomas and lung adenomas *(107).* When *Mgmt*-mutant mice were crossed with *Mlh1* mutants, the resultant double homozygotes were resistant to MNU-induced lethality, although they still developed frequent lymphomas in response to MNU *(108).* These results are consistent with the hypothesis that MMR results in double-strand breaks at sites of alkylation-related mutation, leading to cell death. Cell death, in turn, prevents cells with mutations from giving rise to tumors, hence, the high lymphoma rate in MNU-treated *Mgmt/Mlh1* double mutants, in which apoptotic elimination of cells with alkylation-related mutations does not occur. Mice with mutations in DNA repair genes should continue to serve as valuable models for the mechanistic study of mutagen-induced tumorigenesis.

8. *BRCA 1* AND *2*

BRCA1 and *BRCA2* are TSGs involved in human breast cancer, which have recently received considerable attention. The function of the two encoded proteins remains unclear, although BRCA1 may associate with p53 *(109),* affecting its transcriptional regulatory function; BRCA2 associates with Rad51 *(110),* suggesting a function in DNA repair. Mouse embryos deficient in either *BRCA1* or *2* arrest early in development, around E7.5–8.5, with apparent defects in cell proliferation *(111–113).* In neither case did heterozygous mice develop tumors *(112,113).* *BRCA* mutant mice have been crossed with *p53* mutants, again without the development of new tumors; the absence of *p53* did slightly prolong survival of both *BRCA1* or *2* homozygous mutant embryos, suggesting a possible functional association *(111,114).* Recently, mice carrying a targeted *BRCA2* mutation were derived, which resulted in production of a truncated BRCA2 protein. Most homozygous mutant embryos arrest during development, but, among the few homozygous animals that are born, lethal thymic lymphomas develop at about 12–14 wk of age *(115,116).* *BRCA2–/–* fibroblasts obtained from mutant embryos have defects in proliferation and DNA repair that can be related to overexpression of p53 and *p21^{WAF1/CIP1}* (116), and further study of these cells and animals may provide more insight into the role of *BRCA* genes in mammary tumorigenesis.

9. TUBEROUS SCLEROSIS

The Eker rat develops renal cell carcinomas with dominant pattern of inheritance, which is consistent with the inheritance of a dominantly acting oncogene, or an inactive TSG. The *Tsc2* gene, which is mutated in the human condition, tuberous sclerosis, was found to be the responsible gene *(117–119)*. The additional observation that this locus underwent LOH during tumor development indicated that *Tsc2* was likely to be functioning as a TSG *(117,120)*. Transgenic rats, in which a wt *Tsc2* gene was inserted, did not develop tumors *(121)*. *Tsc2* encodes a membrane-associated guanosine triphosphatase-activating protein designated "tuberin" *(122)*. This large protein has also been found to have transcriptional regulatory activity *(123,124)*, and efforts are underway to identify potential target genes *(125)*. Members of the AP1 family of TFs are overexpressed in renal carcinomas from Eker rats, and may constitute one set of targets for regulation by *Tsc2 (126)*.

10. TSG MUTATIONS NOT RESULTING IN MURINE NEOPLASMS

Several human TSGs, not discussed in previous subheadings, have been targeted in mice, without yielding a model for tumor formation. Some of these genes have been found to be homozygous lethal, including *WT1* and *Vhl*. Mice heterozygous for targeted mutations of these genes do not develop tumors, indicating either that the LOH at their respective loci is rare or that they do not function as TSGs in these animals.

Von Hippel-Landau (VHL) disease predisposes affected individuals to multiple tumor types, including hemangioblastomas and renal cell carcinoma. The *Vhl* gene encodes a protein that associates with transcriptional elongation proteins, although it remains unclear whether this is the major function of this protein. The VHL protein associates intracellularly with the extracellular matrix protein, fibronectin, and fibronectin secretion and matrix assembly appeared to be diminished in *Vhl–/–* cells and embryos *(127)*. Since the abnormal interaction of cells with the extracellular matrix is an important aspect of tumorigenesis, this association with fibronectin assembly provides one possible mode through which the mutation of this gene may be involved in tumor formation. *Vhl–/–* mouse embryos do not survive beyond E12.5, because of defective placental vasculogenesis *(128)*.

WT1 encodes a zinc finger TF, and is deleted in a small fraction of Wilms' tumors, a kidney tumor that occurs in young children *(129)*. Mouse embryos carrying a homozygous mutation in this gene fail to develop kidneys or gonads, with the kidney progenitor mesenchyme becoming completely apoptotic *(65)*. *Wt1+/–* mice, on the other hand, survive to adulthood without the development of kidney or other tumors *(65)*. The WT1 protein has been demonstrated to physically interact with p53 *(129a)*. However, *WT1+/–p53–/–* mice still fail to develop kidney tumors, and, in *WT1/p53* double-mutant embryos, the renal phenotype is identical, with no amelioration of the apoptosis (J. Kreidberg, unpublished results).

11. NONMAMMALIAN MODELS

The discussion thus far has dealt with mammalian systems that have almost exclusively been models utilizing mice carrying targeted mutations. However, it is also important to acknowledge the potential contribution of nonmammalian systems to this field. Although powerful genetic systems have been developed to analyze tumorigenesis in mice and humans, control of cell proliferation in organisms such as *Drosophila* and *Caenorhabditis* can be genetically dissected with greater throughput and at lower cost. Current analysis reveals that the molecular pathways that control cell proliferation are highly conserved between all organisms, justifying the use of nonmammalian models for the study of tumorigenesis. *Drosophila melanogaster* deserves particular mention, because flies develop tumors with important similarities to mammals *(130,131)*, including induction by carcinogens, excessive cellular proliferation, morphological changes in cellular architecture, and invasiveness. Tumors can also be transplanted from fly to fly as a test of malignancy. One genetic approach described by Xu et al. *(131)* in *Drosophila* should be especially useful in the search for new tumor suppressors. Chimeric flies can be made so that some cells contain the previously discussed *Frt* site-specific recombination sites on both homologs of a chr in close proximity to the centromeres. These cells are also able to express the Flp recombinase under the control of the heat-shock promoter, which allows conditional expression at any desired timepoint, with far greater ease than is possible in mammalian systems. Induced expression of Flp in somatic cells results in mitotic recombination between chrs; such daughter cells become homozygous for the chr region distal to the *Frt* sites. If this region carries a mutation that inactivates a TSG, a tumor will develop from the chimeric cells. Further genetic analysis can identify the gene responsible. This approach has been used by Xu et al. *(132)* to identify several novel genes whose inactivation results in excessive cell proliferation.

12. NEW TECHNOLOGIES

Gene targeting and other transgenic technologies are powerful approaches for the study of known TSGs. New technologies are also being developed to identify novel TSGs. This chapter has already discussed the approaches proposed by Balmain and Nagase *(90)*, which will attempt to map and eventually clone genes, based on genetic differences between mouse strains. Approaches based on physical differences between chrs of tumor cells and normal cells are also being developed (*see*, for examples, refs. *133–137*). Spectral karyotyping is one such technology *(138)*, although several others are being developed that also detect differences between karyotypes of normal and neoplastic cells. Sets of probes that paint specific chrs are labeled with single or combinations of chromogens, so that all 19 and X, Y chrs can be distinguished from each other at once, using fluorescent *in situ* hybridization, because the unique set of chromogens will impart a different color to each chr. In tumor cells, translocations between chrs are detected, with immediate identification of both involved chrs *(138)*. This approach is especially useful for mouse chrs, which are all acrocentric and close in size, making karyotyping much more difficult than in humans. Spectral karyotyping, and related approaches that allow analysis of the entire karyotype in a single experiment, should allow for more efficient screening of tumors for cancer-causing mutations.

13. SUMMARY

As the results summarized in this chapter amply demonstrate, the pursuit of animal models of TSGs has thus far yielded several valuable models for the study of tumorigenesis. Obviously, if the only result of these efforts was to demonstrate tumor formation upon loss of a specific gene, one may question the value of such an experiment. However, as evidenced by some of the earliest mutant strains that have been studied in greater detail, it is clear that these animals are providing experimental models that will provide important insights into tumorigenesis. New technologies, such as conditional gene targeting and various procedures for karyotyping tumor cells, combined with the wealth of information provided by the human and mouse genome projects, will allow even more powerful experimental approaches in the near future.

ACKNOWLEDGMENTS

The authors thank Suzanne Newby for excellent editing of this chapter. Work in this laboratory has been supported by grants from the March of Dimes Foundation, the Harcourt General Foundation, the Charles Hood Foundation, the National Kidney Foundation, and the National Institutes of Health.

REFERENCES

1. Knudson A, Jr. Mutation and cancer: statistical study of retinoblastoma. *Proc Natl Acad Sci USA* 1971; 68:820–823.
2. Fung YK, Murphree AL, T'Ang A, Qian J, Hinrichs SH, Benedict WF. Structural evidence for the authenticity of the human retinoblastoma gene. *Science* 1987; 236:1657–1661.
3. Lee WH, Bookstein R, Hong F, Young LJ, Shew JY, Lee EY. Human retinoblastoma susceptibility gene: cloning, identification, and sequence. *Science* 1987; 235:1394–1399.
4. Friend SH, Horowitz JM, Gerber MR, Wang XF, Bogenmann E, Li FP, Weinberg RA. Deletions of a DNA sequence in retinoblastomas and mesenchymal tumors: organization of the sequence and its encoded protein. *Proc Natl Acad Sci USA* 1987; 84:9059–9063.
5. Kuhn R, Schwenk F. Advances in gene targeting methods. *Curr Opin Immunol* 1997; 9:183–188.
6. Sauer B. Inducible gene targeting in mice using the Cre/lox system. *Methods* 1998; 14:381–392.
7. Shastry BS. Gene disruption in mice: models of development and disease. *Mol Cell Biochem* 1998; 181:163–179.
8. Torres M. The use of embryonic stem cells for the genetic manipulation of the mouse. *Curr Topics Dev Biol* 1998; 36:99–114.
9. Capecchi MR. The new mouse genetics: altering the genome by gene targeting. *Trends Genet* 1989; 5:70–76.
10. Donehower LA, Harvey M, Slagle BL, McArthur MJ, Montgomery CA, Jr, Butel JS, Bradley A. Mice deficient for p53 are developmentally normal but susceptible to spontaneous tumours. *Nature* 1992; 356:215–221.
11. Donehower LA. The p53-deficient mouse: a model for basic and applied cancer studies. *Semin Cancer Biol* 1996; 7:269–278.
12. Jacks T, Remington L, Williams BO, Schmitt EM, Halachmi S, Bronson RT, Weinberg RA. Tumor spectrum analysis in p53-mutant mice. *Curr Biol* 1994; 4:1–7.
13. Armstrong JF, Kaufman MH, Harrison DJ, Clarke AR. High-frequency developmental abnormalities in p53-deficient mice. *Curr Biol* 1995; 5:931–936.
14. Malkin D, Li FP, Strong LC, Fraumeni JF, Jr, Nelson CE, Kim DH, et al. Germ line p53 mutations in a familial syndrome of breast cancer, sarcomas, and other neoplasms *Science* 1990; 250:1233–1238.
15. Donehower LA, Harvey M, Vogel H, McArthur MJ, Montgomery CA, Jr, Park SH, et al. Effects of genetic background on tumorigenesis in p53-deficient mice. *Mol Carcinog* 1995; 14:16–22.
16. Harvey M, McArthur MJ, Montgomery CA, Jr, Bradley A, Donehower LA. Genetic background alters the spectrum of tumors that develop in p53-deficient mice. *FASEB J* 1993; 7:938–943.
17. Dietrich WF, Miller J, Steen R, Merchant MA, Damron BD, Husain Z, et al. A comprehensive genetic map of the mouse genome. *Nature* 1996; 380:149–152.
18. Dumaz N, van Kranen HJ, de Vries A, Berg RJ, Wester PW, van Kreijl CF, et al. The role of UV-B light in skin carcinogenesis through the analysis of p53 mutations in squamous cell carcinomas of hairless mice. *Carcinogenesis* 1997; 18:897–904.

19. Harvey M, McArthur MJ, Montgomery CA, Jr, Butel JS, Bradley A, Donehower LA. Spontaneous and carcinogen-induced tumorigenesis in p53-deficient mice. *Nat Genet* 1993; 5:225–229.
20. Hursting SD, Perkins SN, Haines DC, Ward JM, Phang JM. Chemoprevention of spontaneous tumorigenesis in p53-knockout mice. *Cancer Res* 1995; 55:3949–3953.
21. Lowe SW, Bodis S, McClatchey A, Remington L, Ruley HE, Fisher DE, Housman DE, Jacks T. p53 status and the efficacy of cancer therapy in vivo. *Science* 1994; 266:807–810.
22. Sands AT, Suraokar MB, Sanchez A, Marth JE, Donehower LA, Bradley A. p53 deficiency does not affect the accumulation of point mutations in a transgene target. *Proc Natl Acad Sci USA* 1995; 92:8517–8521.
23. Kemp CJ, Donehower LA, Bradley A, Balmain A. Reduction of p53 gene dosage does not increase initiation or promotion but enhances malignant progression of chemically induced skin tumors. *Cell* 1993; 74:813–822.
24. Wang L, Cui Y, Lord BI, Roberts SA, Potten CS, Hendry JH, Scott D. Gamma-ray-induced cell killing and chromosome abnormalities in the bone marrow of p53-deficient mice. *Radiat Res* 1996; 146:259–266.
25. Yuan J, Yeasky TM, Havre PA, Glazer PM. Induction of p53 in mouse cells decreases mutagenesis by UV radiation. *Carcinogenesis* 1995; 16:2295–2300.
26. van Kranen HJ, de Gruijl FR, de Vries A, Sontag Y, Wester PW, Senden HC, Rozemuller E, van Kreijl CF. Frequent p53 alterations but low incidence of ras mutations in UV-B-induced skin tumors of hairless mice. *Carcinogenesis* 1995; 16:1141–1147.
27. Lee JM, Abrahamson JL, Kandel R, Donehower LA, Bernstein A. Susceptibility to radiation-carcinogenesis and accumulation of chromosomal breakage in p53 deficient mice. *Oncogene* 1994; 9:3731–3736.
28. Donehower LA, Godley LA, Aldaz CM, Pyle R, Shi YP, Pinkel D, et al. Deficiency of p53 accelerates mammary tumorigenesis in Wnt-1 transgenic mice and promotes chromosomal instability. *Genes Dev* 1995; 9:882–895.
29. Elson A, Deng C, Campos-Torres J, Donehower LA, Leder P. The MMTV/c-myc transgene and p53 null alleles collaborate to induce T-cell lymphomas, but not mammary carcinomas in transgenic mice. *Oncogene* 1995; 11:181–190.
30. Li B, Murphy KL, Laucirica R, Kittrell F, Medina D, Rosen JM. A transgenic mouse model for mammary carcinogenesis. *Oncogene* 1998; 16:997–1007.
31. Jones JM, Attardi L, Godley LA, Laucirica R, Medina D, Jacks T, Varmus HE, Donehower LA. Absence of p53 in a mouse mammary tumor model promotes tumor cell proliferation without affecting apoptosis. *Cell Growth Differ* 1997; 8:829–838.
32. Putzer BM, Bramson JL, Addison CL, Hitt M, Siegel PM, Muller WJ, Graham FL. Combination therapy with interleukin-2 and wild-type p53 expressed by adenoviral vectors potentiates tumor regression in a murine model of breast cancer. *Hum Gene Ther* 1998; 9:707–918.
33. Agarwal ML, Taylor WR, Chernov MV, Chernova OB, Stark GR. The p53 network. *J Biol Chem* 1998; 273:1–4.
34. Montes de Oca Luna R, Wagner DS, Lozano G. Rescue of early embryonic lethality in mdm2-deficient mice by deletion of p53. *Nature* 1995; 378:203–206.
35. Jones SN, Roe AE, Donehower LA, Bradley A. Rescue of embryonic lethality in Mdm2-deficient mice by absence of p53. *Nature* 1995; 378:206–208.
36. Oliner JD, Kinzler KW, Meltzer PS, George DL, Vogelstein B. Amplification of a gene encoding a p53-associated protein in human sarcomas [see comments]. *Nature* 1992; 358:80–83.
37. Momand J, Zambetti GP, Olson DC, George D, Levine AJ. The mdm-2 oncogene product forms a complex with the p53 protein and inhibits p53-mediated transactivation. *Cell* 1992; 69:1237–1245.
38. Jones SN, Sands AT, Hancock AR, Vogel H, Donehower LA, Linke SP, Wahl GM, Bradley A. Tumorigenic potential and cell growth characteristics of p53-deficient cells are equivalent in the presence or absence of Mdm2. *Proc Natl Acad Sci USA* 1996; 93:14,106–14,111.
39. Nacht M, Strasser A, Chan YR, Harris AW, Schlissel M, Bronson RT, Jacks T. Mutations in the p53 and SCID genes cooperate in tumorigenesis. *Genes Dev* 1996; 10:2055–2066.
40. Weinberg RA. The retinoblastoma gene and gene product. *Cancer Surv* 1992; 12:43–57.
41. Lee EY, Chang CY, Hu N, Wang YC, Lai CC, Herrup K, Lee WH, Bradley A. Mice deficient for Rb are nonviable and show defects in neurogenesis and haematopoiesis [see comments]. *Nature* 1992; 359:288–294.
42. Jacks T, Fazeli A, Schmitt EM, Bronson RT, Goodell MA, Weinberg RA. Effects of an Rb mutation in the mouse [see comments]. *Nature* 1992; 359:295–300.

43. Hu N, Gutsmann A, Herbert DC, Bradley A, Lee WH, Lee EY. Heterozygous Rb-1 delta 20/+mice are predisposed to tumors of the pituitary gland with a nearly complete penetrance. *Oncogene* 1994; 9:1021–1027.

44. Dymecki SM. A modular set of Flp, FRT and lacZ fusion vectors for manipulating genes by site-specific recombination. *Gene* 1996; 171:197–201.

45. Orban PC, Chui D, Marth JD. Tissue- and site-specific DNA recombination in transgenic mice. *Proc Natl Acad Sci USA* 1992; 89:6861–6865.

46. Vooijs M, van der Valk M, te Riele H, Berns A. Flp-mediated tissue-specific inactivation of the retinoblastoma tumor suppressor gene in the mouse. *Oncogene* 1998; 17:1–12.

47. Williams BO, Remington L, Albert DM, Mukai S, Bronson RT, Jacks T. Cooperative tumorigenic effects of germline mutations in Rb and p53. *Nat Genet* 1994; 7:480–484.

48. Harvey M, Vogel H, Lee EY, Bradley A, Donehower LA. Mice deficient in both p53 and Rb develop tumors primarily of endocrine origin. *Cancer Res* 1995; 55:1146–1151.

49. Yamasaki L, Jacks T, Bronson R, Goillot E, Harlow E, Dyson NJ. Tumor induction and tissue atrophy in mice lacking E2F-1. *Cell* 1996; 85:537–548.

50. Yamasaki L, Bronson R, Williams BO, Dyson NJ, Harlow E, Jacks T. Loss of E2F-1 reduces tumorigenesis and extends the lifespan of Rb1(+/–)mice. *Nat Genet* 1998; 18:360–364.

51. Cobrinik D, Lee MH, Hannon G, Mulligan G, Bronson RT, Dyson N, et al. Shared role of the pRB-related p130 and p107 proteins in limb development. *Genes Dev* 1996; 10:1633–1644.

52. Lee MH, Williams BO, Mulligan G, Mukai S, Bronson RT, Dyson N, Harlow E, Jacks T. Targeted disruption of p107: functional overlap between p107 and Rb. *Genes Dev* 1996; 10:1621–1632.

53. Robanus-Maandag E, Dekker M, van der Valk M, Carrozza ML, Jeanny JC, Dannenberg JH, Berns A, te Riele H. p107 is a suppressor of retinoblastoma development in pRb-deficient mice. *Genes Dev* 1998; 12:1599–1609.

54. Riley DJ, Nikitin AY, Lee WH. Adenovirus-mediated retinoblastoma gene therapy suppresses spontaneous pituitary melanotroph tumors in Rb+/– mice. *Nature Med* 1996; 2:1316–1321.

55. Hall M, Peters G. Genetic alterations of cyclins, cyclin-dependent kinases, and Cdk inhibitors in human cancer. *Adv Cancer Res* 1996; 68:67–108.

56. Kiyokawa H, Koff A. Roles of cyclin-dependent kinase inhibitors: lessons from knockout mice. *Curr Topics Microbiol Immunol* 1998; 227:105–120.

57. Deng C, Zhang P, Harper JW, Elledge SJ, Leder P. Mice lacking p21CIP1/WAF1 undergo normal development, but are defective in G1 checkpoint control. *Cell* 1995; 82:675–684.

58. Montes de Oca Luna R, Amelse LL, Chavez-Reyes A, Evans SC, Brugarolas J, Jacks T, Lozano G. Deletion of p21 cannot substitute for p53 loss in rescue of mdm2 null lethality [letter]. *Nat Genet* 1997; 16:336–337.

59. Brugarolas J, Chandrasekaran C, Gordon JI, Beach D, Jacks T, Hannon GJ. Radiation-induced cell cycle arrest compromised by p21 deficiency. *Nature* 1995; 377:552–557.

60. Missero C, Di Cunto F, Kiyokawa H, Koff A, Dotto GP. The absence of p21Cip1/WAF1 alters keratinocyte growth and differentiation and promotes ras-tumor progression. *Genes Dev* 1996; 10:3065–3075.

61. Fero ML, Rivkin M, Tasch M, Porter P, Carow CE, Firpo E, et al. Syndrome of multiorgan hyperplasia with features of gigantism, tumorigenesis, and female sterility in p27(Kip1)-deficient mice. *Cell* 1996; 85:733–744.

62. Kiyokawa H, Kineman RD, Manova-Todorova KO, Soares VC, Hoffman ES, Ono M, et al. Enhanced growth of mice lacking the cyclin-dependent kinase inhibitor function of p27(Kip1). *Cell* 1996; 85:721–732.

63. Fero ML, Randel E, Gurley KE, Roberts JM, Kemp CJ. The murine gene p27Kip1 is haploinsufficient for tumor suppression. *Nature* 1998; 396:177–180.

64. Zhang P, Liegeois NJ, Wong C, Finegold M, Hou H, Thompson JC, et al. Altered cell differentiation and proliferation in mice lacking p57KIP2 indicates a role in Beckwith-Wiedemann syndrome. *Nature* 1997; 387:151–158.

65. Kreidberg JA, Sariola H, Loring JM, Maeda M, Pelletier J, Housman D, Jaenisch R. WT-1 is required for early kidney development. *Cell* 1993; 74:679–691.

66. Serrano M, Lee H, Chin L, Cordon-Cardo C, Beach D, DePinho RA. Role of the INK4a locus in tumor suppression and cell mortality. *Cell* 1996; 85:27–37.

67. Xu Y, Ashley T, Brainerd EE, Bronson RT, Meyn MS, Baltimore D. Targeted disruption of ATM leads to growth retardation, chromosomal fragmentation during meiosis, immune defects, and thymic lymphoma [see comments]. *Genes Dev* 1996; 10:2411–2422.

68. Barlow C, Hirotsune S, Paylor R, Liyanage M, Eckhaus M, Collins F, et al. Atm-deficient mice: a paradigm of ataxia telangiectasia. *Cell* 1996; 86:159–171.

69. Westphal CH, Rowan S, Schmaltz C, Elson A, Fisher DE, Leder P. atm and p53 cooperate in apoptosis and suppression of tumorigenesis, but not in resistance to acute radiation toxicity. *Nat Genet* 1997; 16:397–401.

70. Xu Y, Yang EM, Brugarolas J, Jacks T, Baltimore D. Involvement of p53 and p21 in cellular defects and tumorigenesis in Atm–/– mice. *Mol Cell Biol* 1998; 18:4385–4390.

71. Barlow C, Brown KD, Deng CX, Tagle DA, Wynshaw-Boris A. Atm selectively regulates distinct p53-dependent cell-cycle checkpoint and apoptotic pathways. *Nat Genet* 1997; 17:453–456.

72. Barlow C, Liyanage M, Moens PB, Deng CX, Ried T, Wynshaw-Boris, A. Partial rescue of the prophase I defects of Atm-deficient mice by p53 and p21 null alleles. *Nat Genet* 1997; 17:462–466.

73. Westphal CH, Schmaltz C, Rowan S, Elson A, Fisher DE, Leder P. Genetic interactions between atm and p53 influence cellular proliferation and irradiation-induced cell cycle checkpoints. *Cancer Res* 1997; 57:1664–1667.

74. Moser AR, Pitot HC, Dove WF. A dominant mutation that predisposes to multiple intestinal neoplasia in the mouse. *Science* 1990; 247:322–324.

75. Su LK, Kinzler KW, Vogelstein B, Preisinger AC, Moser AR, Luongo C, Gould KA, Dove WF. Multiple intestinal neoplasia caused by a mutation in the murine homolog of the APC gene. *Science* 1992; 256:668–670.

76. Oshima M, Oshima H, Kitagawa K, Kobayashi M, Itakura C, Taketo M. Loss of Apc heterozygosity and abnormal tissue building in nascent intestinal polyps in mice carrying a truncated Apc gene. *Proc Natl Acad Sci USA* 1995; 92:4482–4486.

77. Jankowski JA, Bruton R, Shepherd N, Sanders DS. Cadherin and catenin biology represent a global mechanism for epithelial cancer progression. *Mol Pathol* 1997; 50:289–290.

78. Luongo C, Moser AR, Gledhill S, Dove WF. Loss of Apc+ in intestinal adenomas from Min mice. *Cancer Res* 1994; 54:5947–5952.

79. Luongo C, Dove WF. Somatic genetic events linked to the Apc locus in intestinal adenomas of the Min mouse. *Genes Chromosomes Cancer* 1996; 17:194–198.

80. Oshima H, Oshima M, Kobayashi M, Tsutsumi M, Taketo MM. Morphological and molecular processes of polyp formation in Apc(delta716) knockout mice. *Cancer Res* 1997; 57:1644–1649.

81. Shibata H, Toyama K, Shioya H, Ito M, Hirota M, Hasegawa S, et al. Rapid colorectal adenoma formation initiated by conditional targeting of the Apc gene. *Science* 1997; 278:120–123.

82. Tsujii M, Kawano S, Tsuji S, Sawaoka H, Hori M, DuBois RN. Cyclooxygenase regulates angiogenesis induced by colon cancer cells. *Cell* 1998; 93:705–716.

83. Smalley WE, DuBois RN. Colorectal cancer and nonsteroidal antiinflammatory drugs. *Adv Pharmacol* 1997; 39:1–20.

84. Rosenberg L, Louik C, Shapiro S. Nonsteroidal antiinflammatory drug use and reduced risk of large bowel carcinoma. *Cancer* 1998; 82:2326–2333.

85. Clarke AR, Cummings MC, Harrison DJ. Interaction between murine germline mutations in p53 and APC predisposes to pancreatic neoplasia but not to increased intestinal malignancy. *Oncogene* 1995; 11:1913–1920.

86. Dobbie Z, Heinimann K, Bishop DT, Muller H, Scott RJ. Identification of a modifier gene locus on chromosome 1p35–36 in familial adenomatous polyposis. *Hum Genet* 1997; 99:653–657.

87. Dietrich WF, Lander ES, Smith JS, Moser AR, Gould KA, Luongo C, Borenstein N, Dove W. Genetic identification of Mom-1, a major modifier locus affecting Min-induced intestinal neoplasia in the mouse. *Cell* 1993; 75:631–639.

88. Gould KA, Luongo C, Moser AR, McNeley MK, Borenstein N, Shedlovsky A, Dove WF, Hong K, Dietrich WF, Lander ES. Genetic evaluation of candidate genes for the Mom1 modifier of intestinal neoplasia in mice. *Genetics* 1996; 144:1777–1785.

89. Keshav S, McKnight AJ, Arora R, Gordon S. Cloning of intestinal phospholipase A2 from intestinal epithelial RNA by differential display PCR. *Cell Proliferation* 1997; 30:369–383.

90. Balmain A, Nagase H. Cancer resistance genes in mice: models for the study of tumour modifiers. *Trends Genet* 1998; 14:139–144.

91. Takaku K, Oshima M, Miyoshi H, Matsui M, Seldin MF, Taketo MM. Intestinal tumorigenesis in compound mutant mice of both Dpc4 (Smad4) and Apc genes. *Cell* 1998; 92:645–656.

92. Zhu Y, Richardson JA, Parada LF, Graff JM. Smad3 mutant mice develop metastatic colorectal cancer. *Cell* 1998; 94:703–714.

93. Reitmair AH, Schmits R, Ewel A, Bapat B, Redston M, Mitri A, et al. MSH2 deficient mice are viable and susceptible to lymphoid tumours. *Nat Genet* 1995; 11:64–70.

94. Reitmair AH, Redston M, Cai JC, Chuang TC, Bjerknes M, Cheng H, et al. Spontaneous intestinal carcinomas and skin neoplasms in Msh2-deficient mice. *Cancer Res* 1996; 56:3842–3849.

95. Reitmair AH, Cai JC, Bjerknes M, Redston M, Cheng H, Pind MT, et al. MSH2 deficiency contributes to accelerated APC-mediated intestinal tumorigenesis. *Cancer Res* 1996; 56:2922–2926.

96. Prolla TA, Baker SM, Harris AC, Tsao JL, Yao X, Bronner CE, et al. Tumour susceptibility and spontaneous mutation in mice deficient in Mlh1, Pms1 and Pms2 DNA mismatch repair. *Nat Genet* 1998; 18:276–279.

97. Baker SM, Bronner CE, Zhang L, Plug AW, Robatzek M, Warren G, et al. Male mice defective in the DNA mismatch repair gene PMS2 exhibit abnormal chromosome synapsis in meiosis. *Cell* 1995; 82:309–319.

98. Baker SM, Harris AC, Tsao JL, Flath TJ, Bronner CE, Gordon M, Shibata D, Liskay RM. Enhanced intestinal adenomatous polyp formation in Pms2–/–;Min mice. *Cancer Res* 1998; 58:1087–1089.

99. Miyaki M, Konishi M, Tanaka K, Kikuchi-Yanoshita R, Muraoka M, Yasuno M, et al. Germline mutation of MSH6 as the cause of hereditary nonpolyposis colorectal cancer [letter]. *Nat Genet* 1997; 17:271–272.

100. Edelmann W, Yang K, Umar A, Heyer J, Lau K, Fan K, et al. Mutation in the mismatch repair gene Msh6 causes cancer susceptibility. *Cell* 1997; 91:467–477.

101. Nakane H, Takeuchi S, Yuba S, Saijo M, Nakatsu Y, Murai H, et al. High incidence of ultraviolet-B-or chemical-carcinogen-induced skin tumours in mice lacking the xeroderma pigmentosum group A gene. *Nature* 1995; 377:165–168.

102. Sands AT, Suraokar MB, Sanchez A, Marth JE, Donehower LA, Bradley A. High susceptibility to ultraviolet-induced carcinogenesis in mice lacking XPC. *Nature* 1995; 377:162–165.

103. de Vries A, van Oostrom CT, Hofhuis FM, Dortant PM, Berg RJ, de Gruijl FR, et al. Increased susceptibility to ultraviolet-B and carcinogens of mice lacking the DNA excision repair gene XPA. *Nature* 1995; 377:169–173.

104. Berg RJ, de Vries A, van Steeg H, de Gruijl FR. Relative susceptibilities of XPA knockout mice and their heterozygous and wild-type littermates to UVB-induced skin cancer. *Cancer Res* 1997; 57:581–584.

105. Cheo DL, Meira LB, Hammer RE, Burns DK, Doughty AT, Friedberg EC. Synergistic interactions between XPC and p53 mutations in double-mutant mice: neural tube abnormalities and accelerated UV radiation-induced skin cancer. *Curr Biol* 1996; 6:1691–1694.

106. Becker K, Dosch J, Gregel CM, Martin BA, Kaina B. Targeted expression of human O(6)-methylguanine-DNA methyltransferase (MGMT) in transgenic mice protects against tumor initiation in two-stage skin carcinogenesis. *Cancer Res* 1996; 56:3244–3249.

107. Sakumi K, Shiraishi A, Shimizu S, Tsuzuki T, Ishikawa T, Sekiguchi M. Methylnitrosourea-induced tumorigenesis in MGMT gene knockout mice. *Cancer Res* 1997; 57:2415–2418.

108. Kawate H, Sakumi K, Tsuzuki T, Nakatsuru Y, Ishikawa T, Takahashi S, et al. Separation of killing and tumorigenic effects of an alkylating agent in mice defective in two of the DNA repair genes. *Proc Natl Acad Sci USA* 1998; 95:5116–120.

109. Zhang H, Somasundaram K, Peng Y, Tian H, Bi D, Weber BL, El-Deiry WS. BRCA1 physically associates with p53 and stimulates its transcriptional activity. *Oncogene* 1998; 16:1713–1721.

110. Sharan SK, Morimatsu M, Albrecht U, Lim DS, Regel E, Dinh C, et al. Embryonic lethality and radiation hypersensitivity mediated by Rad51 in mice lacking Brca2. *Nature* 1997; 386:804–810.

111. Ludwig T, Chapman DL, Papaioannou VE, Efstratiadis A. Targeted mutations of breast cancer susceptibility gene homologs in mice: lethal phenotypes of Brca1, Brca2, Brca1/Brca2, Brca1/p53, and Brca2/p53 nullizygous embryos. *Genes Dev* 1997; 11:1226–1241.

112. Hakem R, de la Pompa JL, Sirard C, Mo R, Woo M, Hakem A, et al. The tumor suppressor gene Brca1 is required for embryonic cellular proliferation in the mouse. *Cell* 1996; 85:1009–1023.

113. Suzuki A, de la Pompa JL, Hakem R, Elia A, Yoshida R, Mo R, et al. Brca2 is required for embryonic cellular proliferation in the mouse. *Genes Dev* 1997; 11:1242–1252.

114. Hakem R, de la Pompa JL, Elia A, Potter J, Mak TW. Partial rescue of Brca1 (5–6) early embryonic lethality by p53 or p21 null mutation. *Nat Genet* 1997; 16:298–302.

115. Friedman LS, Thistlethwaite FC, Patel KJ, Yu VP, Lee H, Venkitaraman AR, et al. Thymic lymphomas in mice with a truncating mutation in Brca2. *Cancer Res* 1998; 58:1338–1343.

116. Connor F, Bertwistle D, Mee PJ, Ross GM, Swift S, Grigorieva E, Tybulewicz VL, Ashworth A. Tumorigenesis and a DNA repair defect in mice with a truncating Brca2 mutation. *Nat Genet* 1997; 17:423–430.

117. Hino O, Kobayashi E, Hirayama Y, Kobayashi T, Kubo Y, Tsuchiya H, Kikuchi Y, Mitani H. Molecular genetic basis of renal carcinogenesis in the Eker rat model of tuberous sclerosis (Tsc2). *Mol Carcinog* 1995; 14:23–27.

118. Kobayashi T, Hirayama Y, Kobayashi E, Kubo Y, Hino O. A germline insertion in the tuberous sclerosis (Tsc2) gene gives rise to the Eker rat model of dominantly inherited cancer. *Nat Genet* 1995; 9:70–74.

119. Yeung RS, Xiao GH, Jin F, Lee WC, Testa JR, Knudson AG. Predisposition to renal carcinoma in the Eker rat is determined by germ-line mutation of the tuberous sclerosis 2 (TSC2) gene. *Proc Natl Acad Sci USA* 1994; 91:11,413–11,416.

120. Yeung RS, Xiao GH, Everitt JI, Jin F, Walker CL. Allelic loss at the tuberous sclerosis 2 locus in spontaneous tumors in the Eker rat. *Mol Carcinog* 1995; 14:28–36.

121. Kobayashi T, Mitani H, Takahashi R, Hirabayashi M, Ueda M, Tamura H, Hino O. Transgenic rescue from embryonic lethality and renal carcinogenesis in the Eker rat model by introduction of a wild-type Tsc2 gene. *Proc Natl Acad Sci USA* 1997; 94:3990–3993.

122. The European Chromosome 16 Tuberous Sclerosis Consortium. Identification and characterization of the tuberous sclerosis gene on chromosome 16. *Cell* 1993; 75:1305–1315.

123. Henry KW, Yuan X, Koszewski NJ, Onda H, Kwiatkowski DJ, Noonan DJ. Tuberous sclerosis gene 2 product modulates transcription mediated by steroid hormone receptor family members. *J Biol Chem* 1998; 273:20,535–20,539.

124. Tsuchiya H, Orimoto K, Kobayashi K, Hino O. Presence of potent transcriptional activation domains in the predisposing tuberous sclerosis (Tsc2) gene product of the Eker rat model. *Cancer Res* 1996; 56:429–433.

125. Orimoto K, Tsuchiya H, Sakurai J, Nishizawa M, Hino O. Identification of cDNAs induced by the tumor suppressor Tsc2 gene using a conditional expression system in Tsc2 mutant (Eker) rat renal carcinoma cells. *Biochem Biophys Res Commun* 1998; 247:728–733.

126. Urakami S, Tsuchiya H, Orimoto K, Kobayashi T, Igawa M, Hino O. Overexpression of members of the AP-1 transcriptional factor family from an early stage of renal carcinogenesis and inhibition of cell growth by Ap-1 gene antisense oligonucleotides in the Tsc2 gene mutant (Eker) rat model. *Biochem Biophys Res Commun* 1997; 241:24–30.

127. Ohh M, Yauch RL, Longergan KM, Whaley JM, Stemmer-Rachamimov AO, Louis DN, et al. The von Hippel-Lindau tumor suppressor protein is required for proper assembly of an extracellular fibronectin matrix. *Mol Cell* 1998; 1:959–968.

128. Gnarra JR, Ward JM, Porter FD, Wagner JR, Devor DE, Grinberg A, et al. Defective placental vasculogenesis causes embryonic lethality in VHL-deficient mice. *Proc Natl Acad Sci USA* 1997; 94:9102–9107.

129. Haber DA, Housman DE. The genetics of Wilms' tumor. *Adv Cancer Res* 1992; 59:41–68.

129a. Maheswaran S, Park S, Bernard A, Morris JF, et al. Physical and functional interaction between WT1 and p53 protein. *Proc Natl Acad Sci USA* 1993; 90:5100–5104.

130. Woodhouse E, Hersperger E, Shearn A. Growth, metastasis, and invasiveness of Drosophila tumors caused by mutations in specific tumor suppressor genes. *Dev Genes Evolution* 1998; 207:542–550.

131. St. John MA, Xu T. Understanding human cancer in a fly? *Am J Hum Genet* 1997; 61:1006–1010.

132. Xu T, Wang W, Zhang S, Stewart RA, Yu W. Identifying tumor suppressors in genetic mosaics: the Drosophila lats gene encodes a putative protein kinase. *Development* 1995; 121:1053–1063.

133. Herzog CR, Lubet RA, You M. Genetic alterations in mouse lung tumors: implications for cancer chemoprevention. *J Cell Biochem Suppl* 1997; 28–29:49–63.

134. Jolicoeur P, Bouchard L, Guimond A, Ste-Marie M, Hanna Z, Dievart A. Use of mouse mammary tumour virus (MMTV)/neu transgenic mice to identify genes collaborating with the c-erbB-2 oncogene in mammary tumour development. *Biochem Soc Symp* 1998; 63:159–165.

135. Ritland SR, Rowse GJ, Chang Y, Gendler SJ. Loss of heterozygosity analysis in primary mammary tumors and lung metastases of MMTV-MTAg and MMTV-neu transgenic mice. *Cancer Res* 1997; 57:3520–3525.

136. Yuan BZ, Miller MJ, Keck CL, Zimonjic DB, Thorgeirsson SS, Popescu NC. Cloning, characterization, and chromosomal localization of a gene frequently deleted in human liver cancer (DLC-1) homologous to rat RhoGAP. *Cancer Res* 1998; 58:2196–2199.

137. Zenklusen JC, Hodges LC, Conti CJ. Loss of heterozygosity on murine chromosome 6 in two-stage carcinogenesis: evidence for a conserved tumor suppressor gene. *Oncogene* 1997; 14:109–114.

138. Liyanage M, Coleman A, du Manoir S, Veldman T, McCormack S, Dickson RB, et al. Multicolour spectral karyotyping of mouse chromosomes. *Nat Genet* 1996; 14:312–315.

139. Moser AR, Shoemaker AR, Connelly CS, Clipson L, Gould KA, Luongo C, et al. Homozygosity for the Min allele of Apc results in disruption of mouse development prior to gastrulation. *Dev Dynamics* 1995; 203:422–433.
140. Riethmacher D, Brinkmann V, Birchmeier C. A targeted mutation in the mouse E-cadherin gene results in defective preimplantation development. *Proc Natl Acad Sci USA* 1995; 92:855–859.
141. Larue L, Ohsugi M, Hirchenhain J, Kemler R. E-cadherin null mutant embryos fail to form a trophectoderm epithelium. *Proc Natl Acad Sci USA* 1994; 91:8263–8267.
142. Rudolph U, Finegold MJ, Rich SS, Harriman GR, Srinivasan Y, Brabet P, et al. Ulcerative colitis and adenocarcinoma of the colon in G alpha i2-deficient mice. *Nat Genet* 1995; 10:143–150.
143. Matzuk MM, Finegold MJ, Su JG, Hsueh AJ, Bradley A. Alpha-inhibit is a tumour-suppressor gene with gonadal specificity in mice. *Nature* 1992; 360:313–319.
144. Tsuzuki T, Sakumi K, Shiraishi A, Kawate H, Igarashi H, Iwakuma T, et al. Targeted disruption of the DNA repair methyltransferase gene renders mice hypersensitive to alkylating agent. *Carcinogenesis* 1996; 17:1215–1220.
145. Baker SM, Plug AW, Prolla TA, Bronner CE, Harris AC, Yao X, et al. Involvement of mouse Mlh1 in DNA mismatch repair and meiotic crossing over [see comments]. *Nat Genet* 1996; 13:336–342.
146. Jacks T, Shih TS, Schmitt EM, Bronson RT, Bernards A, Weinberg RA. Tumour predisposition in mice heterozygous for a targeted mutation in Nf1. *Nat Genet* 1994; 7:353–361.
147. McClatchey AI, Saotome I, Ramesh V, Gusella JF, Jacks T. The Nf2 tumor suppressor gene product is essential for extraembryonic development immediately prior to gastrulation. *Genes Dev* 1997; 11:1253–1265.
148. McClatchey AI, Saotome I, Mercer K, Crowley D, Gusella JF, Bronson RT, Jacks T. Mice heterozygous for a mutation at the Nf2 tumor suppressor locus develop a range of highly metastatic tumors. *Genes Dev* 1998; 12:1121–1133.
149. Hino O, Klein-Szanto AJ, Freed JJ, Testa JR, Brown DQ, Vilensky M, et al. Spontaneous and radiation-induced renal tumors in the Eker rat model of dominantly inherited cancer. *Proc Natl Acad Sci USA* 1993; 90:327–331.
150. Gu H, Marth JD, Orban PC, Mossmann H, Rajewsky K. Deletion of a DNA polymerase b gene segment in T cells using cell type-specific gene targeting. *Science* 1994; 265:103–106.

2

Viral Oncoproteins as Probes for Tumor Suppressor Function

Srikumar P. Chellappan, PhD

CONTENTS

1. INTRODUCTION

Transformation by small DNA tumor viruses requires multiple events that induce normally quiescent cells into a state of proliferation *(1)*. This is necessary because the genomes of such viruses do not carry the machinery required for the replication of its own genome nor the components necessary for the transcription of its genes *(2)*. Hence, the common strategy utilized by this group of viruses is to make use of the transcription and replication machinery used by the host cells for their own purposes. The small DNA tumor viruses such as adenovirus (Ad), simian virus 40 (SV40), and human papillomavirus (HPV), are all capable of inducing a proliferative state in quiescent host cells, by the judicious use of their transforming oncoproteins *(2–5)*.

Apparently, the viral genome is designed so that the products of their early genes are capable of inactivating the major negative regulators of mammalian cell proliferation, namely, Rb and its family members and the p53 tumor suppressor protein *(3)*. Each of the three small DNA tumor viruses have proteins that can interact with these growth-suppressive proteins, and in every situation this interaction results in an inactivation of the tumor suppressor proteins *(6–9)*. The specific interaction between such viral oncoproteins (V-ONC) and the cellular tumor suppressor proteins have been studied in great detail, and this has led to a greater understanding of the biochemical processes involved in oncogenic transformation *(10)*. An important fallout of these detailed studies is that now it is possible to utilize these interactions for diagnostic purposes, because the V-ONC act as specific probes for the functional integrity of tumor suppressor protein.

From: *Tumor Suppressor Genes in Human Cancer*
Edited by: D. E. Fisher © Humana Press Inc., Totowa, NJ

Though the small DNA tumor viruses are not closely related evolutionarily, they all share the ability to inactivate the Rb family proteins, as well as p53 *(2,11)*. In the case of Ad, the *E1A* gene product can physically interact with the functional form of Rb and its family members *(12);* at the same time, products of the *E1B* gene can bind to and inactivate the p53 protein *(13–15)*. Similarly, in the case of HPVs, the E7 protein interacts with the Rb family proteins *(16),* and a separate protein, the product of the *E6* gene, binds to and inactivates p53 *(5)*. The situation is slightly different in the case of SV40: Here, the large T-antigen (T-Ag) is capable of interacting with both Rb family members and the p53 protein *(4,14)*. But, in all cases, the viruses carry genes that can neutralize the Rb- and p53-mediated suppression of cell proliferation.

This review is organized so that first the V-ONCs that interact with the Rb family of tumor suppressors are discussed, followed by those that bind to p53. In the last subheading, the potential use of these interactions in assessing the presence and functional status of the cellular antioncogenes is discussed.

2. V-ONCS INTERACTING WITH RB AND RB FAMILY PROTEINS

2.1. Adenovirus E1A

2.1.1. Historical Perspective

A considerable amount of work was done in the early 1980s to understand the mechanism of oncogenic transformation by the early gene products of DNA tumor viruses, especially the *E1A* gene of Ad. Two major lines of investigation were undertaken by different groups: mutational analysis of the different regions of the *E1A* gene required for transformation, and analysis of different cellular proteins associating with Ad E1A proteins. Results of these studies converged on the potential mechanisms involved in the E1A-mediated cellular transformation, and laid a solid foundation for what is known about viral oncogenesis today.

Ad E1A is a phosphoprotein expressed at the early stage of infection, and the phosphorylation of E1A has been shown to affect its function. Differential splicing gives rise predominantly to two different polypeptides of different sizes, one 243 residues long (12S E1A) and the other 289 residues long (13S E1A) *(17)*. There are additional smaller forms of E1A proteins also generated by differential splicing, but the complete transformation capacity of E1A requires the regions present in the 12S E1A protein. Analysis of the E1A structure revealed three regions highly conserved among different serotypes of Ads. These conserved regions (CR) have been named CR1, CR2, and CR3 (Fig. 1). Of these, CR1 and CR2 are present on both the 12S and 13S E1A, but CR3 is present only on the 13S E1A. As can be seen from the figure, the CR3 region almost perfectly overlaps the domain spliced out in the 12S E1A.

The CR1 and CR2 regions are derived from the exon 1 of the *E1A* gene, and the CR3 region is at the junction of exon 1 and exon 2. exon 2 of the *E1A* gene constitutes the multifunctional carboxy-terminal region, and is present in both the 12S and 13S forms of the protein *(18)*. This C-terminal region is involved in effecting functions like transcriptional repression, suppression of cellular transformation, and mediated susceptibility to the host cytotoxic T-lymphocyte response *(17)*.

It was discovered early that the Ad *E1A* gene could modulate the gene expression of both viral and cellular genes *(2)*. This was based on studies using mutant viruses, which had inactive forms of the *E1A* gene. Further studies revealed that E1A regulates

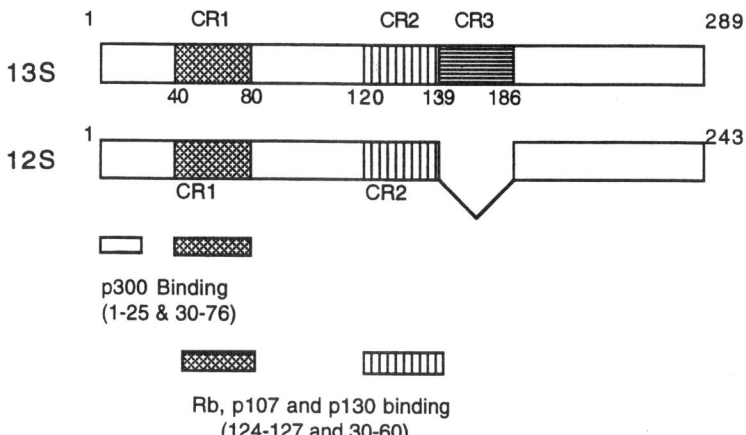

Fig. 1. Schematic of the Ad 5 E1A protein. Note that the CR3 domain present in the 13S protein is absent in the shorter 12S E1A. The conserved regions involved in binding to cellular proteins are also indicated.

gene expression, mostly at the level of transcription, and this is achieved through the mediation of many cellular transcription factors (TFs). E1A can induce, as well as repress, transcription from many cellular promoters, and these functions reside on different regions of the E1A protein. E1A does not bind to DNA directly, and all its cellular effects are mediated through targeting the cellular proteins of the host.

2.1.2. IMMORTALIZATION AND TRANSFORMATION BY E1A

It is well established that *E1A* is an oncogene, and is capable of immortalizing primary rodent cells on its own, and can effectively transform cells in cooperation with a second oncogene, such as *ras (1,2)*. The transformation function of *E1A* has been studied in detail, especially the requirements of different regions necessary for functionally cooperating with other oncogenes. Such studies have revealed that sequences contained in the exon 1 of E1A are capable of cooperating with oncogenic ras to transform primary cells *(17)*. But additional regions are necessary for transformation in collaboration with *E1B* gene or polyoma middle-T-Ag. For example, residues 140–193, overlapping with the CR3 region of the E1A protein, are required for cooperation with polyoma middle-T-Ag, and residues 266–276 in the C-terminal domain are required for cooperating with E1B in transformation.

Such detailed analysis of the regions of the E1A required for transformation helped identify three distinct domains within exon 1 sequences to be essential for transformation function. The CR1 region, as well as the CR2 regions, were absolutely necessary for transformation function, but were not sufficient by themselves. It soon became clear that the amino (N)-terminal residues, 2–25, which fell outside the span of CR1 sequences, were indispensable for transformation. Mutations within CR1 and CR2 significantly reduced the capacity of E1A to transform cells. Although the 12S E1A can bring about complete transformation in most primary cell lines, the CR3 region is essential for transformation of certain specific cell lines *(17)*.

In addition to transforming cells, E1A was found to have certain growth-suppressive properties also *(17,19,20)*. The first indication of this was obtained when it was found that transformation of baby rat kidney cells by the 12S E1A was 100× more efficient than by the 13S E1A molecule. This raised the possibility that the CR3 domain, or other regions encoded by the exon 2, can function to suppress transformation. Further studies *(21–23)* showed that the repressive function resided within residues 237 and 283, especially in the region spanning 256–283. The molecular mechanisms by which this region mediates such suppressive properties are not yet clear, even though certain cellular proteins, such as CtBP, have been found to bind to this region. Repression of transformation by E1A exon 2 region may involve more indirect mechanisms also. For example, it is clear that E1A protein can induce the accumulation of the p53 protein, which has strong tumor-suppressive properties. E1A is thought to enhance p53 levels both by increasing the transcription of its genes, as well as by increasing the stability of the protein *(24,25)*. The detailed molecular mechanisms involved in these processes are emerging now.

2.1.3. E1A BINDING PROTEINS

The identification of different regions of E1A involved in cellular transformation facilitated detailed studies on the underlying molecular mechanisms. As mentioned earlier, one fruitful strategy was to identify cellular proteins that interact with E1A, and assess the potential effects of the interaction *(1)*.

Early studies *(26)* identified a set of six cellular proteins that co-immunoprecipitated with Ad E1A from cellular extracts. They ranged in size from 300 to 33 kDa. The first protein to be positively identified as an E1A-binding protein was the Rb protein, the product of the retinoblastoma tumor suppressor gene. This finding was a milestone in the study of oncogenic mechanisms, because it provided an example of a V-ONC binding to and conceivably neutralizing a cellular tumor-suppressor protein. It further became evident *(12)* that both of the remaining Rb family members, p107 and p130 proteins, also bound to E1A equally well *(27)*. E1A was found to bind to these proteins through the peptide sequence, LXCXE, which is present in the CR2 region. It was further found to be present in HPV E7 protein, as well as SV40 large T-Ag *(18,28)*. As described in subheading 2.1.4., these oncoproteins, although derived from unrelated viruses, all appear to bind to the members of the Rb family. In the case of E1A, there appeared to be two distinct regions that made contacts with the Rb protein: the LXCXE motif in the CR2 region, and residues 30–60 of the CR1 region.

The 300 kDa protein that was found to associate with E1A has now been identified to be the transcriptional co-activator, p300/CBP *(29)*. Many members of this co-activator family have been found to bind to E1A, and E1A can effectively block their transcriptional activity. It has also been reported that E1A can prevent the cyclin cyclin-dependent kinase (CDK)-mediated phosphorylation of p300. The p300/CBP family of transcriptional co-activators has indigenous histone acetylene transferase (HAT) activity, and they are associated with other HAT proteins, such as PCAF-1, in vivo. E1A is believed to affect these interactions, thus modulating the transcriptional activity of p300. Mutational studies of E1A protein had made it clear that the extreme N-terminal residues spanning 1–25 are essential for the transformation function of E1A *(30)*. It is apparent that this region of E1A is involved in binding to p300, and that the subsequent inactivation is indispensable for E1A to transform cells.

The two proteins of 60 and 33 kDa have been identified to be cyclin A and CDK2. The CR2 region is chiefly involved in the binding to these proteins, and the CR1 region plays a secondary role in the interaction. Although it is established that E1A binds to

Table 1
Cellular Proteins Binding to E1A

Protein	Region of E1A required for binding
p400	**1–48**
p300	**1–25,** 30–76
Rb	**121–127,** 30–60
p107	**124–127,** 30–60
p130	**124–139,** 30–60
Cyclin A, p60	**124–127,** 30–60
P33CDK2	**124–127,** 30–60
BS69	**140–185,** 76–120
CtBP	**271–284**

cyclin A directly, it is not yet clear whether the interaction with CDK2 is direct or through cyclin A. Nevertheless, the pattern that is emerging suggests that E1A can bind to and affect the function of critical cellular proteins involved in cell cycle regulation. A list of the proteins known to bind directly to E1A, and the regions involved in the binding, is shown in Table 1.

2.1.4. INACTIVATION OF RB FUNCTION BY E1A

Studies on the mechanisms involved in E1A-mediated transformation have highlighted the inactivation of Rb and its family proteins as an essential step in this process *(12)*. Because Rb is a well-characterized tumor suppressor protein, the interaction between Rb and E1A has garnered maximum attention.

Rb is a nuclear phosphoprotein that plays a critical role in the progression of the mammalian cell cycle *(31)*. Inactivation of Rb by phosphorylation is a necessary step for the transition of proliferating cells from G1 to S phase. It is well-established that it is the hypophosphorylated form of Rb that is functional in arresting cell proliferation, and inactivation of Rb in the mid-to-late G1 phase, by cyclins D and E and their associated kinases, facilitates the G1-to-S transition *(31–33)*. E1A, as well as other V-ONCs, such as HPV E7 and SV40 large T-Ag, all preferentially binds to the functional hypophosphorylated form of Rb *(13)*. It has now been shown that interaction with E1A, E7, or SV40 large T-Ag all bring about an inactivation of Rb that is equivalent to its phosphorylation by CDKs.

All three Rb family members have a central conserved domain, named "pocket domain" *(34)*. The pocket domain imparts the growth-regulatory functions of the Rb protein *(35)*, and almost all point mutations or deletions of the *Rb* gene found in human cancers map to this region *(36–38)*. E1A binds to the functional pocket domain of the Rb protein. Studies in the past few years have established that interaction of E1A with Rb, or the phosphorylation of Rb, leads to its inactivation, which is equivalent to a deletion or mutation of the pocket domain of the gene.

As shown in Fig. 2, the pocket domain comprises two subdomains, named the A and B pockets. In the case of p107 and p130, these subdomains are separated by a spacer region *(39)*. p107 and p130 are known to bind to cyclins A and E through the spacer region, through a sequence that is similar to the cyclin-A-binding domain of the

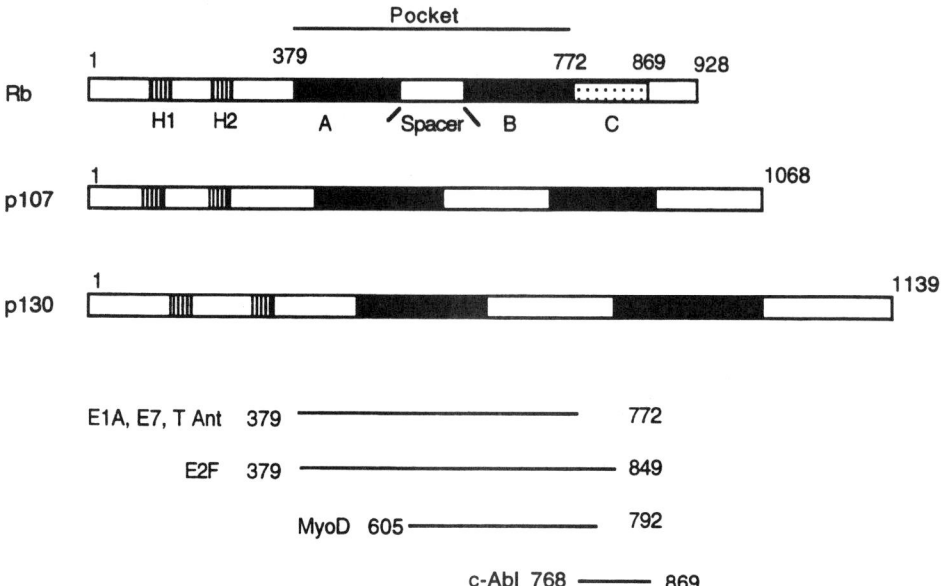

Fig. 2. Structure of the Rb family proteins. The conserved pocket domain is shown by the filled box; the H1 and H2 domains also share significant homology. The region 768–869 of Rb is involved in binding to c-Abl protein, and is referred to as the C box. The spacer region between A and B boxes are considerably larger and functionally distinct in p107 and p130, compared to Rb.

Waf1/cip1/p21 CDK-inhibitory protein *(40)*. The spacer region is almost absent in Rb protein, although Rb can bind to the cyclin D proteins *(41)*. E1A and other V-ONCs associate with Rb and Rb family members through the mediation of the A and B pockets.

The functional consequence of E1A interacting with the Rb family proteins has been elucidated. Studies in the early 1990s showed that Rb is associated with a cellular TF, E2F *(42–44)*, which was originally identified as a factor necessary for the E1A-mediated induction of the Ad *E2* gene, and was found to bind to a sequence element, TTTCGCGC *(18)*. There were two such elements present in the E2 promoter, and it was found that E1A induces cooperative binding of E2F to the sites, in association with the Ad E4 protein. It was also found that E1A can dissociate multiprotein complexes that contain the E2F TF *(45)*, and it was already known that E1A bound to the Rb protein. Attempts were being made to identify the cellular proteins that bind to Rb as well, and such combined efforts revealed that E2F is a target for the Rb protein, and that E1A binding to Rb disrupts the interaction between Rb and E2F. The term E2F now refers to a family of six proteins, E2Fs 1–6. Of these, E2Fs 1–5 are transcriptionally active *(46)*, and E2F6 is repressive in nature *(47)*. Although the transcriptionally active E2Fs have similar DNA-binding abilities, they show a preference in their ability to bind to different Rb family members: E2Fs 1, 2, and 3 bind to the Rb protein *(39)*; E2Fs 4 and 5 preferentially interact with p107 and p130 *(39,48,49)*.

Further analysis of the functional consequences of the interactions involving Rb, E1A, and E2F showed that Rb binds to E2F, and represses its transcriptional activity *(50)*. The binding of E1A could effectively reverse this Rb-mediated repression of E2F to Rb *(50)*. Cloning of the E2F family members revealed that Rb binds to the transcriptional activation domain of E2F1, thus preventing its ability to function as a TF *(51,52)*. Recent studies have shown that Rb can block transcription from promoters containing E2F sites, by recruiting the cellular histone deacetylase enzyme, *HDAC1 (53–55)*.

In the case of E1A, it appeared that induction of E2F activity facilitated the expression of the *E2* gene. As described in later subheadings, HPV E7, as well as SV40 large T-Ag, were also capable of dissociating E2F from Rb, and thus inducing its transcription *(56)*. This was surprising, since none of the genes present in these viruses had E2F sites in their promoters. This raised the possibility that activation of E2F may be contributing to the expression of the *E2* gene in Ad, but it plays a more important role in the regulation of the cell cycle. It is now evident that the V-ONCs activate E2F to achieve a proliferative state, which is conducive for their DNA replication *(13,14)*.

It became apparent that many cellular promoters that were known to be induced by Ad E1A had E2F sites in their promoter *(57)*. Further, an analysis of the genes that are regulated by E2F showed a variety of cell-cycle regulatory genes as downstream targets of E2F. For example, genes for proteins such as DHFR, DNA polymerase α, ribonucleotide reductase, thymidylate kinase, thymidylate synthase, and so on, which are all necessary for DNA synthesis, are regulated by E2F. Further, many additional cell-cycle proteins, such as cyclins A and E, p107 CDK2, and soon were found to be regulated by E2F as well. The current model suggests that inactivation of Rb by phosphorylation, during the G1 phase of the cell cycle, releases free E2F activity, which transcribes this set of genes, required for cell cycle progression. In the case of viral infection, the products of the immediate early genes achieve the same end result, but by inactivating Rb by a direct interaction. Similarly, it may be imagined that, when the *Rb* gene is inactivated by mutation or deletion, there would be an abundance of active free E2F, which would contribute to uncontrolled cell cycle progression, and hence oncogenesis *(57)*.

The regions of E1A necessary for dissociation of E2F–Rb complexes have been worked out in detail. As discussed earlier, it is the CR2 region that chiefly mediates the interaction of 1A with Rb, but the CR1 region, especially residues 30–60 *(58)*, contributes to the binding. Within this region, a tyrosine residue at position 47 is important for stable binding of the human Rb protein. There is no region in E7 or SV40 large T-Ag similar to this E1A CR1 motif. It appears that the same motifs are essential for the disruption of E2F-containing complexes by E1A. One model that has been proposed, based on extensive mutational analysis, as well as on competition experiments, is that the CR2 region would tether E1A to the Rb protein *(58)*. Once E1A is bound to Rb, the CR1 region blocks the region of Rb involved in binding to E2F, thus preventing the formation or existence of a Rb–E2F complex *(59)*. It has been proposed that cyclin D would be functioning in a similar fashion to disrupt Rb–E2F complex. Because cyclin D has an LXCXE motif, it may tether to the Rb protein and bring to its proximity the CDK4/6 kinases, which can phosphorylate Rb and disrupt an E2F–Rb interaction.

2.1.5. Interaction of E1A with p107 and p130

The interaction of E1A with Rb is the best-characterized interaction among all E1A-binding proteins, but all members of the Rb family interact with E1A in essentially the same fashion. It has been shown that interaction of E1A with p107, as well as with p130, can dissociate the E2F proteins associated with them *(60)*. The regions of E1A involved in binding to p130 spans a few additional residues in the CR2 region; residues 121–127 are required for binding to Rb, as well as to p107, and the residues 121–139 are necessary for binding to p130 *(17)*.

The functional consequences of E1A binding to p107 and p130 may be the same, but it is not yet clear whether E1A must bind to and inactivate these proteins to induce cell proliferation. This is a question especially in the case of p107, in which it complexes with E2F in the S phase of the cell cycle *(61)*, and hence the importance of inactivating a protein that functions past the G1–S transition point is questionable. The interaction of E1A with p130 may be more important, since Ad infects mostly quiescent cells, and p130–E2F complexes are prevalent in resting cells *(62)*. In a broader view, the interaction of E1A with either p107 or p130 appears to be less important than its interaction with Rb, simply because the role of these proteins in normal cell cycle regulation is not as significant as that of Rb. This is borne out by the fact that no mutations of the p107 gene has been reported in human cancers, and p130 has been reported to be mutated in a small subset of lung carcinomas. It may be assumed that, for the purposes of this review, the interactions of these proteins with E1A is not relevant.

2.2. HPV E7 Protein

Papilloma viruses are small DNA tumor viruses that have been linked to cancers of the genital tract *(63)*. About 70 types of papilloma viruses have been isolated, and are broadly classified into low-risk and high-risk HPVs, based on the correlation between their presence in benign genital warts or malignant cervical carcinoma *(64)*. HPVs are unique in that, although they share the same pathways of transformation as Ads and SV40, the latter two are not correlated with human cancer.

The genomes of HPV types are different, but the general organization is highly conserved *(3)*. There are three distinct regions within the genome: a region containing the regulatory elements for the transcription of viral genes, a region encoding six early genes, and one encoding two late genes. The most common HPV types found in cervical carcinoma are HPV16 and HPV18; for all practical purposes, their transforming proteins are identical. As in the case of Ads, there are two early genes that are crucial for transformation: the *E7* gene, which is functionally equivalent to Ad E1A; and the *E6* gene, which is comparable to the *E1B* gene *(9)*.

It has been found that integration of the viral genome into mammalian cells causes a loss of many viral genes, but maintains the *E7* and *E6* genes. Transformation experiments have suggested that *E7* gene from the high-risk HPV types can transform cells very efficiently; the *E6* gene has a lesser capacity to do so *(65)*. Studies using the soft-agar colony formation assay have shown that E7 protein, in association with a second oncogene, can transform a variety of primary rodent and mammalian cell lines, and the continued presence of the E7 protein is required for maintenance of the transformed phenotype. The *E6* gene has a lower capacity to transform cells, but it can efficiently cooperate with the *E7* gene to transform primary human keratinocytes, which are the natural hosts for HPV. Both genes, of low-risk HPV type, have weak transformation potential, thus correlating their activity with the tumorigenicity of the HPV type.

As mentioned earlier, the E6 protein is similar to Ad E1B protein functionally, and it can interact with the p53 tumor suppressor protein like E1A. Similarly, the E7 protein, which is functionally analogous to the Ad E1A, can bind to the Rb family of tumor suppressor proteins. This interaction is dealt with here first.

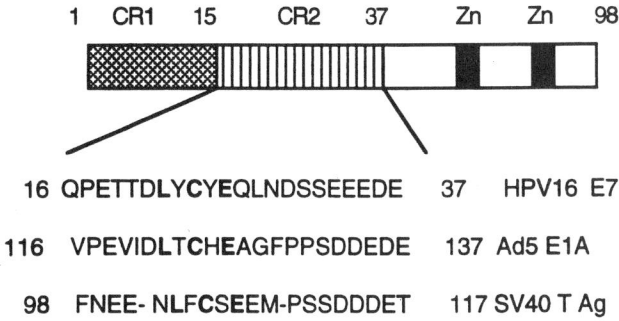

Fig. 3. Schematic of the HPV16 E7 protein. The conserved CR1 and CR2 domains are at the N-terminal half of the protein. The region of high-sequence homology between E7, E1A, and T-Ag is shown below, with the LXCXE motif in bold.

2.2.1. E7 Protein and its Interaction with Rb

The HPV E7 is a small acidic protein that localizes to the nucleus and nuclear matrix. This 98-amino-acid (aa) protein can be divided into three domains: CR1, CR2, and CR3, based on the structural similarity to the Ad *E1A* gene *(66)*. CR1 and CR2 are at the N-terminal region of the protein, and the CR3 at the C-terminal end (Fig. 3). CR2 is essential for the binding to the Rb protein, and both CR1 and CR3 are required for the transformation function of E7. The CR3 region of the protein has two zinc-binding motifs, which are involved in dimerization of E7. Mutations in the zinc-binding region of E7 abolished its ability to transform cells, although it was able to bind to Rb protein very efficiently. There are two potential casein kinase sites at the C-terminus of the protein, and phosphorylation on these sites is believed to be important for the full transformation function of E7 *(3)*.

The interaction of E7 with Rb family of tumor suppressor proteins has been studied in detail *(16,67,68)*. In fact, the crystal structure of the CR2 region of E7, bound to the pocket domain of Rb, has been elucidated. As in the case of E1A, E7 protein binds to the pocket domain of Rb through a conserved LXCXE motif, and the binding to Rb is required for E7 to transform cells *(16)*. The first indication of the correlation between Rb inactivation and the transformation by HPV E7 was obtained in 1991 *(69)*, when an analysis of human cervical carcinoma cell lines revealed that those having an intact *Rb* gene had HPV E7 incorporated in the genome. In contrast, those that had a mutant *Rb* gene contained no E7. This suggested a direct correlation between the inactivation of Rb, either by mutation, or through the binding or HPV E7 protein.

Mutations in the Rb-binding motif of E7 totally abolished the ability of E7 to promote growth of primary cells in soft agar in cooperation with oncogenic *ras*. As already mentioned, low-risk HPV types have low capacity to transform cells, and, supporting this observation, HPV6 and HPV11 E7 can bind to Rb only weakly. Although the Rb-binding moieties of the high- and low-risk HPV types are similar, a comparison showed that HPV 16 E7 has an aspartic acid at position 22; HPV 6 E7 had a glycine *(3,70)*. Substitution of the glycine with aspartic acid, in the HPV 6 E7, enhanced its ability to bind to Rb, and increased its ability to transform cells. This correlation does not appear universal, because certain low-risk types of HPV have an aspartic acid at this position, and can bind to Rb efficiently. The correlation between the ability of E7 to transform cells and to bind Rb is still not clear, especially in the case of low-risk types. It has been proposed that, although the low-risk E7 proteins bind to Rb with comparable affinity, the functional consequences of the interactions may be different in the various types of HPV.

As in the case of Ad E1A, HPV E7 preferentially binds to the hypophosphorylated form of Rb. Similarly, binding of E7 could effectively dissociate Rb–E2F complexes in mammalian cell extracts. The crystal structure of HPV16 E7, bound to the pocket domain of Rb, has been solved recently, and the structure supports the earlier functional observations made on the interaction *(71)*. It was found that a 9-aa E7 peptide, carrying the LXCXE motif, binds to a highly conserved region within the B box of the Rb pocket. Both the A and B boxes seem to have structural similarities to cyclins and TFIIB, in that they all possess a five-helix cyclin fold. The LXCXE sequence was found to bind to a shallow groove in the B pocket, which was formed by three cyclin fold helices. Alternating Leu, Cys, Glu, and Leu side chains of the E7 peptide make intermolecular contacts with the B-box groove. In addition, there is a high density of van der Waals forces and hydrogen bond contacts distributed uniformly between the E7 peptide and the B box, contributing to the strength of the binding. The actual contact of the E7 peptide was to the groove of the B box, but the interphase of the A–B boxes appeared to contribute significantly to the binding. The structure also revealed a high degree of conservation of the B-box site, which binds to LXCXE motifs. Four residues that contact the backbone of the E7 peptide, Tyr709, Tyr756, Asn757, and Lys713 are identical in diverse species of Rb proteins, as well as in p107 and p130. The high level of conservation of this B-box site suggests that it plays a major role in the functioning of the Rb family proteins *(71)*.

It has been shown recently *(72)* that the half-life of the Rb protein is considerably reduced in cells stably transformed with the E7 protein *(72)*. An overexpression system showed that high levels of E7 protein can lead to an increased decay of the Rb protein, and this could be blocked by proteasome inhibitors. The degradation was limited to Rb, since there was no change in the stability of p107 or p130 proteins in response to E7. Further, this function appeared to be a specific feature of the HPV E7 protein, because neither Ad E1A nor SV40 T-Ag could affect the half-life of Rb *(72)*.

2.2.2. BINDING OF E7 TO P107 AND P130 PROTEINS

The binding of E7 to p107 and p130 has been studied in detail. Unlike in the case of E1A, there are apparent differences in the consequences of binding to Rb vs p107 or p130. First, it became apparent that E7 protein cannot dissociate p107–cyclin A–E2F complexes, unlike E1A; instead, it remains associated with the complex *(73)*. The association with the p107–cyclin A–E2F complex was also dependent on the LXCXE motif, and E7 from HPV 6 had a reduced capacity for association. It has been suggested that HPV E7 can target cellular genes like c-*myb* by targeting the p107–cyclin A–E2F complex in NIH 3T3 cells. Apparently, the expression of B-*myb* promoter in these cells correlates with the binding of distinct p107–E2F complexes at the E2F binding site, and Rb–E2F complexes do not appear to play a major role in this regulation *(74)*. It has been found that, although the inactivation of Rb family proteins and the induction of E2F activity correlates with the transformation function of E7, this alone is not sufficient. Despite the suggestion that E7 interacts differently with p107 and p130, it is not yet clear whether the interactions of E7 with these proteins are important for its transformation function.

It may be concluded that the interaction of HPV E7 with Rb has been elucidated more clearly at the structural level, and the conclusions drawn from this can be extended to the other Rb-binding proteins, such as like E1A and SV40 T-Ag. Further, cellular proteins, such as cyclin D, may be targeting Rb through similar interactions *(71)*.

3. V-ONCS INTERACTING IWTH P53
3.1. HPV E6 Protein

The HPV E6 is an 18 kDa 151-aa basic protein that also localizes to the nuclear matrix and cell membranes *(10)*. Its most notable structural feature is the presence of four Cys motifs, which can form two well-defined Zn fingers. These motifs can bind Zn in vitro, and are highly conserved between all serotypes of HPV. The E6 protein has no homology to Ad E1B or SV40 T-Ag, but can function in a similar fashion. The major common feature of these three proteins is their ability to bind and inactivate p53 tumor suppressor protein. In certain high-risk HPV-infected cells, polycistronic E6 messages have been detected, which can give raise to full-length, or to a shorter, protein, E6* *(10)*.

E6 protein from high-risk HPV types associate with p53 with higher efficiency than E6 from low-risk HPV types *(5)*. The binding of E6 to p53 is enhanced by a cellular protein, E6AP (E6-associated protein). It has now been established that E6 protein binding leads to the degradation of p53, thus reducing its half-life. The proteolytic degradation of p53 through the ubiquitin–proteasome pathway has been studied in detail. Consistent with these observations, cell lines that carry a high level of E6 have very low amounts of p53 protein, mimicking situations in which *p53* gene is mutated. In vitro analyses have identified two domains of E6 that are involved in the binding and degradation of p53: the C-terminal end of E6 is required for binding; the N-terminal end is required for effecting degradation. E6 protein can inhibit the transcriptional activity of p53, and this does not require the activation of the proteasome pathway. In addition to p53, a variety of cellular proteins, ranging in mol wt from 33 to 212 kDa, have been found to associate with HPV E6, but their identities are not yet known.

3.2. Ad E1B Protein

The Ad 5 *E1B* gene has been studied with respect to its interaction with the p53 protein *(75)*. Unlike the Rb-binding V-ONCs, there are no structural similarities between the p53-binding proteins *(8)*. Thus, although Ad, HPV, and SV40 all encode proteins that can bind to p53, there are no conserved or shared domains between them. Further, the functional consequence of binding of these proteins to p53 are also different: the Ad E1B binding represses the transcriptional activity of p53; the HPV E6 protein leads to the degradation of p53 *(10)*.

Ad *E1B* gene codes for two distinct protein products, one 55 kDa and the other 19 kDa in size. Only the 55 kDa E1b protein has been found to physically interact with p53. The p53-binding domain of E1B is not conserved, even in different serotypes of Ad, and E1B protein from certain strains, such as Ad12, cannot bind to p53 *(28)*. Because p53 plays a major role in arresting cells in G1, in response to DNA damaging agents, or induces apoptosis if the DNA damage cannot be successfully repaired, it is believed that the V-ONCs that target p53 lead to a suppression of the cell death program *(16)*. E1B proteins are believed to suppress cell death programs initiated by DNA damage, as well as by other viral proteins like E1A *(15,77)*. Thus, the V-ONCs that bind to p53 cause distinct functional effects than those binding to Rb and facilitating G1–S transition.

The functional characterization of p53 has shown that it is a TF, possessing distinct DNA-binding and activation domains. Further, it is very well established that p53 induces a wide variety of cellular genes, while repressing certain other genes. Studies on the functional consequences of E1b binding to p53 revealed that the 55 kDa protein targets the activation domain of p53, and thus inhibits p53-mediated transcriptional induction. This could influence the expression of vital cell cycle genes, such as the p21$^{Waf1/Cip1}$ proteins and Mdm2, which is a regulator of p53 itself. Detailed mutational analysis has shown that the ability of E1B 55 kDa protein to repress p53-mediated transcriptional activation strongly correlated with its ability to transform cells. Ad12 E1B 55 kDa protein, which was unable to bind to p53, was effective in blocking p53-mediated transcriptional activation. Conversely, the interaction between p53 and E1B was necessary, but not sufficient, for transcriptional repression, as well as transformation functions of E1B *(78,79)*. This conclusion is based on the finding that a single aa insertion at residue 443 abolishes the ability of E1B to bind p53, but it was effective in transcriptional repression and transformation. The C-terminal end of E1B outside the p53-binding region was required for its transformation function; phosphorylation at three sites within this region was also essential for the transformation function *(78,80)*.

The region of E1B 55 kDa that binds to p53 also is involved in binding to the Ad E4 protein Orf6 *(78)*. This led to the suggestion that the repressive properties of E1B are facilitated by the presence of E4Orf6. In addition, one study has shown that E1B, in cooperation with E4Orf6, modulates not only the transcriptional activity of p53, but also the levels of p53 protein. Because E4Orf6 has also been shown to bind to p53, it is thought that collective interactions among E1B55kd, E4Orf6, and p53 lead to modulation of the levels, as well as the activity, of p53.

It has been reported that the E1B 19 kDa protein can also affect p53 function, but that this does not require a direct interaction. Unlike E1B 55 kDa protein, the 19 kDa protein was unable to block the transcriptional activity of p53; since p53 is also known to repress the transcription of certain cellular genes, it is believed that E1B19 kDa protein affects the transcriptional repressive properties of p53.

3.3. SV40 T-Antigen

SV40 is an oncogenic DNA tumor virus that was originally discovered in rhesus monkey kidney cells. The oncogenic property of this virus resides in two early gene products, the large (T) and the small (t) tumor Ag *(4)*. The large T-Ag can transform cells on its own, but the small t-Ag cannot; the latter can enhance the transformation potential for the large T-Ag. The transformation function of the T-Ag was found to require its interaction with the Rb family tumor suppressor proteins, as well as the inactivation of p53. SV40 thus differs from Ad and HPV in having one protein that can inactivate Rb, as well as p53 pathways; these functions reside in separate proteins in the latter two. Further, SV40 T-Ag is also capable of binding to DNA, unlike E1A or E7 proteins *(81)*.

The SV40 T-Ag is a polypeptide of 708 aa, and is considerably larger than the other oncoproteins discussed thus far (Fig. 4). There are several distinct functional domains

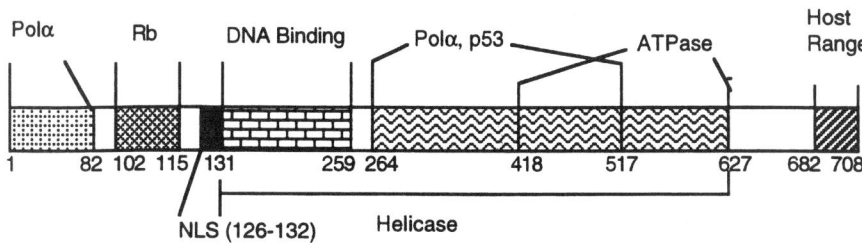

Fig. 4. Structure of the SV40 large T-Ag. The various functional and protein-binding domains are indicated.

that have been extensively characterized. A schematic of the T-Ag domain structure is shown in Fig. 4; as can be seen, the N-terminal end (1–82) and an internal domain is involved in binding to DNA polymerase α-primase. The Rb-binding domain carrying the LXCXE motif spans residues 102–115. Other well-defined domains include, progressively toward the C-terminal end, a NLS (126–132); a DNA-binding domain (131–259); a finger motif (302–320); a second DNA polymerase α-primase-binding domain (259–517), which partly overlaps with the p53-binding domain (275–517); and an adenosine triphosphatase (ATPase)/ATP-binding domain (418–627). A helicase domain extends from residue 131 to 627. There is a cluster of phosphorylation sites on the C-terminal region, which has been reported to be critical for T-Ag function. A small domain that defines the host range of (55)SV40 resides at the extreme C-terminal end, spanning residues 682–708. The ability of T-Ag to perform such a diverse array of functions enables it to facilitate all stages of viral replication and propagation *(4)*.

3.3.1. INTERACTION WITH RB FAMILY MEMBERS

The interaction of SV40 large T-Ag with Rb protein has been well-characterized, and the inactivation of Rb is essential for T-Ag to transform cells *(82)*. As mentioned earlier, the Rb-binding domain of large T-Ag has the conserved canonical LXCXE motif. The integrity of this motif is essential, but not sufficient, for the oncogenic activity of T-Ag. It is clear that interaction of T-Ag with Rb leads to an inactivation of Rb function, as in the case of E1A or HPV E7, leading to a dissociation and activation of E2F TF. Such studies have established that these three DNA tumor viruses all utilize common mechanisms for inducing the cells to enter S phase, creating an environment conducive to the replication of viral DNA. One difference in the case of T-Ag, however, is that it can perform considerably more functions than E1A alone, or even E1A and E1B combined *(7,83)*.

Consistent with the finding that T-Ag can efficiently dissociate Rb–E2F complexes, it was found that a human cell line, WI38-VA13, which stably expresses T-Ag, had no Rb–E2F complexes at all *(84)*. There was a reduced level of the cyclin A–E2F complex, and the loss of these complexes corresponded to an increase in the levels of free, transcriptionally active E2F. Like E1A and HPV E7, T-Ag also targets the pocket domain of hypophosphorylated Rb. Hence, biochemically, as well as functionally, the interaction of T-Ag with Rb is similar to that of E1A or E7.

It has been shown recently *(85)* that a region of T-Ag N-terminal to the conserved Rb-binding region is required for the functional inactivation of the Rb protein. This region has been named the J-domain, because it has sequence homology to the J-domain of the DnaJ family of molecular chaperons of *Escherichia coli*. The J-domains are characterized by conserved histidine–proline–aspartate (HPD) residues, and this tripeptide is present at the N-terminus of the SV40 large T-Ag. Further, the J-domain of SV40 T-Ag, as well as certain other polyomaviruses, can functionally substitute for the J-domain of *E. coli* DnaJ chaperon.

Studies on the J-domain of T-Ag showed that it is required to overcome the G1–S arrest induced by all Rb-family members *(85,86)*. In addition, it was required to reverse the repression of E2F activity brought about by Rb, as well as p130. From these experiments, it appears that, in the case of T-Ag also, the LXCXE motif tethers the protein to Rb, while a N-terminal region functionally inactivates it *(86)*. This general pattern is similar to the one seen in the E1A-mediated inactivation of Rb *(58,59)*.

It has also been reported that the J-domain of T-Ag can affect the phosphorylation status of p107 and p130 proteins. In cells stably expressing T-Ag, there is a reduced amount of phosphorylated forms of p107 and p130. There was also a faster turnover of p130 protein, which could be a result of its aberrant phosphorylation status. These effects of T-Ag appear to be J-domain-dependent, because point mutations in the HPD motif abolished these changes. Further, replacement of the N-terminal J-domain of T-Ag with J-domain motifs from cellular proteins restored this ability. Similarly, the transformation function of T-Ag also appeared to require a functional J-domain. The biochemical basis for the functioning of the J-domain is not yet clear, but it appears to be as important as the LXCXE motif in the Rb-binding domain for full T-Ag function.

3.3.2. INTERACTION OF T-AG WITH P53

As in the case of other p53 binding vONCs, the region of T-Ag that is involved in binding to p53 is large and not conserved. The biochemical effect of the binding of T-Ag to p53 is also different from the binding of E6 or E1B. One interesting facet of this interaction is that phosphorylation of both T-Ag and p53 appear to be necessary for binding in murine cells. In addition, the p53-binding region of T-Ag overlaps with the binding sites for DNA polymerase α and ATP, raising the possibility that T-Ag–p53 complexes may affect the viral functions regulated by T-Ag *(4)*. This notion is supported by the finding that a single point mutation (Pro to Leu), at position 584 of T-Ag, resulted in a loss of p53 binding, along with changes in the ATPase activity of T-Ag, as well as its ability to oligomerize. This mutation also resulted in reduced stability of T-Ag, and resulted in defective replication and reduced transformation functions.

It has been found that wild-type murine p53 can block the binding of T-Ag to DNA polymerase α; in addition, wild-type p53 could effectively reduce the replication of viral origins, but mutant p53 molecules were unable to do this *(6)*. This could possibly result from p53 competing for the binding to T-Ag to DNA polymerase α or another cellular protein involved in replication. One other interesting aspect of the interaction between T-Ag and p53 is that the phosphorylation, as well as the stability, of the latter increase, upon T-Ag-mediated cellular transformation. The relative importance of p53 binding in T-Ag-mediated transformation of primary murine cells was highlighted in one study, in which it was found that the N-terminal region of T-Ag, up to residue 250, was not necessary for this function. This ruled out a role for Rb-binding, nuclear translocation, and DNA binding abilities of T-Ag. In that study, residues 251–626 were found to be vital for immortalization, suggesting that p53 binding and inactivation is indispensable for the transformation function of T-Ag *(87).*

The modulation of p53 by T-Ag extends beyond enhancing the stability of the former. It has been shown that T-Ag can block the DNA-binding activity of p53, which correlated with T-Ag inhibiting the transcriptional activation functions of p53. One recent study showed that the N-terminal domain of T-Ag, which is not involved in DNA binding nor binding to p53, could effectively repress p53-mediated transcription. This suggests that the N-terminal region of T-Ag may be affecting p53-mediated transcription indirectly, through other cellular factors involved *(6).*

Overall, it may be summarized that, although the interaction of V-ONCs with Rb family members has been characterized in great detail, ambiguity still exists as to the nature of their interactions with p53, and its functional consequences, partly because the p53-binding domains of the V-ONCs are large and not very well defined, and they do not share extensive homology. Despite these drawbacks, it appears to be a fruitful endeavor to study the interactions of these proteins with p53 in greater detail.

4. V-ONCS AS PROBES FOR RB FUNCTION

The observations described above make it clear that V-ONCs interact with Rb family members specifically and with high affinity. The most notable aspect of the interaction is that the V-ONCs specifically target the active form of Rb and Rb family members, which leads to an alteration of their normal function. These features make the V-ONCs efficient probes for assessing the functional status of the Rb protein in a given cell.

The functional effects of V-ONCs on Rb have been studied more extensively in Ad E1A, but the recent structural studies on HPV E7 peptide would make it a more adaptable probe for Rb function. The 9-aa peptide derived from E7 appears to bind to Rb efficiently, and may provide a good model for designing custom probes for Rb function *(88,89).* Since V-ONCs bind specifically to the functional, wild-type, hypophosphorylated form of Rb, the binding itself may be considered as a measure of the functional status of Rb. In addition, because the binding of such V-ONCs leads to a perturbation of Rb activity, methods could be designed to evaluate such changes in Rb function. Many novel methods have been developed recently to measure protein–protein interactions in vivo in living cells *(90),* mostly using fluorescent probes *(91,92).* Such methods generally measure changes in the fluorescence properties of the tagged protein or peptide when it interacts with another component. It may be imagined that a tagged E7 peptide, or one derived from Ad E1A, would be able to detect the functional status of the Rb protein, using such methods.

The tagged V-ONCs may be used to detect functional Rb proteins in biopsy samples of human tumors, or to evaluate whether the Rb protein is expressed in its functional form after being introduced into cells for gene therapy purposes *(93).* Although constitutively active, phosphorylation site mutants of Rb are expected to be used for this purpose, the assay using V-ONCs would be of immense help in assessing the amounts, as well as the functional status, of the protein. The advantage of such an assay system is that it would be able to detect functional Rb protein, even in single cells, and the assay may be modified easily to an in vitro diagnostic system very efficiently.

The V-ONCs would be especially useful for detecting functional Rb protein, because, unlike the p53 protein, there are no good antibodies that can distinguish between a functional form of the Rb protein and a mutated protein. Further, because even small peptides derived from the oncoproteins can specifically interact with functional Rb makes this approach feasible and attractive. It may be imagined that such specific biological probes would be of value in assessing the functional status of the vital growth regulatory proteins.

5. V-ONCS AS PROBES FOR P53 FUNCTION

The use of V-ONCs to assess the levels or the functional status of p53 tumor suppressor protein is not as attractive or feasible as in the case of the Rb family proteins. The chief reasons for this are that the regions of V-ONCs interacting with p53 are large, and they do not specifically interact with the functional form of p53 alone. In contrast, there are excellent immunological reagents available that can distinguish between the functional and inactive forms of p53. But it still remains an option, which could be utilized in circumstances in which antibodies may not be effective or accessible. Again, one of the many recent techniques for detecting protein–protein interactions in vivo and in vitro may be modified for this purpose. It would appear that polypeptides derived from the Ad E1B 55 kDa protein or the HPV E6 protein would be more suitable for this purpose.

It appears an exciting possibility that the oncoproteins of DNA tumor viruses may be harnessed to detect and quantitate cellular proteins that can prevent cell proliferation and oncogenesis. This would be a valuable addition to the repertoire of modern techniques to combat cancer.

ACKNOWLEDGMENT

The research in the author's laboratory is supported by the grants CA63136 and CA77301 from the National Cancer Institute.

REFERENCES

1. Nevins JR. Adenovirus E1A: transcription regulation and alteration of cell growth control. 1995.
2. Shenk T, Flint SJ. Transcriptional and transforming activities of the adenovirus E1A proteins. *Adv Cancer Res* 1991; 57:47–85.
3. Barbosa MS. Oncogenic role of human papillomavirus proteins. *Crit Rev Oncog* 1996; 7:1–18.
4. Fanning E. Simian virus 40 large T antigen: the puzzle, the pieces and the emerging picture. *J Virol* 1992; 66:1289–1293.
5. Huibregtse JM, Beaudenon SL. Mechanism of HPV E6 proteins in cellular transformation. *Semin Cancer Biol* 1996; 7:317–326.
6. Ludlow JW. Interactions between SV40 large-tumor antigen and the growth suppressor proteins pRb and p53. *FASEB J* 1993; 7:866–871.
7. Ludlow JW, Skuse GR. Viral oncoprotein binding to pRB, p107, p130, and p300. *Virus Res* 1995; 35:113–121.
8. Moran E. Interaction of adenoviral proteins with pRb and p53. *FASEB J* 1993; 7:880–885.
9. Vousden K. Interactions of human papillomavirus transforming proteins with the products of tumor suppressor genes. *FASEB J* 1993; 7:872–879.
10. Tommasino M, Crawford L. Human papillomavirus E6 and E7: proteins which deregulate the cell cycle. *Bioessays* 1995; 17:509–518.
11. White E. Regulation of p53-dependent apoptosis by E1A and E1B. *Curr Top Microbiol Immunol* 1995; 199:34–58.
12. Whyte P, Buchkovich KJ, Horowitz JM, Friend SH, Raybuck M, Weinberg RA, Harlow E. Association between an oncogene and an anti-oncogene: the adenovirus E1A proteins bind to the retinoblastoma gene product. *Nature* 1988; 334:124–129.

13. Nevins JR. Cell cycle targets of the DNA tumor viruses. *Curr Opin Genet Dev* 1994; 4:130–134.
14. Nevins JR. Disruption of cell-cycle control by viral oncoproteins. *Biochem Soc Trans* 1993; 21:935–938.
15. White E. Regulation of p53-dependent apoptosis by E1A and E1B. *Curr Top Microbiol Immunol* 1995; 199:34–58.
16. Whyte P, Williamson NM, HE. The human papilloma virus-16 E7 oncoprotein is able to bind to the retinoblastoma gene product. *Science* 1989; 243:934–937.
17. Mymryk JS. Tumour suppressive properties of the adenovirus 5 E1A oncogene. *Oncogene* 1996; 13:1581–1589.
18. Chen H, Yu D, Chinnadurai G, Karunagaran D, Hung MC. Mapping of adenovirus 5 E1A domains responsible for suppression of neu-mediated transformation via transcriptional repression of neu. *Oncogene* 1997; 14:1965–1971.
19. Chinnadurai G. Adenovirus E1A as a tumor suppressor gene. *Oncogene* 1992; 7:1255–1258.
20. Nevins JR. Mechanisms of viral-mediated transactivation of transcription. *Adv Virus Res* 1989; 37:35–83.
21. Kuppuswamy M, Chinnadurai G. Cell type dependent transformation by adenovirus 5 E1a proteins. *Oncogene* 1988; 2:567–572.
22. Kuppuswamy M, Subramanian T, Chinnadurai G. Separation of immortalization and T24-ras oncogene cooperative functions of adenovirus E1a. *Oncogene* 1988; 2:613–615.
23. Subramanian T, Kuppuswamy M, Nasr RJ, Chinnadurai G. An N-terminal region of adenovirus E1a essential for cell transformation and induction of an epithelial cell growth factor. *Oncogene* 1988; 2:105–112.
24. Braithwaite A, Nelson C, Skulimowski A, McGovern J, Pigott D, Jenkins J. Transactivation of the p53 oncogene by E1a gene products. *Virology* 1990; 177:595–605.
25. Lowe SW, Ruley HE. Stabilization of the p53 tumor suppressor is induced by adenovirus 5 E1A and accompanies apoptosis. *Genes Dev* 1993; 7:535–545.
26. Harlow E, Whyte P, Franza BRJ, Schley C. Association of adenovirus early-region 1A protein with cellular polypetides. *Mol Cell Biol* 1986; 6:1579–1589.
27. Whyte P, Williamson NM, HE. Cellular targets for transformation by the adenovirus E1A proteins. *Cell* 1989; 56:67–75.
28. Cress WD, Nevins JR. Use of the E2F transcription factor by DNA tumor virus regulatory proteins. *Curr Top Microbiol Immunol* 1996; 208:63–78.
29. Yang XJ, Ogryzko VV, Nishikawa J, Howard BH, Nakatani Y. A p300/CBP-associated factor that competes with the adenoviral oncoprotein E1A. *Nature* 1996; 382:319–324.
30. Barbeau D, Marcellus RC, Bacchetti S, Bayley ST, Branton PE. Quantitative analysis of regions of adenovirus E1A products involved ininteractions with cellular proteins. *Biochem Cell Biol* 1992; 70:1123–1134.
31. Weinberg RA. The retinoblastoma protein and cell cycle control. *Cell* 1995; 81:323–330.
32. Kamb A. Cell cycle regulators and cancer. *Trends Genet* 1995; 11:136–140.
33. Sherr CJ. The ins and outs of RB: coupling gene expression to the cell cycle clock. *Trends Cell Biol* 1994; 4:15–18.
34. Hollingsworth RJ, Hensey CE, Lee WH. Retinoblastoma protein and the cell cycle. *Curr Opin Genet Dev* 1993; 3:55–62.
35. Hu Q, Dyson N, Harlow E. The regions of the retinoblastoma protein needed for binding to adenovirus E1A or or SV40 larget T-antigen are common sites for mutations. *EMBO J* 1990; 9:1147–1155.
36. Horowitz JM, Yandell DW, Park SH, Canning S, Whyte P, Buchkovich K, HE, WRA, Dryja TP. Point mutational inactivation of the retinoblastoma antioncogene. *Science* 1989; 243:937–940.
37. Huang S, Wang NP, Tseng BY, Lee WH, Lee EH. Two distinct and frequently mutated regions of the retinoblastoma protein are required for binding to SV40 T antigen. *EMBO J* 1990; 9:1815–1822.
38. Kaelin WG, Ewen ME, Livingston DM. Definition of the minimal simian virus 40 larget T-antigen and adenovirus E1A-binding domain in the retinoblastoma gene product. *Mol Cell Biol* 1990; 10:3761–3769.
39. Beijersbergen RL, Bernards R. Cell cycle regulation by the retinoblastoma family of growth inhibitory proteins. *Biochim Biophys Acta* 1996; 1287:103–120.
40. Adams PD, Sellers WR, Sharma SK, Wu AD, Nalin CM, Kaelin WG, Jr. Identification of a cyclin-cdk2 recognition motif present in substrates and p21-like cyclin-dependent kinase inhibitors. *Mol Cell Biol* 1996; 16:6623–6633.

41. Kato J, Matsushime H, Hiebert SW, Ewen ME, Sherr CJ. Direct binding of cyclin D to the retinoblastoma gene product (pRb) and pRb phosphorylation by the cyclin D-dependent kinase CDK4. *Genes Dev* 1993; 7:331–342.

42. Bagchi S, Weinmann R, Raychaudhuri P. The retinoblastoma protein copurifies with E2F-I, an E1A-regulated inhibitor of the transcription factor E2F. *Cell* 1991; 65:1073–1082.

43. Chellappan SP, Hiebert S, Mudryj M, Horowitz JM, Nevins JR. The E2F transcription factor is a cellular target for the RB protein. *Cell* 1991; 65:1053–1061.

44. Chittenden T, LDM, Jr KWG. RB associates with an E2F-like, sequence-specific DNA-binding protein. *Cold Spring Harbor Symp Quant Biol* 1991; 56:187–195.

45. Bagchi S, Raychaudhuri P, Nevins JR. Adenovirus E1A proteins can dissociate heteromeric complexes involving the E2F transcription factor: a novel mechanism for E1A trans-activation. *Cell* 1990; 62:659–669.

46. Slansky JE, Farnham PJ. Introduction to the E2F family: protein structure and gene regulation. *Curr Top Microbiol Immunol* 1996; 208:1–30.

47. Trimarchi JM, Fairchild B, Verona R, Moberg K, Andon N, Lees JA. E2F-6, a member of the E2F family that can behave as a transcriptional repressor. *Proc Natl Acad Sci USA* 1998; 95:2850–2855.

48. Cobrinik D. Regulatory interactions among E2Fs and cell cycle control proteins. *Curr Top Microbiol Immunol* 1996; 208:32–63.

49. Dyson N. 1998. The regulation of E2F by pRB-family proteins. *Genes Dev* 12:2245–2262.

50. Hiebert SW, Chellappan SP, Horowitz JM, Nevins JR. The interaction of RB with E2F coincides with an inhibition of the transcriptional activity of E2F. *Genes Dev* 1992; 6:177–185.

51. Helin K, Lees JA, Vidal M, Dyson N, HE, FA. A cDNA encoding a pRB-binding protein with properties of the transcription factor E2F. *Cell* 1992; 70:337–350.

52. Kaelin WG, Krek W, Sellers WR, DeCaprio JA, Ajchenbaum F, Fuchs CS, et al. Expression cloning of a cDNA encoding a retinoblastoma-binding protein with E2F-like properties. *Cell* 1992; 70:351–364.

53. Brehm A, Miska EA, McCance DJ, Reid JL, Bannister AJ, Kouzarides T. Retinoblastoma protein recruits histone deacetylase to repress transcription. *Nature* 1998; 391:597–601.

54. Luo RX, Postigo AA, Dean DC. Rb interacts with histone deacetylase to repress transcription. *Cell* 1998; 92:463–473.

55. Magnaghi-Jaulin L, Groisman R, Naguibneva I, Robin P, Lorain S, Le Villain JP, et al. Retinoblastoma protein represses transcription by recruiting a histone deacetylase. *Nature* 1998; 391:601–605.

56. Dyson N. pRB, p107 and the regulation of the E2F transcription factor. *J Cell Sci Suppl* 1994; 18:81–7.

57. Nevins JR. E2F: a link between the Rb tumor suppressor protein and viral oncoproteins. *Science* 1992; 258:424–429.

58. Fattaey AR, Harlow E, Helin K. Independent regions of adenovirus E1A are required for binding to and dissociation of E2F-protein complexes. *Mol Cell Biol* 1993; 13:7267–7277.

59. Ikeda MA, Nevins JR. Identification of distinct roles for separate E1A domains in disruption of E2F complexes. *Mol Cell Biol* 1993; 13:7029–7035.

60. Chellappan S, Kraus VB, Kroger B, Munger K, Howley PM, Phelps WC, Nevins JR. Adenovirus E1A, simian virus 40 tumor antigen, and human papillomavirus E7 protein share the capacity to disrupt the interaction between transcription factor E2F and the retinoblastoma. *Proc Natl Acad Sci USA* 1992; 89:4549–4553.

61. Devoto SH, Mudryj M, Pines J, Hunter T, Nevins JR. A cyclin A-protein kinase complex possesses sequence-specific DNA binding activity: p33cdk2 is a component of the E2F-cyclin A complex. *Cell* 1992; 68:167–176.

62. Chellappan SP. The E2F transcription factor: role in cell cycle regulation and differentiation. *Mol Cell Differ* 1994; 2:201–220.

63. Park TW, Fujiwara H, Wright TC. Molecular biology of cervical cancer and its precursors. *Cancer* 1995; 76:1902–1913.

64. Galloway DA, McDougall JK. The disruption of cell cycle checkpoints by papillomavirus oncoproteins contributes to anogenital neoplasia. *Semin Cancer Biol* 1996; 7:309–315.

65. Munger K, Phelps WC, Bubb V, Howley PM, Schlegel R. The E6 and E7 genes of the human papillomavirus type 16 together are necessary and sufficient for transformation of primary human keratinocytes. *J Virol* 1989; 63:4417–4421.

66. Phelps WC, Yee CL, Munger K, Howley PM. Functional and sequence similarities between HPV16 E7 and adenovirus E1A. *Curr Top Microbiol Immunol* 1989; 144:153–166.

67. Dyson N, Howley PM, Munger K, Harlow E. The human papilloma virus-16 E7 oncoprotein is able to bind to the retinoblastoma gene product. *Science* 1989; 243:934–937.

68. Jones DL, Munger K. Interactions of the human papillomavirus E7 protein with cell cycle regulators. *Semin Cancer Biol* 1996; 7:327–337.

69. Scheffner M, Munger K, Byrne JC, Howley PM. State of the p53 and retinoblastoma genes in human cervical carcinoma cell lines. *Proc Natl Acad Sci USA* 1991; 88:5523–5527.

70. Howley PM, Scheffner M, Huibregtse J, Munger K. Oncoproteins encoded by the cancer-associated human papillomaviruses target the products of the retinoblastoma and p53 tumor suppressor genes. *Cold Spring Harbor Symp Quant Biol* 1991; 56:149–155.

71. Lee J-O, Russo A, Pavletich NP. Structure of the retinoblastoma tumor-suppressor pocket domain bound to a peptide from HPV E7. *Nature* 1998; 391:859–865.

72. Berezutskaya E, Yu B, Morozov A, Raychaudhuri P, Bagchi S. Differential regulation of the pocket domains of the retinoblastoma family proteins by the HPV16 E7 oncoprotein. *Cell Growth Differ* 1997; 8:1277–1286.

73. Arroyo M, Bagchi S, Raychaudhuri P. Association of the human papillomavirus type 16 E7 protein with the S-phase-specific E2F-cyclin A complex. *Mol Cell Biol* 1993; 13:6537–6546.

74. Lam EW, Morris JD, Davies R, Crook T, Watson RJ, Vousden KH. HPV16 E7 oncoprotein deregulates B-myb expression: correlation with targeting of p107/E2F complexes. *EMBO J* 1994; 13:871–878.

75. Boyd JM, Malstrom S, Subramanian T, Venkatesh LK, Schaeper U, Elangovan B, D'Sa-Eipper C, Chinnadurai G. Adenovirus E1B 19 kDa and Bcl-2 proteins interact with a common set of cellular proteins [see comments] [published erratum appears in *Cell* 1994; 16:79:following 1120]. *Cell* 1994; 79:341–351.

76. Martin ME, Berk AJ. Adenovirus E1B 55K represses p53 activation in vitro. *J Virol* 1998; 72:3146–3154.

77. Sabbatini P, Chiou SK, Rao L, White E. Modulation of p53-mediated transcriptional repression and apoptosis by the adenovirus E1B 19K protein. *Mol Cell Biol* 1995; 15:1060–1070.

78. Querido E, Marcellus RC, Lai A, Charbonneau R, Teodoro JG, Ketner G, Branton PE. Regulation of p53 levels by the E1B 55-kilodalton protein and E4orf6 in adenovirus-infected cells. *J Virol* 1997; 71:3788–3798.

79. Querido E, Teodoro JG, Branton PE. Accumulation of p53 induced by the adenovirus E1A protein requires regions involved in the stimulation of DNA synthesis. *J Virol* 1997; 71:3526–3533.

80. Teodoro JG, Branton PE. Regulation of p53-dependent apoptosis, transcriptional repression, and cell transformation by phosphorylation of the 55-kilodalton E1B protein of human adenovirus type 5. *J Virol* 1997; 71:3620–3627.

81. Flint SJ, Shenk T. Viral transactivating proteins. *Ann Rev Genet* 1997; 31:177–212.

82. DeCaprio J, Ludlow JW, Figge J, Shew JY, Huang CM, Lee WH, et al. SV40 large tumor antigen forms a specific complex with the product of the retinoblastoma susceptibility gene. *Cell* 1988; 54:275–283.

83. Ludlow JW. Use of sequential immunoprecipitation to reveal discrete, separable populations of SV40 T-antigen binding to host cellular proteins. *J Virol Methods* 1996; 59:105–112.

84. Chellappan S, Kraus VB, Kroger B, Munger K, Howley PM, Phelps WC, Nevins JR. Adenovirus E1A, simian virus 40 tumor antigen, and human papillomavirus E7 protein share the capacity to disrupt the interaction between transcription factor E2F and the retinoblastoma gene product. *Proc Natl Acad Sci USA* 1992; 89:4549–4553.

85. Sheng Q, Denis D, Ratnofsky M, Roberts TM, DeCaprio JA, Schaffhausen B. The DnaJ domain of polyomavirus large T antigen is required to regulate Rb family tumor suppressor function. *J Virol* 1997; 71:9410–9416.

86. Zalvide J, Stubdal H, DeCaprio JA. The J domain of simian virus 40 large T antigen is required to functionally inactivate RB family proteins. *Mol Cell Biol* 1998; 18:1408–1415.

87. Zerrahn J, Tiemann F, Deppert W. Simian virus 40 small t antigen activates the carboxyl-terminal transforming p53-binding domain of large T antigen. *J Virol* 1996; 70:6781–6789.

88. Jones RE, Wegrzyn RJ, Patrick DR, Balishin NL, Vuocolo GA, Riemen MW, et al. Identification of HPV-16 E7 peptides that are potent antagonists of E7 binding to the retinoblastoma suppressor protein. *J Biol Chem* 1990; 265:12,782–12,785.

89. Munger K, Werness BA, Dyson N, Phelps WC, Harlow E, Howley PM. Complex formation of human papillomavirus E7 proteins with the retinoblastoma tumor suppressor gene product. *EMBO J* 1989; 8:4099–105.

90. Karlsson R, Falt A. Experimental design for kinetic analysis of protein-protein interactions with surface plasmon resonance biosensors. *J Immunol Methods* 1997; 200:121–133.
91. Garamszegi N, Garamszegi ZP, Rogers MS, DeMarco SJ, Strehler EE. Application of a chimeric green fluorescent protein to study protein–protein interactions. *Biotechniques* 1997; 23:864–6, 868–870, 872.
92. Nishimune A, Nash SR, Nakanishi S, Henley JM. Detection of protein-protein interactions in the nervous system using the two-hybrid system. *Trends Neurosci* 1996; 19:261–266.
93. Riley DJ, Nikitin AY, Lee WH. Adenovirus-mediated retinoblastoma gene therapy suppresses spontaneous pituitary melanotroph tumors in Rb+/– mice. *Nat Med* 1996; 2:1316–1321.

3 Tumor Suppressors in Metastasis

Taoufik Ouatas, PhD,
Melanie T. Hartsough, PhD,
and Patricia S. Steeg, PhD

1. INTRODUCTION

Recent advances in cancer development studies showed tumorigenesis and metastasis to be complex multistep processes. In the past decade, it became evident that cancer cells have multiple genetic alterations, including point mutations, gene amplification, recombination, gene deletion, reduced or overexpression, and loss of heterozygosity (LOH). Those affected genes include tumor suppressors, oncogenes, and a panoply of genes involved in cell cycle, motility, and adhesion. This review emphasizes the tumor suppressor genes (TSGs) and metastasis suppressor genes, and their alteration at different levels in metastasizing cells.

Metastasis is a major cause of cancer patients' deaths, but cures are often possible for patients whose tumors remain localized. Thus, understanding the genetic and physiological alterations leading to metastasis formation should be of value in defining new diagnostic techniques and therapeutic approaches. The invading tumor cells must accomplish a multistage process to metastasize to distant sites. They commonly invade the adjacent tissues, and enter the lymphatic and/or circulatory system. They disseminate, and adhere at a second site, either by binding to specific ligands or by physical limitations. Thereafter, they either proliferate in the vasculature or extravasate the endothelium and proliferate in the host tissue or organ, once conveniently irrigated. Each of the steps included in this pathway must be completed by an originally invading cell, to be able to form a micrometastasis, and a secondary or tertiary tumor may then form.

Many of the TSGs for which a role in metastatic progression was shown are discussed herein.

From: *Tumor Suppressor Genes in Human Cancer*
Edited by: D. E. Fisher © Humana Press Inc., Totowa, NJ

2. TUMOR SUPPRESSOR GENES

TSGs encode proteins in which their absence, inactivation, or mutation promotes oncogenesis, or prevents apoptosis at various levels in the cell, because of their diverse functions. They have the ability to inhibit or reduce tumorigenicity when transfected in transformed cells. Most of the TSGs were originally identified by the occurrence of mutations in families with hereditary predisposition to develop certain cancer types.

2.1. p53 Gene

The *p53* gene product was originally identified based on its ability, when phosphory-lated, to associate with the large T-antigen (T-Ag) in SV40-infected cells *(1)*. Studies then focused on the oncogenic potential of *p53,* because its mutant form cooperates with the *ras* oncogene in transforming rat embryo fibroblasts *(2)*. Later transfection studies showed that *p53* is a TSG *(3–9)* whose expression and structure is altered in several cancers *(6,10–18)*, and that earlier studies were performed with a mutant prod-uct. *p53* was then shown to mediate apoptosis *(19–22)*, DNA damage repair *(23–30)*, and cell cycle control *(31–34)*.

By transcriptionally activating the expression of p21[WAF/CIP1/sdi1], an inhibitory pro-tein to cell cycle kinases, p53 can inhibit the cell cycle in the G1–S transition *(35,36)*. Also, *p53*-dependent upregulation of 14-3-3 σ is proposed to play a role in the inhibi-tion of G2–M phase progression *(37)*. On the other hand, p53 transcriptional regulation and binding to Mdm2 protein has been shown to regulate *p53* activity, and to play a role in uncoupling S phase from mitosis *(38–44)*. The p53-deficient mice were shown to develop normally, but were prone to the spontaneous development of a variety of tumors *(45–51)*. Similar observations were reported for *p53*-deficient cells *(52,53)*.

Several studies showed the mutation of p53 in a variety of cancers. In breast can-cer(BC), *p53* overexpression, mutation, and LOH are common mutations *(54,55)*. Recent cohort studies revealed the presence of overexpression and/or structural alter-ations of *p53* gene in 38.1% of chondrosarcomas of bone *(56)* and 54% of squamous cell carcinoma *(57)*. A statistically significant correlation was observed between overexpression or alteration of the *p53* gene and both the histologic grade of the tumor and the presence of metastasis. The probability of local recurrence-free, metas-tasis-free, and overall survival was shown to be significantly higher for patients with no overexpression or alteration of *p53* than for patients with *p53* overexpression or alteration, in the case of pediatric adrenal cortical tumors, small cell lung cancer, oropharyngeal carcinoma, BC, and several other tumors *(14,54,58–63)*. The observa-tion that *p53* overexpression is associated with a poor outcome for the patients in these studies may be explained by the accumulation of genetic alterations in the downstream targets of *p53,* which mediate cell cycle or apoptosis. Also, the overex-pression of *p53* by different types of tumor cells suggests that *p53* immunity may be useful for tumor immunotherapy.

A recent study *(64)* showed that p53-immunized mice acquired resistance to tumor metastases. Therefore, an anti-idiotypic network built around certain domains of p53 seems to be programmed within the immune system, and immunity to p53 can be asso-ciated with resistance to tumor cells *(64–66)*.

Data are further complicated by the fact that different domains of p53 are involved in the control of apoptosis and cell cyle. A polyproline region in p53 mediates apoptosis induction, but not cell growth arrest *(67)*. Therefore, it is important to investigate the molecular alterations of *p53* in primary and metastatic tumors. In most cases, the description of *p53* mutation in primary tumor is indicative of a poor prognosis *(15,16,57,68–70)*. Mutations in the metastatic sites correlate with those in the primary tumors, even if their frequency is generally higher *(68,71)*, and suggest that *p53* status is indicative of the aggressiveness of the tumors *(58,72)*. A recent study *(73)* characterized *p53* mutations in 39/58 cell lines analyzed from the National Cancer Institute (NCI) anticancer drug screen. Cells containing mutant *p53* were deficient in γ-ray induction of CIP1/WAF1, GADD45, and Mdm2 mRNA and the ability to arrest in G1 following γ-irradiation *(73)*.

Further studies, using transgenic mice that overexpress a mutant *p53* gene, show its ability to promote tumorigenesis *(74,75)* and the cellular resistance of a variety of hematopoietic cell lineages to γ-radiation *(76)*. These observations provide direct evidence that *p53* mutations affect the cellular response to DNA damage, by increasing either DNA repair processes or cellular tolerance to DNA damage. Thus, deletion, overexpression, or mutations of the *p53* gene seem to be important predictors of aggressive clinical behavior.

2.2. Retinoblastoma Gene

The hereditary form of retinoblastoma (RB) is a dominant autosomal disease characterized by an ocular tumor in childhood. Heterozygotes carrying one mutant and one normal *Rb* allele develop RB with 90% penetrance. The Rb protein is a nuclear protein variably phosphorylated during the cell cycle. The unphosphorylated form is characteristic of the G1 phase; the phosphorylated form is present during the S, G2, and M phases *(77)*. *Rb* was recently shown to play an essential role in DNA damage-induced arrest at the G1–S phase checkpoint *(78)*.

The cell cycle inhibition activity of Rb protein was shown by transfection in many different cell lines *(79–82)*, and the phosphorylation of Rb by overexpressed cyclins E and D suppresses its growth-inhibition activity *(80,81)*. Several studies *(83–88)* have shown the tumor suppressor activity of *Rb* in RB, osteosarcoma, and bladder carcinoma cells, either by in vivo or in vitro transfection. Nullizygous mice for Rb are nonviable, and die between 10 and 14 d gestation, and show increased levels of both cell division and cell death by apoptosis in the hematopoietic and nervous systems *(89,90)*. However, heterozygous and chimeric mice are viable, and exhibit several tumors arising from the pituitary gland *(91–93)*. Rb transgenic mice, carrying several copies of the human *Rb* gene, were smaller than the wild-type (wt) ones *(94)*. The degree of dwarfism correlated with the copy number of the transgene and the corresponding level of Rb protein *(94)*.

Mutations and/or LOH of the *Rb* gene were found in several types of cancer, including osteosarcoma, breast carcinoma, prostate carcinoma, and small cell lung carcinoma *(95–97)*. Although reduced expression, LOH, and mutations were shown in several BC cells *(98,99)*, conflicting results have been found concerning the correlation between the loss of *Rb* expression and the presence of lymph node metastasis *(63,100–103)*. Borg et al. *(102)* found that the LOH of *Rb* does not reflect a reduced expression of Rb protein; Spandidos et al. *(101)* and Naka et al. claimed a higher incidence of lymph node metastasis in tumors expressing *Rb*. However, most of these data show that loss or aberrant expression of Rb is one of the most common molecular alterations in invasive tumors.

2.3. E-Cadherin–Catenin Genes

The E-cadherin–catenin complex, an organizer of epithelial structure and function, is commonly disturbed in invasive cancer. E-cadherin, a transmembrane protein with an extracellular and an intracellular domain, is one of the key players involved in cell–cell adhesion. The function of E-cadherin in preventing metastasis in tumor development is believed to be dependent on intracellular catenins *(104–112)*. Transfection studies have shown the ability of E-cadherin to decrease cell growth and invasiveness of human carcinoma cells *(104,112)*, and to suppress the development of osteolytic bone metastases from BC cells in an experimental metastasis model *(113)*. Byers et al. *(109)* showed that E-cadherin-negative BC cells, which had been transfected with E-cadherin, exhibited large increases in adhesion strength, only if the expressed protein was appropriately linked to the cytoskeleton. The data also show that E-cadherin-negative tumor cells, or cells in which the adhesion molecule is inefficiently linked to the cytoskeleton, are more likely than E-cadherin-expressing cells to detach from a tumor mass in response to shear forces equivalent to those exerted in a lymphatic vessel or venule.

Several studies have shown that downregulation of E-cadherin *(106,108,111,114–120)* and catenins *(121–123)* correlate with a poor outcome of the patients. When E-cadherin and α-, β-, and γ-catenins were analyzed as one group, a significant association was seen between reduction in immunoreactivity of at least one of these four proteins and the presence of metastases *(110)*. These data indicate that if one of these proteins is downregulated, the function of the others in suppressing tumors and metastasis is altered. Also, a significant association was seen between lobular invasive tumors and β-catenin expression *(110)*.

Another study *(124)* on laryngeal squamous cell carcinoma (LSCC) suggested that the aberrations in the function of α-catenin, the anchoring protein of E-cadherin, may cause dysfunction of the cadherin–catenin complex, leading to disturbed cell–cell adhesion. Patients with cytoplasmic presence of α-catenin appeared to have a trend toward poor overall survival, which suggests that cytoplasmic localization of α-catenin is associated with aggressive behavior and metastatic phenotype of LSCC *(124)*.

However, specific E-cadherin peptides (histidine-alanine-valine [HAV]) were shown to inhibit cell aggregation and disturb the epithelial morphology, and were able to stimulate invasion of cells expressing E-cadherins *(125)*. This suggests that E-cadherin functions are inhibited by homologous proteolytic HAV-containing fragments, which may promote cancer invasion. Therefore, it seems important to assess E-cadherin–catenin status in primary tumors, as a prognostic factor.

2.4. MMAC/PTEN Gene

The tumor suppressor, phosphatase and tensin homolog deleted on chromosome 10 (PTEN) was isolated by different groups *(126–129)*, based on its mutation in several human cancers. The gene product was shown to possess an intrinsic protein tyrosine phosphatase activity *(128)*. The abundance of its transcription is altered in many transformed cells. In transforming growth factor β (TGF-β)-sensitive cells, PTEN expression is rapidly downregulated by TGF-β *(128)*. The transfection of wt PTEN into glioma cells, expressing endogenous mutant alleles, caused growth suppression, but was without effect in cells containing endogenous wt PTEN *(130)*. Tamura et al. *(131)* have shown that overexpression of PTEN inhibited cell migration; antisense PTEN enhanced migration. Integrin-mediated cell spreading and the formation of focal adhesions were downregulated by wt PTEN, but not by PTEN with an inactive phosphatase domain *(131)*.

PTEN was recently shown to dephosphorylate focal adhesion kinase (FAK) and to inhibit integrin-mediated cell spreading and cell migration *(132)*. This suggests that a general function of PTEN is to downregulate FAK and Shc phosphorylation, Ras activity, downstream mitogen-activated protein kinase activation, and associated focal contact formation and cell spreading *(132)*. Also, a role for PTEN in regulating the activity of the phosphoinositol 3-kinase pathway in malignant human cells was shown *(133–136)*. The phosphoinositol phosphatase activity of PTEN mediates a serum-sensitive G1 growth arrest in glioma cells *(134)* and the expression of PTEN in primary astrocytes reduced the levels of 3′ phosphoinositides and inhibited protein kinase B (PKB/Akt) activity *(135)*. Homozygous PTEN-deficient mice were found to die by d 9.5 of development, and to show abnormal patterning and overgrowth of the cephalic and caudal regions *(137)*. Embryonic fibroblasts from PTEN-deficient mice exhibited decreased sensitivity to apoptosis stimuli and constitutively elevated activity and phosphorylation of PKB/Akt *(137)*.

Several studies gave evidence of LOH, reduced expression, and alteration of PTEN in several types of aggressive cancer, including prostate, breast, endometrium, lung and ovary *(127,138–155)*. Furthermore, Furnari et al. *(130)* showed that ectopic expression of mutant PTEN alleles, which carried mutations found in primary tumors, and have been shown, or are expected, to inactivate its phosphatase activity, caused little growth suppression. Taken together, these observations suggest that PTEN could serve as a prognostic factor in human cancers, and that its mutation or loss of expression is predictive of poor prognosis.

2.5. Deleted in Colorectal Cancer Gene

The deleted in colorectal cancer *(DCC)* gene is one of several genes altered during tumorigenesis. It encodes a receptor for netrin-1, a chemoattractant involved in axon guidance *(156–159)*. When wt cDNA was transfected into nitrosomethylurea (NMU)-transformed tumorigenic human papilloma virus-immortalized human epithelial cells that had allelic loss and reduced expression of *DCC,* it suppressed tumorigenicity, but truncated *DCC* did not *(160)*. However, *DCC*-deficient mice displayed defects in axonal projections, but were not affected in growth, differentiation, morphogenesis, or tumorigenesis of the intestine *(161)*. On the other hand, Mehlen et al. *(162)* have shown that DCC induces apoptosis in the absence of netrin-1 binding, but blocks apoptosis when engaged by the ligand.

Although mutation of *DCC* gene was rare, a high incidence of LOH and decreased mRNA expression in colorectal cancer, especially with hepatic metastasis, was observed *(163)*. Different studies indicate that LOH or decreased *DCC* expression are closely associated with liver and lymph node metastasis of gastric, colorectal, breast, and testicular germ cell cancers *(164–171)*. The farther away the lymph node metasta-

Table 1
Molecular Alterations of Some tumor suppressor Genes in Primary and Invasive
Human Tumors

Gene	Invasive tumors	Molecular alterations	Refs.
BRCA1	BC and ovarian cancers and squamous cell carcinomas of the esophagus	LOH, deletion, reduced expression and/or mutation	(318–324)
BRCA2	Female hereditary and male sporadic BC	LOH, deletion and/or mutation	(320,322,325,326)
Nf1	Neurofibromatosis type 1, neurofibrosarcoma, leukemia, melanoma, shwannoma	Deletion, reduced expression, mutation, splicing alteration	(196,197,327–345)
Nf2	Neurofibromatosis type 2, meningiomas, schwannomas, mesotheliomas	Deletion, reduced expression, mutation, splicing alteration	(198,199,346–358)
MTS1	Melanomas, sarcomas, esophageal, hepatocellular and lung carcinomas, prostate, cervix, breast, and oral cancer.	LOH, deletion, reduced expression, mutation, methylation	(98,187,359–368)
APC	Familial polyposis coli, colorectal tumors, hepatocellular adenoma, oral and ampullary carcinomas, fibromatosis, thyroid tumors	Deletion, LOH, mutation, reduced expression	(210–212,369–380)

sis was from the primary tumor, the higher the frequency of allelic deletions became (165).

Furthermore, DCC is a caspase substrate, and mutation of the site at which caspase-3 cleaves DCC (162) suppresses the proapoptotic effect of DCC completely. This data suggests that DCC may function as a tumor-suppressor protein by inducing apoptosis in physiological conditions in which netrin-1 is unavailable. This function depends on caspase cascades by a mechanism that requires cleavage of DCC. Assessment of this parameter in primary lesions may thus find predictive application.

The p53 gene, the retinoblastoma gene, the E-cadheringene, the MMAC/PTEN gene, and the DCC gene constitute the five most-reported TSGs for which a correlation between mutations and/or deletion and metastasis was reported. Other functional studies characterized additional putative tumor suppressors, including BRCA1 (172–177), BRCA2 (178–185), MTS1 (98,186–191), Nf1 (192–197), Nf2 (198–204), and APC (205–212) genes (Table 1).

However, as discussed above, the metastatic process is a multistep phenomenon, and the causal event of tumorigenicity may be of different sources. Thus, it is important to focus on the latest steps leading to metastasis, and to try to isolate genes, that have the ability to suppress metastasis, and to study their mechanisms of function.

3. METASTASIS SUPPRESSOR GENES

In the past decade, four such genes were identified. Here the authors review one of the most functionally studied among them, *nm23,* which was identified in this lab.

3.1. nm23-H1

Nm23-H1 was identified through a differential colony hybridization between related low and high metastatic potential murine melanoma cell lines, in which *nm23* mRNA expression was found to be reduced in the high-metastatic-potential cell line *(213).* Presently, the human *nm23* family of genes consists of five members: *nm23-H1, nm23-H2, nm23-DR, nm23-H4,* and *nm23-H5 (213–218).* These isoforms are closely related, having identity to Nm23-H1 of 94, 76, 59, and 30%, respectively. Homologs of this family are studied as abnormal wing disks *(awd)* in *Drosophila* development *(219,220)* and nucleoside diphosphate kinase (ndpk) in other species *(221,222),* and are highly conserved through evolution. When compared to Nm23-H1, mouse and rat homologs have 88–95% identity; the *Drosophila awd* has 87% identity; *Xenopus, Arabidopsis,* yeast, and *Dictyostelium* are 82–89, 78, 73–75, and 70–76% identical, respectively. On the other hand, several strains of *Escherichia coli* contain genes displaying more than 60% identity to *nm23.* The evolutionary conservation of this protein suggests that the *nm23* gene is pivotal for cellular survival, and that the signaling pathway involving this protein may be held intact from lower to higher organisms.

Among the family members, the *Nm23-H1* isoform has been repeatedly linked to metastasis, and, therefore, is the focus of the rest of this chapter. Reduced *nm23*-H1 mRNA or protein expression levels have been observed in highly metastatic tumor cells among a variety of *in vitro* monmdel systems, including N-NMU-induced rat and human mammary tumors, mouse mammary tumor virus-induced mammary tumors and Ras or Ras+ adenovirus 2 E1a-transfected rat embryo fibroblasts *(223–226).* In some other model systems, differential *nm23* expression was not apparent *(227).* This inconsistency is indicative of tumor heterogeneity, whereby multiple molecular pathways may underlie the metastatic process.

Regarding human *in vivo* analyses, breast tumor cohort studies showed a correlation of reduced Nm23 expression with indicators of high tumor metastatic potential, such as reduced patient survival *(228–233),* positive lymph node metastases *(234–237),* or poor differentiation state *(238).* No mutations were found in the coding sequence of *Nm23-H1* among 20 tumor specimens sequenced *(239).* Most recently, *Nm23-H1* was reported to be an independent prognostic factor for disease-free survival, upon multivariate analysis of a cohort of patients with node-negative breast tumors *(228).* Similar expression relationships have been reported in cervical *(240–243),* hepatocellular *(244–246),* ovarian carcinomas *(247–250),* and melanoma *(251–255).* Again, not all cohort studies reported this expression association; however, the data, although not unanimous, indicate that reduced *nm23* expression accompanied high tumor metastatic potential in several cancer cell types, which is characteristic of a causal relationship.

Reduced RNA and protein expression can sometimes be the result of chromosomal aberrations. LOH of the *nm23-H1* locus has been detected in breast carcinomas, lung adenomas, and kidney and colon carcinomas *(256,257).* However, an infiltrating breast tumor cohort study, conducted comparing Nm23-H1 protein expression with allelic deletion at the *nm23-H1* locus, and patient disease-free survival, showed that, although allelic deletion of *nm23-H1* was observed, protein levels in these specimens were not uniformly low. In fact, the best correlate of poor patient survival (metastatic potential) was low Nm23-H1 expression, regardless of deletion status *(239)*.

Loss of Nm23-H1 expression may contribute to metastatic progression in many cancers, but overexpression of *nm23* RNA predicts advanced-staged neuroblastoma tumors. In 20% of these advanced cases, a mutated form of Nm23, a serine (Ser) to a gylcine (Gly) change, at amino acid (aa) 120, was present *(258–260)*. A similar occurrence was cited for metastatic colon carcinomas, in which high expression of nm23 was associated with the more aggressive tumors, and these tumors were more susceptible to *nm23* mutations *(256)*. Although possible, inactivating Nm23 protein through mutations seems to be relatively rare. In contrast, loss of expression of this gene is a consistent property of aggressive tumors.

3.1.1. METASTASIS SUPPRESSOR ACTIVITY

Nm23 function was assessed by the transfection of murine *nm23-1* cDNA and subsequently human *nm23-H1* cDNA into the highly metastatic murine K-1735 TK melanoma and human MDA-MB-435 breast carcinoma cell lines, respectively. Clonal cell lines constitutively overexpressing *nm23* exhibited a 50–96% reduction in tumor metastatic potential *in vivo,* with no effect on tumor growth *(261,262). In vitro, nm23* transfectants were deficient in soft agar colonization and motility to serum, insulin-like growth factor, and platelet-derived growth factor, compared to control transfectants, but exhibited comparable proliferation rates *(261–263)*. Four independent laboratories subsequently reported *(264–267)* that transfection of *nm23* into melanoma or mammary carcinoma cell lines resulted in significantly reduced tumor metastatic potential *in vivo* (Table 2). Additionally, *in vitro* suppression of cell motility was replicated in metastatic MDA-MB-231 breast carcinoma cells transfected with *nm23-H1 (264)*. Nm23-H1 seems to control metastasis independent of primary tumor size, thereby suggesting that nm23 is not a tumor suppressor, but a metastasis suppressor.

How could Nm23-H1 be inhibiting metastasis? Based on the observation that reduced or mutated *Drosophila awd* expression correlated with multiple developmental abnormalities postmetamorphosis *(219,220),* it is reasonable to postulate that *nm23* may function in differentiation in higher organisms. Indeed, *nm23-H1* transfectants of the MDA-MB-435 human breast carcinoma cell line were found to recapitulate aspects of normal mammary differentiation in three-dimensional culture, including formation of ascinar structures, production of the basement membrane proteins and sialomucin, and growth arrest *(268)*. Independent laboratories reported *(269,270)* that transfection of PC12 cells with *nm23* induced neuronal differentiation, and antisense to *nm23* reversed TGF-β—induced differentiation of colon carcinoma cells. Taken together, these results suggest that Nm23-H1 plays a role in the development/differentiation aspect of cells, which may be an important underlying biological mechanism affecting cell motility and overall metastasis behavior.

3.1.2. MECHANISMS OF NM23

The biochemical mechanism whereby Nm23 overexpression suppresses tumor metastasis, or induces differentiation, is still under investigation. Site-directed mutagenesis was utilized to correlate Nm23-H1 sequence and biological function in the highly metastatic human MDA-MB-435 breast carcinoma cell line *(271)*. Mutations were pro-

Table 2
Transfection Experiments Reporting and in vivo Metastasis Suppressor Activity for Nm23-H1

Cell line	% inhibition	Ref.
K-1735 TK Melanoma	58–96	Cell 65:25, 1991 (261)
MDA-MB-435 Breast carcinoma	65–90	Oncogene 8:2325, 1993 (262)
B16F10 Melanoma	93	Int. J. Cancer 60:204, 1995 (265)
B16FE7 Melanoma	80–87	Cancer Res. 55:1977, 1995 (267)
MDA-MB-231 Breast carcinoma	44–46	Br. J. Cancer 78:710, 1998 (264)
MTLn3 Mammary carcinoma	44–52	Int. J. Cancer 65:531, 1996 (266)
DU 145 Prostate carcinoma	60–90	Cancer Lett. 133:143, 1998 (381)

duced at various aas, and the control, wt, and mutant constructs were transfected into MDA-MB-435 cells, and assayed for one aspect of the tumor metastatic process, motility in Boyden chambers *in vitro* toward serum, or a defined chemoattractant (ATX) *(272)*. MDA-MB-435 clonal cell lines, transfected with a control vector, exhibited dose-dependent chemotaxis; clones expressing wt *nm23-H1* were 44–98% less motile to serum, and 86–99% less motile to ATX. Three mutations at two aa positions failed to inhibit chemotaxis: Mutation of proline 96 to Ser (P96S) is the conditional dominant killer of prune *(k-pn)* mutation in *Drosophila awd,* which causes developmental defects; serine 120 has been reported to be mutated to glycine (Gly) in 6/28 stage IV human neuroblastomas *(258),* and may be subject to autophosphorylation *(273).* Additionally, serine 120 lies two aas carboxy to a histidine (His) autophosphorylation site involved in the nucleoside diphosphate kinase (NDPK) and His PK activities of Nm23, and contributes to the structure of the active enzymatic pocket *(274).* Mutation of Ser 120 to Gly (S120G), or, to a lesser extent, to an alanine (Ala) (S120A), resulted in high levels of motility. In contrast, mutation of the autophosphorylation site Ser 44 to Ala (S44A) resulted in motility comparable to that of wt *nm23-H1.* The significance of this data is twofold: It indicates that the two sites of Nm23-H1, P96 and S120, are essential domains for Nm23-dependent regulation of cell motility; and it provides the first structure–function correlation for Nm23 in tumor metastasis.

How does Nm23-H1 signal a cell? Although this question has not been definitively answered, the Nm23 proteins have been reported to possess many biochemical functions, including a nonspecific (NDPK) activity *(222,275),* His PK activity *(276–279),* Ser PK activity *(280,281),* autophosphorylation on both Ser and His residues *(273,282–284),* and binding to multiple G proteins, *(285),* RZR orphan receptors *(286),* β-tubulin *(287),* glyceraldehyde-3-phosphate dehydrogenase *(281),* and vimentin *(288).* Several of these activities are debated in the literature. In order to determine if any known biochemical activity correlated with biological function in breast carcinoma motility suppression, the wt and site-directed mutant Nm23-H1 proteins were produced in *E. coli,* and purified to apparent homogeneity *(279).* The S120G/A and P96S mutations exhibited unique deficiencies in aspects of His dependent protein phosphotransferase activity: The S120 mutants exhibited reduced His autophosphorylation levels, which resulted in reduced downstream transfer to Sers and reduced His PK activity to three in vitro substrates; the P96 mutation exhibited normal levels of autophosphorylation, but reduced transfer of phosphate to three substrates in a His PK assay. The

Table 3
Site-Directed Mutagenesis Studies of Nm23-H1

Mutation	Autophosphorylation		NDPK	His PK	Cell migration (Mean ± SEM)
	His	Ser			
Wt	+	+	+	+	1.0+/–0.7
P96S	+	+	+	–	17.5+/–3.7
S120G	±	±	+	±	42.2+/–6.6
S44A	+	±	+	+	1.0+/–0.6

+ = activity; – = no activity; ± = reduced activity.

mutant Nm23-H1 proteins exhibited levels of NDPK activity comparable to wt. The data permit the hypothesis that a His dependent protein phosphotransferase activity is responsible for the motility suppressive effect of *nm23-H1* (Table 3; *279*). The importance of the Nm23 His phosphotransferase activity, but not its NDPK activity, was replicated in *Drosophila* development *(289)*. To date, however, the nm23-dependent signal transduction pathway has not yet been defined.

3.1.3. TRANSLATION STUDIES

In most cases, cancer patients succumb to complications arising from metastatic disease. One way to lengthen a patient's life-span may be to halt or eliminate the metastatic capacity of the tumor. Utilizing the knowledge acquired about *nm23-H1,* two clinical possibilities arise: First, based on the evidence that many metastatically competent tumors exhibit reduced Nm23 expression, and that overexpression of Nm23 results in decreased tumor metastatic potential *in vivo,* it is possible that elevation of the Nm23 expression of overt or micrometastatic tumor deposits in BC or other cancers may be of therapeutic benefit to limit metastatic colonization and dissemination. Second, Nm23 could be used as a marker of tumor metastatic potential, to identify compounds with preferential antiproliferative activity against metastatic breast carcinoma and melanoma cells.

The first strategy utilizes the observations that low nm23-H1 expression seems to be a limiting feature in the regulation of metastasis and cell motility. These findings lead to the hypothesis that transcriptional and translational regulation of *nm23-H1* expression may be important determinants of high vs low expression levels. Thus, if one could turn on the transcription of nm23, thereby increasing RNA and subsequent protein levels, the tumor cells' metastatic capability may be hindered or reversed. A study *(290),* substantiating the clinical potential of increasing nm23 expression, reported that Nm23 transfectants of four different cell lines exhibited increased sensitivity to cisplatin (a clinical chemotherapy [CT] drug) to inhibition of proliferation *in vitro,* and that a more pronounced reduction of metastatic colonization occurred upon cisplatin administration in *in vivo* experiments using control and *nm23* transfectants of the K-1735 TK murine melanoma cell line. In addition, a recent ovarian cohort study *(247)* supported this finding. This suggests that, if agents are identified that can increase the Nm23 expression of micrometastatic breast carcinoma or other cancer cells *in vivo,* the combination of these agents with alkylating agents may be a potential therapeutic regime.

In order to identify factors that could elevate Nm23-H1 expression, the transcriptional regulation of *nm23-H1* promoter is being studied. The *nm23-H1* promoter has been cloned by several laboratories *(291–293)*. Deletion constructs of the *nm23-H1* promoter identified a 248-bp fragment that mediates high vs low expression levels (unpublished data). Current research is attempting to define the transcriptional elements (i.e., transcription factors, methylation) that are responsible for its higher expression in nonmetastatic cells.

The second translational approach also utilizes the fact that nm23 expression is reduced in highly metastatic cells; however, in this case, *nm23-H1* expression was used as a marker for metastatic potential. In this analysis, Nm23 expression levels in the breast carcinoma and melanoma subset of the panel were determined by densitometry of Western blots, and correlated to *in vitro* sensitivity to CT agents. Through a screen of antiproliferative activity of compounds on a 55-cell-line panel, performed by NCI's Developmental Therapeutic Program, growth-inhibitory potential of drugs can be correlated to a molecular event that varies among the cell lines (in this case, Nm23-H1), using the COMPARE program (in this case, Nm23-H1). Among 177 standard CT compounds that are currently in clinical use, no agents were identified that were preferentially growth-inhibitory to low Nm23-expressing, high metastatic breast carcinoma and melanoma cell lines. However, a compound known as NSC 645306 (resynthesized as NSC680718) was identified from a 30,000 compound repository, with such *in vitro* activity *(279)*.

Tumor-growth-inhibitory potential was demonstrated *in vivo* by a hollow fiber assay in five tumor cell lines historically difficult to treat by conventional CT: two metastatically competent breast carcinoma cell lines, a melanoma cell line, and colon carcinoma and non-small cell lung carcinoma cell lines *(279)*. NSC680718 also possesses antiangiogenic activity in vivo, using a rat aorta assay, with 50 and 98% inhibition of angiogenesis at 5 and 10 µM, respectively (unpublished data). The compound did not elevate Nm23 expression, indicating that this was simply a marker in these studies. Taken together, the data define an approach for identifying novel compounds that may exhibit preferential inhibitory capacity for metastatically competent breast and other tumor cells, and which may become important standard CT in the future *(279)*.

3.2. KAI1 Gene

KAI1/CD82 is a metastasis suppressor gene on human chromosome 11p11.2 that encodes a glycoprotein of the transmembrane four superfamily *(294)*. Reduced KAI1 expression associates with malignant progression of human tumors, including prostate, breast, lung, bladder, ovary, and pancreatic cancers *(69,295–299)*. The transfection of KAI1 into invasive melanoma or colon cancer cells reduced or suppressed their invasive and metastatic potential *(300,301)*. Reduced KAI1 mRNA in tumor cells seems to influence their metastatic ability, and thereby enhances their malignant potential *(296,297,302–305)*.

In addition, several studies suggest that KAI protein may undergo posttranslational modifications (N-linked glycosylation) in invasive cells. Other members of the transmembrane four superfamily were cloned, and their expression seems also to inversely correlate with the aggressive behavior of tumor cells. The expressions of *MRP-1/CD9*, *CD81*, and *KAI1/CD82* genes were found to be useful indicators of a poor prognosis in BC patients *(299,306)*. It was recently suggested that KAI1 expression may be directly activated by p53 *(69)*. A cohort study of prostate tumors showed a direct correlation between p53 and KAI1 expression. This suggests that the loss of p53 function, which is commonly observed in many types of cancer, leads to the downregulation of the *KAI1* gene, which may result in the progression of metastasis *(69)*. Therefore, a simultaneous survey of p53 and KAI1 expression in human cancers should be of interest.

3.3. KiSS-1 Gene

The *KiSS-1* gene encodes a predominantly hydrophilic, 164-aa protein with a polyproline-rich domain indicative of an SH3 ligand (binds to the homology 3 domain of the oncoprotein Src) and a putative PK C-α phosphorylation site *(307,308)*. When transfected into C8161 melanoma cells or MDA-MB435 cells, the full-length *KiSS-1* cDNA suppressed metastasis in an expression-dependent manner *(307,309,310)*. Its expression seems to occur only in nonmetastatic cells *(307)*. The data suggest a mechanism whereby *KiSS-1* regulates events downstream of cell–matrix adhesion, perhaps involving cytoskeletal reorganization. However, more studies are needed to address the function of *KiSS-1* and its expression in human cancers.

3.4. TSP1 Gene

Another example of a metastasis suppressor gene is provided by the thrombospondin 1 *(TSP1)* gene *(311–314)*, encoding an extracellular matrix protein. Transfection of *TSP1* into human breast carcinoma cells reduces their metastatic potential and angiogenesis *(315)*. However, *TSP1* also inhibits cell growth by a different mechanism *(315–317)*, and additional mechanistic studies are needed to address the role of TSP1 in cell growth and motility. Therefore, this gene may be a good prognostic factor, when combined with other TSGs, for human cancers.

In addition to those genes, numerous growth factor and hormone receptor families have been shown to affect the tumor progression and metastasis. Despite these findings, it is still unclear whether all those factors independently or synergistically affect the metastatic progression in human cancers. Further studies on the cooperation between TSGs and/or metastasis suppressor genes, in controlling the metastatic phenotype, will raise new insights into human cancer diagnosis and gene therapy.

REFERENCES

1. Lane D, Crawford L. T antigen is bound to a host protein in SV40-transformed cells. *Nature* 1979; 278:261–263.
2. Eliyahu D, Raz A, Gruss P, Givol D, Oren M. Participation of p53 cellular tumour antigen in transformation of normal embryonic cells. *Nature* 1984; 312:646–649.
3. Johnson P, Gray D, Mowat M, Benchimol S. Expression of wild-type p53 is not compatible with continued growth of p53-negative tumor cells. *Mol Cell Biol* 1991; 11:1–11.
4. Lu X, Park SH, Thompson TC, Lane DP. Ras-induced hyperplasia occurs with mutation of p53, but activated ras and myc together can induce carcinoma without p53 mutation. *Cell* 1992; 70:153–161.
5. Kern SE, Pietenpol JA, Thiagalingam S, Seymour A, Kinzler KW, Vogelstein B. Oncogenic forms of p53 inhibit p53-regulated gene expression. *Science* 1992; 256:827–830.
6. Takahashi T, Carbone D, Nau MM, Hida T, Linnoila I, Ueda R, Minna JD. Wild-type but not mutant p53 suppresses the growth of human lung cancer cells bearing multiple genetic lesions. *Cancer Res* 1992; 52:2340–2343.
7. Ginsberg D, Mechta F, Yaniv M, Oren M. Wild-type p53 can down-modulate the activity of various promoters. *Proc Natl Acad Sci USA* 1991; 88:9979–9983.
8. Hicks GG, Egan SE, Greenberg AH, Mowat M. Mutant p53 tumor suppressor alleles release ras-induced cell cycle growth arrest. *Mol Cell Biol* 1991; 11:1344–1352.

9. Milner J, Medcalf EA, Cook AC. Tumor suppressor p53: analysis of wild-type and mutant p53 complexes. *Mol Cell Biol* 1991; 11:12–19.

10. Hinds P, Finlay C, Quartin R, Baker S, Fearon E, Vogelstein B, Levine A. Mutant p53 DNA clones from human colon carcinomas cooperate with ras in transforming primary rat cells: a comparison of the "hot spot" mutant phenotypes. *Cell Growth Differ* 1990; 1:571–580.

11. Hollstein M, Sidransky D, Vogelstein B, Harris C. p53 mutations in human cancers. *Science* 1991; 253:49–53.

12. Tokunaga T, Nakamura M, Oshika Y, Tsuchida T, Kazuno M, Fukushima Y, et al. Alterations in tumour suppressor gene p53 correlate with inhibition of thrombospondin-1 gene expression in colon cancer cells. *Virchows Arch* 1998; 433:415–418.

13. Oshiro Y, Chaturvedi V, Hayden D, Nazeer T, Johnson M, Johnston DA, et al. Altered p53 is associated with aggressive behavior of chondrosarcoma: a long term follow-up study. *Cancer* 1998; 83:2324–2334.

14. Levesque MA, D'Costa M, Spratt EH, Yaman MM, Diamandis EP. Quantitative analysis of p53 protein in non-small cell lung cancer and its prognostic value. *Int J Cancer* 1998; 79:494–501.

15. Yokoyama R, Schneider-Stock R, Radig K, Wex T, Roessner A. Clinicopathologic implications of MDM2, p53 and K-ras gene alterations in osteosarcomas: MDM2 amplification and p53 mutations found in progressive tumors. *Pathol Res Pract* 1998; 194:615–621.

16. Wang LS, Chow KC, Liu CC, Chiu JH. p53 gene alternation in squamous cell carcinoma of the esophagus detected by PCR-cold SSCP analysis. *Proc Natl Sci Counc Repub China B* 1998; 22:114–121.

17. Esteve P, Embade N, Perona R, Jimenez B, del Peso L, Leon J, Arends M, Miki T, Lacal JC. Rho-regulated signals induce apoptosis in vitro and in vivo by a p53-independent, but Bcl2 dependent pathway. *Oncogene* 1998; 17:1855–1869.

18. Indinnimeo M, Cicchini C, Stazi A, Limiti MR, Giarnieri E, Ghini C, Vecchione A. The prevalence of p53 immunoreactivity in anal canal carcinoma. *Oncol Rep* 1998; 5:1455–1457.

19. Shaw P, Bovey R, Tardy S, Sahli R, Sordat B, Costa J. Induction of apoptosis by wild-type p53 in a human colon tumor-derived cell line. *Proc Natl Acad Sci USA* 1992; 89:4495–4499.

20. Yonish-Rouach E, Grunwald D, Wilder S, Kimchi A, May E, Lawrence JJ, May P, Oren M. p53-mediated cell death: relationship to cell cycle control. *Mol Cell Biol* 1993; 13:1415–1423.

21. Debbas M, White E. Wild-type p53 mediates apoptosis by E1A, which is inhibited by E1B. *Genes Dev* 1993; 7:546–554.

22. Ramqvist T, Magnusson KP, Wang Y, Szekely L, Klein G, Wiman KG. Wild-type p53 induces apoptosis in a Burkitt lymphoma (BL) line that carries mutant p53. *Oncogene* 1993; 8:1495–1500.

23. Kastan MB, Onyekwere O, Sidransky D, Vogelstein B, Craig RW. Participation of p53 protein in the cellular response to DNA damage. *Cancer Res* 1991; 51:6304–6311.

24. Lane DP. Cancer. p53, guardian of the genome. *Nature* 1992; 358:15–16.

25. Ananthaswamy HN, Pierceall WE. Molecular alterations in human skin tumors. *Prog Clin Biol Res* 1992; 376:61–84.

26. Reznikoff CA, Belair C, Savelieva E, Zhai Y, Pfeifer K, Yeager T, et al. Long-term genome stability and minimal genotypic and phenotypic alterations in HPV16 E7-, but not E6-, immortalized human uroepithelial cells. *Genes Dev* 1994; 8:2227–2240.

27. Pogribny IP, Basnakian AG, Miller BJ, Lopatina NG, Poirier LA, James SJ. Breaks in genomic DNA and within the p53 gene are associated with hypomethylation in livers of folate/methyl-deficient rats [published erratum appears in *Cancer Res* 1995; 55:2711]. *Cancer Res* 1995; 55:1894–1901.

28. Moro F, Ottagio L, Bonatti S, Simili M, Miele M, Bozzo S, Abbondandolo A. p53 expression in normal versus transformed mammalian cells. *Carcinogenesis* 1995; 16:2435–2440.

29. Bertrand P, Rouillard D, Boulet A, Levalois C, Soussi T, Lopez BS. Increase of spontaneous intrachromosomal homologous recombination in mammalian cells expressing a mutant p53 protein. *Oncogene* 1997; 14:1117–1122.

30. Mekeel KL, Tang W, Kachnic LA, Luo CM, DeFrank JS, Powell SN. Inactivation of p53 results in high rates of homologous recombination. *Oncogene* 1997; 14:1847–1857.

31. Kuerbitz SJ, Plunkett BS, Walsh WV, Kastan MB. Wild-type p53 is a cell cycle checkpoint determinant following irradiation. *Proc Natl Acad Sci USA* 1992; 89:7491–7495.

32. Kastan MB, Zhan Q, el-Deiry WS, Carrier F, Jacks T, Walsh WV, et al. A mammalian cell cycle checkpoint pathway utilizing p53 and GADD45 is defective in ataxia-telangiéctasia. *Cell* 1992; 71:587–597.

33. Carr AM, Green MH, Lehmann AR. Checkpoint policing by p53 [letter; comment]. *Nature* 1992; 359:486–487.
34. Ryan JJ, Danish R, Gottlieb CA, Clarke MF. Cell cycle analysis of p53-induced cell death in murine erythroleukemia cells. *Mol Cell Biol* 1993; 13:711–719.
35. El-Deiry W, Tokino T, Velculescu V, Levy D, Parsons R, Trent J, et al. WAF1, a potential mediator of p53 tumor suppression. *Cell* 1993; 75:817–825.
36. Harper J, Adami G, Wei N, Keyomarsi K, Elledge S. The p21 Cdk-interacting protein Cip1 is a potent inhibitor of G1 cyclin-dependent kinases. *Cell* 1993; 75:805–816.
37. Hermeking H, Lengauer C, Polyak K, He TC, Zhang L, Thiagalingam S, Kinzler KW, Vogelstein B. 14-3-3 sigma is a p53-regulated inhibitor of G2/M progression. *Mol Cell* 1997; 1:3–11.
38. Juven T, Barak Y, Zauberman A, George DL, Oren, M. Wild type p53 can mediate sequence-specific transactivation of an internal promoter within the mdm2 gene. *Oncogene* 1993; 8:3411–3416.
39. Barak Y, Juven T, Haffner R, Oren M. mdm2 expression is induced by wild type p53 activity. *EMBO J* 1993; 12:461–468.
40. Oliner JD, Pietenpol JA, Thiagalingam S, Gyuris J, Kinzler KW, Vogelstein B. Oncoprotein MDM2 conceals the activation domain of tumour suppressor p53. *Nature* 1993; 362:857–860.
41. Leveillard T, Wasylyk B. The MDM2 C-terminal region binds to TAFII250 and is required for MDM2 regulation of the cyclin A promoter. *J Biol Chem* 1997; 272:30,651–30,661.
42. Kubbutat MH, Jones SN, Vousden KH. Regulation of p53 stability by Mdm2. *Nature* 1997; 387:299–303.
43. Nozaki T, Masutani M, Sugimura T, Takato T, Wakabayashi K. Abrogation of G1 arrest after DNA damage is associated with constitutive overexpression of Mdm2, Cdk4, and Irf1 mRNAs in the BALB/c 3T3 A31 variant 1–1 clone. *Biochem Biophys Res Commun* 1997; 233:216–220.
44. Lundgren K, Montes de Oca Luna R, McNeill YB, Emerick EP, Spencer B, Barfield OR, et al. Targeted expression of MDM2 uncouples S phase from mitosis and inhibits mammary gland development independent of p53. *Genes Dev* 1997; 11:714–725.
45. Donehower LA, Harvey M, Slagle BL, McArthur MJ, Montgomery CA Jr, Butel JS, Bradley A. Mice deficient for p53 are developmentally normal but susceptible to spontaneous tumours. *Nature* 1992; 356:215–221.
46. Lee JM, Abrahamson JL, Kandel R, Donehower LA, Bernstein A. Susceptibility to radiation-carcinogenesis and accumulation of chromosomal breakage in p53 deficient mice. *Oncogene* 1994; 9:3731–3736.
47. Jacks T, Remington L, Williams BO, Schmitt EM, Halachmi S, Bronson RT, Weinberg RA. Tumor spectrum analysis in p53-mutant mice. *Curr Biol* 1994; 4:1–7.
48. Symonds H, Krall L, Remington L, Saenz-Robles M, Lowe S, Jacks T, Van Dyke T. p53-dependent apoptosis suppresses tumor growth and progression in vivo. *Cell* 1994; 78:703–711.
49. Hursting SD, Perkins SN, Phang JM. Calorie restriction delays spontaneous tumorigenesis in p53-knockout transgenic mice. *Proc Natl Acad Sci USA* 1994; 91:7036–7040.
50. Howes KA, Ransom N, Papermaster DS, Lasudry JG, Albert DM, Windle JJ. Apoptosis or retinoblastoma: alternative fates of photoreceptors expressing the HPV-16 E7 gene in the presence or absence of p53 [published erratum appears in *Genes Dev* 1994; 8:1738]. *Genes Dev* 1994; 8:1300–1310.
51. Bowman T, Symonds H, Gu, L, Yin C, Oren M, Van Dyke T. Tissue-specific inactivation of p53 tumor suppression in the mouse. *Genes Dev* 1996; 10:826–835.
52. Livingstone LR, White A, Sprouse J, Livanos E, Jacks T, Tlsty TD. Altered cell cycle arrest and gene amplification potential accompany loss of wild-type p53. *Cell* 1992; 70:923–935.
53. Nikiforov MA, Hagen K, Ossovskaya VS, Connor TM, Lowe SW, Deichman GI, Gudkov AV. p53 modulation of anchorage independent growth and experimental metastasis. *Oncogene* 1996; 13:1709–1719.
54. Levesque MA, Yu H, Clark GM, Diamandis EP. Enzyme-linked immunoabsorbent assay-detected p53 protein accumulation: a prognostic factor in a large breast cancer cohort. *J Clin Oncol* 1998; 16:2641–2650.
55. Tsuda H, Sakamaki C, Tsugane S, Fukutomi T, Hirohashi S. A prospective study of the significance of gene and chromosome alterations as prognostic indicators of breast cancer patients with lymph node metastases. *Breast Cancer Res Treat* 1998; 48:21–32.
56. Oshiro Y, Chaturved V, Hayden D, Nazeer T, Johnson M, Johnston D, Ordonez N, Ayala A, Czerniak B. Altered p53 is associated with aggressive behavior of chondrosarcoma: a long term follow-up study. *Cancer* 1998; 83:2324–2334.

57. Atula S, Kurvinen K, Grenman R, Syrjanen S. SSCP pattern indicative for p53 mutation is related to advanced stage and high-grade of tongue cancer. *Eur J Cancer B Oral Oncol* 1996; 32B:222–229.

58. Rohlke P, Milde-Langosch K, Weyland C, Pichlmeier U, Jonat W, Loning T. p53 is a persistent and predictive marker in advanced ovarian carcinomas: multivariate analysis including comparison with Ki67 immunoreactivity. *J Cancer Res Clin Oncol* 1997; 123:496–501.

59. Sato Y, Nio Y, Song MM, Sumi S, Hirahara N, Minari Y, Tamura K. p53 protein expression as prognostic factor in human pancreatic cancer. *Anticancer Res* 1997; 17: 2779–2788.

60. Caminero MJ, Nunez F, Suarez C, Ablanedo P, Riera JR, Dominguez F. Detection of p53 protein in oropharyngeal carcinoma. Prognostic implications. *Arch Otolaryngol Head Neck Surg* 1996; 122:769–772.

61. Venara M, Marull R, Bergada I, Gamboni M, Chemes H. Functional adrenal cortical tumors in childhood: a study of ploidy, p53- protein and nucleolar organizer regions (AgNORs) as prognostic markers. *J Pediatr Endocrinol Metab* 1998; 11:597–605.

62. Silvestrini R, Daidone M, Benini E, Faranda A, Tomasic G, Boracchi P, Salvadori B, Veronesi U. Validation of p53 accumulation as a predictor of distant metastasis at 10 years of follow-up in 1400 node-negative breast cancers. *Clin Cancer Res* 1996; 12:2007–2013.

63. Naka T, Toyota N, Kaneko T, Kaibara N. Protein expression of p53, p21WAF1, and Rb as prognostic indicators in patients with surgically treated hepatocellular carcinoma. *Anticancer Res* 1998; 18:555–564.

64. Erez-Alon N, Herkel J, Wolkowicz R, Ruiz PJ, Waisman A, Rotter V, Cohen IR. Immunity to p53 induced by an idiotypic network of anti-p53 antibodies: generation of sequence-specific anti-DNA antibodies and protection from tumor metastasis. *Cancer Res* 1998; 58:5447–5452.

65. Hallak R, Mueller J, Lotter O, Gansauge S, Gansauge F, el-Deen Jumma M, et al. p53 genetic alterations, protein expression and autoantibodies in human colorectal carcinoma: a comparative study. *Int J Oncol* 1998; 12:785–791.

66. Shiota G, Kishimoto Y, Suyama A, Okubo M, Katayama S, Harada K, Ishida M, Hori K, Suou T, Kawasaki H. Prognostic significance of serum anti-p53 antibody in patients with hepatocellular carcinoma. *J Hepatol* 1997; 27:661–668.

67. Sakamuro D, Sabbatini P, White E, Prendergast GC. The polyproline region of p53 is required to activate apoptosis but not growth arrest. *Oncogene* 1997; 15:887–898.

68. Meyers FJ, Gumerlock PH, Chi SG, Borchers H, Deitch AD, deVere White RW. Very frequent p53 mutations in metastatic prostate carcinoma and in matched primary tumors. *Cancer* 1998; 83:2534–2539.

69. Mashimo T, Watabe M, Hirota S, Hosobe S, Miura K, Tegtmeyer P, Rinker-Shaeffer C, Watabe K. The expression of the KAI1 gene, a tumor metastasis suppressor, is directly activated by p53. *Proc Natl Acad Sci USA* 1998; 95:11,307–11,311.

70. Campomenosi P, Assereto P, Bogliolo M, Fronza G, Abbondandolo A, Capasso A, et al. p53 mutations and DNA ploidy in colorectal adenocarcinomas. *Anal Cell Pathol* 1998; 17:1–12.

71. Stoll C, Baretton G, Lohrs U. The influence of p53 and associated factors on the outcome of patients with oral squamous cell carcinoma. *Virchows Arch* 1998; 433:427–433.

72. Kappes S, Milde-Langosch K, Kressin P, Passlack B, Dockhorn-Dworniczak B, Rohlke P, Loning T. p53 mutations in ovarian tumors, detected by temperature-gradient gel electrophoresis, direct sequencing and immunohistochemistry. *Int J Cancer* 1995; 64:52–59.

73. O'Connor PM, Jackman J, Bae I, Myers TG, Fan S, Mutoh M, et al. Characterization of the p53 tumor suppressor pathway in cell lines of the National Cancer Institute anticancer drug screen and correlations with the growth-inhibitory potency of 123 anticancer agents. *Cancer Res* 1997; 57:4285–4300.

74. Lavigueur A, Maltby V, Mock D, Rossant J, Pawson T, Bernstein A. High incidence of lung, bone, and lymphoid tumors in transgenic mice overexpressing mutant alleles of the p53 oncogene. *Mol Cell Biol* 1989; 9:3982–3991.

75. Lavigueur A, Bernstein A. p53 transgenic mice: accelerated erythroleukemia induction by Friend virus. *Oncogene* 1991; 6:2197–2201.

76. Lee JM, Bernstein A. p53 mutations increase resistance to ionizing radiation. *Proc Natl Acad Sci USA* 1993; 90:5742–5746.

77. Cox L, Chen G, Lee E. Tumor suppressor genes and their role in breast cancer. *Breast Cancer Res Treat* 1994; 32:19–38.

78. Harrington EA, Bruce JL, Harlow E, Dyson N. pRB plays an essential role in cell cycle arrest induced by DNA damage. *Proc Natl Acad Sci USA* 1998; 95:11,945–11,950.
79. Goodrich D, Wang N, Qian Y, Lee E, Lee W. The retinoblastoma gene product regulates progression through the G1 phase of the cell cycle. *Cell* 1991; 67:293–302.
80. Hinds P, Mittnacht S, Dulic V, Arnold A, Reed S, Weinberg R. Regulation of retinoblastoma protein functions by ectopic expression of human cyclins. *Cell* 1992; 70:993–1006.
81. Qin X, Chittenden T, Livingston D, Kaelin WJ. Identification of a growth suppression domain within the retinoblastoma gene product. *Genes Dev* 1992; 6:953–964.
82. Awazu S, Nakata K, Hida D, Sakamoto T, Nagata K, Ishii N, Kanematsu T. Stable transfection of retinoblastoma gene promotes contact inhibition of cell growth and hepatocyte nuclear factor-1-mediated transcription in human hepatoma cells. *Biochem Biophys Res Commun* 1998; 252:269–273.
83. Huang H, Yee J, Shew J, Chen P, Bookstein R, Friedmann T, Lee E, Lee W. Suppression of the neoplastic phenotype by replacement of the RB gene in human cancer cells. *Science* 1988; 242:1563–1566.
84. Sumegi J, Uzvolgyi E, Klein G. Expression of the RB gene under the control of MuLV-LTR suppresses tumorigenicity of WERI-Rb-27 retinoblastoma cells in immunodefective mice. *Cell Growth Differ* 1990; 1:247–250.
85. Madreperla S, Whittum-Hudson J, Prendergast R, Chen P, Lee W. Intraocular tumor suppression of retinoblastoma gene-reconstituted retinoblastoma cells. *Cancer Res* 1991; 51:6381–6384.
86. Takahashi R, Hashimoto T, Xu HJ, Hu SX, Matsui T, Miki T, et al. The retinoblastoma gene functions as a growth and tumor suppressor in human bladder carcinoma cells. *Proc Natl Acad Sci USA* 1991; 88:5257–5261.
87. Chen P, Chen Y, Shan B, Bookstein R, Lee W. Stability of retinoblastoma gene expression determines the tumorigenicity of reconstituted retinoblastoma cells. *Cell Growth Differ* 1992; 3:119–125.
88. Lefebvre D, Gala JL, Heusterspreute M, Delhez H, Philippe M. Introduction of a normal retinoblastoma (Rb) gene into Rb-deficient lymphoblastoid cells delays tumorigenicity in immunodefective mice. *Leuk Res* 1998; 22:905–912.
89. Lee EY, Chang CY, Hu N, Wang YC, Lai CC, Herrup K, Lee WH, Bradley A. Mice deficient for Rb are nonviable and show defects in neurogenesis and haematopoiesis. *Nature* 1992; 359:288–294.
90. Hooper ML. The role of the p53 and Rb-1 genes in cancer, development and apoptosis. *J Cell Sci Suppl* 1994; 18:13–17.
91. Williams BO, Schmitt EM, Remington L, Bronson RT, Albert DM, Weinberg RA, Jacks T. Extensive contribution of Rb-deficient cells to adult chimeric mice with limited histopathological consequences. *Embo J* 1994; 13:4251–4259.
92. Harrison DJ, Hooper ML, Armstrong JF, Clarke AR. Effects of heterozygosity for the Rb-lt19neo allele in the mouse. *Oncogene* 1995; 10:1615–1620.
93. Harvey M, Vogel H, Lee EY, Bradley A, Donehower LA. Mice deficient in both p53 and Rb develop tumors primarily of endocrine origin. *Cancer Res* 1995; 55:1146–1151.
94. Bignon YJ, Chen Y, Chang CY, Riley DJ, Windle JJ, Mellon PL, Lee WH. Expression of a retinoblastoma transgene results in dwarf mice. *Genes Dev* 1993; 7:1654–1662.
95. Abramson D, Ellsworth R, Kitchin F, Tung G. Second nonocular tumors in retinoblastoma survivors. Are they radiation-induced? *Ophthalmology* 1984; 91:1351–1355.
96. Goodrich D, Lee W. Molecular characterization of the retinoblastoma susceptibility gene. *Biochim Biophys Acta* 1993; 1155:43–61.
97. Welch D, Wei L. Genetic and epigenetic regulation of human breast cancer progression and metastasis. *Endocr-Related Cancer* 1998; 5:155–197.
98. Cohen J, Geradts J. Loss of RB and MTS1/CDKN2 (p16) expression in human sarcomas. *Hum Pathol* 1997; 28:893–898.
99. Ishikawa T, Furihata M, Ohtsuki Y, Murakami H, Inoue A, Ogoshi S. Cyclin D1 overexpression related to retinoblastoma protein expression as a prognostic marker in human oesophageal squamous cell carcinoma. *Br J Cancer* 1998; 77:92–97.
100. Andersen T, Gaustad A, Ottestad L, Farrants G, Nesland J, Tveit K, Borresen A. Genetic alterations of the tumour suppressor gene regions 3p, 11p, 13q, 17p, and 17q in human breast carcinomas. *Genes Chromosomes Cancer* 1992; 4:113–121.
101. Spandidos D, Karaiossifidi H, Malliri A, Linardopoulos S, Vassilaros S, Tsikkinis A, Field J. Expression of ras Rb1 and p53 proteins in human breast cancer. *Anticancer Res* 1992; 12:81–89.

102. Borg A, Zhang Q, Alm P, Olsson H, Sellberg G. The retinoblastoma gene in breast cancer: allele loss is not correlated with loss of gene protein expression. *Cancer Res* 1992; 52:2991–2994.

103. Varley J, Armour J, Swallow J, Jeffreys A, Ponder B, T'Ang A, Fung Y, Brammar W, Walker R. The retinoblastoma gene is frequently altered leading to loss of expression in primary breast tumours [published erratum appears in *Oncogene* 1990; 5:245]. *Oncogene* 1989; 4:725–729.

104. Miyaki M, Tanaka K, Kikuchi-Yanoshita R, Muraoka M, Konishi M, Takeichi M. Increased cell-substratum adhesion, and decreased gelatinase secretion and cell growth, induced by E-cadherin transfection of human colon carcinoma cells. *Oncogene* 1995; 11:2547–2552.

105. Maruyama K, Ochiai A, Nakamura S, Baba S, Hirohashi S. Dysfunction of E-cadherin-catenin system in invasion and metastasis of colorectal cancer. *Nippon Geka Gakkai Zasshi* 1998; 99:402–408.

106. Shimoyama Y, Hirohashi S. Cadherin intercellular adhesion molecule in hepatocellular carcinomas: loss of E-cadherin expression in an undifferentiated carcinoma. *Cancer Lett* 1991; 57:131–135.

107. Tahara E, Kuniyasu H, Nakayama H, Yasui W, Yokozaki H. Metastasis related genes and malignancy in human esophageal, gastric and colorectal cancers. *Gan To Kagaku Ryoho* 1993; 20:326–331.

108. Imao T, Koshida K, Endo Y, Uchibayashi T, Sasaki T, Namiki M. Dominant role of E-cadherin in the progression of bladder cancer. *J Urol* 1999; 161:692–698.

109. Byers SW, Sommers CL, Hoxter B, Mercurio AM, Tozeren A. Role of E-cadherin in the response of tumor cell aggregates to lymphatic, venous and arterial flow: measurement of cell-cell adhesion strength. *J Cell Sci* 1995; 108:2053–2064.

110. Bukholm I, Nesland J, Karesen R, Jacobsen U, Borresen-Dale A. E-cadherin and alpha-, beta-, and gamma-catenin protein expression in relation to metastasis in human breast carcinoma. *J Pathol* 1998; 185:262–266.

111. Bussemakers MJ, van Moorselaar RJ, Giroldi LA, Ichikawa T, Isaacs JT, Takeichi M, Debruyne FM, Schalken JA. Decreased expression of E-cadherin in the progression of rat prostatic cancer. *Cancer Res* 1992; 52:2916–2922.

112. Frixen UH, Behrens J, Sachs M, Eberle G, Voss B, Warda A, Lochner D, Birchmeier W. E-cadherin-mediated cell-cell adhesion prevents invasiveness of human carcinoma cells. *J Cell Biol* 1991; 113:173–185.

113. Mbalaviele G, Dunstan CR, Sasaki A, Williams PJ, Mundy GR, Yoneda T. E-cadherin expression in human breast cancer cells suppresses the development of osteolytic bone metastases in an experimental metastasis model. *Cancer Res* 1996; 56:4063–4070.

114. Mareel MM, Behrens J, Birchmeier W, De Bruyne GK, Vleminckx K, Hoogewijs A, Fiers WC, Van Roy FM. Down-regulation of E-cadherin expression in Madin Darby canine kidney (MDCK) cells inside tumors of nude mice. *Int J Cancer* 1991; 47:922–928.

115. Schipper JH, Frixen UH, Behrens J, Unger A, Jahnke K, Birchmeier W. E-cadherin expression in squamous cell carcinomas of head and neck: inverse correlation with tumor dedifferentiation and lymph node metastasis. *Cancer Res* 1991; 51:6328–6337.

116. Inoue M, Ogawa H, Miyata M, Shiozaki H, Tanizawa O. Expression of E-cadherin in normal, benign, and malignant tissues of female genital organs. *Am J Clin Pathol* 1992; 98:76–80.

117. Hashimoto M, Niwa O, Nitta Y, Takeichi M, Yokoro K. Unstable expression of E-cadherin adhesion molecules in metastatic ovarian tumor cells. *Jpn J Cancer Res* 1989; 80:459–463.

118. Franchi A, Gallo O, Boddi V, Santucci M. Prediction of occult neck metastases in laryngeal carcinoma: role of proliferating cell nuclear antigen, MIB-1, and E-cadherin immunohistochemical determination. *Clin Cancer Res* 1996; 2:1801–1808.

119. Shun CT, Wu MS, Lin JT, Wang HP, Houng RL, Lee WJ, Wang TH, Chuang SM. An immunohistochemical study of E-cadherin expression with correlations to clinicopathological features in gastric cancer. *Hepatogastroenterology* 1998; 45:944–949.

120. Kuniyasu H, Ellis LM, Evans DB, Abbruzzese JL, Fenoglio CJ, Bucana CD, Cleary KR, Tahara E, Fidler IJ. Relative expression of E-cadherin and type IV collagenase genes predicts disease outcome in patients with resectable pancreatic carcinoma. *Clin Cancer Res* 1999; 5:25–33.

121. Washington K, Chiappori A, Hamilton K, Shyr Y, Blanke C, Johnson D, Sawyers J, Beauchamp D. Expression of beta-catenin, alpha-catenin, and E-cadherin in Barrett's esophagus and esophageal adenocarcinomas. *Modern Pathol* 1998; 11:805–813.

122. Karayiannakis AJ, Syrigos KN, Chatzigianni E, Papanikolaou S, Alexiou D, Kalahanis N, Rosenberg T, Bastounis E. Aberrant E-cadherin expression associated with loss of differentiation and advanced stage in human pancreatic cancer. *Anticancer Res* 1998; 18:4177–4180.

123. Gofuku J, Shiozaki H, Tsujinaka T, Inoue M, Tamura S, Doki Y, Matsui S, Tsukita S, Kikkawa N, Monden M. Expression of E-cadherin and alpha-catenin in patients with colorectal carcinoma. Correlation with cancer invasion and metastasis. *Am J Clin Pathol* 1999; 111:29–37.

124. Hirvikoski P, Kumpulainen E, Virtaniemi J, Helin H, Rantala I, Johansson R, Juhola M, Kosma V. Cytoplasmic accumulation of alpha-catenin is associated with aggressive features in laryngeal squamous-cell carcinoma. *Int J Cancer* 1998; 79:546–550.

125. No V, Willems J, Vandekerckhove J, Roy F, Bruyneel E, Mareel M. Inhibition of adhesion and induction of epithelial cell invasion by HAV-containing E-cadherin-specific peptides. *J Cell Sci* 1998; 112:127–135.

126. Steck PA, Pershouse MA, Jasser SA, Yung WK, Lin H, Ligon AH, et al. Identification of a candidate tumour suppressor gene, MMAC1, at chromosome 10q23.3 that is mutated in multiple advanced cancers. *Nat Genet* 1997; 15:356–362.

127. Li J, Yen C, Liaw D, Podsypanina K, Bose S, Wang S, et al. R. PTEN, a putative protein tyrosine phosphatase gene mutated in human brain, breast, and prostate cancer. *Science* 1997; 275:1943–1947.

128. Li DM, Sun H. TEP1, encoded by a candidate tumor suppressor locus, is a novel protein tyrosine phosphatase regulated by transforming growth factor beta. *Cancer Res* 1997; 57:2124–2129.

129. Liaw D, Marsh DJ, Li J, Dahia PL, Wang SI, Zheng Z, et al. Germline mutations of the PTEN gene in Cowden disease, an inherited breast and thyroid cancer syndrome. *Nat Genet* 1997; 16:64–67.

130. Furnari FB, Lin H, Huang HS, Cavenee WK. Growth suppression of glioma cells by PTEN requires a functional phosphatase catalytic domain. *Proc Natl Acad Sci USA* 1997; 94:12,479–12,484.

131. Tamura M, Gu J, Matsumoto K, Aota S, Parsons R, Yamada KM. Inhibition of cell migration, spreading, and focal adhesions by tumor suppressor PTEN. *Science* 1998; 280:1614–1617.

132. Gu J, Tamura M, Yamada K. Tumor suppressor PTEN inhibits integrin- and growth factor-mediated mitogen-activated protein (MAP) kinase signaling pathways. *J Cell Biol* 1998; 143:1375–1383.

133. Wu X, Senechal K, Neshat MS, Whang YE, Sawyers CL. The PTEN/MMAC1 tumor suppressor phosphatase functions as a negative regulator of the phosphoinositide 3-kinase/Akt pathway. *Proc Natl Acad Sci USA* 1998; 95:15,587–15,691.

134. Furnari FB, Huang HJ, Cavenee WK. The phosphoinositol phosphatase activity of PTEN mediates a serum-sensitive G1 growth arrest in glioma cells. *Cancer Res* 1998; 58:5002–5008.

135. Haas-Kogan D, Shalev N, Wong M, Mills G, G, Y, Stokoe D. Protein kinase B (PKB/Akt) activity is elevated in glioblastoma cells due to mutation of the tumor suppressor PTEN/MMAC. *Curr Biol* 1998; 8:1195–1198.

136. Maehama T, Dixon JE. The tumor suppressor, PTEN/MMAC1, dephosphorylates the lipid second messenger, phosphatidylinositol 3,4,5-trisphosphate. *J Biol Chem* 1998; 273:13,375–13,378.

137. Stambolic V, Suzuki A, de la Pompa JL, Brothers GM, Mirtsos C, Sasaki T, et al. Negative regulation of PKB/Akt-dependent cell survival by the tumor suppressor PTEN. *Cell* 1998; 95:29–39.

138. Dahia PL, Marsh DJ, Zheng Z, Zedenius J, Komminoth P, Frisk T, et al. Somatic deletions and mutations in the Cowden disease gene, PTEN, in sporadic thyroid tumors. *Cancer Res* 1997; 57:4710–4713.

139. Tsou HC, Teng DH, Ping XL, Brancolini V, Davis T, Hu R, et al. The role of MMAC1 mutations in early-onset breast cancer: causative in association with Cowden syndrome and excluded in BRCA1-negative cases. *Am J Hum Genet* 1997; 61:1036–1043.

140. Tashiro H, Blazes MS, Wu R, Cho KR, Bose S, Wang SI, Li J, Parsons R, Ellenson LH. Mutations in PTEN are frequent in endometrial carcinoma but rare in other common gynecological malignancies. *Cancer Res* 1997; 57:3935–3940.

141. Rasheed BK, Stenzel TT, McLendon RE, Parsons R, Friedman AH, Friedman HS, Bigner DD, Bigner SH. PTEN gene mutations are seen in high-grade but not in low-grade gliomas. *Cancer Res* 1997; 57:4187–4190.

142. Wang SI, Puc J, Li J, Bruce JN, Cairns P, Sidransky D, Parsons R. Somatic mutations of PTEN in glioblastoma multiforme. *Cancer Res* 1997; 57:4183–4186.

143. Guldberg P, thor Straten P, Birck A, Ahrenkiel V, Kirkin AF, Zeuthen J. Disruption of the MMAC1/PTEN gene by deletion or mutation is a frequent event in malignant melanoma. *Cancer Res* 1997; 57:3660–3663.

144. Rhei E, Kang L, Bogomolniy F, Federici MG, Borgen PI, Boyd J. Mutation analysis of the putative tumor suppressor gene PTEN/MMAC1 in primary breast carcinomas. *Cancer Res* 1997; 57:3657–3659.

145. Nelen MR, van Staveren WC, Peeters EA, Hassel MB, Gorlin RJ, Hamm H, et al. Germline mutations in the PTEN/MMAC1 gene in patients with Cowden disease. *Hum Mol Genet* 1997; 6:1383–1387.

146. Marsh DJ, Dahia PL, Zheng Z, Liaw D, Parsons R, Gorlin RJ, Eng C. Germline mutations in PTEN are present in Bannayan-Zonana syndrome. *Nat Genet* 1997; 16:333–334.

147. Lynch ED, Ostermeyer EA, Lee MK, Arena JF, Ji H, Dann J, et al. Inherited mutations in PTEN that are associated with breast cancer, cowden disease, and juvenile polyposis. *Am J Hum Genet* 1997; 61:1254–1260.

148. Teng DH, Hu R, Lin H, Davis T, Iliev D, Frye C, et al. MMAC1/PTEN mutations in primary tumor specimens and tumor cell lines. *Cancer Res* 1997; 57:5221–5225.

149. Cairns P, Okami K, Halachmi S, Halachmi N, Esteller M, Herman JG, et al. Frequent inactivation of PTEN/MMAC1 in primary prostate cancer. *Cancer Res* 1997; 57:4997–5000.

150. Gray I, Stewart L, Phillips S, Hamilton J, Gray N, Watson G, Spurr N, Snary D. Mutation and expression analysis of the putative prostate tumour-suppressor gene PTEN. *Br J Cancer* 1998; 78:1296–1300.

151. Lin WM, Forgacs E, Warshal DP, Yeh IT, Martin JS, Ashfaq R, Muller CY. Loss of heterozygosity and mutational analysis of the PTEN/MMAC1 gene in synchronous endometrial and ovarian carcinomas. *Clin Cancer Res* 1998; 4:2577–2583.

152. Shao X, Tandon R, Samara G, Kanki H, Yano H, Close L, Parsons R, Sato T. Mutational analysis of the PTEN gene in head and neck squamous cell carcinoma. *Int J Cancer* 1998; 77:684–688.

153. Kim SK, Su LK, Oh Y, Kemp BL, Hong WK, Mao L. Alterations of PTEN/MMAC1, a candidate tumor suppressor gene, and its homologue, PTH2, in small cell lung cancer cell lines. *Oncogene* 1998; 16:89–93.

154. Okami K, Wu L, Riggins G, Cairns P, Goggins M, Evron E, et al. Analysis of PTEN/MMAC1 alterations in aerodigestive tract tumors. *Cancer Res* 1998; 58:509–511.

155. Bostrom J, Cobbers JM, Wolter M, Tabatabai G, Weber RG, Lichter P, Collins VP, Reifenberger G. Mutation of the PTEN (MMAC1) tumor suppressor gene in a subset of glioblastomas but not in meningiomas with loss of chromosome arm 10q. *Cancer Res* 1998; 58:29–33.

156. Keino-Masu K, Masu M, Hinck L, Leonardo ED, Chan SS, Culotti JG, Tessier-Lavigne M. Deleted in colorectal cancer (DCC) encodes a netrin receptor. *Cell* 1996; 87:175–185.

157. Drescher U. Netrins find their receptor. *Nature* 1996; 384:416–417.

158. Deiner MS, Kennedy TE, Fazeli A, Serafini T, Tessier-Lavigne M, Sretavan DW. Netrin-1 and DCC mediate axon guidance locally at the optic disc: loss of function leads to optic nerve hypoplasia. *Neuron* 1997; 19:575–589.

159. de la Torre JR, Hopker VH, Ming GL, Poo MM, Tessier-Lavigne M, Hemmati-Brivanlou A, Holt CE. Turning of retinal growth cones in a netrin-1 gradient mediated by the netrin receptor DCC. *Neuron* 1997; 19:1211–1224.

160. Klingelhutz A, Hedrick L, Cho K, McDougall J. The DCC gene suppresses the malignant phenotype of transformed human epithelial cells. *Oncogene* 1995; 10:1581–1586.

161. Fazeli A, Dickinson SL, Hermiston ML, Tighe RV, Steen RG, Small CG, et al. Phenotype of mice lacking functional Deleted in colorectal cancer (Dcc) gene. *Nature* 1997; 386:796–804.

162. Mehlen P, Rabizadeh S, Snipas SJ, Assa-Munt N, Salvesen GS, Bredesen DE. The DCC gene product induces apoptosis by a mechanism requiring receptor proteolysis. *Nature* 1998; 395:801–804.

163. Aoyama N, Minami R, Fujimori T, Maeda S. Structure and function of DCC (deleted in colorectal cancer) gene and its product. *Nippon Rinsho* 1996; 54:972–980.

164. Ookawa K, Sakamoto M, Hirohashi S, Yoshida YTS, Terada M, Yokota J. Concordant p53 and DCC alterations and allelic losses on chromosomes 13q and 14q associated with liver metastases of colorectal carcinoma. *Int J Cancer* 1993; 53:382–387.

165. Miyake S, Nagai K, Yoshino K, Oto M, Endo M, Yuasa Y. Point mutations and allelic deletion of tumor suppressor gene DCC in human esophageal squamous cell carcinomas and their relation to metastasis. *Cancer Res* 1994; 54:3007–3010.

166. Wakita K, Kohno N, Sakoda Y, Ishikawa Y, Sakaue M. Decreased expression of the DCC gene in human breast carcinoma. *Surg Today* 1996; 26:900–903.

167. Strohmeyer D, Langenhof S, Ackermann R, Hartmann M, Strohmeyer T, Schmidt B. Analysis of the DCC tumor suppressor gene in testicular germ cell tumors: mutations and loss of expression. *J Urol* 1997; 157:1973–1976.

168. Yoshida Y, Itoh F, Endo T, Hinoda Y, Imai K. Decreased DCC mRNA expression in human gastric cancers is clinicopathologically significant. *Int J Cancer* 1998; 79:634–639.

169. Fang DC, Jass JR, Wang DX. Loss of heterozygosity and loss of expression of the DCC gene in gastric cancer. *J Clin Pathol* 1998; 51:593–596.

170. Reymond MA, Dworak O, Remke S, Hohenberger W, Kirchner T, Kockerling F. DCC protein as a predictor of distant metastases after curative surgery for rectal cancer. *Dis Colon Rectum* 1998; 41:755–760.

171. Goi T, Yamaguchi A, Nakagawara G, Urano T, Shiku H, Furukawa K. Reduced expression of deleted colorectal carcinoma (DCC) protein in established colon cancers. *Br J Cancer* 1998; 77:466–471.

172. Chen J, Silver DP, Walpita D, Cantor SB, Gazdar AF, Tomlinson G, et al. Stable interaction between the products of the BRCA1 and BRCA2 tumor suppressor genes in mitotic and meiotic cells. *Mol Cell* 1998; 2:317–328.

173. Fan S, Wang JA, Yuan RQ, Ma YX, Meng Q, Erdos, MR, Brody LC, Goldberg ID, Rosen EM. BRCA1 as a potential human prostate tumor suppressor: modulation of proliferation, damage responses and expression of cell regulatory proteins. *Oncogene* 1998; 16:3069–3082.

174. Jensen DE, Proctor M, Marquis ST, Gardner HP, Ha SI, Chodosh LA, et al. BAP1: a novel ubiquitin hydrolase which binds to the BRCA1 RING finger and enhances BRCA1-mediated cell growth suppression. *Oncogene* 1998; 16:1097–1112.

175. Rohlfs EM, Learning WG, Friedman KJ, Couch FJ, Weber BL, Silverman LM. Direct detection of mutations in the breast and ovarian cancer susceptibility gene BRCA1 by PCR-mediated site-directed mutagenesis. *Clin Chem* 1997; 43:24–29.

176. Shao N, Chai YL, Shyam E, Reddy P, Rao VN. Induction of apoptosis by the tumor suppressor protein BRCA1. *Oncogene* 1996; 13:1–7.

177. Rao VN, Shao N, Ahmad M, Reddy ES. Antisense RNA to the putative tumor suppressor gene BRCA1 transforms mouse fibroblasts. *Oncogene* 1996; 12:523–528.

178. Vaughn JP, Cirisano FD, Huper G, Berchuck A, Futreal PA, Marks JR, Iglehart JD. Cell cycle control of BRCA2. *Cancer Res* 1996; 56:4590–4594.

179. Gudmundsson J, Johannesdottir G, Bergthorsson JT, Arason A, Ingvarsson S, Egilsson V, Barkardottir RB. Different tumor types from BRCA2 carriers show wild-type chromosome deletions on 13q12–q13. *Cancer Res* 1995; 55:4830–4832.

180. Wong AKC, Pero R, Ormonde PA, Tavtigian SV, Bartel PL. RAD51 interacts with the evolutionarily conserved BRC motifs in the human breast cancer susceptibility gene brca2. *J Biol Chem* 1997; 272:31,941–31,944.

181. Bertwistle D, Swift S, Marston NJ, Jackson LE, Crossland S, Crompton MR, Marshall CJ, Ashworth A. Nuclear location and cell cycle regulation of the BRCA2 protein. *Cancer Res* 1997; 57:5485–5488.

182. Suzuki A, de la Pompa JL, Hakem R, Elia A, Yoshida R, et al. Brca2 is required for embryonic cellular proliferation in the mouse. *Genes Dev* 1997; 11:1242–1252.

183. Rajan JV, Marquis ST, Gardner HP, Chodosh LA. Developmental expression of Brca2 colocalizes with Brcal and is associated with proliferation and differentiation in multiple tissues. *Dev Biol* 1997; 184:385–401.

184. Marmorstein LY, Ouchi T, Aaronson SA. The BRCA2 gene product functionally interacts with p53 and RAD51. *Proc Natl Acad Sci USA* 1998; 95:13,869–13,874.

185. Rio PG, Pernin D, Bay JO, Albuisson E, Kwiatkowski F, De Latour M, Bernard-Gallon DJ, Bignon YJ. Loss of heterozygosity of BRCA1, BRCA2 and ATM genes in sporadic invasive ductal breast carcinoma. *Int J Oncol* 1998; 13:849–853.

186. Kamb A, Gruis NA, Weaver-Feldhaus J, Liu Q, Harshman K, Tavtigian SV, et al. A cell cycle regulator potentially involved in genesis of many tumor types. *Science* 1994; 264:436–440.

187. Mori T, Miura K, Aoki T, Nishihira T, Mori S, Nakamura Y. Frequent somatic mutation of the MTS1/CDK4I (multiple tumor suppressor/cyclin-dependent kinase 4 inhibitor) gene in esophageal squamous cell carcinoma. *Cancer Res* 1994; 54:3396–3397.

188. Arap W, Knudsen E, Sewell DA, Sidransky D, Wang JY, Huang HJ, Cavenee WK. Functional analysis of wild-type and malignant glioma derived CDKN2Abeta alleles: evidence for an RB-independent growth suppressive pathway. *Oncogene* 1997; 15:2013–2020.

189. Wu Q, Possati L, Montesi M, Gualandi F, Rimessi P, Morelli C, Trabanelli C, Barbanti-Brodano G. Growth arrest and suppression of tumorigenicity of bladder-carcinoma cell lines induced by the P16/CDKN2 (p16INK4A, MTS1) gene and other loci on human chromosome 9. *Int J Cancer* 1996; 65:840–846.

190. Tenan M, Benedetti S, Finocchiaro G. Deletion and transfection analysis of the p15/MTS2 gene in malignant gliomas. *Biochem Biophys Res Commun* 1995; 217:195–202.

191. Jin X, Nguyen D, Zhang WW, Kyritsis AP, Roth JA. Cell cycle arrest and inhibition of tumor cell proliferation by the p16INK4 gene mediated by an adenovirus vector. *Cancer Res* 1995; 55:3250–3253.

192. Gutmann DH, Wood DL, Collins FS. Identification of the neurofibromatosis type 1 gene product. *Proc Natl Acad Sci USA* 1991; 88:9658–9662.

193. Wallace MR, Marchuk DA, Andersen LB, Letcher R, Odeh HM, Saulino AM, et al. Type 1 neurofibromatosis gene: identification of a large transcript disrupted in three NF1 patients [published erratum appears in *Science* 1990; 250:1749]. *Science* 1990; 249:181–186.

194. Cawthon RM, Weiss R, Xu GF, Viskochil D, Culver M, Stevens J, et al. A major segment of the neurofibromatosis type 1 gene: cDNA sequence, genomic structure, and point mutations [published erratum appears in *Cell* 1990; 62:following 608]. *Cell* 1990; 62:193–201.

195. Viskochil D, Buchberg AM, Xu G, Cawthon RM, Stevens J, Wolff RK, et al. Deletions and a translocation interrupt a cloned gene at the neurofibromatosis type 1 locus. *Cell* 1990; 62:187–192.

196. Johnson MR, DeClue JE, Felzmann S, Vass WC, Xu G, White R, Lowy DR. Neurofibromin can inhibit Ras-dependent growth by a mechanism independent of its GTPase-accelerating function. *Mol Cell Biol* 1994; 14:641–645.

197. Nur EKMS, Varga M, Maruta H. The GTPase-activating NF1 fragment of 91 amino acids reverses v-Ha-Ras-induced malignant phenotype. *J Biol Chem* 1993; 268:22,331–22,337.

198. Gutmann DH, Sherman L, Seftor L, Haipek C, Hoang Lu K, Hendrix M. Increased expression of the NF2 tumor suppressor gene product, merlin, impairs cell motility, adhesionand spreading. *Hum Mol Genet* 1999; 8:267–275.

199. McClatchey AI, Saotome I, Mercer K, Crowley D, Gusella JF, Bronson RT, Jacks T. Mice heterozygous for a mutation at the Nf2 tumor suppressor locus develop a range of highly metastatic tumors. *Genes Dev* 1998; 12:1121–1133.

200. Shaw RJ, McClatchey AI, Jacks T. Localization and functional domains of the neurofibromatosis type II tumor suppressor, merlin. *Cell Growth Differ* 1998; 9:287–296.

201. Sherman L, Xu HM, Geist RT, Saporito-Irwin S, Howells N, Ponta H, Herrlich P, Gutmann DH. Interdomain binding mediates tumor growth suppression by the NF2 gene product. *Oncogene* 1997; 15:2505–2509.

202. Lutchman M, Rouleau GA. The neurofibromatosis type 2 gene product, schwannomin, suppresses growth of NIH 3T3 cells. *Cancer Res* 1995; 55:2270–2274.

203. Rouleau GA, Seizinger BR, Wertelecki W, Haines JL, Superneau DW, Martuza RL, Gusella JF. Flanking markers bracket the neurofibromatosis type 2 (NF2) gene on chromosome 22. *Am J Hum Genet* 1990; 46:323–328.

204. Trofatter JA, MacCollin MM, Rutter JL, Murrell JR, Duyao MP, Parry DM, et al. A novel moesin-, ezrin-, radixin-like gene is a candidate for the neurofibromatosis 2 tumor suppressor [published erratum appears in *Cell* 1993; 75:826]. *Cell* 1993; 72:791–800.

205. Rubinfeld B, Albert I, Porfiri E, Munemitsu S, Polakis P. Loss of beta-catenin regulation by the APC tumor suppressor protein correlates with loss of structure due to common somatic mutations of the gene. *Cancer Res* 1997; 57:4624–4630.

206. Korinek V, Barker N, Morin PJ, van Wichen D, de Weger R, Kinzler KW, Vogelstein B, Clevers H. Constitutive transcriptional activation by a beta-catenin-Tcf complex in APC–/– colon carcinoma. *Science* 1997; 275:1784–1787.

207. Morin PJ, Vogelstein B. Kinzler KW. Apoptosis and APC in colorectal tumorigenesis. *Proc Natl Acad Sci U S A* 1996; 93:7950–7954.

208. Groden J, Joslyn G, Samowitz W, Jones D, Bhattacharyya N, Spirio L, et al. Response of colon cancer cell lines to the introduction of APC, a colon-specific tumor suppressor gene. *Cancer Res* 1995; 55:1531–1539.

209. Sulekova Z, Ballhausen WG. A novel coding exon of the human adenomatous polyposis coli gene. *Hum Genet* 1995; 96:469–471.

210. Joslyn G, Carlson M, Thliveris A, Albertsen H, Gelbert L, Samowitz W, et al. Identification of deletion mutations and three new genes at the familial polyposis locus. *Cell* 1991; 66:601–613.

211. Nakamura Y, Nishisho I, Kinzler KW, Vogelstein B, Miyoshi Y, Miki Y, Ando H, Horii A. Mutations of the APC (adenomatous polyposis coli) gene in FAP (familial polyposis coli) patients and in sporadic colorectal tumors. *Tohoku J Exp Med* 1992; 168:141–7.

212. Ichii S, Horii A, Nakatsuru S, Furuyama J, Utsunomiya J, Nakamura Y. Inactivation of both APC alleles in an early stage of colon adenomas in a patient with familial adenomatous polyposis (FAP). *Hum Mol Genet* 1992; 1:387–390.

213. Steeg PS, Bevilacqua G, Kopper L, Thorgeirsson UP, Talmadge JE, Liotta LA, Sobel ME. Evidence for a novel gene associated with low tumor metastatic potential. *J Natl Cancer Inst* 1988; 80:200–204.

214. Rosengard AM, Krutzsch HC, Shearn A, Biggs JR, Barker E, Margulies IM, et al. Reduced Nm23/Awd protein in tumour metastasis and aberrant Drosophila development. *Nature* 1989; 342:177–180.

215. Stahl JA, Leone A, Rosengard AM, Porter L, King CR, Steeg PS. Identification of a second human nm23 gene, nm23-H2. *Cancer Res* 1991; 51:445–449.

216. Venturelli D, Martinez R, Melotti P, Casella I, Peschle C, Cucco C, et al. Overexpression of DR-nm23, a protein encoded by a member of the nm23 gene family, inhibits granulocyte differentiation and induces apoptosis in 32Dc13 myeloid cells. *Proc Natl Acad Sci USA* 1995; 92:7435–7439.

217. Milon L, Rousseau-Merck MF, Munier A, Erent M, Lascu I, Capeau J. Lacombe ML. nm23-H4, a new member of the family of human nm23/nucleoside diphosphate kinase genes localised on chromosome 16p13. *Hum Genet* 1997; 99:550–557.

218. Munier A, Feral C, Milon L, Pinon VP, Gyapay G, Capeau F, Guellaen G, Lacombe ML. A new human nm23 homologue (nm23-H5) specifically expressed in testis germinal cells. *FEBS Lett* 1998; 434:289–294.

219. Dearolf CR, Hersperger E, Shearn A. Developmental consequences of awdb3, a cell-autonomous lethal mutation of Drosophila induced by hybrid dysgenesis. *Dev Biol* 1988; 129:159–168.

220. Dearolf CR, Tripoulas N, Biggs J, Shearn A. Molecular consequences of awdb3, a cell-autonomous lethal mutation of Drosophila induced by hybrid dysgenesis. *Dev Biol* 1988; 129:169–178.

221. Ouatas T, Abdallah B, Gasmi L, Bourdais J, Postel E, Mazabraud A. Three different genes encode NM23/nucleoside diphosphate kinases in Xenopus laevis. *Gene* 1997; 194:215–225.

222. Wallet V, Mutzel R, Troll H, Barzu O, Wurster B, Veron M, Lacombe ML. Dictyostelium nucleoside diphosphate kinase highly homologous to Nm23 and Awd proteins involved in mammalian tumor metastasis and Drosophila development. *J Natl Cancer Inst* 1990; 82:1199–1202.

223. Lakshmi MS, Parker C, Sherbet GV. Metastasis associated MTS1 and NM23 genes affect tubulin polymerisation in B16 melanomas: a possible mechanism of their regulation of metastatic behaviour of tumours. *Anticancer Res* 1993; 13:299–303.

224. Caligo MA, Cipollini G, Cope Di Valromita A, Bistocchi M, Bevilacqua G. Decreasing expression of NM23 gene in metastatic murine mammary tumors of viral etiology (MMTV). *Anticancer Res* 1992; 12:969–973.

225. Steeg PS, Bevilacqua G, Pozzatti R, Liotta LA, Sobel ME. Altered expression of NM23, a gene associated with low tumor metastatic potential, during adenovirus 2 Ela inhibition of experimental metastasis. *Cancer Res* 1988; 48:6550–6554.

226. Su ZZ, Austin VN, Zimmer SG, Fisher PB. Defining the critical gene expression changes associated with expression and suppression of the tumorigenic and metastatic phenotype in Ha-ras-transformed cloned rat embryo fibroblast cells. *Oncogene* 1993; 8:1211–1219.

227. Radinsky R, Weisberg HZ, Staroselsky AN, Fidler IJ. Expression level of the nm23 gene in clonal populations of metastatic murine and human neoplasms. *Cancer Res* 1992; 52:5808–5814.

228. Heimann R, Ferguson DJ, Hellman S. The relationship between nm23, angiogenesis, and the metastatic proclivity of node-negative breast cancer. *Cancer Res* 1998; 58:2766–2771.

229. Han S, Yun IJ, Noh DY, Choe KJ, Song SY, Chi JG. Abnormal expression of four novel molecular markers represents a highly aggressive phenotype in breast cancer. Immunohistochemical assay of p53, nm23, erbB-2, and cathepsin D protein. *J Surg Oncol* 1997; 65:22–27.

230. Hennessy C, Henry JA, May FE, Westley BR, Angus B, Lennard TW. Expression of anti-metastatic gene nm23. *Br J Cancer* 1991; 63:1024.

231. Charpin C, Bouvier C, Garcia S, Martini F, Andrac L, Lavaut MN, Allasia C. Automated and quantitative immunocytochemical assays of Nm23/NDPK protein in breast carcinomas. *Int J Cancer* 1997; 74:416–420.

232. Barnes R, Masood S, Barker E, Rosengard AM, Coggin DL, Crowell T, et al. Low nm23 protein expression in infiltrating ductal breast carcinomas correlates with reduced patient survival. *Am J Pathol* 1991; 139:245–250.

233. Tokunaga Y, Urano T, Furukawa K, Kondo H, Kanematsu T, Shiku H. Reduced expression of nm23-H1, but not of nm23-H2, is concordant with the frequency of lymph-node metastasis of human breast cancer. *Int J Cancer* 1993; 55:66–71.

234. Noguchi M, Earashi M, Ohnishi I, Kinoshita K, Thomas M, Fusida S, Miyazaki I, Mizukami Y. Nm23 expression versus *Helix pomatia* lectin binding in human breast cancer metastases. *Int J Oncol* 1994; 4:1353–1358.

235. Sauer T, Furu I, Beraki K, Jebsen PW, Ormerod E, Naess O. nm23 protein expression in fine-needle aspirates from breast carcinoma: inverse correlation with cytologic grading, lymph node status, and ploidy. *Cancer* 1998; 84:109–114.

236. Duenas-Gonzalez A, Abad-Hernandez MM, Garcia-Mata J, Paz-Bouza JI, Cruz-Hernandez JJ. Gonzalez-Sarmiento R. Analysis of nm23-H1 expression in breast cancer. Correlation with p53 expression and clinicopathologic findings. *Cancer Lett* 1996; 101:137–142.

237. Bertheau P, Steinberg SM, Merino MJ. C-erbB-2, p53, and nm23 gene product expression in breast cancer in young women: immunohistochemical analysis and clinicopathologic correlation. *Hum Pathol* 1998; 29:323–329.

238. Russo A, Bazan V, Morello V, Valli C, Giarnieri E, Dardanoni G, et al. Nm23-H1 protein immunohistochemical expression in human breast cancer-Relationship to prognostic factors and risk of relapse. *Oncol Rep* 1996; 3:183–189.

239. Cropp CS, Lidereau R, Leone A, Liscia D, Cappa AP, Campbell G, et al. NME1 protein expression and loss of heterozygosity mutations in primary human breast tumors. *J Natl Cancer Inst* 1994; 86:1167–1169.

240. Sarac E, Ayhan A, Ertoy D, Tuncer ZS, Yasui W, Tahara E. nm23 expression in carcinoma of the uterine cervix. *Eur J Gynaecol Oncol* 1998; 19:312–315.

241. Marone M, Scambia G, Ferrandina G, Giannitelli C, Benedetti-Panici P, Iacovella S, Leone A, Mancuso S. Nm23 expression in endometrial and cervical cancer: inverse correlation with lymph node involvement and myometrial invasion. *Br J Cancer* 1996; 74:1063–1068.

242. Mandai M, Konishi I, Koshiyama M, Komatsu T, Yamamoto S, Nanbu K, Mori T, Fukumoto M. Altered expression of nm23-H1 and c-erbB-2 proteins have prognostic significance in adenocarcinoma but not in squamous cell carcinoma of the uterine cervix. *Cancer* 1995; 75:2523–2529.

243. Ilijas M, Pavelic K, Sarcevic B, Kapitanovic S, Kurjak A, Stambrook P, Gluckman J, Pavelic Z. Expression of nm23-H1 gene in squamous cell carcinoma of the cervix correlates with 5-year survival. *Int J Oncol* 1994; 5:1455–1457.

244. Boix L, Bruix J, Campo E, Sole M, Castells A, Fuster J, Rivera F, Cardesa A, Rodes J. nm23-H1 expression and disease recurrence after surgical resection of small hepatocellular carcinoma. *Gastroenterology* 1994; 107:486–491.

245. Nakayama T, Ohtsuru A, Nakao K, Shima M, Nakata K, Watanabe K, et al. Expression in human hepatocellular carcinoma of nucleoside diphosphate kinase, a homologue of the nm23 gene product. *J Natl Cancer Inst* 1992; 84:1349–1354.

246. Yamaguchi A, Urano T, Goi T, Takeuchi K, Niimoto S, Nakagawara G, Furukawa K, Shiku H. Expression of human nm23-H1 and nm23-H2 proteins in hepatocellular carcinoma. *Cancer* 1994; 73:2280–2284.

247. Scambia G, Ferrandina G, Marone M, Benedetti Panici P, Giannitelli C, Piantelli M, Leone A, Mancuso S. nm23 in ovarian cancer: correlation with clinical outcome and other clinicopathologic and biochemical prognostic parameters. *J Clin Oncol* 1996; 14:334–342.

248. Veil A, Dall'Agnese L, Canzonieri V, Sopracordevole F, Capozzi E, Carbone A, Visentin M, Boiocchi M. Suppressive role of the metastasis-related nm23-H1 gene in human ovarian carcinomas: association of high messenger RNA expression with lack of lymph node metastasis. *Cancer Res* 1995; 55:2645–2650.

249. Mandai M, Konishi I, Komatsu T, Mori T, Arao S, Nomura H, Kanda Y, Hiai H, Fukumoto M. Mutation of the nm23 gene, loss of heterozygosity at the nm23 locus and K-ras mutation in ovarian carcinoma: correlation with tumour progression and nm23 gene expression. *Br J Cancer* 1995; 72:691–695.

250. Kapitonovic S, Spaventi R, Vujsic S, Petrovic Z, Kurjak A, Pavelic Z, Gluckman J, Stambrook P, Pavelic K. Nm23-H1 gene expression in ovarian tumors: a potential tumor marker. *Anticancer Res* 1995; 15:587–590.

251. Bodey B, Kaiser HE, Goldfarb RH. Immunophenotypically varied cell subpopulations in primary and metastatic human melanomas. Monoclonal antibodies for diagnosis, detection of neoplastic progression and receptor directed immunotherapy. *Anticancer Res* 1996; 16:517–531.

252. Florenes VA, Aamdal S, Myklebost O, Maelandsmo GM, Bruland OS, Fodstad O. Levels of nm23 messenger RNA in metastatic malignant melanomas: inverse correlation to disease progression. *Cancer Res* 1992; 52:6088–6091.

253. Greco IM, Calvisi G, Ventura L, Cerrito F. An immunohistochemical analysis of nm23 gene product expression in uveal melanoma. *Melanoma Res* 1997; 7:231–236.

254. Lee CS, Pirdas A, Lee MW. Immunohistochemical demonstration of the nm23-H1 gene product in human malignant melanoma and Spitz nevi. *Pathology* 1996; 28:220–224.

255. Xerri L, Grob JJ, Battyani Z, Gouvernet J, Hassoun J, Bonerandi JJ. NM23 expression in metastasis of malignant melanoma is a predictive prognostic parameter correlated with survival. *Br J Cancer* 1994; 70:1224–1228.

256. Wang L, Patel U, Ghosh L, Chen HC, Banerjee S. Mutation in the nm23 gene is associated with metastasis in colorectal cancer [published erratum appears in *Cancer Res* 1993; 53:3652]. *Cancer Res* 1993; 53:717–720.

257. Leone A, McBride OW, Weston A, Wang MG, Anglard P, Cropp CS, et al. Somatic allelic deletion of nm23 in human cancer. *Cancer Res* 1991; 51:2490–2493.

258. Chang CL, Zhu XX, Thoraval DH, Ungar D, Rawwas J, Hora N, et al. Nm23-H1 mutation in neuroblastoma [letter]. *Nature* 1994; 370:335–336.

259. Hailat N, Keim DR, Melhem RF, Zhu XX, Eckerskorn C, Brodeur GM, et al. High levels of p19/nm23 protein in neuroblastoma are associated with advanced stage disease and with N-myc gene amplification. *J Clin Invest* 1991; 88:341–345.

260. Leone A, Seeger RC, Hong CM, Hu YY, Arboleda MJ, Brodeur GM, et al. Evidence for nm23 RNA overexpression, DNA amplification and mutation in aggressive childhood neuroblastomas. *Oncogene* 1993; 8:855–865.

261. Leone A, Flatow U, King CR, Sandeen MA, Margulies IM, Liotta LA, Steeg PS. Reduced tumor incidence, metastatic potential, and cytokine responsiveness of nm23-transfected melanoma cells. *Cell* 1991; 65:25–35.

262. Leone A, Flatow U, VanHoutte K, Steeg PS. Transfection of human nm23-H1 in to the human MDA-MB-435 breast carcinoma cell line: effects on tumor metastatic potential, colonization and enzymatic activity. *Oncogene* 1993; 8:2325–2333.

263. Kantor JD, McCormick B, Steeg PS, Zetter BR. Inhibition of cell motility after nm23 transfection of human and murine tumor cells. *Cancer Res* 1993; 53:1971–1973.

264. Russell R, Pedersen A, Kantor J, Geisinger K, Long R, Zbieranski N, et al. Relationship of nm23 to proteolytic factors, proliferation and motility in breast cancer tissues and cell lines. *Br J Cancer* 1998; 78:710–717.

265. Parhar RS, Shi Y, Zou M, Farid NR, Ernst P, al-Sedairy ST. Effects of cytokine-mediated modulation of nm23 expression on the invasion and metastatic behavior of B16F10 melanoma cells. *Int J Cancer* 1995; 60:204–210.

266. Fukuda M, Ishii A, Yasutomo Y, Shimada N, Ishikawa N, Hanai N, et al. Decreased expression of nucleoside diphosphate kinase alpha isoform, an nm23-H2 gene homolog, is associated with metastatic potential of rat mammary-adenocarcinoma cells. *Int J Cancer* 1996; 65:531–537.

267. Baba H, Urano T, Okada K, Furukawa K, Nakayama E, Tanaka H, Iwasaki K, Shiku H. Two isotypes of murine nm23/nucleoside diphosphate kinase, nm23-M1 and nm23-M2, are involved in metastatic suppression of a murine melanoma line. *Cancer Res* 1995; 55:1977–1981.

268. Howlett AR, Petersen OW, Steeg PS, Bissell MJ. A novel function for the nm23-H1 gene: overexpression in human breast carcinoma cells leads to the formation of basement membrane and growth arrest. *J Natl Cancer Inst* 1994; 86:1838–1844.

269. Hsu S, Huang F, Wang L, Banerjee S, Winawer S, Friedman E. The role of nm23 in transforming growth factor beta 1-mediated adherence and growth arrest. *Cell Growth Differ* 1994; 5:909–917.

270. Gervasi F, D'Agnano I, Vossio S, Zupi G, Sacchi A, Lombardi D. nm23 influences proliferation and differentiation of PC12 cells in response to nerve growth factor. *Cell Growth Differ* 1996; 7:1689–1695.

271. MacDonald NJ, Freije JMP, Stracke ML, Manrow RE, Steeg PS. Site-directed mutagenesis of nm23-H1. Mutation of proline 96 or serine 120 abrogates its motility inhibitory activity upon transection into human breast carcinoma cells. *J Biol Chem* 1996; 271:25,107–25,116.

272. Stracke ML, Krutzsch HC, Unsworth EJ, Arestad A, Cioce V, Schiffmann E, Liotta LA. Identification purification and partial sequence analysis of autotaxin, a novel motility-stimulating protein. *J Biol Chem* 1992; 267:2524–2529.

273. MacDonald NJ, De la Rosa A, Benedict MA, Freije JM, Krutsch H, Steeg PS. A serine phosphorylation of Nm23, and not its nucleoside diphosphate kinase activity, correlates with suppression of tumor metastatic potential. *J Biol Chem* 1993; 268:25,780–25,789.

274. Tepper AD, Dammann H, Bominaar AA, Veron M. Investigation of the active site and the conformational stability of nucleoside diphosphate kinase by site-directed mutagenesis. *J Biol Chem* 1994; 269:32,175–32,180.

275. Biggs J, Hersperger E, Steeg PS, Liotta, LA, Shearn A. A Drosophila gene that is homologous to a mammalian gene associated with tumor metastasis codes for a nucleoside diphosphate kinase. *Cell* 1990; 63:933–940.

276. Wagner PD, Vu ND. Phosphorylation of ATP-citrate lyase by nucleoside diphosphate kinase. *J Biol Chem* 1995; 270:21758–21764.

277. Wagner PD, Steeg PS, Vu ND. Two-component kinase-like activity of nm23 correlates with its motility-suppressing activity. *Proc Natl Acad Sci USA* 1997; 94:9000–9005.

278. Lu Q, Park H, Egger LA, Inouye M. Nucleoside-diphosphate kinase-mediated signal transduction via histidyl-aspartyl phosphorelay systems in Escherichia coli. *J Biol Chem* 1996; 271:32,886–32,893.

279. Freije JM, Blay P, MacDonald NJ, Manrow RE, Steeg PS. Site-directed mutation of Nm23-H1. Mutations lacking motility suppressive capacity upon transfection are deficient in histidine-dependent protein phosphotransferase pathways in vitro. *J Biol Chem* 1997; 272:5525–5532.

280. Engel M, Veron M, Theisinger B, Lacombe ML, Seib T, Dooley S, Welter C. A novel serine/threonine-specific protein phosphotransferase activity of Nm23/nucleoside-diphosphate kinase. *Eur J Biochem* 1995; 234:200–207.

281. Engel M, Seifert M, Theisinger B, Seyfert U, Welter C. Glyceraldehyde-3-phosphate dehydrogenase and Nm23-H1/nucleoside diphosphate kinase A. Two old enzymes combine for the novel Nm23 protein phosphotransferase function. *J Biol Chem* 1998; 273:20,058–20,065.

282. Biondi RM, Engel M, Sauane M, Welter C, Issinger OG, Jimenez de Asua L, Passeron S. Inhibition of nucleoside diphosphate kinase activity by in vitro phosphorylation by protein kinase CK2. Differential phosphorylation of NDP kinases in HeLa cells in culture. *FEBS Lett* 1996; 399:183–187.

283. Hemmerich S, Pecht I. Oligomeric structure and autophosphorylation of nucleoside diphosphate kinase from rat mucosal mast cells. *Biochemistry* 1992; 31:4580–4587.

284. Munoz-Dorado J, Almaula N, Inouye S, Inouye M. Autophosphorylation of nucleoside diphosphate kinase from Myxococcus xanthus. *J Bacteriol* 1993; 175:1176–1181.

285. Leung SM, Hightower LE. A 16-kDa protein functions as a new regulatory protein for Hsc70 molecular chaperone and is identified as a member of the Nm23/nucleoside diphosphate kinase family [published erratum appears in *J Biol Chem* 1997; 272:12,248]. *J Biol Chem* 1997; 272:2607–2014.

286. Paravicini G, Steinmayr M, Andre E, Becker-Andre M. The metastasis suppressor candidate nucleotide diphosphate kinase NM23 specifically interacts with members of the ROR/RZR nuclear orphan receptor subfamily. *Biochem Biophys Res Commun* 1996; 227:82–87.

287. Lombardi D, Sacchi A, D'Agostino G, Tibursi G. The association of the Nm23-M1 protein and beta-tubulin correlates with cell differentiation. *Exp Cell Res* 1995; 217:267–271.

288. Otero AS. Copurification of vimentin, energy metabolism enzymes, and a MER5 homolog with nucleoside diphosphate kinase. Identification of tissue-specific interactions. *J Biol Chem* 1997; 272:14,690–14,694.

289. Xu J, Liu LZ, Deng XF, Timmons L, Hersperger E, Steeg PS, Veron M, Shearn A. The enzymatic activity of Drosophila AWD/NDP kinase is necessary but not sufficient for its biological function. *Dev Biol* 1996; 177:544–557.

290. Ferguson AW, Flatow U, MacDonald NJ, Larminat F, Bohr VA, Steeg PS. Increased sensitivity to cisplatin by nm23-transfected tumor cell lines. *Cancer Res* 1996; 56:2931–2935.

291. De la Rosa A, Mikhak B, Steeg PS. Identification and characterization of the promoter for the human metastasis suppressor gene nm23-H1. *Arch Med Res* 1996; 27:395–401.

292. Chen HC, Wang L, Banerjee S. Isolation and characterization of the promoter region of human nm23-H1, a metastasis suppressor gene. *Oncogene* 1994; 9:2905–2912.

293. Okada K, Urano T, Baba H, Furukawa K, Shiku H. Independent and differential expression of two isotypes of human Nm23: analysis of the promoter regions of the nm23-H1 and H2 genes. *Oncogene* 1996; 13:1937–1943.

294. Dong J, Lamb P, Rinker-Schaeffer C, Vukanovic J, Ichikawa T, Isaacs J, Barrett J. KAI1, a metastasis suppressor gene for prostate cancer on human chromosome 11p11.2 *Science* 1995; 268:884–886.

295. Dong J, Suzuki H, Pin S, Bova G, Schalken J, Isaacs W, Barrett J, Isaacs J. Down regulation of the KAI1 metastasis suppressor gene during the progression of human prostatic cancer infrequently involves gene mutation or allelic loss. *Cancer Res* 1996; 56:4387–4390.

296. Sun HC, Tang ZY, Zhou G, Li XM. KAI1 gene expression in hepatocellular carcinoma and its relationship with intrahepatic metastases *J Exp Clin Cancer Res* 1998; 17:307–311.

297. Friess H, Guo XZ, Berberat P, Graber HU, Zimmermann A, Korc M, Buchler MW. Reduced KAI1 expression in pancreatic cancer is associated with lymph node and distant metastasis. *Int J Cancer* 1998; 79:349–355.

298. Higashiyama M, Kodama K, Yokouchi H, Takami K, Adachi M, Taki T, et al. KAI1/CD82 expression in nonsmall cell lung carcinoma is a novel, favorable prognostic factor: an immunohistochemical analysis. *Cancer* 1998; 83:466–474.

299. White A, Lamb P, Barrett J. Frequent downregulation of the KAI1 (CD82) metastasis suppressor protein in human cancer cell lines. *Oncogene* 1998; 16:3143–3149.

300. Takaoka A, Hinoda Y, Satoh S, Adachi Y, Itoh F, Adachi M, Imai K. Suppression of invasive properties of colon cancer cells by a metastasis suppressor KAI1 gene. *Oncogene* 1998; 16:1443–1453.

301. Takaoka A, Hinoda Y, Sato S, Itoh F, Adachi M, Hareyama M, Imai K. Reduced invasive and metastatic potentials of KAI1-transfected melanoma cells. *Jpn J Cancer Res* 1998; 89:397–404.

302. Yu Y, Yang J, Markovic B, Jackson P, Yardley G, Barrett J, Russell P. Loss of KAI1 messenger RNA expression in both high-grade and invasive human bladder cancers. *Clin Cancer Res* 1997; 3:1045–1049.

303. Yang X, Welch DR, Phillips KK, Weissman BE, Wei LL. KAI1, a putative marker for metastatic potential in human breast cancer. *Cancer Lett* 1997; 119:149–155.

304. Guo X, Friess H, Di Mola F, Heinicke J, Abou-Shady M, Graber H, et al. KAI1, a new metastasis suppressor gene, is reduced in metastatic hepatocellular carcinoma. *Hepatology* 1998; 28:1481–1488.

305. Hinoda Y, Adachi Y, Takaoka A, Mitsuuchi H, Satoh Y, Itoh F, Kondoh Y, Imai K. Decreased expression of the metastasis suppressor gene KAI1 in gastric cancer. *Cancer Lett* 1998; 129:229–234.

306. Huang C, Kohno N, Ogawa E, Adachi M, Taki T, Miyake M. Correlation of reduction in MRP-1/CD9 and KAI1/CD82 expression with recurrences in breast cancer patients. *Am J Pathol* 1998; 153:973–983.

307. Lee J, Miele M, Hicks D, Phillips K, Trent J, Weissman B, Welch D. KiSS-1, a novel human malignant melanoma metastasis-suppressor gene [published erratum appears in *J Natl Cancer Inst* 1997; 89:1549]. *J Natl Cancer Inst* 1996; 88:1731–1737.

308. West A, Vojta P, Welch D, Weissman B. Chromosome localization and genomic structure of the KiSS-1 metastasis suppressor gene (KISS1). *Genomics* 1998; 54:145–148.

309. Lee J, Welch D. Identification of highly expressed genes in metastasis-suppressed chromosome 6/human malignant melanoma hybrid cells using subtractive hybridization and differential display. *Int J Cancer* 1997; 71:1035–1044.

310. Lee J, Welch D. Suppression of metastasis in human breast carcinoma MDA-MB-435 cells after transfection with the metastasis suppressor gene, KiSS-1. *Cancer Res* 1997; 57:2384–2387.

311. Bertin N, Clezardin P, Kubiak R, Frappart L. Thrombospondin-1 and-2 messenger RNA expression in normal, benign, and neoplastic human breast tissues: correlation with prognostic factors, tumor angiogenesis, and fibroblastic desmoplasia. *Cancer Res* 1997; 57:396–399.

312. Grant S, Kyshtoobayeva A, Kurosaki T, Jakowatz J, Fruehauf J. Mutant p53 correlates with reduced expression of thrombospondin-1, increased angiogenesis, and metastatic progression in melanoma. *Cancer Detect Prev* 1998; 22:185–194.

313. Volpert O, Lawler J, Bouck N. A human fibrosarcoma inhibits systemic angiogenesis and the growth of experimental metastases via thrombospondin-1. *Proc Natl Acad Sci USA* 1998; 95:6343–6348.

314. Xu M, Kumar D, Stass S, Mixson A. Gene therapy with p53 and a fragment of thrombospondin I inhibits human breast cancer in vivo. *Mol Genet Metab* 1998; 63:103–109.

315. Weinstat-Saslow DL, Zabrenetzky VS, VanHoutte K, Frazier WA, Roberts DD, Steeg PS. Transfection of thrombospondin 1 complementary DNA into a human breast carcinoma cell line reduces primary tumor growth, metastatic potential, and angiogenesis. *Cancer Res* 1994; 54:6504–6511.

316. Roberts DD. Regulation of tumor growth and metastasis by thrombospondin-1. *FASEB J* 1996; 10:1183–1191.

317. Guo N, Zabrenetzky V, Chandrasekaran L, Sipes J, Lawler J, Krutzsch H, Roberts D. Differential roles of protein kinase C and pertussis toxin-sensitive G-binding proteins in modulation of melanoma cell proliferation and motility by thrombospondin 1. *Cancer Res* 1998; 58:3154–3162.

318. Wagner TM, Moslinger RA, Muhr D, Langbauer G, Hirtenlehner K, Concin H, et al. BRCA1-related breast cancer in Austrian breast and ovarian cancer families: specific BRCA1 mutations and pathological characteristics. *Int J Cancer* 1998; 77:354–360.

319. Eisinger F, Nogues C, Birnbaum D, Jacquemier J, Sobol H. Low frequency of lymph-node metastasis in BRCA1-associated breast cancer. *Lancet* 1998; 351:1633–1634.

320. Robson M, Gilewski T, Haas B, Levin D, Borgen P, Rajan P, et al. BRCA-associated breast cancer in young women. *J Clin Oncol* 1998; 16:1642–1649.

321. Karp SE, Tonin PN, Begin LR, Martinez JJ, Zhang JC, Pollak MN, Foulkes WD. Influence of BRCA1 mutations on nuclear grade and estrogen receptor status of breast carcinoma in Ashkenazi Jewish women. *Cancer* 1997; 80:435–441.

322. Beckmann MW, Picard F, An HX, van Roeyen CR, Dominik SI, Mosny DS, Schnurch HG, Bender HG, Niederacher D. Clinical impact of detection of loss of heterozygosity of BRCA1 and BRCA2 markers in sporadic breast cancer. *Br J Cancer* 1996; 73:1220–1226.

323. Mori T, Aoki T, Matsubara T, Iida F, Du X, Nishihira T, Mori S, Nakamura Y. Frequent loss of heterozygosity in the region including BRCA1 on chromosome 17q in squamous cell carcinomas of the esophagus. *Cancer Res* 1994; 54:1638–1640.

324. Schmutzler RK, Homann A, Bierhoff E, Wiestler OD, von Daimling A, Krebs D. Detection of genetic alterations in sporadic breast tumors. *Gynakol Geburtshilfliche Rundsch* 1995; 35:(Suppl 1) 63–67.

325. Prechtel D, Werenskiold AK, Prechtel K, Keller G, Hofler H. Frequent loss of heterozygosity at chromosome 13q12–13 with BRCA2 markers in sporadic male breast cancer. *Diagn Mol Pathol* 1998;7:57–62.

326. Bieche I, Nogues C, Rivoilan S, Khodja A, Latil A, Lidereau R. Prognostic value of loss of heterozygosity at BRCA2 in human breast carcinoma. *Br J Cancer* 1997; 76:1416–1418.

327. Lothe RA, Saeter G, Danielsen HE, Stenwig AE, Hoyheim B, O'Connell P, Borresen AL. Genetic alterations in a malignant schwannoma from a patient with neurofibromatosis (NF1). *Pathol Res Pract* 1993; 189:465–471 discussion 471–474.

328. Legius E, Marchuk DA, Collins FS, Glover TW. Somatic deletion of the neurofibromatosis type 1 gene in a neurofibrosarcoma supports a tumour suppressor gene hypothesis. *Nat Genet* 1993; 3:122–126.

329. Andersen LB, Fountain JW, Gutmann DH, Tarle SA, Glover TW, Dracopoli NC, Housman DE, Collins FS. Mutations in the neurofibromatosis 1 gene in sporadic malignant melanoma cell lines. *Nat Genet* 1993; 3:118–121.

330. Shannon KM, O'Connell P, Martin GA, Paderanga D, Olson K, Dinndorf P, McCormick F. Loss of the normal NF1 allele from the bone marrow of children with type 1 neurofibromatosis and malignant myeloid disorders. *N Engl J Med* 1994; 330:597–601.

331. Miles DK, Freedman MH, Stephens K, Pallavicini M, Sievers EL, Weaver M, et al. Patterns of hematopoietic lineage involvement in children with neurofibromatosis type 1 and malignant myeloid disorders. *Blood* 1996; 88:4314–4120.

332. Colman SD, Williams CA, Wallace MR. Benign neurofibromas in type 1 neurofibromatosis (NF1) show somatic deletions of the NF1 gene. *Nat Genet* 1995; 11:90–92.

333. Jensen S, Paderanga DC, Chen P, Olson K, Edwards M, Iavorone A, Israel MA, Shannon K. Molecular analysis at the NF1 locus in astrocytic brain tumors. *Cancer* 1995; 76:674–677.

334. Fridman M, Tikoo A, Varga M, Murphy A, Nur EKMS, Maruta H. The minimal fragments of c-Raf-1 and NF1 that can suppress v-Ha-Ras-induced malignant phenotype. *J Biol Chem* 1994; 269:30105–30108.

335. Kalra R, Paderanga DC, Olson K, Shannon KM. Genetic analysis is consistent with the hypothesis that NF1 limits myeloid cell growth through p21ras. *Blood* 1994; 84:3435–3439.

336. Mangues R, Corral T, Lu S, Symmans WF, Liu L, Pellicer A. NF1 inactivation cooperates with N-ras in in vivo lymphogenesis activating Erk by a mechanism independent of its Ras-GTPase accelerating activity. *Oncogene* 1998; 17:1705–1716.

337. Side LE, Emanuel PD, Taylor B, Franklin J, Thompson P, Castleberry RP, Shannon KM. Mutations of the NF1 gene in children with juvenile myelomonocytic leukemia without clinical evidence of neurofibromatosis, type 1. *Blood* 1998; 92:267–272.

338. Zhang YY, Vik TA, Ryder JW, Srour EF, Jacks T, Shannon K, Clapp DW. Nf1 regulates hematopoietic progenitor cell growth and ras signaling in response to multiple cytokines. *J Exp Med* 1998; 187:1893–1902.

339. Serra E, Puig S, Otero D, Gaona A, Kruyer H, Ars E, Estivill X, Lazaro C. Confirmation of a double-hit model for the NF1 gene in benign neurofibromas. *Am J Hum Genet* 1997; 61:512–519.

340. Side L, Taylor B, Cayouette M, Conner E, Thompson P, Luce M, Shannon K. Homozygous inactivation of the NF1 gene in bone marrow cells from children with neurofibromatosis type 1 and malignant myeloid disorders. *N Engl J Med* 1997; 336:1713–1720.

341. Park VM, Pivnick EK. Neurofibromatosis type 1 (NF1): a protein truncation assay yielding identification of mutations in 73% of patients. *J Med Genet* 1998; 35:813–820.
342. Kai S, Sumita H, Fujioka K, Takahashi H, Hanzawa N, Funabiki T, Ikuta K, Sasaki H. Loss of heterozygosity of NF1 gene in juvenile chronic myelogenous leukemia with neurofibromatosis type 1. *Int J Hematol* 1998; 68:53–60.
343. Klose A, Ahmadian MR, Schuelke M, Scheffzek K, Hoffmeyer S, Gewies A, et al. Selective disactivation of neurofibromin GAP activity in neurofibromatosis type 1. *Hum Mol Genet* 1998; 7:1261–1268.
344. Rasmussen SA, Colman SD, Ho VT, Abernathy CR, Arn PH, Weiss L, et al. Constitutional and mosaic large NF1 gene deletions in neurofibromatosis type 1. *J Med Genet* 1998; 35:468–471.
345. Ars E, Kruyer H, Gaona A, Casquero P, Rosell J, Volpini V, Serra E, Lazaro C, Estivill X. A clinical variant of neurofibromatosis type 1: familial spinal neurofibromatosis with a frameshift mutation in the NF1 gene. *Am J Hum Genet* 1998; 62:834–841.
346. Wellenreuther R, Kraus JA, Lenartz D, Menon AG, Schramm J, Louis DN, et al. Analysis of the neurofibromatosis 2 gene reveals molecular variants of meningioma. *Am J Pathol* 1995; 146:827–832.
347. Papi L, De Vitis LR, Vitelli F, Ammannati F, Mennonna P, Montali E, Bigozzi U. Somatic mutations in the neurofibromatosis type 2 gene in sporadic meningiomas. *Hum Genet* 1995; 95:347–351.
348. Merel P, Khe HX, Sanson M, Bijlsma E, Rouleau G, Laurent-Puig P, et al. Screening for germ-line mutations in the NF2 gene. *Genes Chromosomes Cancer* 1995; 12:117–127.
349. Jacoby LB, MacCollin M, Barone R, Ramesh V, Gusella JF. Frequency and distribution of NF2 mutations in schwannomas. *Genes Chromosomes Cancer* 1996; 17:45–55.
350. Mautner VF, Baser ME, Kluwe L. Phenotypic variablity in two families with novel splice-site and frameshift NF2 mutations. *Hum Genet* 1996; 98:203–206.
351. Huynh DP, Pulst SM. Neurofibromatosis 2 antisense oligodeoxynucleotides induce reversible inhibition of schwannomin synthesis and cell adhesion in STS26T and T98G cells. *Oncogene* 1996; 13:73–84.
352. De Vitis LR, Tedde A, Vitelli F, Ammannati F, Mennonna P, Bigozzi U, Montali E, Papi L. Screening for mutations in the neurofibromatosis type 2 (NF2) gene in sporadic meningiomas. *Hum Genet* 1996; 97:632–637.
353. Harada T, Irving RM, Xuereb JH, Barton DE, Hardy DG, Moffat DA, Maher ER. Molecular genetic investigation of the neurofibromatosis type 2 tumor suppressor gene in sporadic meningioma. *J Neurosurg* 1996; 84:847–51.
354. Kluwe L, Mautner VF. A missense mutation in the NF2 gene results in moderate and mild clinical phenotypes of neurofibromatosis type 2. *Hum Genet* 1996; 97:224–7.
355. Bianchi AB, Mitsunaga SI, Cheng JQ, Klein WM, Jhanwar SC, Seizinger B, et al. High frequency of inactivating mutations in the neurofibromatosis type 2 gene (NF2) in primary malignant mesotheliomas. *Proc Natl Acad Sci USA* 1995; 92:10854–10858.
356. Hitotsumatsu T, Iwaki T, Kitamoto T, Mizoguchi M, Suzuki SO, Hamada Y, Fukui M, Tateishi J. Expression of neurofibromatosis 2 protein in human brain tumors: an immunohistochemical study. *Acta Neuropathol (Ber)* 1997; 93:225–232.
357. Welling DB. Clinical manifestations of mutations in the neurofibromatosis type 2 gene in vestibular schwannomas (acoustic neuromas). *Laryngoscope* 1998; 108:178–189.
358. Jacoby LB, Jones D, Davis K, Kronn D, Short MP, Gusella J, MacCollin M. Molecular analysis of the NF2 tumor-suppressor gene in schwannomatosis. *Am J Hum Genet* 1997; 61:1293–1302.
359. Luca M, Xie S, Gutman M, Huang S, Bar-Eli M. Abnormalities in the CDKN2 (p16INK4/MTS-1) gene in human melanoma cells: relevance to tumor growth and metastasis. *Oncogene* 1995; 11:1399–1402.
360. Roncalli M, Bosari S, Marchetti A, Buttitta F, Bossi P, Graziani D, et al. Cell cycle-related gene abnormalities and product expression in esophageal carcinoma. *Lab Invest* 1998; 78:1049–1057.
361. Takeuchi H, Ozawa S, Ando N, Shih CH, Koyanagi K, Ueda M, Kitajima M. Altered p16/MTS1/CDKN2 and cyclin D1/PRAD-1 gene expression is associated with the prognosis of squamous cell carcinoma of the esophagus. *Clin Cancer Res* 1997; 3:2229–2236.
362. Jarrard DF, Bova GS, Ewing CM, Pin SS, Nguyen SH, Baylin SB, et al. Deletional, mutational, and methylation analyses of CDKN2 (p16/MTS1) in primary and metastatic prostate cancer. *Genes Chromosomes Cancer* 1997; 19:90–96.
363. Marchetti A, Buttitta F, Pellegrini S, Bertacca G, Chella A, Carnicelli V, et al. Alterations of P16 (MTS1) in node-positive non-small cell lung carcinomas. *J Pathol* 1997; 181:178–182.

364. Liu L, Dilworth D, Gao L, Monzon J, Summers A, Lassam N, Hogg D. Mutation of the CDKN2A 5′ UTR creates an aberrant initiation codon and predisposes to melanoma. *Nat Genet* 1999; 21:128–132.

365. Kim JR, Kim SY, Kim MJ, Kim JH. Alterations of CDKN2 (MTS1/p16INK4A) gene in paraffin-embedded tumor tissues of human stomach, lung, cervix and liver cancers. *Exp Mol Med* 1998; 30:109–114.

366. Pande P, Mathur M, Shukla NK, Ralhan R. pRb and p16 protein alterations in human oral tumorigenesis. *Oral Oncol* 1998; 34:396–403.

367. Chi SG, deVere White RW, Muenzer JT, Gumerlock PH. Frequent alteration of CDKN2 (p16(INK4A)/MTS1) expression in human primary prostate carcinomas. *Clin Cancer Res* 1997; 3:1889–1897.

368. Gazzeri S, Della Valle V, Chaussade L, Brambilla C, Larsen CJ, Brambilla E. The human p19ARF protein encoded by the beta transcript of the p16INK4a gene is frequently lost in small cell lung cancer. *Cancer Res* 1998; 58:3926–3931.

369. Hao Y, Zhang J, Yi C, Qian W. Abnormal change of p53 gene in gastric and precancerous lesions and APC gene deletion in gastric carcinoma and near tissues. *J Tongji Med Univ* 1997; 17:75–78.

370. Imai Y, Oda H, Tsurutani N, Nakatsuru Y, Inoue T, Ishikawa T. Frequent somatic mutations of the APC and p53 genes in sporadic ampullary carcinomas. *Jpn J Cancer Res* 1997; 88:846–854.

371. Bala S, Sulekova Z, Ballhausen WG. Constitutive APC exon 14 skipping in early-onset familial adenomatous polyposis reveals a dramatic quantitative distortion of APC gene-specific isoforms. *Hum Mutat* 1997; 10:201–206.

372. Bala S, Wunsch PH, Ballhausen WG. Childhood hepatocellular adenoma in familial adenomatous polyposis: mutations in adenomatous polyposis coli gene and p53. *Gastroenterology* 1997; 112:919–922.

373. Alman BA, Li C, Pajerski ME, Diaz-Cano S, Wolfe HJ. Increased beta-catenin protein and somatic APC mutations in sporadic aggressive fibromatoses (desmoid tumors). *Am J Pathol* 1997; 151:329–34.

374. Zhuang Z, Vortmeyer AO, Mark EJ, Odze R, Emmert-Buck MR, Merino MJ, et al. Barrett's esophagus: metaplastic cells with loss of heterozygosity at the APC gene locus are clonal precursors to invasive adenocarcinoma. *Cancer Res* 1996; 56:1961–1964.

375. Mao EJ, Oda D, Haigh WG, Beckmann AM. Loss of the adenomatous polyposis coli gene and human papillomavirus infection in oral carcinogenesis. *Eur J Cancer B Oral Oncol* 1996; 32B:260–263.

376. Miyaki M, Konishi M, Kikuchi-Yanoshita R, Enomoto M, Igari T, Tanaka K, et al. Characteristics of somatic mutation of the adenomatous polyposis coli gene in colorectal tumors. *Cancer Res* 1994; 54:3011–3020.

377. Varesco L, Gismondi V, Presciuttini S, Groden J, Spirio L, Sala P, et al. Mutation in a splice-donor site of the APC gene in a family with polyposis and late age of colonic cancer death. *Hum Genet* 1994; 93:281–286.

378. Harach HR, Williams GT, Williams ED. Familial adenomatous polyposis associated thyroid carcinoma: A distinct type of follicular cell neoplasm. *Histopathology* 1994; 25:549–561.

379. Cawkwell L, Lewis FA, Quirke P. Frequency of allele loss of DCC, p53, RBI, WT1, NF1, NM23 and APC/MCC in colorectal cancer assayed by fluorescent multiplex polymerase chain reaction. *Br J Cancer* 1994; 70:813–818.

380. Miyaki M, Konishi M, Kikuchi-Yanoshita R, Enomoto M, Tanaka K, Takashashi H, et al. Coexistence of somatic and germ-line mutations of APC gene in desmoid tumors from patients with familial adenomatous polyposis. *Cancer Res* 1993; 53:5079–5082.

381. Lim S, Lee HY, Lee M. Inhibition of colonization and cell-matrix adhesion after *nm23* transfection of human prostate carcinoma cells. Cancer Lett. 1998; 133:143–149.

4

Apoptosis

Machinery of Cell Death in Development and Cancer

Jinyan Du and David E. Fisher, MD, PhD

CONTENTS

1. INTRODUCTION: CELL DEATH AND THE GOAL OF CANCER THERAPEUTICS

Apoptosis, or programmed cell death, plays a critical role in development, tissue homeostasis, and disease. Characterized by nuclear condensation, membrane blebbing, and DNA fragmentation *(1)*, this suicide program allows cells to die without inciting a local inflammatory response, and is thought to play a central role in both cell death during normal homeostasis and in response to stress. One of the most important observations with respect to apoptosis is the recognition that this death pathway is commonly disrupted within cancer cells. This observation led to the realization that cancer is more than a disease of uncontrolled proliferation: It is also usually a disease of ineffective cell death.

Through the study of cell death in cancer, insights have been made that directly impact not only on tumorigenesis, but also on cancer treatment. For example, signifi-

From: *Tumor Suppressor Genes in Human Cancer*
Edited by: D. E. Fisher © Humana Press Inc., Totowa, NJ

cant information has been learned about the manner in which p53 regulates the apoptotic response to DNA damage. This in turn has been found to correlate with clinical prognosis (i.e., virtually all malignancies currently curable with chemotherapy remain wild-type for p53). Dismantling of the apoptosis cascade appears to be remarkably common in cancer, e.g., loss of p53, overexpression of bcl-2 or family members, and loss of phosphatase and tensin homolog deleted on chromosome 10 (PTEN) (upregulation of the Akt survival pathway). Although it is too soon to see direct clinical benefits from understanding of apoptotic machinery, these investigations may permit the construction of productive and well-targeted screens for cancer drug discovery.

2. Apoptotic Machinery

The basic apoptosis machinery appears to be evolutionarily conserved. The first detailed genetic description of this death pathway came from studies of Horvitz et al. *(2–3)*, who described cell death mutants in the nematode *Cenorhabditis elegans* during development. These mutants, termed CED genes (for *C. elegans* death), were identified as positive or negative death regulators. In *C. elegans,* the core cell death machinery consists of three proteins, CED-9, CED-4, and CED-3. CED-9 protects cells from apoptosis; CED-4 and CED-3 promote cell death *(2–3)*. Under physiological conditions, CED-9 forms a complex with CED-4 and CED-3, and prevents the activation of CED-3 by CED-4 *(4–6)*. With apoptotic stimuli, CED-9 dissociates from the complex, and CED-3 is activated to execute the cell death program. Recently, a novel protein, Egl-1, was found to antagonize CED-9's function *(7)*: It binds to CED-9 through its BH3 domain, disrupts the interactions between CED-9 and CED-4, and thereby promotes CED-4-mediated CED-3 activation *(8)*.

Molecular counterparts of all these gene products have also been identified in mammals. CED-3 is homologous to mammalian caspases, a cysteine protease family that cleaves after aspartic acid residues *(9–10)*. Apaf-1 is a mammalian counterpart to CED-4 *(11)*. Caspases exist in proenzyme forms (zymogens), which may themselves be activated by other caspases. Following release of cytochrome-c from mitochondria, cytochrome-*c* binds to Apaf-1, and Apaf-1 recruits procaspase-9, stimulating its proteolytic processing *(12–13)*. Both CED-9 and Egl-1 are related to Bcl-2 family proteins, which are alternatively activators or inhibitors of apoptosis *(3,7,14,15)*.

Multiple stimuli can trigger apoptosis. Several of the apoptotic pathways have been well studied at a mechanistic level. In one pathway, DNA damage and other stresses activate p53 through posttranslational protein stabilization, which triggers apoptosis through both transcription-dependent and -independent mechanisms *(16–17)*. In the transcription-dependent pathway, p53 stimulates a variety of target genes, which include certain ones capable of producing apoptosis. For example, p53 can upregulate Bax expression *(18–20)*. Bax drives cytochrome-*c* release from mitochondria *(21–22)*, and thereby caspase-9 activation *(12)*. Caspase-9 activates caspase-3, and stimulates subsequent cell death *(12)*. Another pathway is initiated by death receptor ligation. Adaptor proteins bind to oligomerized Fas/CD95 *(23)* and recruit procaspase-8 to form (DISC) *(24)*, a protein complex that modulates death signaling from this receptor. At high local concentration, the intrinsic protease activity of procaspase-8 is sufficient for

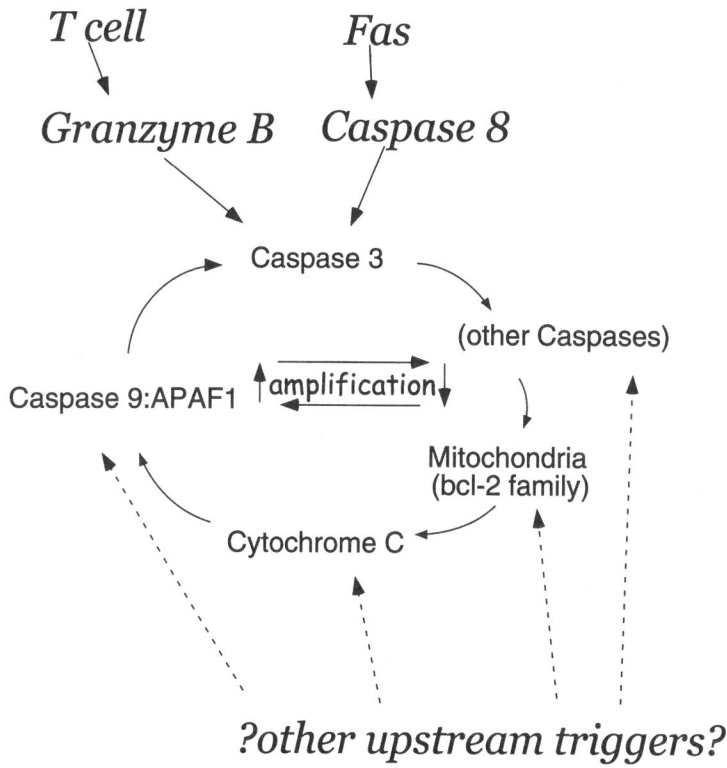

Fig. 1. Apoptosis triggers and amplification loop. Several pathways of apoptosis activate caspases at known stages. Others remain to be discovered, and could potentially target multiple sites within the caspase cascade. Because of interactions among caspases and mitochondria, an amplification loop may contribute to the strength of the death signal.

its activation by intermolecular proteolyic cleavage *(25)*. Subsequently, caspase-8 activates the caspase cascade.

Like many signaling pathways, apoptotic pathways may undergo crosstalk, amplification, or homeostasis. One important example appears to be a role for mitochondrial cytochrome-*c* release as a means of amplifying an apoptotic signal. For example, mitochondrial damage downstream of caspase activation may result in further cytochrome-*c* release, Apaf-1 caspase-9 activation, and caspase-3 stimulation. In addition, caspase-8 can directly cleave Bid *(26,27)*, a bcl-2 family member located in mitochondria, thereby causing cytochrome-*c* release, caspase-9 activation, and caspase-3 induction (Fig. 1).

Deregulated apoptosis may lead to diseases. Too little apoptosis may occur in cancer or autoimmune disease (related to incomplete removal of autoreactive lymphocyte clones), and too much apoptosis in neurodegenerative disease. Efficacy of cancer therapy can be dramatically affected by the status of apoptotic regulators. For example, tumors with p53 mutations, or wild-type p53 inactivated by oncogenes, tend to respond poorly to treatments, but highly curable cancers, such as childhood acute lymphoblastic leukemias, typically retain wild-type p53, which can be activated by irradiation or chemotherapy drugs *(17,28–30)*.

3. SIGNALING FOR DEATH (31)

Fas-induced apoptosis and the perforin–granzyme degranulation pathway are the two major pathways in cytotoxic T-cell-mediated target cell killing (32).

3.1. Fas and FasL

Fas-mediated apoptosis helps to eliminate autoreactive T- and B-cells in the immune system (33). FasL binds to Fas, a transmembrane receptor, and induces receptor oligomerization (33,34). (FADD) and caspase-8 are recruited into the multiprotein DISC complex (23,24) through protein–protein interactions involving a discrete death domain. Recruitment of caspase-8 into such complexes is thought to lead to its auto-proteolytic cleavage, and, thereby to activation (25). Activated caspase-8, in turn, activates downstream caspases, which execute apoptosis (Fig. 1). Alternatively, weaker activation of caspase-8 may cause proteolytic activation of Bid in mitochondria, which subsequently stimulates cytochrome-c release from mitochondria (26,27). Cytochrome-c interacts with Apaf-1, and procaspase-9 is recruited into the complex for activation (12,13).

3.2. Tumor Necrosis Factor

Paradoxically, Tumor necrosis factor (TNF) typically promotes cell survival, except when protein synthesis is inhibited (31). One of its receptors, TNFRI, trimerizes upon TNF binding (34). (TRADD), an adaptor protein that interacts with the TNFR death domain, is recruited to the receptor (35). TRADD may recruit FADD and caspase-8 to induce apoptosis (36–38). Alternatively, it may recruit receptor interacting protein (RIP) (39,40) and TNF receptor-associated factor 2 (TRAF2) (36,41), which activate jun-N-terminal kinase (JNK) or nuclear factor-κB (NF-κB) pathways to promote cell survival. The blockage of NF-κB makes primary glomerular mesangial cells more sensitive to TNF-α-induced apoptosis, but it has no effect on Fas-stimulated cell death (42). A feedback loop has been suggested, in which activated TNFRI upregulates TRAF1 transcription through NF-κB, and, conversely, TRAF1 modulates TNFRI-induced NF-κB activation (43).

Another TNFR, TNFRII, can promote cell death or survival by recruiting TRAFs (44). Association of TRAFs to TNFRII activates JNK and NF-κB, and induces the survival or apoptosis pathways (45,46). TNFRII lacks a death domain. It may function in apoptosis by producing endogenous TNF, which can activate TNFRI (47).

TR6, a new member of TNFR family, is expressed in lung tissues and colon adenocarcinoma. TR6 specifically interacts with Fas ligand and LIGHT (herpes virus entry mediator), and may negatively regulate apoptosis induced by these two ligands (48).

3.3. Granzyme B

Granzyme B, a serine protease, plays a critical role in cytotoxic T-lymphocyte killing. In murine granzyme B, arginine 226 is the critical determinant of its substrate specificity (49). Its translocation into a target cell is perforin-dependent and adenosine triphosphate (ATP) independent. The translocation precedes nuclear apoptotic events, such as DNA fragmentation and nuclear envelop breakdown. Bcl-2 and caspase inhibitors can block these events (50). The unique feature of granzyme B is its preference for proteolytic cleavage after aspartate, which mimics caspase substrate specificity. In this way, it is thought that granzyme B may directly activate caspases to induce apoptotic death of target cells (51; Fig. 1).

4. SIGNALING FOR SURVIVAL

MMAC/PTEN, a phosphatidylinositol phosphatase, is implicated in the oncogenesis of prostate cancer. It may regulate apoptosis and growth arrest through separate pathways. In PTEN-deficient LNCaP cells, expression of exogenous PTEN blocks Akt activation, and induces growth arrest or apoptosis. Overexpression of Bcl-2 only rescues cells from PTEN-mediated apoptosis, but has no effect on PTEN-induced growth suppression (52).

The kinase, protein kinase B PKB/Akt, protects cells from apoptosis, following a variety of triggers. Mechanisms for the protective effect appear to include phosphorylation of Bcl-2 proteins, caspases, and forkhead transcription factors. PKB/Akt controls association/dissociation of Bcl-2 proteins by phosphorylation. Phosphorylation of Bad at Ser 136 ablates its apoptosis-promoting function, because phosphorylated Bad is translocated from mitochondria to cytosol, and can no longer interact with Bcl-XL. TNF induces Akt-mediated Bad phosphorylation through the PI3K pathway (53–55). Akt can also phosphorylate caspase-9, to block its proapoptotic function (56).

The forkhead transcription factors, FKHR, is translocated in the pediatric tumor, alveolar rhabdomyosarcoma. It is found that PKB/Akt can phosphorylate FKHR1 at Thryonine 24 and Serine 253 in vitro (57). Upon phosphorylation by activated Akt, FKHR becomes associated with 14-3-3 proteins, and is retained in cytoplasm. FKHR is dephosphorylated after survival factor withdrawal. Its dephosphorylation leads to its nuclear translocation, to target gene activation, and, thereby, to cell demise. Thus, Akt may promote cell survival by inhibiting FKHR (58).

It has been shown that several growth factors can suppress apoptosis by activating the Akt pathway. For example, epidermal growth factor can rescue T47D breast adenocarcinoma and HEK293 embryonic kidney epithelial cells from Fas-induced apoptosis, in an Akt-dependent manner (59). Also, vascular endothelial growth factor is able to activate Akt and protect endothelial cells from loss of anchorage-dependent survival (60).

5. BCL-2 FAMILY PROTEINS (14,61)

Bcl-2 family proteins are either proapoptotic or antiapoptotic. To date, at least 16 members have been identified (62–64). Bcl-2, the first family member identified, was found to be misexpressed by t(14;18) chromosome translocation in low-grade B-cell lymphomas (65–67). It is homologous to C. elegans CED-9 (3,68). Overexpression of Bcl-2 extends cell life-span; antisense treatment accelerates apoptosis (15,69–73). Bcl-2 docks on the mitochondrial outer membrane, as well as endoplasmic reticulum and nuclear envelope, through its C-terminal hydrophobic domain (64,74–76), although the membrane-docking is not required for its survival-promoting function (77,78). The Bcl-2 family is divided into two subfamilies, based on their activities: prosurvival and proapoptotic (14).

All Bcl-2 proteins contain at least one of four Bcl-2 homology domains (BH1–BH4) (14). BH3 appears to be essential for cell-killing functions of the proapoptotic family members (7,79,80). Proapoptotic Bcl-2 proteins are further divided into the Bax family, whose members contain BH1, BH2, and BH3 domains, and BH3-only proteins, which possess only BH3 (but no other BH) domain *(14)*.

Bcl-2 proteins may execute their functions by dimerization with homologous proteins, association with nonhomologous proteins, or formation of ion channels/pores *(61)*. It has been reported that prosurvival Bcl-2 can heterodimerize with proapoptotic Bax *(81)*, and their ratio determines the cell fate *(82)*. Bcl-2 and Bax can also function independently as regulators of cell death and survival *(61)*. There are mutants of both proteins that cannot heterodimerize with each other, but are still able to suppress apoptosis and cell survival, respectively *(83–88)*. Also, in Bcl-2 or Bax knockout mice, each protein is functional in the absence of the other *(89)*. Nevertheless, the dimerization between Bcl-2 and Bax provides an important generic mechanism by which numerous members of this family may regulate their activity. The BH3 domain is critical for the dimerization among Bcl-2 proteins. Mutagenesis and structural study of Bcl-XL elucidated the structural basis of the dimerization. The amphipathic α-helix in the BH3 domain inserts into the hydrophobic cleft created by the BH1, BH2, and BH3 domains, and binds to the pocket through the hydrophobic surface of the helix *(79,90–92)*.

In addition, Bcl-2 proteins may regulate apoptosis by interacting with nonhomologous proteins *(61)*. Bcl-2 and Bcl-XL can interact with a wide range of cellular proteins. Their partners include Apaf-1 *(93,94)*; death effector domain (DED) containing proteins Bap31 and MRIT *(95,96)*; Raf-1 *(97,98)*; the guanosine triphosphatase, R-Ras *(99)*; calcineurin *(100)*; chaperone regulator BAG-1 *(101)*; p53-binding protein, 53BP-2 *(102)*; prion protein *(103)*; membrane protein BI-1 *(104)*; the spinal muscular atrophy protein, SMN *(105)*; and adenine nucleotide translocator located at the mitochondrial inner membrane *(106)*. Interaction of Bcl-XL with Apaf-1 prevents the association of Apaf-1 with pro caspase-9, suppressing procaspase-9 proteolytic activation *(14,94,93)*.

Bcl-2 family proteins may also insert into intracellular membranes, and function without dimerization with other family members *(61)*. The three dimensional structure of Bcl-XL resembles pore-forming bacterial proteins, diphtheria toxin, and colicins *(91)*. It has been reported that Bcl-2, Bax, and Bcl-XL are able to form ion channels with different conformations and ion selectivity in synthetic lipid bilayers *(107–111)*. The membrane-docking domain of proapoptotic Nip3 is essential for its activity, but its BH3 domain is dispensable *(112–113)*. Other indirect evidence supports the importance of channel-forming capacity for Bcl-2 proteins. Upon exposure to apoptotic stimuli, cytoplasmic localized Bax becomes resistant to alkaline or high-salt extraction, consistent with translocation to intracellular membranes *(108)*.

More than one mechanism has been proposed to explain cytochrome-*c* release from mitochondria during apoptosis. Normally, cytochrome-*c* resides between the inner and outer mitochondrial membranes. Apoptosis signals promote the translocation of Bax from cytosol to the outer mitochondrial membrane. Channels formed by Bax may allow cytochrome-*c* to escape into the cytoplasm *(61)*. Alternatively, loss of the electrochemical gradient (Dψ) across the inner mitochondrial membrane during apoptosis may promote the opening of mitochondrial PT pores, which consist of ANT and voltage-dependent anion channel (VDAC). The opening of PT pores leads to the loss of an ion gradient, influx of water into mitochondria, and cytochrome-*c* release. Expanding of the matrix space eventually results in organelle swelling and membrane rupture *(61)*.

It is known that Bcl-2 proteins can interact with mitochondrial PT pores, and control their conformation. Bax has been co-purified with the PT pore complex. Bcl-2 has been shown to be able to close PT pores in reconstituted liposomes. Bcl-2 family proteins may interact with mitochondrial VDAC, to modulate cytochrome-*c* release during apoptosis. PT pore opening seems to cause cytochrome-*c* release, rather than being a result of cytochrome-*c* release. A study using reconstituted liposomes shows that proapoptotic Bax and Bak promote VDAC opening and cytochrome-*c* passage through the channel; antiapoptotic Bcl-XL suppresses VDAC opening and blocks cytochrome-*c* release *(114)*.

Cellular localization of Bcl-2 proteins is regulated by the association/dissociation between family members. Hypophosphorylated Bad can interact with Bcl-XL to promote cell death. Upon phosphorylation by PKB/Akt or Raf-1, Bad is no longer able to bind to Bcl-XL, thereby abolishing its proapoptotic function *(53,54,115)*. The locations of Bcl-2 proteins can also be controlled by proteolysis. Removal of N-terminus of Bid by caspase-8 promotes its translocation to the mitochondrial outer membrane *(26,27)*.

6. MITOCHONDRIA AND CYTOCHROME-*C* *(76)*

Mitochondria may promote cell death through multiple mechanisms. Apoptotic stimuli disrupt electron transport *(116)* and ATP-production mitochondria *(117)*. In addition, caspase activators, such as cytochrome-*c* *(118,119)* and apoptosis-inducing factor are released into the cytoplasm *(120,121)*. Also, reactive oxygen species are produced by mitochondria during apoptosis *(122)*. Release of mitochondrial cytochrome-*c* is thought to activate a proapoptotic complex consisting of caspase-9 and Apaf-1 (mammalian homolog of *C. elegans* CED-4), resulting in stimulation or amplification of downstream caspase activity (Fig. 1).

7. CASPASES *(123,124)*

Caspases belong to a family of cysteine proteases that cleave after aspartate residues *(125)*. Fourteen caspases are now known, and many have been shown to be involved in apoptosis *(123)*. Caspases are synthesized as inactive precursors, and their activation is tightly regulated. A procaspase consists of four domains: an N-terminal prodomain, a large subunit, a linker, and a small subunit. Proteolytic activation removes the prodomain and linker *(126,127)*, then two large subunits and two small subunits form the active enzyme. Each tetramer contains two active sites, which are composed of residues from both the large subunits and small subunits *(128–130)*.

Caspases are classified as initiators and effectors. Initiators become activated earlier in the cascade. They have long prodomains that contain either DED or caspase recruitment domains (CARDs). Initiators can cleave themselves when clustering (autoactivation), and process downstream caspases as well. Effectors are activated downstream in the cascade, and have short prodomains. The variability of prodomains may facilitate the regulation of caspase activation by different upstream pathways *(123,124,126)*.

Only a subset of cellular proteins is inactivated or processed during apoptosis. Thus, caspases have strict specificity. Caspases cleave after aspartate residues at what is called the substrate P1 site. Four amino acids N-terminal to the P1 site specify substrate preferences. The P4-site amino acid is the primary determinant of substrate specificity *(131,132)*.

Cellular targets of caspases fall into two categories: apoptosis regulators and structural proteins *(123)*. Multiple cytoprotective proteins are inactivated by caspases during apoptosis. DNA endonuclease, DFF40/CAD, is responsible for chromatin collapse and DNA degradation during apoptosis. An inhibitor of DFF 40/CAD (ICAD), binds to CAD to repress its activity in nonapoptotic cells. In apoptotic cells, ICAD is cleaved by caspases, and dissociates from CAD. The freed CAD is now able to function as a nuclease *(133,134)*. Bcl-2 proteins are also targets of caspases *(14,135,136)*. Caspases can directly disassemble cell structures as well. Nuclear lamina is a cytoskeletal structure involved in chromosome organization. Its major component, nuclear lamin, is cleaved by caspase-6 *(137)*. Actin, as well as some actin-regulatory proteins, are also targeted by caspases during apoptosis *(138–142)*.

Inactive procaspases are constitutively expressed in cells. Caspase activation must be regulated in a stringent manner. Effector caspases are usually processed by initiator caspases. Initiator caspases are typically activated by a different strategy: induced proximity. When caspase co-factors become active upon exposure to apoptotic signals, these co-factors may recruit caspases into complexes, producing high local concentrations. Low intrinsic proteolytic activity of caspase precursors is sufficient for their processing under conditions of close proximity. Alternatively, initiators may exist in a conformation that forbids autocleavage. Binding to co-factors may then cause conformational changes in caspases, to permit autoprocessing *(123,143)*.

Ordering of the cytochrome-*c*-initiated caspase cascade (Fig. 1) has been examined in cell-free systems, using immunodepletion. Cytochrome-*c* released from mitochrondria binds to Apaf-1. The complex recruits procaspase-9, and clustering of caspase-9 stimulates its autoproteolytic cleavage. A recent study showed that cytochrome-*c*-mediated caspase-9 activation requires dATP-dependent apaf-1 oligomerization. In the absence of dATP, apaf-1 interacts with cytochrome-*c* as a monomer. dATP hydrolysis drives apaf-1 oligomerization. After oligomerization, the apaf-1–cytochrome-*c* complex can recruit and activate caspase-9, then mature caspase-9 is released from the complex to further activate downstream caspases *(144)*. Another study showed that procaspase-9 may recruit procaspase-3 to the apaf-1–caspase-9 complex through the WD-40 repeats in apaf-1 *(145)*.

Following activation, caspase-9 may dissociate from the complex to initiate the caspase cascade, or, alternatively, downstream caspases may be recruited to the complex through adaptor proteins. In either case, caspases-7 and -3 become activated next in the cascade. Caspase-3 further activates caspases-6 and 2. Finally, caspase-8 and caspase-10 are activated by caspase-6. Caspase-1, -4, and -5 failed to be processed in this system. A positive feedback loop was revealed in the cascade. Caspase-9 initiates the processing of caspase-3, and, conversely, caspase-3 activates caspase-9 *(147)*. In the context of living cells, it is possible that this entire pathway reflects an amplification loop. By this model, upstream apoptotic signals may trigger mitochrondrial injury (possibly through caspase action), resulting in cytochrome-*c* release, apaf-1/caspase-9 activation, caspase-3 activation, further mitochondrial damage, further cytochrome-*c* release and so on. The precise role of such an amplification loop (Fig. 1) in the overall life–death decision of a cell remains to be determined.

Fas has been shown to activate caspase-3, not only by promoting autoproteolytic activation of caspase-8, but also by stimulating denitrosylation of the catalytic cysteine of caspase-3. Procaspase-3 is S-nitrosylated in unstimulated cells. Its denitrosylation upon Fas activation correlates with an increase in intracellular caspase activity. Together, nitrosylation/denitrosylation may provide an additional mechanism to control caspase activity *(146)*. Caspase activity is also regulated by inhibitors. A subset of inhibitors of apoptosis (IAPs) suppress apoptosis by blocking caspase activation.

Caspase-independent apoptosis has also been described in the literature. For example, granzymes A and B are loaded into target cells through perforin in cytotoxic T-lymphocyte-killing pathway. Granzyme B-mediated apoptosis depends on caspase activation; granzyme A-induced cell death is not impaired by caspase inhibitors *(148)*. Overexpression of Bax-like proteins has also been suggested to induce cell death without caspase activation *(149–151)*.

8. *P53 (17)*

The *p53* gene encodes the most commonly mutated tumor suppressor in human cancer. The protein contains 393 amino acids, and displays transcription factor function. Through a domain located at the N-terminus, *p53* interacts with additional factors to stimulate transcription *(152,153)*. The region following the transactivation domain contains five SH3-binding motifs, and is crucial to the tumor-suppressing function of *p53* *(154)*. The central domain functions as both a DNA-binding and a protein-binding motif *(155–163)*. The C-terminal region is involved in tetramerization and nonsequence-specific DNA binding *(164–166)*.

p53 is thought to modulate two discrete functions, both in response to stress: cell cycle arrest and apoptosis (*see* Fig. 2): Both activities may play significant roles in its tumor-suppressive function. Considerable information has been learned about the pathways leading from stress (particularly DNA damage) to *p53* upregulation. This pathway involves a series of kinase activities, thought to include DNA-dependent protein kinase and the product of the ataxia telangiectasia gene *(ATM)*. The consequence of this signaling cascade is thought to be phosphorylation and stabilization of p53 protein. p53 stability may also be regulated by viral products. For example, human papillomavirus E6 protein triggers ubiquitin-dependent proteolysis of p53 *(167)*. E1A is thought to upregulate p53 protein levels, through a pathway in which it induces expression of the alternative reading frame (ARF) at the p16[Ink4a] locus, which in turns inhibits the action of Mdm2, an oncoprotein that triggers degradation of p53. Via this pathway, p53 levels can be downregulated by loss of ARF or amplification of Mdm2, two events that occur in a significant number of human cancers.

p53 upregulation produces either cell cycle arrest or apoptosis, depending on cellular context. In fibroblasts, oncogenic transformation is associated with, p53 inducing apoptosis; the untransformed state is associated with p53 inducing cell cycle arrest. These alternative outcomes (arrest vs death) represent a potential therapeutic index in the clinical scenario, permitting a potential treatment or stress to produce selective toxicity (Fig. 2). Many human cancers are associated with mutations in p53 or p53 regulators *(17,28–30,168)*. In this light, it is noteworthy that p53 loss is common in adult, often incurable, tumors, but it is much less commonly mutated or deleted in the more highly curable pediatric malignancies *(169)*.

p53 is thought to regulate apoptosis through both transcription-dependent and -independent mechanisms. It may control the expression of several key regulators of apoptosis, such as bax *(20)*, although the in vivo significance of this regulation to

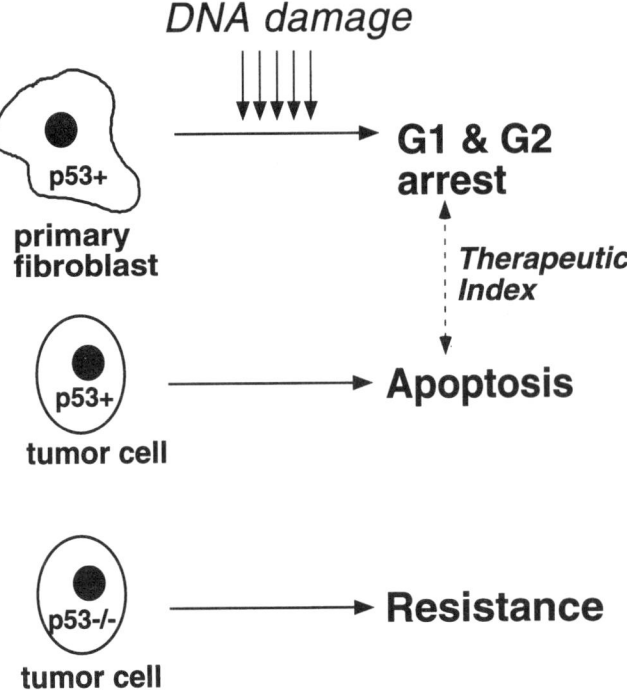

Fig. 2. Response to DNA damage is modulated by p53. Induction of p53 results in growth arrest at G1 and G2-M in untransformed (primary) fibroblasts; identical treatment of oncogene-transformed fibroblasts (tumor cells) results in apoptotic death. The differential responses of arrest vs. death produce a potential therapeutic index. Loss of p53 in tumor cells is associated with treatment resistance.

p53-dependent apoptosis remains unclear. It has been found that 14 p53-induced genes are activated by overexpression of p53 in the colorectal cancer line, DLD-1 *(170)*. These proteins probably function in response to oxidative stress. p53 may also transcriptionally target the death receptors, Fas *(171–174)* and DR5 *(175)*, insulin-like growth factor binding protein-3 *(176)*, and p53-activated gene 608 *(177)*, to repress antiapoptotic pathways and promote cell death. The human homolog of *Drosophila sina*, SIAH-1, was recently identified, and its expression is induced by p53 during apoptosis. SIAH-1 may function as an apoptosis inducer and tumor suppressor *(178)*.

In some contexts, however, the transactivation function of p53 may not be required for p53-mediated apoptosis. In transcription-independent apoptosis, p53 may be able to act as a transcriptional repressor, or p53 may induce cell death through protein–protein interactions *(17)*.

9. IAPS AND DEATH-ASSOCIATED PROTEINS *(179,180)*

IAPs were first found in baculoviruses *(181,182)*. Two defining motifs exist in almost all IAPs: The first is the (baculoviral inhibitor of apoptosis repeat domain *(181,182)*, which consists of a putative Ser/Thr phosphorylation site and arrays of DX3CX2C and HX6C, which may be metal coordination sites; the second is the interesting new gene zinc finger motif *(183)*, which contains two Zn coordination sites

(184). This domain may be involved in DNA–protein interactions, and is critical to the antiapoptotic function of baculoviral IAPs *(185).* The CARD motif has been found in two human IAPs *(186).*

IAPs can protect cells from diversity of apoptotic triggers, including death receptor ligation, viral infection, growth factor withdrawal, chemotherapy drugs, and irradiation *(179).* The ability of IAPs to block apoptosis is cell-type-specific *(179).* IAPs may inhibit cell death by blocking caspases or activating NF-κB. It has been shown that IAPs can directly inhibit caspase-3 and -7 *(187–189).* Alternatively, they may block caspase-9 activation by cytochrome-*c (190).* Some IAPs are able to activate NF-κB. Conversely, NF-κB can induce IAPs in multiple cell lines *(179).* One of the human IAPs, survivin, may be involved in malignancy. Survivin, which is normally expressed only in fetal tissue, has been found in most cancer cells tested *(191–193),* where its expression has been suggested to correlate with shorter disease survival *(194–179).*

The technical knockout strategy is an ingenious method that was designed to identify rate-limiting genes in apoptosis. HeLa cells are transfected with an antisense cDNA library. Transfected cells are selected against apoptosis-triggering agents, such as interferon γ. cDNA fragments are rescued from interferons γ-resistant cells, and used to identify the genes that are responsible for the resistance *(195).*

Five death-associated proteins (DAPs) have been identified through this approach *(180).* Among them, DAP-2 (DAP-kinase [DAPK]) is a Ca$^+$/calmodulin (CaM)-dependent protein kinase. Its N-terminus contains a kinase domain followed by a CaM-binding domain. The 3′-end of the protein is analogous to the death domain *(196).* DAPK can autophosphorylate or phosphorylate other substrates in vitro. It is associated with microfilaments in vivo *(197).* Overexpression of DAPK induces apoptosis in a kinase-activity-dependent manner. Moreover CpG islands of the DAPK gene are hypermethylated in many B-cell malignancies *(198).* This kinase may be involved in Fas-mediated apoptosis, since DAPK antisense treatment or overexpression of the DAP-kinase death domain protects cells from Fas-induced apoptosis. Thus, DAPK may function downstream of the Fas receptor, and its apoptosis-inducing activity requires the death domain *(199).* Recently, DAPK2, a homolog of DAPK, was found in mice. Localized in the cytoplasm, DAPK2 also stimulates apoptosis, when overexpressed *(200).*

10. MYC *(201–203)*

The proto-oncogene c-*myc* not only signals proliferation, but also sensitizes cells to apoptosis. Deregulated c-*myc* expression leads to apoptosis in the presence of cell growth arrest signals, cell differentiation signals, or cytotoxic agents. A large number of proteins have been found to associate with c-*myc,* in the process of attempting to uncover the mechanism(s) underlying the ability of this oncoprotein to regulate proliferation and death.

The C-terminus of c-*myc* contains a basic/helix-loop-helix/leucine-zipper (b/HLH/Zip) domain. HLH and leucine zipper motifs mediate protein–protein interactions, and the adjacent region rich in basic residues recognizes E-box DNA sequences in the major groove of DNA *(204–207).* A transcriptional activation domain (TAD) is located at the N-terminus of Myc *(208).* The TAD contains two Myc boxes: MB1 and MB2. MB2 is crucial to the transcription repression function of myc *(209–212).* Max *(213–216),* YY-1 *(217),* AP-2 *(218),* BRCA-1 *(219),* TFII-I *(211,220,221),* and Miz-1

(222) interact with the C-terminal b/HLH/Zip domain; p107 *(223,224)*, Bin1 *(225)*, Myc modulation 1 (MM-1) *(226)*, protein associated with Myc (Pam) *(227)*, transformation/transcription domain-associated protein (TRRAP) *(228)*, and associate of c-Myc-1 AMY-1 *(229)* can associate at or near the N-terminal TAD.

Max possesses a b/HLH/Zip domain highly homologous to Myc, and a C-terminal nuclear localization signal. Max dimerizes either with itself or with Myc, under physiological conditions, and their heterodimerization appears to be essential for Myc-dependent function, including DNA binding, transactivation, transformation, and apoptosis *(214–216)*. Max's dimerization and DNA-binding activity are regulated by phosphorylation and alternative splicing *(58,230–233)*. Max resides in the center of two competing pathways: one in which it dimerizes with Myc to produce proliferation, and one in which it dimerizes with Mad proteins to promote differentiation. Max:Mad dimers repress transcription by recruiting histone deacetylases through the co-repressor, mSin3 *(234,235)*.

Association of YY-1 with c-*myc* prevents Max binding *(217)*. Myc antagonizes transcriptional activation and repression of YY-1 by interfering with its binding to TBP and TFII-B *(236)*. Correspondingly, YY-1 may be able to affect c-*myc* activity indirectly *(237)*. Transcription factor AP-2 inhibits c-*myc* function by competitively binding to the same DNA sequence as the Myc/Max complex, and AP-2 has been found to be capable of suppressing c-*myc*-mediated apoptosis *(203)*. The tumor suppressor, BRCA-1, can associate with c-*myc* in vitro and in tissue culture cells. Overexpression of BRCA-1 inhibits c-*myc*-mediated transactivation *(203)*. Transcription factor TFII-I recognizes both E-box and the pyrimidine-rich Inr sites *(238)*. The binding of the c-*myc*/TFII-I complex to both promotor elements results in transcriptional repression of various target genes *(220,221)*. Miz-1 is a Zn finger protein. c-*myc*/Miz-1 association is required for c-*myc* to inhibit Miz-1-mediated growth arrest *(203)*.

p107, a member of the Rb family, interacts with MB2 of c-*myc*, through its pocket domain, to inhibit c-*myc*-mediated transactivation *(223,224,239)*. Bin-1 is an adaptor protein that may act as a tumor suppressor *(203)*. It may be a significant regulator of c-*myc*-induced apoptosis. MM-1 is a nucleocytoplasmic protein that inhibits transactivation by c-*myc* *(203)*. TRRAP may be a co-activator of c-*myc*-mediated transactivation *(203)*. The yeast homolog of TRRAP, Tra1, is a component of the chromatin remodeling complex, SAGA *(240)*. Pam binds to c-*myc*'s N-terminal domain, both in vitro and in vivo *(227)*, but the physiological role of Pam/c-*myc* remains uncertain. Amy-1 interacts with the MB2 motif of c-*myc* *(229)*, but only binds to certain phosphorylated isoforms of c-*myc*, in a cell-cycle dependent manner. This association may regulate Amy-1 translocation into the nucleus *(203)*.

c-*myc* targets various genes, although their products are not yet known to be essential for Myc-mediated apoptosis. Its genes, identified as potential transcriptional targets, include *cdc25A (241)*; α-prothymosin *(242)*; lactate dehydrogenase A *(243)*; *MrDb (244)*; *DEAD* box family RNA helicase, which functions in RNA processing and translational control; ornithine decarboxylase *(245,246)*; and growth arrest and DNA damage-inducible gene *(gadd45) (247)*.

Inappropriate c-*myc* expression under restricted growth conditions induces apoptosis in interleukin-3-dependent 32Dcl3 cells *(248)*, Rat-1 fibroblast cell lines, primary rat embryo fibroblasts *(210)*, mouse embryo fibroblasts *(249)*, quiescent renal epithelial cells *(250)*, hepatocytes, and lymphoid cells *(249,251)*. Deregulated c-*myc* expression also stimulates apoptosis in M1-Myc cells in the presence of differentiation signals *(252)*. On the other hand, Myc expression is not required for apoptosis in 32Dcl3 cells and TGFβ/p53ts myeloid leukemic M1 cells *(253,254)*, suggesting that Myc overexpression is sufficient, but not necessary for tumor cell apoptosis.

Release of cytochrome-*c* from mitochondria appears to occur downstream of Myc-induced apoptosis. A recent study showed that anticytochrome-*c* antibody can block c-*myc*-mediated apoptosis *(255)*. Currently, it is unclear how direct this regulation of mitochondrial function is within cells.

Two models have been proposed for c-*myc*-mediated apoptosis: the conflict model, and the dual signal model *(249,256)*. According to the conflict model, c-*myc* directly regulates only proliferative signals. Conflicts between such proliferative signals and distinct growth arrest signals, presented simultaneously in the cell, generates the unique biochemical context resulting in caspase activation and apoptosis. By the dual signal model, c-*myc* fundamentally triggers both proliferative and apoptotic signals. Apoptosis occurs in the absence of survival factors; proliferation dominates in the presence of such factors. There is evidence for and against each model. Abundant experimental data have produced their own set of conflicting support for these models. For example, in certain Myc-deregulated cell systems, there is little correlation between growth arrest and apoptosis *(257)*. In addition, insulin-like growth factor 1 and platelet-derived growth factor can protect cells against Myc-mediated apoptosis, without inducing cell proliferation *(258,259)*. Clearly, more mechanistic detail is needed regarding Myc's biochemical activities in regulating both proliferation and apoptosis in vivo.

11. PROSPECTS

The discovery that death can be more than chaotic disruption of cellular metabolism has dramatically changed modern biology. Adding to the examination of growth and differentiation in terms of cell cycle and tissue-specific gene expression, survival signals have gained prominence in regulating key steps in development. Diseases, especially cancer, have been mostly recategorized, not only in terms of growth dysregulation, but also in terms of death dysregulation. For cancer, this probably is important both during the process of tumorigenesis and in the context of cancer therapy. One of the striking features of apoptosis is the recognition of how commonly it occurs under both physiologic and pathologic conditions. Indeed, it is rare to observe descriptions of nonapoptotic death in the current biological literature. Nonetheless, it is clear that nonapoptotic death plays an important role in much of human pathology (in contrast to apoptosis, which also occurs in the course of normal cellular homeostasis). It is likely that the explosion of information regarding the ordered, efficient events in apoptosis will permit the identification of therapeutic targets of importance, and, hopefully, novel treatments. In this manner, manipulation of apoptotic events may permit both up- or downregulation of cellular survival in a manner that could impact on an enormous number of human diseases.

REFERENCES

1. Kerr JF, Wyllie AH, Currie AR. Apoptosis: a basic biological phenomenon with wide-ranging implications in tissue kinetics. *Br J Cancer* 1972; 26:239–257.
2. Ellis HM. Genetic control of programmed cell death in the nematode C. elegans. *Cell* 1986; 44:817–829.

3. Hengartner MO. C. elegans cell survival gene ced-9 encodes a functional homolog of the mammalian proto-oncogene bcl-2. *Cell* 1994; 76:665–676.
4. Chinnaiyan AM, Lane BR, Dixit VM. Interaction of CED-4 with CED-3 and CED-9: a molecular framework for cell death. *Science* 1997; 275:1122–1126.
5. Spector MS, Hoeppner DJ, Hengartner MO. Interaction between the C. elegans cell-death regulators CED-9 and CED-4. *Nature* 1997; 385:653–656.
6. Chinnaiyan AM, O'Rourke K, Koonin EV, Dixit VM. Role of CED-4 in the activation of CED-3. *Nature* 1997; 388:728–729.
7. Conradt B. The C. elegans protein EGL-1 is required for programmed cell death and interacts with the Bcl-2-like protein CED-9. *Cell* 1998; 93:519–529.
8. del Peso L, Nunez G. Caenorhabditis elegans EGL-1 disrupts the interaction of CED-9 with CED-4 and promotes CED-3 activation. *J Biol Chem* 1998; 273:33,495–500.
9. Yuan J, Ledoux S, Ellis HM, Horvitz HR. The C. elegans cell death gene ced-3 encodes a protein similar to mammalian interleukin-1 beta-converting enzyme. *Cell* 1993; 75:641–652.
10. Thornberry NA, Calaycay JR, Chapman KT, Howard AD, Kostura MJ, Miller DK, et al. A novel heterodimeric cysteine protease is required for interleukin-1 beta processing in monocytes. *Nature* 1992; 356:768–774.
11. Zou H, Liu X, Lutschg A, Wang X. Apaf-1, a human protein homologous to C. elegans CED-4, participates in cytochrome c-dependent activation of caspase-3. *Cell* 1997; 90:405–413.
12. Li P, Budihardjo I, Srinivasula SM, Ahmad M, Alnemri ES, Wang X. Cytochrome c and dATP-dependent formation of Apaf-1/caspase-9 complex initiates an apoptotic protease cascade. *Cell* 1997; 91:479–489.
13. Srinivasula SM, Fernandes-Alnemri T, Alnemri ES. Autoactivation of procaspase-9 by Apaf-1-mediated oligomerization. *Mol Cell* 1998; 1:949–957.
14. Adams JM. The Bcl-2 protein family: arbiters of cell survival. *Science* 1998; 281:1322–1326.
15. Chao DT. BCL-2 family: regulators of cell death. *Annu Rev Immunol* 1998; 16:395–419.
16. Levine A. The tumor suppressor genes. *Annu Rev Biochem* 1993; 62:623–651.
17. Ding HF, Fisher FE. Mechanisms of p53-mediated apoptosis. *Crit Rev Oncog* 1998; 9:83–98.
18. Miyashita T, Krajewska M, Wang HG, Lin HK, Liebermann DA, Hoffman B, Reed JC. Tumor suppressor p53 is a regulator of bcl-2 and bax gene expression in vitro and in vivo. *Oncogene* 1994; 9:1799–1805.
19. Selvakumaran M, Miyashita T, Wang HG, Krajewski S, Reed JC, Hoffman B, Liebermann D. Immediate early up-regulation of bax expression by p53 but not TGF beta 1: a paradigm for distinct apoptotic pathways. *Oncogene* 1994; 9:1791–1798.
20. Miyashita T. Tumor suppressor p53 is a direct transcriptional activator of the human bax gene. *Cell* 1995; 80:293–299.
21. Manon S, Guerin M. Release of cytochrome c and decrease of cytochrome c oxidase in Bax-expressing yeast cells, and prevention of these effects by coexpression of Bcl-xL. *FEBS Lett* 1997; 415:29–32.
22. Rosse T, Monney L, Rager M, Conus S, Fellay I, Jansen B, Borner C. Bcl-2 prolongs cell survival after Bax-induced release of cytochrome c. *Nature* 1998; 391:496–499.
23. Chinnaiyan AM, Tewari M, Dixit VM. FADD, a novel death domain-containing protein, interacts with the death domain of Fas and initiates apoptosis. *Cell* 1995; 81:505–512.
24. Boldin MP, Goltsev YV, Wallach D. Involvement of MACH, a novel MORT1/FADD-interacting protease, in Fas/APO-1- and TNF receptor-induced cell death. *Cell* 1996; 85:803–815.
25. Muzio M, Stennicke HR, Salvesen GS, Dixit VM. An induced proximity model for caspase-8 activation. *J Biol Chem* 1998; 273:2926–2930.
26. Li H, Xu CJ, Yuan J. Cleavage of BID by caspase 8 mediates the mitochondrial damage in the Fas pathway of apoptosis. *Cell* 1998; 94:491–501.
27. Luo X, Zou H, Slaughter C, Wang X. Bid, a Bcl2 interacting protein, mediates cytochrome c release from mitochondria in response to activation of cell surface death receptors. *Cell* 1998; 94:481–490.
28. Harris CC, Clinical implications of the p53 tumor-suppressor gene. *N Engl J Med* 1993; 329:1318–1327.
29. Fisher D. Apoptosis in cancer therapy: crossing the threshold. *Cell* 1994; 78:539–542.
30. Ruley H. p53 and response to chemotherapy and radiotherapy. *Important Adv Oncol* 1996.
31. Ashkenazi A. Death receptors: signaling and modulation. *Science* 1998; 281:1305–1308.

32. Olive C, Falk MC. Apoptosis and expression of cytotoxic T lymphocyte effector molecules in renal allografts. *Transplant Immunol* 1999; 7:27–36.

33. Nagata S. Apoptosis by death factor. *Cell* 1997; 88:355–365.

34. Smith CA, Goodwin RG. The TNF receptor superfamily of cellular and viral proteins: activation, costimulation, and death. *Cell* 1994; 76:959–962.

35. Hsu H, Goeddel DV. The TNF receptor 1-associated protein TRADD signals cell death and NF-kappa B activation. *Cell* 1995; 81:495–504.

36. Hsu H, Pan MG, Goeddel DV. TRADD-TRAF2 and TRADD-FADD interactions define two distinct TNF receptor 1 signal transduction pathways. *Cell* 1996; 84:299–308.

37. Chinnaiyan AM, Seldin MF, O'Rourke K, Kischkel FC, Hellbardt S, Krammer PH, Peter ME, Dixit and VM. FADD/MORT1 is a common mediator of CD95 (Fas/APO-1) and tumor necrosis factor receptor-induced apoptosis. *J Biol Chem* 1996; 271:4961–4965.

38. Varfolomeev EE, Goncharov TM, Wallach D. A potential mechanism of "cross-talk" between the p55 tumor necrosis factor receptor and Fas/APO1: proteins binding to the death domains of the two receptors also bind to each other. *J Exp Med* 1996; 183:1271–1275.

39. Hsu H, Shu HB, Baichwal V, Goeddel DV. TNF-dependent recruitment of the protein kinase RIP to the TNF receptor-1 signaling complex. *Immunity* 1996; 4:387–396.

40. Ting AT, Seed B. RIP mediates tumor necrosis factor receptor 1 activation of NF-kappaB but not Fas/APO-1-initiated apoptosis. *EMBO J* 1996; 15:6189–6196.

41. Rothe M, Henzel WJ, Ayres TM, Goeddel DV. The TNFR2-TRAF signaling complex contains two novel proteins related to baculoviral inhibitor of apoptosis proteins. *Cell* 1995; 83:1243–1252.

42. Sugiyama H, Kitamura M, Zhao L, Stylianou E. Selective sensitization to tumor necrosis factor-alpha-induced apoptosis by blockade of NF-kappaB in primary glomerular mesangial cells. *J Biol Chem* 1999; 274:19,532–537.

43. Schwenzer R, Liptay S, Schubert G, Peters N, Scheurich P, Schmid RM, Wajant H. The human tumor necrosis factor (TNF) receptor-associated factor 1 gene (TRAF1) is up-regulated by cytokines of the TNF ligand family and modulates TNF-induced activation of NF-kappaB and c-Jun N-terminal kinase. *J Biol Chem* 1999; 274:19,368–19,374.

44. Rothe M, Henzel WJ, Goeddel DV. A novel family of putative signal transducers associated with the cytoplasmic domain of the 75 kDa tumor necrosis factor receptor. *Cell* 1994; 78:681–692.

45. Rothe M, Dixit VM, Goeddel DV. TRAF2-mediated activation of NF-kappa B by TNF receptor 2 and CD40. *Science* 1995; 269:1424–1427.

46. Dadgostar H. An intact zinc ring finger is required for tumor necrosis factor receptor-associated factor-mediated nuclear factor-kappaB activation but is dispensable for c-Jun N-terminal kinase signaling. *J Biol Chem* 1998; 273:24775–24780.

47. Grell M, Gottfried E, Chen CM, Grunwald U, Huang DC, Lee YH, et al. Induction of cell death by tumour necrosis factor (TNF) receptor 2, CD40 and CD30: a role for TNF-R1 activation by endogenous membrane-anchored TNF. *EMBO J* 1999; 18:3034–3043.

48. Yu KY, Ni J, Zhai Y, Ebner R, Kwon BS. A newly identified member of tumor necrosis factor receptor superfamily (TR6) suppresses LIGHT-mediated apoptosis. *J Biol Chem* 1999; 274:13733–13736.

49. Caputo A, James MN, Powers JC, Bleackley RC. Electrostatic reversal of serine proteinase substrate specificity. *Proteins* 1999; 35:415–424.

50. Blink EJ, Jans DA. Perforin-dependent nuclear targeting of granzymes: a central role in the nuclear events of granule-exocytosis-mediated apoptosis? *Immunol Cell Biol* 1999; 77:206–215.

51. Darmon AJ, Nicholson DW, and Bleackley RC. Activation of the apoptotic protease CPP32 by cytotoxic T-cell-derived granzyme B. *Nature* 1995; 377:446–448.

52. Davies MA, Dhesi H, Berman R, McDonnell TJ, McConkey D, Yung WK, Steck PA. Regulation of Akt/PKB activity, cellular growth, and apoptosis in prostate carcinoma cells by MMAC/PTEN. *Cancer Res* 1999; 59:2551–2556.

53. Datta SR, Tao X, Masters S, Fu H, Gotoh Y, Greenberg ME. Akt phosphorylation of BAD couples survival signals to the cell-intrinsic death machinery. *Cell* 1997; 91:231–241.

54. del Peso L, Page C, Herrera R, Nunez G. Interleukin-3-induced phosphorylation of BAD through the protein kinase Akt. *Science* 1997; 278:687–689.

55. Pastorino JG, Farber JL. Tumor necrosis factor induces phosphorylation and translocation of BAD through a phosphatidylinositide-3-OH kinase-dependent pathway. *J Biol Chem* 1999; 274:19411–19416.

56. Cardone MH, Stennicke HR, Salvesen GS, Franke TF, Stanbridge E, Frisch S, et al. Regulation of cell death protease caspase-9 by phosphorylation. *Science* 1998; 282:1318–1321.

57. Biggs WH 3rd, Hunter T, Cavenee WK, Arden KC. Protein kinase B/Akt-mediated phosphorylation promotes nuclear exclusion of the winged helix transcription factor FKHR1. *Proc Natl Acad Sci USA* 1999; 96:7421–7426.

58. Brunet A, Zigmond MJ, Lin MZ, Juo P, Hu LS, Anderson MJ, et al. Akt promotes cell survival by phosphorylating and inhibiting a Forkhead transcription factor. *Cell* 1999; 96:857–868.

59. Gibson S, Oyer R, Anderson SM, Johnson GL. Epidermal growth factor protects epithelial cells against Fas-induced apoptosis. Requirement for Akt activation. *J Biol Chem* 1999; 274:17612–17618.

60. Fujio Y. Akt mediates cytoprotection of endothelial cells by vascular endothelial growth factor in an anchorage-dependent manner. *J Biol Chem* 1999; 274:16,349–16,354.

61. Reed J. Bcl-2 family proteins. *Oncogene* 1998; 17:3225–3236.

62. Chao DT. BCL-2 family: regulators of cell death. *Annu Rev Immunol* 1998; 16:395–419.

63. Reed J. Bcl-2 and the regulation of programmed cell death. *J Cell Biol* 1994; 124:1–6.

64. Zamzami N, Marzo I, Susin SA, Kroemer G. Subcellular and submitochondrial mode of action of Bcl-2-like oncoproteins. *Oncogene* 1998; 16:2265–2282.

65. Tsujimoto Y, Yunis J, Nowell PC, Croce CM. Cloning of the chromosome breakpoint of neoplastic B cells with the t(14;18) chromosome translocation. *Science* 1984; 226:1097–1099.

66. Bakhshi A, Goldman P, Wright JJ, McBride OW, Epstein AL, Korsmeyer SJ. Cloning the chromosomal breakpoint of t(14;18) human lymphomas: clustering around JH on chromosome 14 and near a transcriptional unit on 18. *Cell* 1985; 41:899–906.

67. Cleary ML, Sklar J. Cloning and structural analysis of cDNAs for bcl-2 and a hybrid bcl-2/immunoglobulin transcript resulting from the t(14;18) translocation. *Cell* 1996; 47:19–28.

68. Vaux DL, Kim SK. Prevention of programmed cell death in Caenorhabditis elegans by human bcl-2. *Science* 1992; 258:1955–1957.

69. Cory S, Jacks T, Corcoran LM, Metz T, Harris AW, Adams JM. Enhanced cell survival and tumorigenesis. *Cold Spring Harbor Symp Quant Biol* 1994; 59:365–375.

70. Cory S. Regulation of lymphocyte survival by the bcl-2 gene family. *Annu Rev Immunol* 1995; 13:513–543.

71. Strasser A, Vaux DL. The role of the bcl-2/ced-9 gene family in cancer and general implications of defects in cell death control for tumourigenesis and resistance to chemotherapy. *Biochim Biophys Acta* 1997; 1333:F151–F178.

72. Yang E. Molecular thanatopsis: a discourse on the BCL2 family and cell death. *Blood* 1996; 88:386–401.

73. Vaux DL, Adams JM. Bcl-2 gene promotes haemopoietic cell survival and cooperates with c-*myc* to immortalize pre-B cells. *Nature* 1998; 335:440–442.

74. Krajewski S, Takayama S, Schibler MJ, Fenton W, Reed JC. Investigation of the subcellular distribution of the bcl-2 oncoprotein: residence in the nuclear envelope, endoplasmic reticulum, and outer mitochondrial membranes. *Cancer Res* 1993; 53:4701–4714.

75. Kroemer G. The proto-oncogene Bcl-2 and its role in regulating apoptosis. *Nat Med* 1997; 3:614–620.

76. Green DR. Mitochondria and apoptosis. *Science* 1998; 281:1309–1312.

77. Nguyen M, Walton PA, Oltvai ZN, Korsmeyer SJ, Shore GC. Role of membrane anchor domain of Bcl-2 in suppression of apoptosis caused by E1B-defective adenovirus. *J Biol Chem* 1994; 269:16521–16524.

78. Borner C, Mattmann C, Irmler M, Schaerer E, Martinou JC, Tschopp J. The protein bcl-2 alpha does not require membrane attachment, but two conserved domains to suppress apoptosis. *J Cell Biol* 1994; 126:1059–1068.

79. Chittenden T, Houghton AB, Ebb RG, Gallo GJ, Elangovan B, Chinnadurai G, Lutz RJ. A conserved domain in Bak, distinct from BH1 and BH2, mediates cell death and protein binding functions. *EMBO J* 1995; 14:5589–5596.

80. Kelekar A. Bcl-2-family proteins: the role of the BH3 domain in apoptosis. *Trends Cell Biol* 1998; 8:324–330.

81. Oltvai ZN, Korsmeyer SJ. Bcl-2 heterodimerizes in vivo with a conserved homolog, Bax, that accelerates programmed cell death. *Cell* 1993; 74:609–619.

82. Oltvai ZN. Checkpoints of dueling dimers foil death wishes. *Cell* 1994; 79:189–192.

83. Cheng EH, Boise LH, Thompson CB, Hardwick JM. Bax-independent inhibition of apoptosis by Bcl-XL. *Nature* 1996; 379:554–556.

84. Simonian PL, Merino R, Nunez G. Bax can antagonize Bcl-XL during etoposide and cisplatin-induced cell death independently of its heterodimerization with Bcl-XL. *J Biol Chem* 1996; 271:22,764–22,772.

85. Simonian PL, Nunez G Bak can accelerate chemotherapy-induced cell death independently of its heterodimerization with Bcl-XL and Bcl-2. *Oncogene* 1997; 15:1871–1875.

86. Tao W, Morgan JI. Modulation of cell death in yeast by the Bcl-2 family of proteins. *J Biol Chem* 1997; 272:15547–15552.

87. Wang K, Waksman G, Korsmeyer SJ. Mutagenesis of the BH3 domain of BAX identifies residues critical for dimerization and killing. *Mol Cell Biol* 1998; 18:6083–6089.

88. Zha H. Heterodimerization-independent functions of cell death regulatory proteins Bax and Bcl-2 in yeast and mammalian cells. *J Biol Chem* 1997; 272:31,482–31,488.

89. Knudson CM. Bcl-2 and Bax function independently to regulate cell death. *Nat Genet* 1997; 16:358–363.

90. Yin XM, Korsmeyer SJ. BH1 and BH2 domains of Bcl-2 are required for inhibition of apoptosis and heterodimerization with Bax. *Nature* 1994; 369:321–323.

91. Muchmore SW, Liang H, Meadows RP, Harlan JE, Yoon HS, Nettesheim D, et al. X-ray and NMR structure of human Bcl-xL, an inhibitor of programmed cell death. *Nature* 1996; 381:335–341.

92. Sattler M, Nettesheim D, Meadows RP, Harlan JE, Eberstadt M, Yoon HS, et al. Structure of Bcl-xL–Bak peptide complex: recognition between regulators of apoptosis. *Science* 1997; 275:983–986.

93. Hu Y, Wu D, Inohara N, Nunez G. Bcl-XL interacts with Apaf-1 and inhibits Apaf-1-dependent caspase-9 activation. *Proc Natl Acad Sci USA* 1998; 95:4386–4391.

94. Pan G, Dixit VM. Caspase-9, Bcl-XL, and Apaf-1 form a ternary complex. *J Biol Chem* 1998; 273:5841–5845.

95. Han DK, Wright ME, Friedman C, Trask BJ, Riedel RT, Baskin DG, Schwartz SM, Hood L. MRIT, a novel death-effector domain-containing protein, interacts with caspases and BclXL and initiates cell death. *Proc Natl Acad Sci USA* 1997; 94:11,333–11,338.

96. Ng FW, Kwan T, Branton PE, Nicholson DW, Cromlish JA, Shore GC. p28 Bap31, a Bcl-2/Bcl-XL- and procaspase-8-associated protein in the endoplasmic reticulum. *J Cell Biol* 1997; 139:327–338.

97. Wang HG, Takayama S, Sato T, Torigoe T, Krajewski S, Tanaka S, et al. Apoptosis regulation by interaction of Bcl-2 protein and Raf-1 kinase. *Oncogene* 1994; 9:2751–2756.

98. Wang HG, Reed JC. Bcl-2 targets the protein kinase Raf-1 to mitochondria. *Cell* 1996; 87:629–638.

99. Fernandez-Sarabia MJ. Bcl-2 associates with the ras-related protein R-ras p23. *Nature* 1993; 366:274–275.

100. Shibasaki F, Akagi T, McKeon F. Suppression of signalling through transcription factor NF-AT by interactions between calcineurin and Bcl-2. *Nature* 1997; 386:728–731.

101. Takayama S, Krajewski S, Kochel K, Irie S, Millan JA, Reed JC. Cloning and functional analysis of BAG-1: a novel Bcl-2-binding protein with anti-cell death activity. *Cell* 1995; 80:279–284.

102. Naumovski L. The p53-binding protein 53BP2 also interacts with Bc12 and impedes cell cycle progression at G2/M. *Mol Cell Biol* 1996; 16:3884–3892.

103. Kurschner C. The cellular prion protein (PrP) selectively binds to Bcl-2 in the yeast two-hybrid system. *Brain Res Mol Brain Res* 1995; 30:165–168.

104. Xu Q. Bax inhibitor-1, a mammalian apoptosis suppressor identified by functional screening in yeast. *Mol Cell* 1998; 1:337–346.

105. Iwahashi H, Yasuhara N, Hanafusa T, Matsuzawa Y, Tsujimoto Y. Synergistic anti-apoptotic activity between Bcl-2 and SMN implicated in spinal muscular atrophy. *Nature* 1997; 390:413–417.

106. Marzo I, Zamzami N, Jurgensmeier JM, Susin SA, Vieira HL, Prevost MC, et al. Bax and adenine nucleotide translocator cooperate in the mitochondrial control of apoptosis. *Science* 1998; 281:2027–2031.

107. Minn AJ, Schendel SL, Liang H, Muchmore SW, Fesik SW, Fill M, Thompson CB. Bcl-x(L) forms an ion channel in synthetic lipid membranes. *Nature* 1997; 385:353–357.

108. Antonsson B, Ciavatta A, Montessuit S, Lewis S, Martinou I, Bernasconi L, et al. Inhibition of Bax channel-forming activity by Bcl-2. *Science* 1997; 277:370–372.

109. Schendel SL, Montal MO, Matsuyama S, Montal M, Reed JC. Channel formation by antiapoptotic protein Bcl-2. *Proc Natl Acad Sci USA* 1997; 94:5113–5118.

110. Schlesinger PH, Yin XM, Yamamoto K, Saito M, Waksman G, Korsmeyer SJ. Comparison of the ion channel characteristics of proapoptotic BAX and antiapoptotic BCL-2. *Proc Natl Acad Sci USA* 1997; 94:11,357–11,362.

111. Lam M, Nunez G, Ma J, Distelhorst CW. Regulation of Bcl-xl channel activity by calcium. *J Biol Chem* 1998; 273:17,307–17,310.

112. Chen G, Dubik D, Shi L, Cizeau J, Bleackley RC, Saxena S, Gietz RD, Greenberg AH. The E1B 19K/Bcl-2-binding protein Nip3 is a dimeric mitochondrial protein that activates apoptosis. *J Exp Med* 1997; 186:1975–1983.

113. Yasuda M, Subramanian T, Chinnadurai G. Adenovirus E1B-19K/BCL-2 interacting protein BNIP3 contains a BH3 domain and a mitochondrial targeting sequence. *J Biol Chem* 1998; 273:12,415–12,421.

114. Shimizu S, Tsujimoto Y. Bcl-2 family proteins regulate the release of apoptogenic cytochrome c by the mitochondrial channel VDAC. *Nature* 1999; 399:483–487.

115. Zha J, Yang E, Jockel J, Korsmeyer SJ. Serine phosphorylation of death agonist BAD in response to survival factor results in binding to 14-3-3 not BCL-X. *Cell* 1996; 87:619–628.

116. Garcia-Ruiz C, Mari M, Morales A, Fernandez-Checa JC. Direct effect of ceramide on the mitochondrial electron transport chain leads to generation of reactive oxygen species. Role of mitochondrial glutathione. *J Biol Chem* 1997; 272:11,369–11,377.

117. Bossy-Wetzel E, Green DR. Mitochondrial cytochrome c release in apoptosis occurs upstream of DEVD-specific caspase activation and independently of mitochondrial transmembrane depolarization. *EMBO J* 1998; 17:37–49.

118. Yang J, Bhalla K, Kim CN, Ibrado AM, Cai J, Peng TI, Jones DP, Wang X. Prevention of apoptosis by Bcl-2: release of cytochrome c from mitochondria blocked. *Science* 1997; 275:1129–1132.

119. Kluck RM, Green DR, Newmeyer DD. The release of cytochrome c from mitochondria: a primary site for Bcl-2 regulation of apoptosis. *Science* 1997; 275:1132–1136.

120. Susin SA, Castedo M, Hirsch T, Marchetti P, Macho A, Daugas E, Geuskens M, Kroemer G. Bcl-2 inhibits the mitochondrial release of an apoptogenic protease. *J Exp Med* 1996; 184:1331–1341.

121. Susin SA, Castedo M, Daugas E, Wang HG, Geley S, Fassy F, Reed JC, Kroemer G. The central executioner of apoptosis: multiple connections between protease activation and mitochondria in Fas/APO-1/CD95- and ceramide-induced apoptosis. *J Exp Med* 1997; 186:25–37.

122. Bredesen D. Neural apoptosis. *Ann Neurol* 1995; 38:839–851.

123. Nunez G, Hu Y, Inohara N. Caspases: the proteases of the apoptotic pathway. *Oncogene* 1998; 17:3237–3245.

124. Thornberry NA. Caspases: enemies within. *Science* 1998; 281:1312–1316.

125. Alnemri ES, Nicholson DW, Salvesen G, Thornberry NA, Wong WW, Yuan J. Human ICE/CED-3 protease nomenclature. *Cell* 1996; 87:171.

126. Nicholson DW. Caspases: killer proteases. *Trends Biochem Sci* 1997; 22:299–306.

127. Salvesen GS. Caspases: intracellular signaling by proteolysis. *Cell* 1997; 91:443–446.

128. Walker NP, Brady KD, Dang LC, Bump NJ, Ferenz CR, Franklin S, et al. Crystal structure of the cysteine protease interleukin-1 beta-converting enzyme: a (p20/p10)2 homodimer. *Cell* 1994; 78:343–352.

129. Wilson KP, Thomson JA, Kim EE, Griffith JP, Navia MA, Murcko MA, et al. Structure and mechanism of interleukin-1 beta converting enzyme. *Nature* 1994; 370:270–275.

130. Rotonda J, Fazil KM, Gallant M, Gareau Y, Labelle M, Peterson EP, et al. The three-dimensional structure of apopain/CPP32, a key mediator of apoptosis. *Nat Struct Biol* 1996; 3:619–625.

131. Talanian RV, Trautz S, Hackett MC, Mankovich JA, Banach D, Ghayur T, Brady KD, Wong WW. Substrate specificities of caspase family proteases. *J Biol Chem* 1997; 272:9677–9682.

132. Thornberry NA, Peterson EP, Rasper DM, Timkey T, Garcia-Calvo M, Houtzager VM, et al. A combinatorial approach defines specificities of members of the caspase family and granzyme B. Functional relationships established for key mediators of apoptosis. *J Biol Chem* 1997; 272:17,907–17,911.

133. Liu X, Slaughter C, Wang X. DFF, a heterodimeric protein that functions downstream of caspase-3 to trigger DNA fragmentation during apoptosis. *Cell* 1997; 89:175–184.

134. Enari M, Yokoyama H, Okawa K, Iwamatsu A, Nagata S. A caspase-activated DNase that degrades DNA during apoptosis, and its inhibitor ICAD. *Nature* 1998; 391:43–50.

135. Cheng EH, Clem RJ, Ravi R, Kastan MB, Bedi A, Ueno K, Hardwick JM. Conversion of Bcl-2 to a Bax-like death effector by caspases. *Science* 1997; 278:1966–1968.

136. Xue D. Caenorhabditis elegans CED-9 protein is a bifunctional cell-death inhibitor. *Nature* 1997; 390:305–308.

137. Takahashi A, Lazebnik YA, Fernandes-Alnemri T, Litwack G, Moir RD, Goldman RD, et al. Cleavage of lamin A by Mch2 alpha but not CPP32: multiple interleukin 1 beta-converting enzyme-related

proteases with distinct substrate recognition properties are active in apoptosis. *Proc Natl Acad Sci USA* 1996; 93:8395–8400.

138. Cryns VL, Zhu H, Li H, Yuan J. Specific cleavage of alpha-fodrin during Fas- and tumor necrosis factor-induced apoptosis is mediated by an interleukin-1beta-converting enzyme/Ced-3 protease distinct from the poly(ADP-ribose) polymerase protease. *J Biol Chem* 1996; 271:31277–31282.

139. Nath R, Stafford D, Hajimohammadreza I, Posner A, Allen H, Talanian RV, et al. Non-erythroid alpha-spectrin breakdown by calpain and interleukin 1 beta-converting-enzyme-like protease(s) in apoptotic cells: contributory roles of both protease families in neuronal apoptosis. *Biochem J* 1996; 319:683–690.

140. Mashima T, Noguchi K, Miller DK, Nicholson DW, Tsuruo T. Actin cleavage by CPP-32/apopain during the development of apoptosis. *Oncogene* 1997; 14:1007–1012.

141. Brockstedt E, Kostka S, Laubersheimer A, Dorken B, Wittmann-Liebold B, Bommert K, Otto A. Identification of apoptosis-associated proteins in a human Burkitt lymphoma cell line. Cleavage of heterogeneous nuclear ribonucleoprotein A1 by caspase 3. *J Biol Chem* 1998; 273:28,057–28,064.

142. Janicke RU, Sprengart ML, Porter AG. Caspase-3 is required for alpha-fodrin cleavage but dispensable for cleavage of other death substrates in apoptosis. *J Biol Chem* 1998; 273:15,540–15,545.

143. Hengartner M. Apoptosis. Death by crowd control. *Science* 1998; 281:1298–1299.

144. Saleh A, Acharya S, Fishel R, Alnemri ES. Cytochrome c and dATP-mediated oligomerization of Apaf-1 is a prerequisite for procaspase-9 activation. *J Biol Chem* 1999; 274:17,941–17,945.

145. Hu Y, Ding L, Nunez G. Role of cytochrome c and dATP/ATP hydrolysis in Apaf-1-mediated caspase-9 activation and apoptosis. *EMBO J* 1999; 18:3586–3595.

146. Mannick JB, Liu L, Hess DT, Zeng M, Miao QX, et al. Fas-induced caspase denitrosylation. *Science* 1999; 284:651–654.

147. Slee EA, Kluck RM, Wolf BB, Casiano CA, Newmeyer DD, Wang HG, et al. Ordering the cytochrome c-initiated caspase cascade: hierarchical activation of caspases-2, -3, -6, -7, -8, and -10 in a caspase-9-dependent manner. *J Cell Biol* 1999; 144:281–292.

148. Beresford PJ, Greenberg AH, Lieberman J. Granzyme A loading induces rapid cytolysis and a novel form of DNA damage independently of caspase activation. *Immunity* 1999; 10:585–594.

149. Xiang J, Korsmeyer SJ. BAX-induced cell death may not require interleukin 1 beta-converting enzyme-like proteases. *Proc Natl Acad Sci USA* 1996; 93:14,559–14,563.

150. McCarthy NJ, Gilbert CS, Evan GI. Inhibition of Ced-3/ICE-related proteases does not prevent cell death induced by oncogenes, DNA damage, or the Bcl-2 homologue Bak. *J Cell Biol* 1997; 136:215–227.

151. Gross A, Wei MC, Korsmeyer SJ. Enforced dimerization of BAX results in its translocation, mitochondrial dysfunction and apoptosis. *EMBO J* 1998; 17:3878–3885.

152. Fields S. Presence of a potent transcription activating sequence in the p53 protein. *Science* 1990; 249:1046–1049.

153. Unger T, Segal S, Minna JD. p53: a transdominant regulator of transcription whose function is ablated by mutations occurring in human cancer. *EMBO J* 1992; 11:1383–1390.

154. Walker KK. Identification of a novel p53 functional domain that is necessary for efficient growth suppression. *Proc Natl Acad Sci USA* 1996; 93:15,335–15,340.

155. el-Deiry WS, Pietenpol JA, Kinzler KW, Vogelstein B. Definition of a consensus binding site for p53. *Nat Genet* 1992; 1:45–49.

156. Funk WD, PD, Karas RH, Wright WE, Shay JW. A transcriptionally active DNA-binding site for human p53 protein complexes. *Mol Cell Biol* 1992; 12:2866–2871.

157. Bargonetti J, Chen X, Marshak DR, Prives C. A proteolytic fragment from the central region of p53 has marked sequence-specific DNA-binding activity when generated from wild-type but not from oncogenic mutant p53 protein. *Genes Dev* 1993; 7:2565–2574.

158. Halazonetis TD. Conformational shifts propagate from the oligomerization domain of p53 to its tetrameric DNA binding domain and restore DNA binding to select p53 mutants. *EMBO J* 1993; 12:5057–5064.

159. Pavletich NP, Pabo CO. The DNA-binding domain of p53 contains the four conserved regions and the major mutation hot spots. *Genes Dev* 1993; 7:2556–2564.

160. Ruppert JM. Analysis of a protein-binding domain of p53. *Mol Cell Biol* 1993; 13:3811–3820.

161. Wang Y, Wang P, Stenger JE, Mayr G, Anderson ME, Schwedes JF, Tegtmeyer P. p53 domains: identification and characterization of two autonomous DNA-binding regions. *Genes Dev* 1993; 7:2575–2586.

162. Iwabuchi K, Li B, Marraccino R, Fields S. Two cellular proteins that bind to wild-type but not mutant p53. *Proc Natl Acad Sci USA* 1994; 91:6098–6102.

163. Gorina S. Structure of the p53 tumor suppressor bound to the ankyrin and SH3 domains of 53BP2. *Science* 1996; 274:1001–1005.

164. Clore GM, Sakaguchi K, Zambrano N, Sakamoto H, Appella E, Gronenborn AM. High-resolution structure of the oligomerization domain of p53 by multidimensional NMR. *Science* 1994; 265:386–391.

165. Lee W, Yin Y, Yau P, Litchfield D, Arrowsmith CH. Solution structure of the tetrameric minimum transforming domain of p53. *Nat Struct Biol* 1994; 1:877–890.

166. Jeffrey PD, Pavletich NP. Crystal structure of the tetramerization domain of the p53 tumor suppressor at 1.7 angstroms. *Science* 1995; 267:1498–1502.

167. Scheffner M, Huibregtse JM, Vierstra RD, Howley PM. The HPV-16 E6 and E6-AP complex functions as a ubiquitin-protein ligase in the ubiquitination of p53. *Cell* 1993; 5:495–505.

168. Bardeesy N, Pelletier J. Clonal expansion and attenuated apoptosis in Wilms' tumors are associated with p53 gene mutations. *Cancer Res* 1995; 55:215–219.

169. McGill G, Fisher DE. p53 and apoptosis: a double edged sword. *J Clin Invest* 1999; 104:223–225.

170. Polyak K, Zweier JL, Kinzler KW, Vogelstein B. A model for p53-induced apoptosis. *Nature* 1997; 389:300–305.

171. Owen-Schaub LB, Cusack JC, Angelo LS, Santee SM, Fujiwara T, Roth JA, et al. Wild-type human p53 and a temperature-sensitive mutant induce Fas/APO-1 expression. *Mol Cell Biol* 1995; 15:3032–3040.

172. Tamura T, Saya H, Haga H, Futami S, Miyamoto M, Koh T, et al. Induction of Fas-mediated apoptosis in p53-transfected human colon carcinoma cells. *Oncogene* 1995; 11:1939–1946.

173. Muller M, Hug H, Heinemann EM, Walczak H, Hofmann WJ, Stremmel W, Krammer PH, Galle PR. Drug-induced apoptosis in hepatoma cells is mediated by the CD95 (APO-1/Fas) receptor/ligand system and involves activation of wild-type p53. *J Clin Invest* 1997; 99:403–413.

174. Reinke V. The p53 targets mdm2 and Fas are not required as mediators of apoptosis in vivo. *Oncogene* 1997; 15:1527–1534.

175. Wu GS, McDonald ER 3rd, Jiang W, Meng R, Krantz ID, Kao G, et al. KILLER/DR5 is a DNA damage-inducible p53-regulated death receptor gene. *Nat Genet* 1997; 17:141–143.

176. Buckbinder L, Velasco-Miguel S, Takenaka I, Faha B, Seizinger BR, Kley N. Induction of the growth inhibitor IGF-binding protein 3 by p53. *Nature* 1995; 377:646–649.

177. Israeli D, Haupt Y, Elkeles A, Wilder S, Amson R, Telerman A, Oren M. A novel p53-inducible gene, PAG608, encodes a nuclear zinc finger protein whose overexpression promotes apoptosis. *EMBO J* 1997; 16:4384–4392.

178. Roperch JP, Prieur S, Piouffre L, Israeli D, Tuynder M, Nemani M, et al. SIAH-1 promotes apoptosis and tumor suppression through a network involving the regulation of protein folding, unfolding, and trafficking: Identification of common effectors with p53 and p21 (Waf1). *Proc Natl Acad Sci USA* 1999; 96:8070–8073.

179. LaCasse EC, Korneluk RG, MacKenzie AE. The inhibitors of apoptosis (IAPs) and their emerging role in cancer. *Oncogene* 1998; 17:3247–3259.

180. Levy-Strumpf N. Death associated proteins (DAPs): from gene identification to the analysis of their apoptotic and tumor suppressive functions. *Oncogene* 1998; 17:3331–3340.

181. Crook NE, Miller LK. An apoptosis-inhibiting baculovirus gene with a zinc finger-like motif. *J Virol* 1993; 67:2168–2174.

182. Birnbaum MJ, Miller LK. An apoptosis-inhibiting gene from a nuclear polyhedrosis virus encoding a polypeptide with Cys/His sequence motifs. *J Virol* 1994; 68:2521–2528.

183. Freemont PS, Trowsdale J. A novel cysteine-rich sequence motif. *Cell* 1991; 64:483–484.

184. Barlow PN, Milner A, Elliott M, Everett R. Structure of the C3HC4 domain by 1H-nuclear magnetic resonance spectroscopy. A new structural class of zinc-finger. *J Mol Biol* 1994; 237:201–211.

185. Borden KL, FP. The RING finger domain: a recent example of a sequence-structure family. *Curr Opin Struct Biol* 1996; 6:395–401.

186. Hofmann K, Tschopp J. The CARD domain: a new apoptotic signalling motif. *Trends Biochem Sci* 1997; 22:155–156.

187. Deveraux QL, Salvesen GS, Reed JC. X-linked IAP is a direct inhibitor of cell-death proteases. *Nature* 1997; 388:300–304.

188. Roy N, Takahashi R, Salvesen GS, Reed JC. The c-IAP-1 and c-IAP-2 proteins are direct inhibitors of specific caspases. *EMBO J* 1997; 16:6914–6925.

189. Tamm I, Sausville E, Scudiero DA, Vigna N, Oltersdorf T, Reed JC. IAP-family protein survivin inhibits caspase activity and apoptosis induced by Fas (CD95), Bax, caspases, and anticancer drugs. *Cancer Res* 1998; 58:5315–5320.

190. Deveraux QL, Stennicke HR, Van Arsdale T, Zhou Q, Srinivasula SM, Alnemri ES, Salvesen GS, Reed JC. IAPs block apoptotic events induced by caspase-8 and cytochrome c by direct inhibition of distinct caspases. *EMBO J* 1998; 17:2215–2223.

191. Ambrosini G, Altieri DC. A novel anti-apoptosis gene, survivin, expressed in cancer and lymphoma. *Nat Med* 1997; 3:917–921.

192. Adida C, BD, Peuchmaur M, Reyes-Mugica M, Altieri DC. Anti-apoptosis gene, survivin, and prognosis of neuroblastoma. *Lancet* 1998; 351:882–883.

193. Lu CD, Tanigawa N. Expression of a novel antiapoptosis gene, survivin, correlated with tumor cell apoptosis and p53 accumulation in gastric carcinomas. *Cancer Res* 1998; 58:1808–1812.

194. Kawasaki H, Lu CD, Toyoda M, Tenjo T, Tanigawa N. Inhibition of apoptosis by survivin predicts shorter survival rates in colorectal cancer. *Cancer Res* 1998; 58:5071–5074.

195. Deiss LP. A genetic tool used to identify thioredoxin as a mediator of a growth inhibitory signal. *Science* 1991; 252:117–120.

196. Feinstein E, Wallach D, Boldin M, Varfolomeev E. The death domain: a module shared by proteins with diverse cellular functions. *Trends Biochem Sci* 1995; 20:342–344.

197. Cohen O, Kimchi A. DAP-kinase is a Ca2+/calmodulin-dependent, cytoskeletal-associated protein kinase, with cell death-inducing functions that depend on its catalytic activity. *EMBO J* 1997; 16:998–1008.

198. Katzenellenbogen RA, Herman JG. Hypermethylation of the DAP-kinase CpG island is a common alteration in B-cell malignancies. *Blood* 1999; 93:4347–4353.

199. Cohen O, Kissil JL, Raveh T, Berissi H, Spivak-Kroizaman T, Feinstein E, Kimchi A. DAP-kinase participates in TNF-alpha- and Fas-induced apoptosis and its function requires the death domain. *J Cell Biol* 1999; 146:141–148.

200. Kawai T, Hoshino K, Copeland NG, Gilbert DJ, Jenkins NA, Akira S. Death-associated protein kinase 2 is a new calcium/calmodulin-dependent protein kinase that signals apoptosis through its catalytic activity. *Oncogene* 1999; 18:3471–3480.

201. Hoffman B. The proto-oncogene c-myc and apoptosis. *Oncogene* 1998; 17:3351–3357.

202. Prendergast, G. Mechanisms of apoptosis by c-Myc. *Oncogene* 1999; 18:2967–2987.

203. Sakamuro D. New Myc-interacting proteins: a second Myc network emerges. *Oncogene* 1999; 18:2942–2954.

204. Landschulz WH, McKnight SL. The leucine zipper: a hypothetical structure common to a new class of DNA binding proteins. *Science* 1988; 240:1759–1764.

205. Murre C, Baltimore D. A new DNA binding and dimerization motif in immunoglobulin enhancer binding, daughterless, MyoD, and myc proteins. *Cell* 1989; 56:777–783.

206. Blackwell TK, Blackwood EM, Eisenman RN, Weintraub H. Sequence-specific DNA binding by the c-Myc protein. *Science* 1990; 250:1149–1151.

207. Fisher DE, Parent L, Sharp PA. High affinity DNA binding Myc analogs: recognition by an α helix. *Cell* 1993; 72:467–476.

208. Henriksson M. Proteins of the Myc network: essential regulators of cell growth and differentiation. *Adv Cancer Res* 1996; 68:109–182.

209. Stone J, Ramsay G, Jakobovits E, Bishop JM, Varmus H, Lee W. Definition of regions in human c-myc that are involved in transformation and nuclear localization. *Mol Cell Biol* 1987; 7:1697–1709.

210. Evan GI, Gilbert CS, Littlewood TD, Land H, Brooks M, Waters CM, Penn LZ, Hancock DC. Induction of apoptosis in fibroblasts by c-myc protein. *Cell* 1992; 69:119–128.

211. Li LH, Prendergast G, MacGregor D, Ziff EB. c-Myc represses transcription in vivo by a novel mechanism dependent on the initiator element and Myc box II. *EMBO J* 1994; 13:4070–4079.

212. Brough DE, Ellwood KB, Townley RA, Cole MD. An essential domain of the c-myc protein interacts with a nuclear factor that is also required for E1A-mediated transformation. *Mol Cell Biol* 1995; 15:1536–1544.

213. Blackwood EM, Eisenman RN. Max: a helix-loop-helix zipper protein that forms a sequence-specific DNA-binding complex with Myc. *Science* 1991; 251:1211–1217.

214. Mukherjee B, DePinho RA. Myc family oncoproteins function through a common pathway to transform normal cells in culture: cross-interference by Max and trans-acting dominant mutants. *Genes Dev* 1992; 6:1480–1492.

215. Amati B, Levy N, Littlewood TD, Evan GI, Land H. Oncogenic activity of the c-Myc protein requires dimerization with Max. *Cell* 1993; 72:233–245.

216. Amati B, Evan GI, Land H. The c-Myc protein induces cell cycle progression and apoptosis through dimerization with Max. *EMBO J* 1993; 12:5083–5087.

217. Shrivastava A, SS, Kalpana GV, Artandi S, Goff SP, Calame K. Inhibition of transcriptional regulator Yin-Yang-1 by association with c-Myc. *Science* 1993; 262:1889–1892.

218. Gaubatz S, Dosch R, Werner O, Mitchell P, Buettner R, Eilers M. Transcriptional activation by Myc is under negative control by the transcription factor AP-2. *EMBO J* 1995; 14:1508–1519.

219. Wang Q, Kajino K, Greene MI. BRCA1 binds c-Myc and inhibits its transcriptional and transforming activity in cells. *Oncogene* 1998; 17:1939–1948.

220. Roy AL, Gutjahr T, Roeder RG. Direct role for Myc in transcription initiation mediated by interactions with TFII-I. *Nature* 1993; 365:359–361.

221. Roy AL, Meisterernst M, Roeder RG. An alternative pathway for transcription initiation involving TFII-I. *Nature* 1993; 365:355–359.

222. Peukert K, Schneider A, Carmichael G, Hanel F, Eilers M. An alternative pathway for gene regulation by Myc. *EMBO J* 1997; 16:5672–5686.

223. Beijersbergen RL, Zhu L, Bernards R. Interaction of c-Myc with the pRb-related protein p107 results in inhibition of c-Myc-mediated transactivation. *EMBO J* 1994; 13:4080–4086.

224. Gu W, Magrath IT, Dang CV, Dalla-Favera R. Binding and suppression of the Myc transcriptional activation domain by p107. *Science* 1994; 264:251–254.

225. Sakamuro D, Wechsler-Reya R, Prendergast GC. BIN1 is a novel MYC-interacting protein with features of a tumour suppressor. *Nat Genet* 1996; 14:69–77.

226. Mori K, Kitaura H, Taira T, Iguchi-Ariga SM, Ariga H. MM-1, a novel c-Myc-associating protein that represses transcriptional activity of c-Myc. *J Biol Chem* 1998; 273:29,794–29,800.

227. Guo Q, Dang CV, Liu ET, Bishop JM. Identification of a large Myc-binding protein that contains RCC1-like repeats. *Proc Natl Acad Sci USA* 1998; 95:9172–9177.

228. McMahon SB, Dugan KA, Copeland TD, Cole MD. The novel ATM-related protein TRRAP is an essential cofactor for the c-Myc and E2F oncoproteins. *Cell* 1998; 94:363–374.

229. Taira T, Onishi T, Kitaura H, Yoshida S, Kato H, Ikeda M, Tamai K, Iguchi-Ariga SM, Ariga H. AMY-1, a novel C-MYC binding protein that stimulates transcription activity of C-MYC. *Genes Cells* 1998; 3:549–565.

230. Arsura M, Hann SR, Sonenshein GE. Variant Max protein, derived by alternative splicing, associates with c-Myc in vivo and inhibits transactivation. *Mol Cell Biol* 1995; 15:6702–6709.

231. Makela TP, Vastrik I, Alitalo K. Alternative forms of Max as enhancers or suppressors of Myc-ras cotransformation. *Science* 1992; 256:373–377.

232. Prochownik EV. Differential patterns of DNA binding by myc and max proteins. *Proc Natl Acad Sci USA* 1993; 90:960–964.

233. Vastrik I, Koskinen PJ, Alitalo K. Determination of sequences responsible for the differential regulation of Myc function by delta Max and Max. *Oncogene* 1995; 11:553–560.

234. Ayer DE, Eisenman RN. Mad-Max transcriptional repression is mediated by ternary complex formation with mammalian homologs of yeast repressor Sin3. *Cell* 1995; 80:767–776.

235. Harper SE, Sharp PA. Sin3 corepressor function in Myc-induced transcription and transformation. *Proc Natl Acad Sci USA* 1996; 93:8536–8540.

236. Shrivastava A, Artandi S, Calame K. YY1 and c-Myc associate in vivo in a manner that depends on c-Myc levels. *Proc Natl Acad Sci USA* 1996; 93:10638–10641.

237. Austen M, Luscher-Firzlaff JM, Luscher B. YY1 can inhibit c-Myc function through a mechanism requiring DNA binding of YY1 but neither its transactivation domain nor direct interaction with c-Myc. *Oncogene* 1998; 17:511–520.

238. Roy AL, Gregor PD, Novina CD, Martinez E, Roeder RG. Cloning of an inr- and E-box-binding protein, TFII-I, that interacts physically and functionally with USF1. *EMBO J* 1997; 16:7091–7104.

239. Hoang AT, LB, Lewis BC, Yano T, Chou TY, Barrett JF, Raffeld M, Hann SR, Dang CV. A link between increased transforming activity of lymphoma-derived MYC mutant alleles, their defective regulation by p107, and altered phosphorylation of the c-Myc transactivation domain. *Mol Cell Biol* 1995; 15:4031–4042.

240. Saleh A, Ting N, McMahon SB, Litchfield DW, Yates JR 3rd, Lees-Miller SP, Cole MD, Brandl CJ. Tralp is a component of the yeast Ada.Spt transcriptional regulatory complexes. *J Biol Chem* 1998; 273:26,559–26,565.

241. Galaktionov K, Beach D. Cdc25 cell-cycle phosphatase as a target of c-myc. *Nature* 1996; 382:511–517.
242. Diaz-Jullien C, Covelo G, Freire M. Prothymosin alpha binds histones in vitro and shows activity in nucleosome assembly assay. *Biochim Biophys Acta* 1996; 1296:219–227.
243. Shim H, CY, Lewis BC, Dang CV. A unique glucose-dependent apoptotic pathway induced by c-Myc. *Proc Natl Acad Sci USA* 1998; 95:1511–1516.
244. Grandori C, Siebelt F, Ayer DE, Eisenman RN. Myc-Max heterodimers activate a DEAD box gene and interact with multiple E box-related sites in vivo. *EMBO J* 1996; 15:4344–4357.
245. Bello-Fernandez C, Cleveland JL. The ornithine decarboxylase gene is a transcriptional target of c-Myc. *Proc Natl Acad Sci USA* 1993; 90:7804–7808.
246. Wagner AJ, Laimins LA, Hay N. c-Myc induces the expression and activity of ornithine decarboxylase. *Cell Growth Differ* 1993; 4:879–883.
247. Marhin WW, Facchini LM, Fornace AJ Jr, Penn LZ. Myc represses the growth arrest gene gadd45. *Oncogene* 1997; 14:2825–2834.
248. Askew DS, Simmons BC, Cleveland JL. Constitutive c-myc expression in an IL-3-dependent myeloid cell line suppresses cell cycle arrest and accelerates apoptosis. *Oncogene* 1991; 6:1915–1922.
249. Packham G. c-Myc and apoptosis. *Biochim Biophys Acta* 1995; 1242:11–28.
250. Zhan Y, Stevens JL. A role for c-myc in chemically induced renal-cell death. *Mol Cell Biol* 1997; 17:6755–6764.
251. Thompson E. The many roles of c-Myc in apoptosis. *Annu Rev Physiol* 1998; 60:575–600.
252. Hoffman-Liebermann B. Interleukin-6- and leukemia inhibitory factor-induced terminal differentiation of myeloid leukemia cells is blocked at an intermediate stage by constitutive c-myc. *Mol Cell Biol* 1991; 11:2375–2381.
253. Guillouf C, Selvakumaran M, De Luca A, Giordano A, Hoffman B, Liebermann DA. Dissection of the genetic programs of p53-mediated G1 growth arrest and apoptosis: blocking p53-induced apoptosis unmasks G1 arrest. *Blood* 1995; 85:2691–2698.
254. Selvakumaran M, Sjin RT, Reed JC, Liebermann DA, Hoffman B. The novel primary response gene MyD118 and the proto-oncogenes myb, myc, and bcl-2 modulate transforming growth factor beta 1-induced apoptosis of myeloid leukemia cells. *Mol Cell Biol* 1994; 14:2352–2360.
255. Juin P, Littlewood T, Evan G. c-myc-induced sensitization to apoptosis is mediated through cytochrome c release. *Genes Dev* 1999; 13:1367–1381.
256. Harrington EA, Evan GI. Oncogenes and cell death. *Curr Opin Genet Dev* 1994; 4:120–129.
257. Gibson AW, Johnston RN. Apoptosis induced by c-myc overexpression is dependent on growth conditions. *Exp Cell Res* 1995; 218:351–358.
258. Harrington EA, Fanidi A, Evan GI. c-Myc-induced apoptosis in fibroblasts is inhibited by specific cytokines. *EMBO J* 1994; 13:3286–3295.
259. Shichiri M, Marumo F, Hirata Y. Endothelin-1 is a potent survival factor for c-Myc-dependent apoptosis. *Mol Endocrinol* 1998; 12:172–180.

5 Drug Discovery in Oncology

Alex Matter, MD

CONTENTS

1. INTRODUCTION

Drug discovery in oncology has undergone profound changes over the past 20 yr; the rate of change has markedly accelerated over the last 5 yr, and it is therefore appropriate to take stock of these changes, and to ask what the next steps in this evolving landscape of concepts, skills, and technologies are likely to be. More than ever, drug discovery in oncology finds itself at the crossroads of academic research, industrial research and development (with a growing share by the biotech industry), clinical research, regulatory authorities, and public health, including major partners, such as the National Cancer Institute (NCI). All of these partners are driven, more than ever, by the forces related to productivity, i.e., a relentless drive for quality at manageable cost, within minimal time frames. These forces are behind the technological revolution that is still taking place, the drive to secure competitive patent positions, the drive to be faster on the market through streamlined R&D processes, the drive for a more efficient approval process and flexible handling of market access by

From: *Tumor Suppressor Genes in Human Cancer*
Edited by: D. E. Fisher © Humana Press Inc., Totowa, NJ

health authorities, and the importance of pharmacoeconomic aspects for the payors at large, even in disease states, such as advanced cancer, in which the medical need is undisputed.

This chapter concentrates mostly on technical aspects of drug discovery; keep in mind, however, that a almost invisible force field is exerting a powerful effect on all aspects of this complex endeavor. Considering the wide field of drug discovery in oncology, the chapter limits itself, for reasons of space, to the discussion of research and some early development activities, excluding late-stage development aspects, and therapeutic modalities directed against cancer itself, to the exclusion of chemopreventive regimens, as well as supportive therapy.

Drug discovery in oncology has undergone changes in two major and separate aspects:

- Development of novel technologies and processes: How does one approach a particular drug target? The technological revolution and the impact of novel technologies, processes, concepts, and rationales on the whole drug discovery process is briefly summarized in Subheading 3.
- Development of concepts and rationales: Which target should one tackle? Molecular epidemiology and a much deeper understanding of the pathophysiology of cancer at the molecular level are at the roots of new concepts and rationales. An evaluation of possible targets, taking these elements into account, is proposed in subheading 4.

Some examples of modern approaches in anticancer drug discovery are illustrated in subheading 5. Key parameters for the selection of one or another treatment modality are briefly explained in subheading 6. Subheading 7 deals with some of the issues related to the predictive quality of drug screens. Subheading 8 proposes some principles for the organization of efficient drug screens. Subheading 9–11 list key success factors for various stages of drug discovery, and Subheading 12 briefly deals with a few major elements of project management.

2. DEVELOPMENT OF NOVEL TECHNOLOGIES AND PROCESSES

Rapid technological change has occurred on many levels.

2.1. Chemistry (1–4)

One major development in synthetic chemistry over the last 10 yr is the emergence of combinatorial techniques to produce large number of compounds in a relatively short time frame. Combinatorial chemistry comprises many different approaches, requiring either deconvolution or allowing direct or indirect analysis, based on an encoding strategy of single compounds (one bead–one compound). Vigorous efforts are underway to make these libraries more informative, using more efficient synthetic strategies and more sophisticated analytic techniques.

Computer-assisted molecular modeling, structural chemistry, two-dimensional (2-D)- and three-dimensional (3-D)-quantitative structure activity relationship (2-D- and 3-D-QSAR) analysis have become commonplace tools of the trade of the modern synthetic chemist. Comparative molecular field analysis is a 3-D-QSAR technique in which the steric and electrostatic fields surrounding a set of molecules are sampled at discrete grid points, then correlated with biological activity. These techniques have acquired a high degree of predictive quality, i.e., in many cases, the predictions of the modeler are borne out by the hard-copy synthetic compound. Combining the above technologies, intelligent combinatorial libraries, based on multivariate design and multivariate (QSAR) analysis, are expected to provide high chemical diversity with a reasonable number of compounds *(3)*.

Another important development in chemistry is the fact that more and more complex structures have become feasible: complex chiral compounds, antisense chemistry (including, e.g., the phosphorothioate and ribonucleotide variations), polysaccharide synthesis, and the elaborate synthetic pathways of natural compounds, are some prominent examples *(77,84)*. Bioreactions (enzymatic reactions) are often introduced to complement classical synthetic chemistry (e.g., for enantioselective hydroxylation or the synthesis of other enantiomerically pure products).

2.2. Analytics (5–29)

The bewildering variety of established and emerging analytical technologies can be grouped conveniently into the following:

1. Nuclear magnetic resonance (NMR) technologies have developed into a panel of highly versatile techniques for the analysis of compounds in solution or attached to solid supports. Both structure and ligand–receptor interactions can be studied simultaneously, in automated fashion. A variety of refinements has evolved for various applications: magic angle spinning-NMR, nuclear Overhauser effect, ^{13}C/^{1}H-NMR, affinity NMR for the observation of ligand binding to a macromolecule (for review, *see* ref. *20*). These technologies are now more and more combined with high-throughput analytics needed in combinatorial chemistry, genomics, and proteomics. Bio-NMR (in tissues and animals; *see* subheading 2.12) is a further addition measuring e.g., ^{31}P-NMR spectra (markers of cellular energetics and metabolism).
2. Mass spectroscopy (MS) technologies: liquid chromatography coupled with MS or tandem-MS, matrix-assisted laser desorption ionization time-of-flight MS (MALDI-TOF MS) *(14)*, electrospray/MS/MS for the rapid and precise measurement of molecular mass of (macro)molecules; high-throughput MS using fully automated MS equipment *(11)*, affinity selection MS in combination with combinatorial chemistry (looking, e.g., at peptide substrate specificity of enzymes/receptors), capillary electrophoresis coupled with MS to study protein–protein and protein–ligand interactions.
3. Optical spectroscopy (OS) comprising technologies such as circular dichroism, infrared-OS, Raman, UV, and fluorescence; fluorescence polarization (FP) for the study of ligand–receptor interactions.
4. X-ray crystallography for structural studies (30–33), more and more frequently utilized also for the analysis of the interaction between drug candidates and drug targets at the atomic level.

This list is by no means exhaustive. In many instances, these technologies are applied in concert, in order to reap maximal information.

2.3. Assay Technologies

Measurement of ligand–receptor, enzyme–substrate, and antigen–antibody (Ab) interactions has been performed by classical methods, including radioimmunoassays, enzyme-linked immunosorbent assay, and many variations thereof. More recently, a number of novel assay technologies have been developed in order to cope with the need to reduce time and cost of assays, while preserving their robustness, simplicity, and sensitivity. New assay techniques that reduce the number of processing steps and the time intervals of incubation and analysis are needed, particularly to cope with the vast number of compounds produced by combinatorial chemistry.

Homogeneous assays (in which components are all in the same solution) and non separation assays (in which the components are bound to a solid phase) are increasingly introduced in high-throughput screens. A widely used example of the latter is the scintillation proximity assay (SPA) of Amersham. SPA is a convenient technology that uses scintillant-containing microspheres (fluoromicrospheres or SPA beads), which are chemically treated to enable the coupling of macromolecules. The assay uses radioisotopes with low-energy radiation (^{3}H-emitting β particles, ^{125}I-emitting Auger electrons with very short path lengths). If a molecule labeled with one of these isotopes is bound to the bead surface, it is in close proximity for the emitted radiation to produce a measurable light signal; the energy of isotopes that are not bound to the beads (attached to unbound ligand) is dissipated in the aqueous environment. This leads to a simple, homogeneous assay format, avoiding separation steps, and reducing manipulations to a minimum, which is particularly suitable for high throughput screening (HTS).

A major drive toward homogeneous assays, avoiding radioisotopes, has led to assays based on FP (*see* below), and to new assay formats such as homogeneous, time-resolved fluorescence resonance transfer (FRET) *(15)*. This latter assay format combines the advantages of radioisotope assays (sensitivity) with the advantages of fluorescence (safety, stability of labels). The principle of this technology entails the use of europium (Eu^{+3}), a rare lanthanide that is known to demonstrate a slow decay of its fluorescent signal. The trapping of Eu^{+3} in a macropolycyclic complex, such as the cryptates, allows formation of a highly efficient fluorescent complex ([Eu]K), which avoids fluorescence quenching by common assay components, such as oxygen, water, and proteins. (Eu)K represents therefore an efficient donor that can be used for labeling macromolecules. A further refinement of this assay format is the use of an acceptor molecule, such as a modified allophycocyanine (XL665), for the efficient energy transfer between donor and acceptor in the same assay. With the combined use of these two fluorescent labels, it is possible to analyze fluorescent signals without quenching, in the homogeneous assay format, and numerous applications have already been designed to make full use of this attractive technology. Numerous variations of this technology have been devised that cannot discussed here.

As mentioned, FP is used as an alternative in modern HTS, to measure an increase of polarization in labeled ligands, upon binding to relevant receptors. A number of applications have been described (reviewed in ref. *22*), which show the feasibility of this approach and its suitability for HTS.

2.4. Biosensors

Biosensors constitute a new and rapidly growing family of established and emerging technologies. Established technologies are based on surface plasmon resonance, which is valuable for measuring protein oligomerization and receptor–ligand-binding affinities. Novel applications, such as planar wave guide sensors and evanescent-field fluorimetry, are emerging as useful components in current drug screens. Future trends comprise, e.g., miniaturization of sensors, detection by Ab capturing of analytes, development of sensor chips, and so on. This field is rapidly broadening into a major technology bundle.

2.5. Vectors

Plasmids, and retroviral, adenoviral, and other vectors are now routinely used to transform cells.

2.6. Genomics/Pharmacogenomics

Description of the changes in gene expression following the administration of drug candidates or drugs in cellular and animal screens, as well as in clinical trials, has certainly become a major activity in drug discovery. Five different methods are commonly used: expressed sequence tag sequencing, subtractive cloning, differential display/representational difference analysis cDNA microarray hybridization, and serial analysis of gene expression (SAGE).

cDNA microarray hybridization technologies for the detection of differential gene expression are expected to become dominant technologies for the measurement of relative mRNA levels of multiple genes in a tissue or cell line. mRNA expression levels are determined after isolation of polyA$^+$ mRNA and synthesis of the cDNA label through hybridization to oligos on different chips (experimental, control). Quantitative fluorescence measurements of both chips are then compared, and the ratio on each spot yields the relative expression level. Expression levels of up to 6500 genes can be analyzed simultaneously, and it is to be expected that performance of these chips will increase dramatically. Affymetrix and Synteni appear to be in the lead at present in a field of about 20 companies directly involved in these technologies *(21)*.

An alternative approach, the SAGE technology *(6,26,62)* has also become a broadly applicable approach for the detection and comparison of thousands of transcripts, particularly low-abundance transcripts. It is estimated that a transcript expressed at an average of three copies per cell can be detected with 92% probability *(75)*. In this publication, studying approx 45,000 different genes of gastrointestinal tumors, transcripts of about 1% of the genes were expressed at significantly different levels in normal and neoplastic cells.

Methods for high-throughput target identification exploit modified yeast two-hybrid systems, which allow screening of any newly identified protein (bait) against a cDNA library (prey), in order to find the native binding partner. The prey cDNA library is expressed in a large panel of individually cloned yeast strains that are formatted into microtiter plates, and are used repeatedly against the different baits, similar to a compound collection that can be used in different HTS assays. For each new screen, the bait code is inserted next to a DNA-binding domain, and this yeast strain is then mated individually with the panel of prey clones. In case of an interaction between bait and prey, transcription is activated and a positive readout generated by the β-galactosidase reporter enzyme. This system has been refined and automated in some places to a high efficiency, allowing rapid and successful identification of novel bait proteins.

An alternative method for rapid detection of relevant drug targets in yeast *(Saccharomyces cerevisiae)* has been devised and called synthetic lethal screening *(9)*. This method screens for second site mutations that, by themselves, are not lethal, but, in combination with a primary defect (such as a defect in the DNA repair machinery), produce lethality. Such gene products that produce lethality in cells with primary genetic lesions would be attractive anticancer drug targets.

A highly sophisticated approach, aiming also at high-throughput target identification, has been developed in mammalian cells. This approach consists of a flow chart of technologies, starting with a combinatorial retroviral gene library to infect cells that are transfected with a green fluorescence protein *(GFP)*/BFP reporter gene, under the control of a promoter that is giving a readout related to a cellular function of interest. Peptide sequences with inhibitory function can then be isolated and tested for their function in secondary cellular screens. Inhibitory peptides can then be used as baits in a yeast two-hybrid screen to detect their intracellular binding partners. The feasibility of this approach has been documented *(10)*. This approach has also been used to identify inhibitors of enzymes, such as caspases, in living cells, in a combinatorial mode *(88)*.

2.7. Proteomics (12)

This is usually understood to represent high-resolution 2-D gel electrophoresis coupled with computerized image and data analysis, followed by MALDI-TOF MS or nanoelectrospray. Several gels of the same preparation can be run to average the spot images and improve the signal-to-noise ratio. Changes in abundance (≥ 2) or position of proteins in the 2-D electrophoresis pattern are then correlated with functional changes induced by altering experimental variables. A 2000 protein/gel resolution is achievable; for higher resolution, so-called "zoom gels", focused on narrow pH zones, are proposed. Estimates are that proteins with more than 400 copies/cell will then become detectable, a level of detection that far outstrips the currently achievable sensitivity of >10,000 copies/cell that are detectable.

2.8. (Engineered) Cell Lines and Transgenic Cell Lines

Cell-based assays have become an integral part of modern drug screens. Classically, these were used in a relatively simple manner to measure proliferation rates, growth, and cytotoxicity of drug candidates; in a more sophisticated way, biochemical markers have been studied, such as autophosphorylation patterns of surface receptors, enzymatic activity of members of signal transduction pathways, and expression of transcription factors by Northern, or Western blots.

More recently, with the introduction of reporter gene technologies, the output signal to any drug exposure has become more refined. In such systems, transcriptional activation or inhibition of a particular reporter gene can be measured following, e.g., ligand binding to a surface receptor or interference with a particular element of a signal transduction pathway leading to the reporter gene. These systems are amenable to high-throughput analysis and automation, and several readout techniques are now available *(22)*: β-galactosidase (bacterial), luciferase (from the firefly), alkaline phosphatase (human placenta), β-lactamase (also in living cells, by loading a substrate intracellularly as a membrane-permeant ester *(29)*, and (jellyfish).

Another invention concerns the introduction *of fusion genes, such as (GFP)*, fused to any protein of interest; the readout is green fluorescence, but different mutant proteins, with blue and yellow emission spectra, are now also available. This technology can also be used to study protein–protein interaction by FRET.

Transgenic cell lines are becoming more and more available, in parallel with the emergence of transgenic animals *(see* subheading 2.10.); gene targeting via homologous recombination (exploiting the Cre/loxP system) is now widely available, in order to produce knockout animals in which the gene of interest is missing *(43,44)*. These animals can then be used to generate cell lines containing the identical genetic defect. A further refinement allows replacement of the knocked-out gene by other genetic material, such as the human homolog of the gene under study. These animals and the derivative cell lines are gradually introduced into drug screens, with obvious benefit in terms of relevance of the model situation.

The logic of knockout animals/cell lines has been carried to the level of mRNA in generating cell lines expressing ribozymes mediating a knockdown of the mRNA of interest. Last, a functional, knockout can be achieved via intracellular single-chain Abs. (sFv), which are directed against the gene product of interest *(38)*.

2.9. Cell and Organ Culture Techniques

These methods have seen gradual refinements, with the introduction of matrices, feeder layers, defined tissue culture media, soft agar cell culture, and spheroid cultures.

2.10. Novel Animal Screens

Among many developments, a few are particularly striking and relevant for the drug discovery process in oncology:

1. Xenografts employing human tumor tissue or cell lines.
2. Orthotopic murine tumor models, which are particularly useful in the study of metastatic processes.
3. Transgenic mice with mutations in relevant oncogenes (such as *ras (37), myc (42), E2F1 (45),* or *erbB-2/neu (47),* or tumor suppressor genes *(p53) (49),* which have seen many important improvements with the introduction of the (inbred-strain-derived) embryonic stem cell technology, conditional knockouts and targeted transgenesis (e.g., introduction of dominant mutant tumor suppressor genes or viral inactivators of p53/Rb (SV40 LT) *(41).* For more information, the reader may wish to consult the database of the Jackson labs (Transgenic and Target Mutation Database, http://tbase.jax.org) *(46,48).*
4. Quantitative angiogenesis assays *(40).*

2.11. Microscopy

Cell sorting and confocal microscopy are now introduced in most drug discovery units. Classical histological techniques, autoradiography, and immunohistochemistry have been complemented more recently by new methods, such as *in situ, hybridization techniques* comprising comparative genomic hybridization and fluorescence *in situ* hybridization (FISH) *(25,28).* FISH has become a very widely applied technique. An interesting development concerns laser-assisted microdissection *(5),* allowing the molecular biology study of isolated cells in mixed tissue samples.

2.12. In Vivo Imaging

Two groups of technologies are particularly worth mention:

1. *Magnetic resonance spectroscopy and magnetic resonance imaging (MRI)* are now well established tools in the drug discovery process. Besides a merely anatomical, noninvasive high-resolution imaging technique, MRI also allows study of blood flow and tissue perfusion, using an intravascular contrast agent (Endorem); the use of extracellular contrast agents (GdDOTA, Dotarem) provides information on vascular permeability. Functional MRI (fMRI) is a further step whereby a control image is subtracted from the image after an experimental intervention, allowing one to map precisely in 3-D the changes that occurred after a stimulus (e.g., drug administration, electrical stimulation, and soon).

2. Fluorescence imaging of whole animals, using GFP technology (*see* subheading 2.8.), or using the luciferase reporter gene system coupled with advanced CCD cameras to detect gene expression with high sensitivity *(35,36)*, yields images of high resolution, and allows noninvasive imaging of whole animals over the time-course of an experiment. These techniques may offer great value in the future, and effectively compete with MRI technologies.

2.13. Tracking of High-affinity Drugs In Vivo

Prerequisites for application of such technologies are high potency of the drug under study, possibility of labeling the drug, and that the signal is not entirely absorbed by the tissue. Currently, only techniques of nuclear medicine have this type of sensitivity:

1. Planar γ camera: 2-D image.
2. Single photon emission computer tomography (SPECT): 3-D image.
3. Positron emission tomography (PET): requires a proton source (cyclotron), and produces 3-D images.

2.13.1. PET

Drug concentrations of about 10^{-12} M may be measured, i.e., <1 μg are usually sufficient for PET studies in humans (tracer dose). Frequently used radionuclides are carbon-11 (with a half-life of 20 min) and fluorine-18 (half-life 110 min). Biodistribution and pharmacokinetic studies are fields of application for this expensive and not user-friendly technique.

2.13.2. SPECT

Technetium-99m (half-life, 6 h) for perfusion studies, iodine-123 (half-life, 13 h), and iodine-131 (half-life, 8 d), are useful for drug labeling, and are among the radionuclides that can be used for SPECT. This is a technology that more easily allows labeling of drugs; it has, however, less spatial resolution (5 mm), and is less sensitive than PET.

2.14. Engineering

Automation, robotization, and miniaturization have been developed to a degree at which efficient handling of large sample numbers has become commonplace. We are now at a stage at which ultrahigh-throughput screening technologies (e.g., 9600-well format) in very small assay volumes (1–2 μL or less) seems feasible, and at which introduction of these technologies into routine screening activities may be achievable over the next few years.

2.15. Information Technology

High-volume data handling and the management of large databases, including networks, have become major challenges. Vigorous efforts are underway to develop efficient tools for data mining that are based on automated applications of a set of algorithms, to explore and model complex relationships between variables and objects. The challenge is to identify interesting compounds. Several tools have been developed that are able to analyze patterns in very large databases. In the opinion of experts, the identification of domain-specific, interesting patterns is still in its infancy, and a completely unsupervised automatic application of data mining tools without domain knowledge and without statistical background knowledge is not feasible at present.

Many data-mining tools attempt to extract frequent patterns from data; the identification of interesting rare patterns is not yet achieved.

Integration of these technologies and processes in an efficient manner, at reasonable cost, into a logical flow of drug screens, remains a major challenge, even for the most experienced drug discovery organizations.

3. SELECTION OF DRUG TARGETS (50–75)

Three major factors appear to play a crucial role in the increasingly refined process of target selection: relevance, validity, and feasibility. Usually, the selection of a novel drug target is triggered primarily by its discovery in biology, usually at an academic center. Only subsequently is its relevance and validity challenged by epidemiological and clinical considerations. Often, the excitement that is generated by the discovery of an apparently important new target molecule is a powerful motor to trigger investments; the slower and less exciting work to substantiate its relevance and validity is often neglected or postponed, many times at a substantial cost, and sometimes at the cost of failure.

Relevance of a drug target means primarily its relative frequency in various cancer types, realizing that high frequency is not necessarily linked with causality in terms of tumorigenesis or maintenance of the tumorigenic state/progression; careful distinction should also be made between familial and sporadic cancers, because relevance of a drug target in familial cancer (found by any positional cloning efforts) may not translate into relevance for sporadic cancer: one classic example is *BRCA1*. Determination of its relevance will allow determination of patient populations with respect to age, sex, race, geography, and the residual medical need after treatment with the best available therapy. These are minimal parameters that are most important for forecasting the market value of any prospective drug candidates, and are also the basis for an assessment of the risk/reward structure of a project.

Validity of a drug target relates to the fact that any target manipulation by drugs or other modalities will lead to a sustainable and desired change in the biology of a given cancer, and, ultimately, after appropriate development, to a favorable outcome for the cancer patient. One is dealing here with issues such as redundancy of pathways and resistance of cancer to manipulation of any kind, in the broadest sense. An example of such difficulties is, for instance, the finding that bcl-2 overexpression in biological terms is deemed to be a target for inhibition, since its well-documented antiapoptotic effects would appear to indicate an important role in tumor progression. In epidemiological terms, the contrary seems to be true (74), and bcl-2 overexpression in breast carcinomas correlates with a number of favorable prognostic factors, such as ER/PgR expression, low-grade histology, and other parameters. Another example is seen with farnesyl transferase inhibitors (FTIs), which inhibit Ras processing. K-Ras processing can proceed via farnesylation (inhibited by FTIs) or geranylgeranylation (which is not inhibited by FTIs). In A549 cells (90), for instance, K-Ras processing is resistant to both FTIs and inhibitors of geranylgeranylation. Yet, either type of compound is able to effectively inhibit tumor growth of A549 tumor cells in mice. It must then be speculated that alternative, unknown targets are hit by these compounds. These two examples illustrate the complexity of the biological systems, and explain to some extent why so many rational hypotheses are not substantiated in the context of a living animal, not to speak of heterogeneous patient populations.

An issue that is also important to consider is genetic instability of cancers, as highlighted in a recent review *(60)*. Clearly, any anticancer drug could become easily ineffective through resistance mechanisms that are driven by genetic instability. In practice, this may be particularly disturbing in fast-dividing tumors, such as leukemias; the more slowly dividing epithelial tumor types may be less quickly able to develop effective resistance mechanisms. Drug candidates that are directed against elements of the host (e.g., angiogenesis, immune mechanisms) are unlikely to be susceptible to resistance caused by genetic changes: One should bear in mind, however, that numerous resistance types (e.g., MDR-1, P450 enzyme induction) do not require genetic alterations to develop. The suggestion of Lengauer et al. to consider genetic instability as a field in which novel drug targets may be found, is very interesting, and the hypothesis that numerous, if not most, cytotoxic agents may be affective because of genetic instability of the cancer cells, merits further analysis.

The validation process comprises both the relevance and validity of a given drug target, and starts with the molecular epidemiology of a given drug target and its role in the pathophysiology of the type of cancer that is under study. In many instances, tool compounds, Abs antisense molecules, and viral vectors, and so on can be used to verify the validity of a drug target at the biochemical, cellular, and intact animal level, even in the absence of any good lead or reference compounds. This work should, at the very least, accompany the drug screen; preferably, it is done prior to large investments in structural biology, lead optimization, and laborious animal screens. However, in many instances, this is hardly possibly, and risk must be carried foreward until a test compound shows its merit in advanced clinical trials. Reducing risk by early validation of a drug target must certainly be one of the primary goals in any drug screening effort. Major consideration in this respect is the predictive quality of any drug screen that must be assessed, whenever possible, using clinically used reference compounds or systems that have shown in other situations a measure of predictive quality. It is fair to say that, in oncological drug discovery, predictive quality is generally low, definitely much lower than is usually thought (*see* below).

The third major factor in drug target selection, *its feasibility,* is dependent on many parameters that are beginning to be understood quite well. It is obvious from decades of pharmaceutical research that, in general, enzymes and receptors for small molecule ligands are almost ideal targets for drug discovery efforts. This is particularly facilitated when the molecular structure of the interaction site is well described at high resolution, when there are high-affinity natural ligands, and when there are assays available that are amenable for high-throughput screens. A particular problem is posed by drug targets that are characterized by protein–protein interactions, such as cytokine receptors and transcription factors. In a few instances, there are natural compounds (e.g., cyclosporin A, rapamycin) that pave the way for a drug discovery effort, and, recently, many examples have been pioneered of total syntheses of seemingly unapproachable molecules (e.g., taxol, epothilones, discodermolide, and so on). When such natural compounds do not exist, and when high throughput screens do not provide high-affinity leads, there is only the route of *de novo* design open to obtain useful lead compounds. It is then important whether a relatively small interaction site can be defined by molecular modeling (corresponding to a few, probably not more than eight, amino acids), whether there are highly charged residues in the vicinity of the binding site (such as the phosphorylated tyrosine recognized by SH2 regions), and whether the binding site is flat or comprises one, or possibly two grooves. In general, results of such *de novo* designs have not been spectacular, despite tremendous efforts (e.g., SH2 regions, STATs, ras-raf, ras-GAP, p53/Mdm2 [*see* Fig. 1]). It may well be that, with the advance of artificial intelligence, combined with high-speed, powerful computers, this situation may change rapidly. On the other hand, low chances of success of such screens have led some to speculate that such targets should then, and only then, be approached when strong, submicromolar leads emerge from high-throughput screening. Thus, detailed validation of targets would only be done later, when principal feasibility of the target has been established. This is a new combinatorial approach to target selection.

4. CANCER DRUG TARGETS WITH ESTABLISHED RELEVANCE

Drug targets can be usefully classified in two ways: first, into targets that are uniquely linked to the genotype of cancer (more or less stable mutations, such as the Philadelphia chromosome in [CML]) that occur in all, or at least in the vast majority of, cancer cells of a given cancer type; a range of oncogenes and tumor suppressor genes have been defined governing key biological parameters, such as the regulation of the cell cycle, apoptosis, differentiation; the genomic instability of cancer cells (driven, e.g., by myc, cyclin D or ras activation, loss of *p53* or ATM function, and so on) is one of the most recent fields in which genetic alterations can now be identified with sufficient precision to build new concepts for drug discovery. Second, drug targets may be defined by host–tumor relationships (hormone-responsive cancers, angiogenesis, invasive properties, osteolysis). This type of classification will have immediate bearing on the level of molecular definition and nature of the drug target, feasibility criteria (*see* Table 2), likelihood of preexistent resistance and/or resistance induction, and size of patient population.

In Table 1, the epidemiology of some of the currently most actively investigated cancer drug targets is summarized, including some examples of drug candidates and their approximate stage of development.

At the earliest possible time-point, a putative target profile for a drug candidate shoud be established. Such a profile would indicate cancer type(s)/patient population, dosage and route of application, treatment schedule and treatment duration, possible partner drugs for combination regimens, and expected patient benefit. It is clear that at the outset of any research project these elements are not available with any degree of precision. A gradual refinement of thinking regarding these parameters is, however, important, and a steady stream of crosschecks with experts from other disciplines, particularly clinical and regulatory, is mandatory, to ensure buy-in from organizations and outside opinion leaders.

5. SELECTION OF TREATMENT MODALITIES

In Table 2, some major advantages and disadvantages of current treatment modalities are compared.

6. PREDICTIVE QUALITY OF DRUG DISCOVERY SCREENS

The following discusses the predictive quality of preclinical screens in the broad sense of clinical outcome, i.e., how predictive, e.g., a particular receptor binding test is regarding clinical outcome in patient populations, when administering the same receptor agonist.

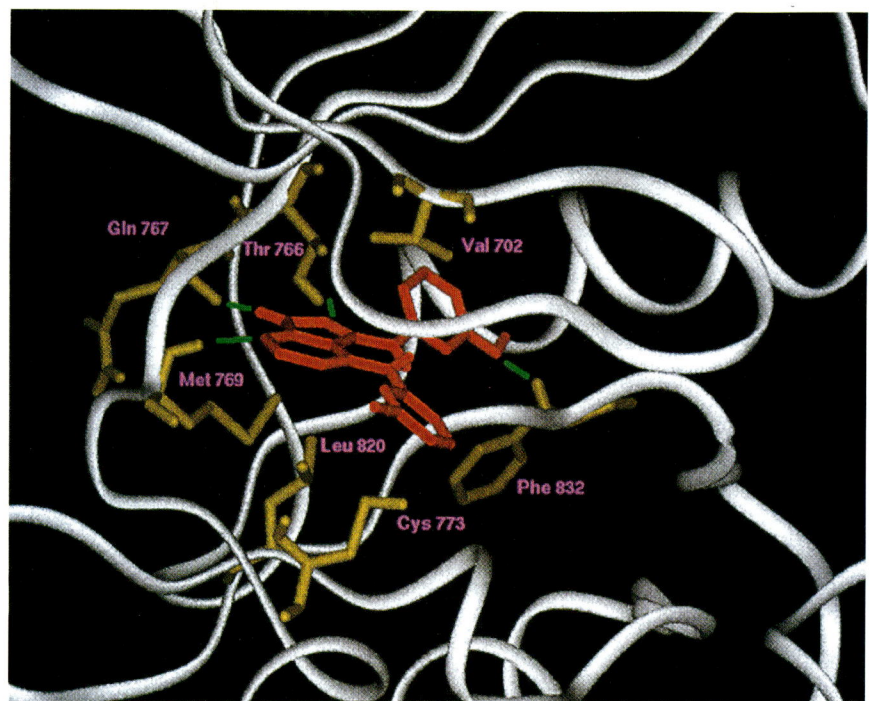

Fig. 1. Tyrosine kinase inhibitor (CGP 59326) modeled into the adenosine, triphosphate-binding pocket of the catalytic region of eadermal growth factor (EGF) receptor tyrosine kinase.

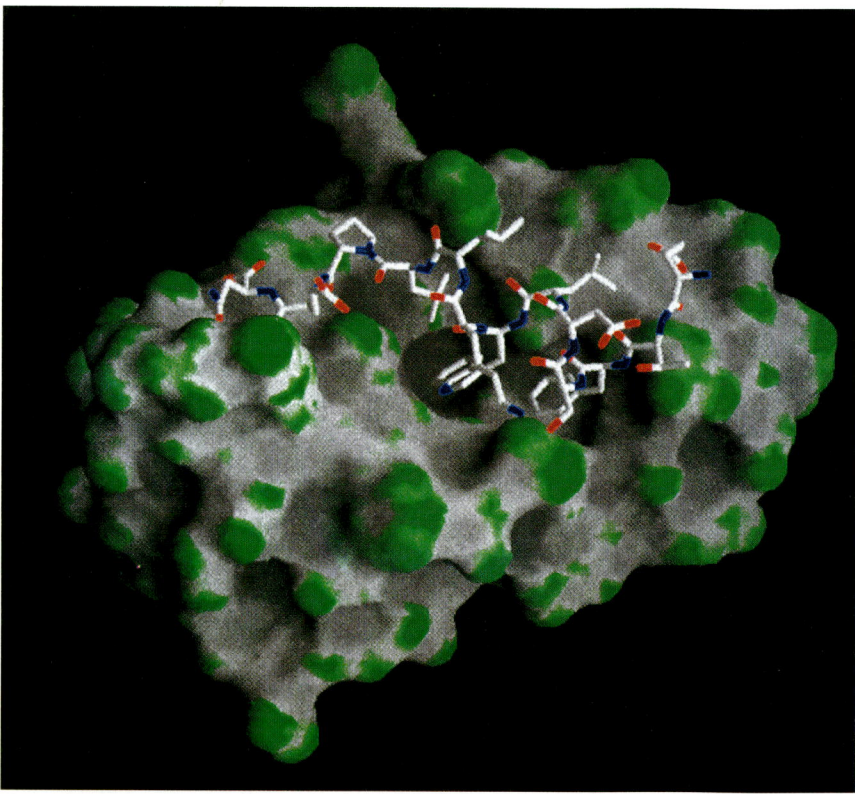

Fig. 2. Deep hydrophobic cleft of Mdm2 in the p53-binding region *(31)*.

Table 1

Target	Epidemiological findings in human cancer	Examples for drugs/drug candidates	Current state of development
Farnesyl transferase (activated Ras)	Mutations of *ras* found in 90% of pancreatic, 50% of colon, and 30% of lung cancer	L-744,832; L-745,631 (Merck), B956 (Shionogi), SCH56580; SCH58450 (Schering-Plough), and many others (78,89)	Phase I/II and preclinical
EGF-receptor/erbB-2 in tumor disease	Overexpression of EGF-R Constitutive activation of mutant receptor in gliomas, breast, lung	CP-358,774 (Pfizer), ZD1839 (Zeneca), SU 5271/PD 153,035 (Sugen), PD 165,557 and PD 158,780 (Parke-Davis), PKI166 (Novartis), many other compounds (79,91,92)	Preclinical–phase I–II
Bcr-abl mutations	100% in CML, 25% of AML	STI571 (Novartis), FCE 28436 (Pharmacia-Up ohn), AG1112 (Yssum)	Preclinical–phase I
(PDGF)-driven tumorigenesis/angiogenesis	Brain, ovarian, lung, prostate	SU 101 (Sugen)	Phase I/II
Role of protein kinase C (PKC)- and c-raf in cancer		ISIS 3521 (antisense against PKCα), ISIS 5132 (antisense against c-raf) (both ISIS-Novartis), PKC412 (Novartis)	Phase II
Cyclin-dependent kinases in cell cycle events		Flavopiridol (HMR) Staurosporine, UCN-01 (NCI-Kyowa Hakko), Butyrolactone-I, Olomoucine, Roscovitine (81)	Phase II Preclinical
Metalloproteases in processes such as invasion and metastasis (94)	Metastatic process, invasion of tumor cells into adjacent tissue	Marimastat (BB 2516, British Biotech), AG 3340 (Agouron), Bay 12-9566 (Bayer) MMI270 (Novartis)	Phase III Phase I Phase I Phase I
Angiogenesis	Tumor angiogenesis in most solid tumors	Fumagillin and TNP-470 (Takeda), ovalicin, αvβ₃ ligand, angiostatin and endostatin (EntreMed)	Preclinical–phase I–II
Vascular endothelial growth factor (VEGF)-dependent angiogenesis	Tumor angiogenesis in most solid tumors	ZD6474 (Zeneca), SU 5416 (Sugen) PTK787 (Novartis-Schering AG)	Phase I trials ongoing

(continues)

Table 1 (*Continued*)

Target	Epidemiological findings in human cancer	Examples for drugs/drug candidates	Current state of development
PDGF-R/VEGF-R-dependent angiogenesis	Tumor angiogenesis in most solid tumors	PD 166,285 (Parke-Davis), SU 6668 (Sugen) (79,91,92)	Preclinical–phase I
Steroid receptors	Acute promyelocytic leukemia liposarcoma, other cancers?	All-*trans*-retinoic acid, Vesanoid (Roche) Troglitazone (PPARγ agonist, Parke-Davis), GW 1929 (Glaxo-Wellcome)	On the market On the market/preclinical Preclinical
Transcriptional regulation of endocrine pathways (estrogen-R, androgen-R)	Hormone-insensitive prostate and breast cancer	Distamycin-related compounds, minor groove-binding compounds, hairpin polyamide design (86)	Preclinical
Apoptotic pathways	1) p53/hdm2 antagonists 2) Bcl-2/Bcl-xL antagonists 3) Histone deacetylase inhibitors 4) Proteolytic activity of the proteasome	(1) No compounds published (2) G 3139 (antisense, Genta); epothilones (indirectly) (3) Trapoxin A, Trichostatin A (4) Lactacystin A, PS 341 (Proscript/Millenium)	(1) Preclinical (2) Phase I/II (3) Preclinical (4) Phase I
P-glycoprotein-mediated multidrug resistance (MDR-1 and many others) (48)	Pre-existant oder drug treatment-induced	PSC833 (Novartis), VX-710 and VX-853 (Vertex), GF 120918 (Glaxo-Wellcome)	Phase I–II–III
Telomerase	Activated in up to 90% of all cancers	Antisense molecules, ribozymes (93), reverse transcriptase inhibitors (e.g., ddGTP, 7-deaza-dGTP, anthraquinones, prophyrin derivatives, and so on) (87)	Preclinical

Information has been adapted, with permissions from refs. 76–94.

Table 2

Modality	Advantages	Disadvantages
Small molecules	Feasibility, ease of manufacturing, purity, stability, patentability	Cost of manufacturing sometimes very high, often insufficient selectivity resulting in nonmechanism-related side effects, risk of mutagenicity/teratogenicity, risk of metabolism-related lack of efficacy or delayed organ toxicity, risk of lack of bioavailability, difficulties in formulation
Endocrine agentss	As above, plus the fact that usually endocrine pathways are well worked-out, and hypotheses are easily testable; in general, endocrine surrogate markers available; high selectivitys	As above, but side effect profile mostly mechanism-related
Natural compounds (77,80)	Feasibility largely dominated by the availability of expression systems and/or natural sources that are easily accessible.	Total synthesis in most cases highly or extremely difficult (sometimes impossible) and very costly; otherwise, as above (small molecules)
Antisense	Exquisite target specificity, at least in theory no nonmechanism-related toxicity; clearcut surrogate markers for biochemical drug effects (mRNA, protein levels)	Cost of goods, difficulties in manufacturing; risk of nonmechanism-related side effects (complement activation, coagulation pathway, proinflammatory effects); parenteral application route, lack of clinically validated models of systemic efficacy of antisense molecules
Peptides	Relative ease of manufacturing; purity, stability as powder, ease of formulation, instability in the bloodstream can be circumvented by unnatural amino acids/cyclization, in general high selectivity, lack of mutagenicity/teratogenicity	Parenteral drugs; can be immunogenic/allergenic; instability in the blood stream needs to be dealt with; in general, **do not cross cell membranes nor the blood–brain barrier;** limited applicability
Cytokines	Production mostly relies on recombinant organisms; Exquisite selectivity at the level of cytokine receptors; however, pleiotropic effects in most cases according to biology of the cytokine in question	Parenteral drugs; in some cases, very serious toxicity (**e.g., vascular leakage syndrome**), metabolic instability in the bloodstream; patent situation often complex
Antibodies	Humanized antibodies now generally available; high selectivity through antigen-binding sites; lack of mutagenicity/teratogenicity, good pharmacokinetic profiles in the case of humanized antibodies; Purity/stability readily achievable	Manufacturing relatively complex, mostly in eukaryotic cells; cost of goods relatively high (on a per gram basis); patent situation often complex

(continues)

Table 2 (*Continued*)

Modality	Advantages	Disadvantages
Radioligands	Technology established; imaging technology available using alternative isotopes	Clinical efficacy still unproven; complex procedure; marked organ toxicity possible, whole body exposure to high irradiation levels
Vaccines	More and more tumor antigens characterized; relevant epitopes becoming available; highly efficient antigen presentation modes and adjuvants available; immunological characterization of immune response types well in hand	Clinical efficacy still unproven; lack of predictive animal models; mostly effects in preclinical models based on a prophylactic regimen; clinical trial plans Based most often on adjuvant setting, resulting in large trials with high risk of failure
Targeted toxins by ligands, antibodies, or viruses	In theory, absence of side effects because of targeting of toxic principles; manufacturing issues addressable by fusion gene approaches; in the case of viruses, topical/local approaches possible	So far, this approach is clinically not validated; sometimes surprising amount of toxicity caused by nonspecific trapping of complexes
Photodynamic therapy	In principle, exquisite selectivity attainable through selective accumulation of the photosensitizer in the tumor and the geometry of excitation at the tumor level; availability of a number of photosensitizers with favorable pharmacokinetic profile and adequate accumulation in the tumor; availability of relatively cheap and user-friendly solid state lasers; clearly demonstrated patient benefit, mostly in the palliative setting	Treatment limited to superficially positioned tumors in hollow organs (brain, necessitating craniotomy) and also limited, at least with currently available sensitizers and lasers, in penetration to a few millimeters. Treatment procedure surgical in nature. Phototoxicity with first generation photosensitizers. Treatment limited to (endoscopically) visible tumors
Gene therapy modalities	At the outset, based on the principle of a once-and-for all correction of genetic disorders, most attractive in cancer research because of the genetic character of the disease. In principle, feasibility established based on efficient integration of new genes by retroviruses and some other viruses	Clinically, this approach is not validated, despite extensive efforts in many top laboratories and clinical institutions. The major hurdles are: low rate of transduction, low rate of integration of desired genetic information, rapid clearance of newly introduced genetic material, immunogenicity of the vector, and manufacturing issues. Nevertheless, efforts continue at a feverish rate

It is a fair assumption that the predictive quality of any test or assay increases with the level of complexity of the screen; thus, biochemical screens, although highly accurate and reproducible, are far removed from any relevant predictive value; complex screens in the intact (tumor-bearing) animal, although less accurate and more expensive, appear to mimic much more closely the clinical situation. To a great extent, however, the notion of highly predictive animal screens has been discredited, as witnessed by several decades of experimental research. This is true for pharmacokinetic, toxicological, and therapeutic efficacy parameters. The pitfalls of predicting pharmacokinetic and toxicological parameters in man, based on animal results, are well documented, and are not be discussed further.

Therapeutic efficacy screens merit close attention because of their experimental complexity, cost, and the pain inflicted on animals. Any cancer researcher in the field will know of instances in which animal screens have not been able to predict clinical outcome. Apart from the fact that negative experiences are always more disturbing and more easily retained in one's memory, there is a consensus that predicting efficacy of a novel anticancer agent for any given cancer type, in the absence of clinically validated predecessor compounds, cannot be attained with any degree of accuracy. The lack of tissue-type predictability of the efficacy of novel anticancer agents has been amply documented by the NCI, exploiting results obtained over several decades of research. It would be wrong, however, to ascribe no predictive quality whatsoever of animal screens for the clinical setting. There are many examples of screens that are almost perfectly predictive for clinical results. Such examples comprise, for instance, cytotoxic agents, such as tubulin stabilizers, antimetabolites, or alkylating agents, tested in xenograft models; endocrine agents, such as estrogen-lowering agents (aromatase inhibitors) tested in the carcinogen (DMBA)-induced rat mammary tumor; and photodynamic therapy, tested in subcutaneous murine tumors.

There are also situations in which negative or unconvincing test results should have prompted a negative decision regarding further development, and in which, only with the benefit of hindsight, the predictive quality of efficacy screens (with negative results) was finally borne out by clinical results. Such examples comprise, for instance, immunostimulants tested in murine metastatic tumor models. There is always a tendency, also, among experienced cancer researchers, to believe more easily positive results, and to disbelieve negative results, when these contradict a personal bias.

Robust assessment of the predictive quality is only possible when solid clinical treatment results become available, usually only after pivotal phase IIb/phase III trials. There are, however, warning signs that should be taken seriously in alerting preclinical researchers and clinicians to possible failure of a drug candidate. Such signs comprise a (very) narrow therapeutic window (<2) in rodent screens; small treatment effects, such as a decrease in tumor growth rate (in the absence of complete tumor stasis or regression); lack of reproducibility of treatment results; marked activity in special tumor models only; and use of demanding routes and schedules of application.

7. BUILDING A FLOW CHART FOR SCREENING ACTIVITIES

Historically, anticancer drug screening was a hit-or-miss activity, testing compounds more or less at random in some murine leukemia models. Generally speaking, this approach led to only marginal successes, and was replaced or complemented by much broader panels of tumor cell lines. With the advent of concepts based on molecular genetics, it became possible to build target-oriented drug screens that were much more efficient in coping with a sizable number of compounds. This trend has continued to this day, and it is now commonplace to test 10^5 to 5×10^5 compounds in early lead-finding screens, in a matter of a several weeks. Bottlenecks remain in complex animal screens, and every effort is therefore made to weed out inactive compounds in less demanding screens. The principle applied is a hierarchical succession of drug discovery screens, going stepwise from a biochemical level to a cellular level, and from there to animal screens. Each screen is then adapted to the planned throughput, and quantitative decision criteria govern the flow of compounds. The parameters that minimally need to be dealt with are:

- Potency
- Selectivity
- Cellular activity/selectivity (for intracellular targets)
- Pharmacokinetic parameters (C_{max}, $t_{1/2}$, area under the curve, oral bioavailability)
- Efficacy
- Toxicity
- Spectrum of activity
- Scheduling
- Combinations with partner compounds
- Resistance (preexistent resistant phenotype, P-glycoprotein induction, bcl-2 overexpression, tubulin isotypes/mutations, gene amplification).

In a very simplified way, for chemistry-driven projects, such flow charts appear in Flow Chart 1.

For projects based on biologicals, such as cytokines, monoclonal antibodies, or gene therapy approaches, all high-throughput screens would be replaced by careful biological characterization of the biological entity in relevant cellular and animal systems. Usually, these projects are characterized by an early emphasis on safety and manufacturing issues, careful characterization of producer cell lines, and so on.

As mentioned above, the validation process of the drug target accompanies the entire process; early validation, even in the absence of any good leads or reference compounds, is sought by all means; a valid drug candidate is later used to characterize therapeutic efficacy at tolerated doses in animals. The next major hurdle is the PoC studies in man, and final validation is obtained, at least in the majority of the cases, at the time of submission.

The underlying principle throughout this process is early attrition, i.e. rejection of nonvalid drug targets and treatment modalities at the earliest possible time. In order to do this, the following points must be watched carefully at all times:

- Quality of screens (accuracy, reproducibility, predictive quality).
- Clearcut decision criteria with clearcut decision-making if success becomes improbable or the competitive situation becomes untenable.
- Efficiency of screening, as can and should be measured by parameters such as throughput, turnaround time, and cost.
- Creative and rapid responses to obstacles that arise during screening.
- Integrated approach to drug screening, applying state-of-the-art technologies at the right time.

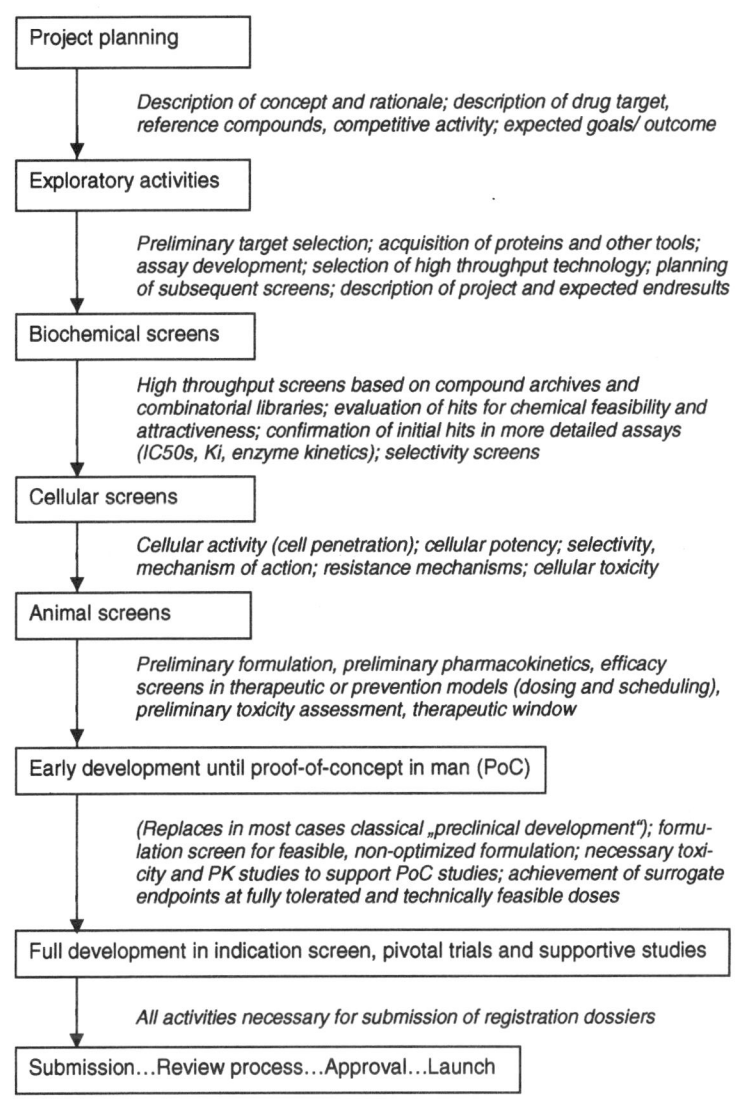

Flow Chart 1 Activities/Measurements

- Adequate planning and preparation of downstream activities, in order to minimize interface problems.

A description of a more detailed flow chart, as used in actual drug screening efforts (again, in the context of a chemistry-driven project) is depicted as Flow Chart 2 (comprising only the research activities in a more narrow sense).

This drug candidate would then be ready for entry into early development activities until PoC in man. One activity, the development of surrogate marker technology, is missing on Flow Chart 2. This is an activity that is pursued at almost every level of the drug screen, and in which an imaginative approach is particularly important, both for early attrition rates and for the success of the PoC studies in man.

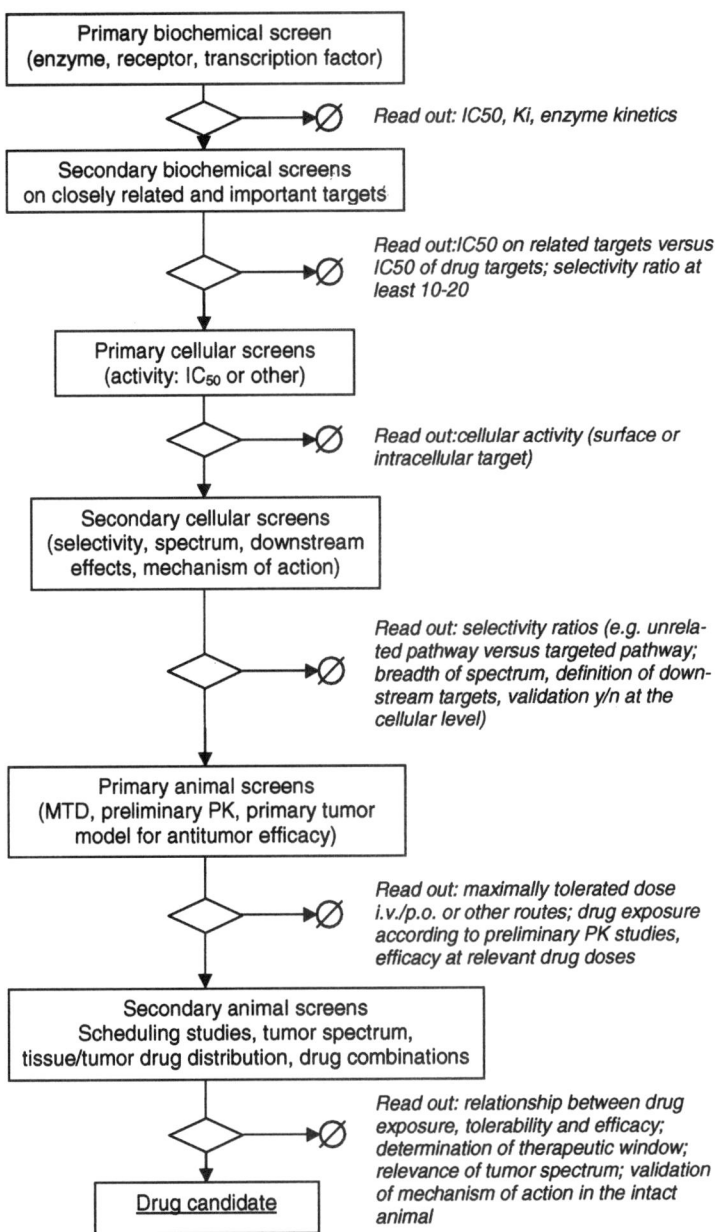

Flow Chart 2 Preclinical Research Activities

8. KEY SUCCESS FACTORS FOR EARLY-STAGE DRUG DISCOVERY SCREENS

High-quality execution and effective bundling of the following technologies, putting aside the overall feasibility of a given approach, greatly determine the success rate of early research projects. A few key attributes should be mentioned.

- High-diversity compound archives, including natural compound libraries: size and diversity of the libraries.
- Cloning/expression/purification of recombinant proteins: immediate availability of a variety of expression systems, state-of-the-art protein chemistry.
- Efficient, high-throughput screens/ultrahigh-throughput screens: assay development by experts, in stages, to the point at which the assay can cope, at reasonable cost, with the desired throughput. Approximate benchmark: 300,000–500,000 single compounds should be screened, and hits confirmed in single-point assays within 2–3 mo; top performance is around 1 mo.
- Structural biology: a series of engineered expression vectors should be rapidly cloned, expressed, and purified proteins checked for propensity to crystallize, in a batch mode, in order to heighten chances of crystal formation; crystallography data are particularly (only?) useful at the time of lead optimization.
- Modeling and medicinal chemistry, including combinatorial chemistry: Although combinatorial chemistry is many times not productive in lead finding, it is most valuable in lead optimization. Creative ways of directed multiparallel synthesis greatly enhance the efficiency of the drug optimization process, provided that screens can cope with the number of compounds, and are sufficiently accurate and able to discriminate between very similar compounds.

9. KEY SUCCESS FACTORS FOR LATE-STAGE COMPOUND DEVELOPMENT

Pharmacological characterization of compounds is only possible if the following technologies are put in place: preliminary formulation, also for difficult-to-handle compounds; preliminary pharmacokinetics, measurements of drug exposure relating to drug action in target tissues; range and predictive quality of available in vivo models.

10. IMPORTANCE OF SURROGATE MARKERS

The availability of surrogate markers with high predictive quality can change entirely the risk profile of a development project, and therefore has eminent practical significance. Very determined efforts should be made to make these available, at least in a prototype assay format, at the time of the first clinical trials.

Surrogate markers can be divided into five categories:

- Mechanistic markers reflecting the mechanism of action of the candidate drug, e.g., EGF receptor downregulation/decrease in autophosphorylation after administration of an EGF receptor tyrosine kinase inhibitor, or decrease of relevant mRNA species in target tissue after administration of antisense compounds.
- Downstream biochemical/endocrine markers of drug action, e.g., decrease of estrogen levels in the bloodstream after administration of aromatase inhibitors, or induction of S-adenosyl methionine decarboxylase (SAMDC) in the target tissue after administration of a SAMDC-inhibitor.
- Noninvasive methods (*see* subheading 2.12.) measuring anatomical parameters (e.g., tumor size) or functional markers, such as tumor tissue blood flow or vascular permeability of the tumor blood vessels using PMRI techniques.
- Tumor markers for early signs of efficacy; examples include CA125 in ovarian cancer or Prostate specific antigen in prostate cancer.

- Toxicity markers identifying target organs and signaling tolerability problems; the most frequently occurring examples are depression of white blood cells as a consequence of bone marrow depression, liver (ALAT, ASAT) and kidney (creatinine clearance) biochemical markers, neurological markers (gait, spasms, convulsions, and so on).

11. TEAM MANAGEMENT IN DRUG DISCOVERY PROCESS

One of the most challenging issues is the management of a multidisciplinary process that is in itself dynamic, i.e., in which the partners are constantly changing, and the nature of the required expertise and the approaches to problem-solving are distinct for each phase of R&D. The challenges occur at different levels. There are conceptual difficulties, particularly for highly innovative projects with a speculative aspect; there are the technological hurdles, almost always including informatics as a key component; personal as well as organizational aspects; and last, the financial implications. In many cases, a geographic component complicates interactions between team members (remote teams).

Given the complexity of the tasks, it is evident that only teams of experts are able to cope with these effectively. These teams must be able to speak a common language, despite the heterogeneity of expertise and the frequently occurring geographical hurdle. This means that teams in different parts of the world need to interact efficiently; in many cases, this also means that language barriers must be overcome, which, for many technical people, can be high. The coordination of these activities is so demanding that most organizations have chosen to entrust professional project managers with the coordination of all the development activities. It is also evident that only organizations with up-to-date IT environments can remain competitive.

One major hurdle is the fact that such teams are subject to two contradictory forces: on the one hand, these teams require a certain stability over time, and team members cannot be changed too frequently without incurring delays; on the other hand, the type of work that is required is constantly changing over the normal research and development time. In these cases, it is highly profitable to minimize interfaces by establishing a sort of fluidity of the team, which means that at least the core team is never changing in its entirety, but gradually evolves over time. Such a process can be graphically depicted as in Fig. 3.

12. CONCLUSIONS

Drug discovery has become an exceedingly complex endeavor. The eruption of revolutionary technologies, mostly linked in some way with miniaturization, robotization, and major IT components, is posing difficult problems in terms of adequate choice of technologies for the target at hand, cost, and overall management of resources. The choice of the drug target in terms of relevance, validity, and feasibility remains tricky, despite the availability of molecular epidemiological data for most important tumor types. Nevertheless, problems can now be tackled that were unassailable just a few years ago. The speed of HTS, the complex chemical syntheses, and the sophisticated molecular screens, coupled with large databases, are examples of tremendous technical progress. On the other hand, it is worthwhile to remember that many problems are still unyielding, even to the most dedicated groups of scientists. Crystallography has many times hit high hurdles, as have efforts to produce high-affinity leads for protein–protein interaction drug targets. These and many other hurdles still exist, but the pressure

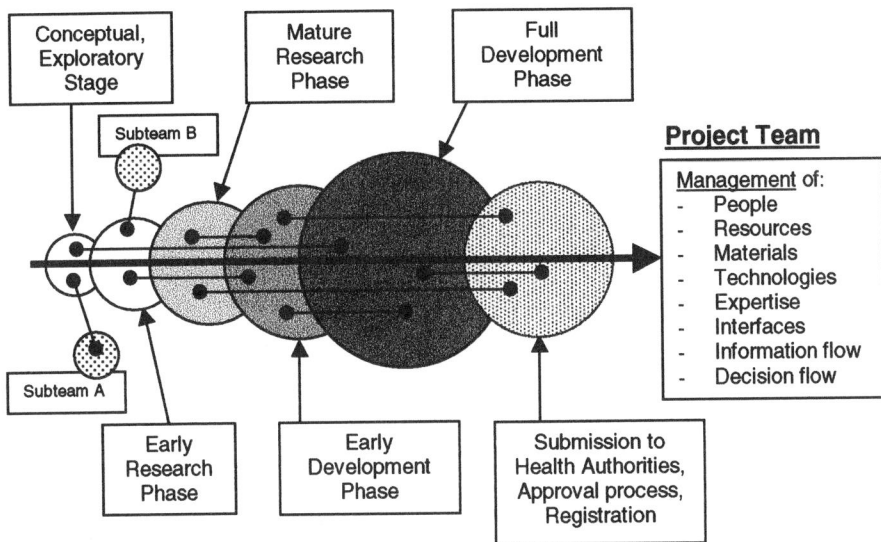

Fig. 3. Project team organization over time. The overlapping circles represent the size of the project team, with mostly changing team members, but some fairly constant ones (such as, e.g., a research or a clinical pharmacology representative, symbolized by a line with a bullet at either end); subteams with specialized missions (e.g., HTS, crystallography, fermentation, and so on) complement the main team. Most bigger teams, particularly in development, work as a collection of subteams.

toward increased productivity steadily increases, and with it the urgency to find technical solutions for ever-more-difficult problems.

ACKNOWLEDGMENTS

The author am greatly indebted to colleagues at Novartis for the countless discussions regarding aspects of this chapter; the author also indebted to Dr. P. Traxler and Dr. C. Garcia-Echeverria for donating the two color slides.

REFERENCES

1. Connolly Martin Y, 3D QSAR. Current state, scope, and limitations, *Perspect Drug Discovery Des* 1998; 12/13/14:3–23.
2. Lam KS. Application of combinatorial library methods in cancer research and drug discovery. *Anti-Cancer Drug Des* 1997; 12:145–167.
3. Van de Waterbeernd H, et al., Lundstedt T, et al. Intelligent combinatorial libraries. In: Comput-Assisted Lead. Optim., [Eur. Symp. Quant. Struct.-Act. Relat.], 11th, Basel, Switzerland: Verlag Chimica Acta. 1997; 191–208.
4. Tominaga Y. Novel 3D Descriptors using excluded volume 2: application to drug classification. *J Chem Inf Comput Sci* 1998; 38:1157–1160.
5. Bernson MR, et al. Identification of multiple mRNA and DNA sequences from small tissue samples isolated by laser-assisted microdissection. *Lab Invest* 1998; 78:1267–1273
6. Carulli JP, et al. High throuput analysis of differential gene expressionk. *J Cell Biochem* 1998; Suppl. 30/31, 286–296
7. Cook ND. Scintillation proximity assay: a versatile high-throughput screening technology. *Drug Discovery Today* 1996; 1:287–294
8. Gosnell PA, et al. Compound library management in high throughput screening. *J Biomol Screening* 1997; 2:99–102

9. Hartwell LH, et al. Integrating genetic approaches into the discovery of anticancer drugs. *Science* 1997; 278:1064–1068

10. Hitoshi Y, et al. Toso, a cell surface, specific regulator of Fas-induced apoptosis in T cells. *Immunity* 1988; 8:461–471

11. Hsieh F, et al. Automated high throughput multiple target screening of molecular libraries by microfluidic MALDI-TOF MS. *J Biomol Screening* 1998; 3:189–198

12. Humphrey-Smith I, et al. Proteome research: complementarity and limitations with respect to the RNA and DNA worlds. *Electrophoresis* 1997; 18:1217–1242.

13. Jayawickreme CK, et al. Gene expression systems in the development of high-throughput screens. *Curr Opin Biotechnol* 1997; 8:629–634

14. Jensen ON, et al. Automation of matrix-assisted laser desorption/ionization mass spectrometry using fuzzy logic feedback control. *Anal Chem* 1997; 69:1706–1714

15. Kolb AJ, et al. Homogeneous, time-resolved fluorescence method for drug discovery. *High Throughput Screening* 1997; 1:345–360

16. Lutz MW, et al. Experimental design for high-throughput screening. *Drug Discovery Today* 1996; 1:277–286

17. Oldenburg KR. Development of an ultra-high throughput screening system: plate design, liquid handling, and image analysis. *Proc SPIE-Int Soc Opt Eng* 1988; 3259:197–208

18. Oldenburg KR, et al. Assay miniaturization for ultra-high throughput screening of combinatorial and discrete compound libraries: a 9600-well (0.2 microliter) assay system. *J Biomol Screening* 1998; 3:55–62

19. Picardo M, et al. Scintillation proximity assays. In: Devlin JP, ed. *High Throughput Screening* New York: Dekker. 1997; 307–316.

20. Shapiro MS, et al. NMR methods in combinatorial chemistry. *Curr Opin Chem Biol* 1998; 2:372–375

21. Schena M, et al. Microarrays: biotechnology's discovery platform for functional genomics. *Trends Biotechnol* 1998; 16:301–306

22. Silverman L, et al. New assay technologies for high-throughput screening. *Curr Opin Chem Biol* 1998; 2:397–403

23. Sittampalam GS, et al. High-throughput screening: advances in assay technologies. *Curr Opin Chem Biol* 1997; 1:384–391

24. Sterrer S, et al. Fluorescence correlation spectroscopy (FCS) – a highly sensitive method to analyze drug/target interactions. *J Recept Signal Transduction Res* 1997; 17:511–520

25. Van de Corput MPC, et al. Sensitive mRNA detection by fluorescence in situ hybridization using horseradish peroxidase-labeled oligodeoxynucleotides and tyramide signal amplification. *J Histochem Cytochem* 1998; 46:1249–1259

26. Velculescu VE, et al. Serial analysis of gene expression. *Science* 1995; 270:484–487

27. Welford SM, et al. Detection of differentially expressed genes in primary tumor tissues using representational differences analysis coupled to microarray hybridization. *Nucleic Acids Res* 1998; 26:3059–3065

28. Wilkens L, et al. Analysis of hematologic diseases using conventional karyotyping, fluorescence in situ hybridization (FISH) and comparative genomic hybridization (CGH). *Hum Pathol* 1998; 29:833–839

29. Zlokarnik G, et al. Quantitation of transcription and clonal selection of single living cells with β-lactamase as reporter. *Science* 1988; 279:84–88

30. Hof P. Crystal structure of the tyrosine phosphatase SHP-2. *Cell* (Cambridge, MA) 1998; 92:441–450.

31. Kussie PH, et al. Structure of the MDM2 oncoprotein bound to the p53 tumor suppressor transactivation domain. *Science* (Washington, DC) 1996; 274:948–953

32. Russo AA, et al. Structural basis for inhibition of the cyclin-dependent kinase Cdk6 by the tumor suppressor p16INK4a. *Nature* (London) 1998; 395:237–243

33. Xu W, et al. Three-dimensional structure of the tyrosine kinase c-Src. *Nature* (London) 1997; 385:595–602

34. Blasco MA, et al. Mouse models for the study of telomerase. *CIBA Found Symp* 1997; 211(Telomeres and Telomerase), 160–176

35. Contag CH, et al. Visualizing gene expression in living mammals using a bioluminescent reporter. *Photochem Photobiol* 1997; 66:523–531

36. Contag PR, et al. Bioluminescent indicators in living mammals. *Nature Med* 1998; 4:245–247

37. Greenhalgh DA, et al. Multistage epidermal carcinogenesis in transgenic mice: cooperativity and paradox. *J Invest Dermatol Symp Proc* 1996; 1:162–176

38. Grim J, et al. erbB-2 knockout employing an intracellular single-chain antibody (sFv) accomplishes specific toxicity in erbB-2-expressing lung cancer cells. *Am J Respir Cell Mol Biol* 1996; 15:348–354

39. Hsu CX, et al. Longitudinal cohort analysis of lethal prostate cancer progression in transgenic mice. *J Urol* 1998; 160:1500–1505

40. Jain RK, et al. Quantitative angiogenesis assays: progress and problems. *Nature Med* 1997; 3:1203–1208

41. Kasper S, et al. Development, progression, and androgen-dependence of prostate tumors in probasin-Large T antigen transgenic mice: a model for prostate cancer. *Lab Invest* 1998; 78:319–333

42. McCormack SJ. Myc/p53 interactions in transgenic mouse mammary development, tumorigenesis and chromosomal instability. *Oncogene* 1998; 16:2755–2766

43. Moreadith RW, et al. Gene targeting in embryonic stem cells. The new physiology and metabolism. *J Mol Med* (Berlin) 1997; 75:208–216

44. Moore RC, et al. Gene targeting. In: Rosenberg RN, ed. The molecular and genetic basis of neurological disease. *Mol Genet Basis Neurol Dis* 2nd ed. Boston, MA: Butterworth-Heinemann. 1997; 33–48

45. Pierce AM, et al. Increased E2F1 activity induces skin tumors in mice heterozygous and nullizygous for p53. *Proc Natl Acad Sci USA* 1998; 95:8858–8863

46. Polites HG. Transgenic model applications to drug discovery. *Int J Exp Path* 1996; 77:257–262

47. Rowse GJ, et al. Genetic modulation of neu proto-oncogene-induced mammary tumorigenesis. *Cancer Res* 1998; 58:2675–2679

48. Van de Vrie W, et al. In vivo model systems in P-glycoprotein-mediated multidrug resistance. *Crit Rev Clin Lab Sci* 1998; 35:1–57

49. Viney JL. Transgenic and gene knockout mice in cancer research. *Cancer Metastasis Rev* 1995; 14:77–90.

50. Bardelli A, et al. Invasive-growth signaling by the Met/HGF receptor. The hereditary renal carcinoma connection. *Biochim Biophys Acta* 1997; 1333:M41–M51

51. Brandt-Rauf PW, et al. The c-erbB-2 protein in oncogenesis: molecular structure to molecular epidemiology. *Crit Rev Oncol* 1994; 5:313–329

52. Bertelson AH, et al. High-throughput gene expression analysis using SAGE. *Drug Discovery Today* 1998; 3:152–159

53. Goldman R, et al. Molecular epidemiology of breast cancer. *In Vivo* 1998; 12:43–48

54. Guha A, et al. Proliferation of human malignant astrocytomas is dependent on ras activation. *Oncogene* 1997; 15:2755–2765

55. Harada N, et al. Molecular and epidemiological analyses of abnormal expression of aromatase in breast cancer. *Pharmacogenetics* 1995; 5:S59–S64

56. Hussain SP, et al. Molecular epidemiology of human cancer. *Recent Results Cancer Res* 1998; 154:22–36

57. Izzotti A, et al. Molecular epidemiology in cancer research. Int. *J Oncol* 1997; 11:1053–1069

58. Lalani E, et al. Molecular and cellular biology of prostate cancer. *Cancer Metast Rev* 1997; 16:29–66

59. Latil A, et al. Genetic aspects of prostate cancer. *Virchows Arch* 1998; 432:389–406

60. Lengauer C, et al. Genetic instabilities in human cancers. *Nature* (London) 1988; 396:643–649

61. Lopez-Otin C, et al. Breast and prostate cancer: an analysis of common epidemiological, genetic, and biochemical features. *Endocr Rev* 1998; 19:365–396

62. Madden SL, et al. SAGE transcript profiles for p53-dependent growth regulation. *Oncogene* 1997; 15:1079–1085

63. Moskaluk CA, et al. Molecular genetics of pancreatic carcinoma. In: Reber HA ed., *Pancreatic Cancer* Totowa, NJ: Humana. 1998; 3–20

64. Rao RN. Targets for cancer therapy in the cell cycle pathway. *Curr Opin Oncol* 1996; 8:516–524

65. Sekido Y, et al. Progress in understanding the molecular pathogenesis of human lung cancer. *Biochim Biophys Acta* 1998; 1378:F21–F59

66. Spivack SD, et al. The molecular epidemiology of lung cancer. *Crit Rev Toxicol* 1997; 27:319–365

67. Sweeney KJ, et al. Lack of relationship between CDK activity and G_1 cyclin expression in breast cancer cells. *Oncogene* 1998; 16:2865–2878

68. Takahashi C, et al. Detection of telomerase activity in prostate cancer by needle biopsy. *Eur Urol* 1997; 32:494–498.

69. Urquidi V, et al. Telomerase in cancer: clinical applications. *Ann Med* (Helsinki) 1998; 30:419–430

70. Van de Woude GF, et al. Met-HGF/SF: tumorigenesis, invasion and metastasis. *Ciba Found Symp* (Plasminogen-Related Growth Factors) 1997; 212:119–132.

71. Weinstein IB. Contributions of molecular biology to cancer epidemiology. *Ann NY Acad Sci* 1995; 768:30–40

72. Welch DR, et al. Genetic and epigenetic regulation of human breast cancer progression and metastasis. *Endocr-Relat Cancer* 1998; 5:155–197

73. Qureshi KN, et al. Molecular biological changes in bladder cancer. *Cancer Surv* (Bladder Cancer) 1998; 31:77–97.

74. Zhang G, et al. Role of bcl-2 expression in breast carcinomas. *Oncol Rep* 1998; 5:1211–1216

75. Zhang L, et al. Gene expression profiles in normal and cancer cells. *Science* 1997; 276:1268–1272

76. Bisoffi M, et al. Inhibition of human telomerase by a retrovirus expressing telomeric antisense RNA. *Eur J Cancer* 1998; 34:1242–1249

77. Cragg GM, et al. Natural products in drug discovery and development. *J Nat Prod* 1997; 60:52–60

78. Gibbs JB, et al. Farnesyltransferase inhibitors versus Ras inhibitors. *Curr Opin Chem Biol* 1997; 1:197–203

79. Klohs WD, et al. Inhibitors of tyrosine kinase. *Curr Opin Oncol* 1997; 9:562–568

80. Long BH, et al. Eleutherobin, a novel cytotoxic agent that induces tubulin polymerization is similar to paclitaxel (Taxol®). *Cancer Res* 1998; 58:1111–1115

81. Meijer L. Chemical inhibitors of cyclin-dependent kinases. *Trends Cell Biol* 1996; 6:393–397

82. Moasser MM, et al. Farnesyl transferase inhibitors cause enhanced mitotic sensitivity to taxol and epothilones. *Proc Natl Acad Sci USA* 1998; 95:1369–1374

83. Moyer JD, et al. Induction of apoptosis and cell cycle arrest by CP-358,774, an inhibitor of epidermal growth factor receptor tyrosine kinase. *Cancer Res* 1997; 57:4838–4848

84. Nicolaou KC, et al. Chemical biology of epothilones. *Angew Chem Int Ed* 1998; 37:2014–2045

85. Njoroge FG, et al. Structure-activity relationship of 3-substituted *N*-(pyridinylacetyl)-4-(8-chloro-5,6-dihydro-11 *H*-benzo[5,6]cycloheptal[1,2-*b*] pyridin-11-ylidene)-piperidine inhibitors of farnesyl-protein transferase: design and synthesis of in vivo active compounds. *J Med Chem* 1997; 40:4290–4301

86. Parks ME, et al. Optimization of the hairpin polyamide design for recognition of the minor groove of DNA. *J Am Chem Soc* 1996; 118:6147–6152

87. Perry PJ, et al. Telomeres and telomerase: targets for cancer chemotherapy? *Expert Opin Ther Pat* 1998; 8:1567–1586

88. Rothenberg SM, et al. Intracellular combinatorial chemistry with peptides in selection of caspase-like inhibitors. *NATO ASI Ser* Ser. H (Gene therapy) 1998; 105:171–183.

89. Sugita K, et al. Inhibitors of Ras-transformation. *Curr Pharm Des* 1997; 3:323–334

90. Sun J, et al. Both farnesyltransferase and geranylgeranyltransferase I inhibitors are required for inhibition of oncogenic K-Ras prenylation but each alone is sufficient to suppress human tumor growth in nude mouse xenografts. *Oncogene* 1998; 16:1467–1473

91. Traxler P. Tyrosine kinase inhibitors in cancer treatment (part II). *Expert Opin Ther Pat* 1998; 8:1599–1625

92. Traxler P, et al. Design and synthesis of novel tyrosine kinase inhibitors using a pharmacophore model of the ATP-binding site of the EGF-R. *J Pharm Belg* 1997; 52:88–96

93. Yokoyama Y, et al. Attenuation of telomerase activity by a hammerhead ribozyme targeting the template region of telomerase RNA in endometrial carcinoma cells. *Cancer Res* 1998; 58:5406–5410

94. Yu AE, et al. Matrix metalloproteinases. Novel targets for directed cancer therapy. *Drugs Aging* 1997; 11:229–244

II TUMOR SUPPRESSOR PATHWAYS

6 Retinoblastoma Protein in Growth Control and Differentiation

Lilia Stepanova, PhD and J. Wade Harper, PhD

1. INTRODUCTION

Proper development of multicellular organisms from a single cell requires precise control of cellular proliferation, differentiation, and apoptosis. This multistep control involves a complex interplay between numerous genes regulating spatial and temporal aspects of development and proliferation. Defects in these control mechanisms can lead to developmental defects or tumor formation. Research performed over the past decade has brought us closer to an understanding of how a family of critical growth regulators, typified by the retinoblastoma (RB) tumor suppressor gene *(Rb),* function to control cell proliferation and differentiation.

From: *Tumor Suppressor Genes in Human Cancer*
Edited by: D. E. Fisher © Humana Press Inc., Totowa, NJ

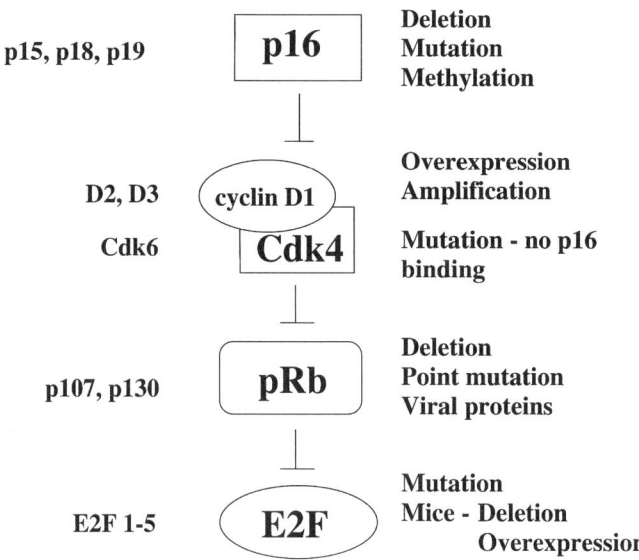

Fig. 1. The *Rb* signaling pathway. Protein components of the Rb pathway are shown alongside their homologs on the left. The types of mutations found in human tumors are indicated on the right. In mice, both deletion and overexpression of E2F-1 were shown to be oncogenic. In humans, frequent mutations of E2F-4 were found in gastrointestinal tumors *(230,231)*.

The *Rb* gene was cloned in 1988 as a gene that, when mutant, predisposes individuals to childhood RB *(1,2)*. Subsequent studies revealed that *Rb* is frequently mutated in various types of human cancers, including small cell lung carcinomas; osteosarcomas; sarcomas; breast, bladder, and prostate carcinomas; myelomas, and leukemias *(1–11)*. Initial insight into the function of Rb protein came with the observation that three unrelated DNA tumor viruses had evolved a similar strategy to transform cells by binding a set of cellular proteins, including Rb *(12–15)*. The central role Rb plays in the transformation by these viruses became apparent after the observation that mutations in these viral proteins, rendering them unable to bind Rb, eliminate their transforming ability *(15–20)*. Since then, significant progress has been made in understanding Rb function, its upstream regulators, and downstream targets. Through analysis of *Rb*-deficient mice, as well as mice deficient in the *Rb* homologs *p107* and *p130,* the complex role played by the *Rb* family in coordinating proliferative control with the processes of differentiation and apoptosis during development is beginning to be understood. Moreover, it is now known that *Rb* is a component of a G1 signaling pathway (Fig. 1), and mutations in a number of components of this pathway have been linked to transformation. It is clear that Rb functions primarily through its ability to bind and modulate the actions of various transcription factors (TF). Negative control of transcription by *Rb* may take several forms, including passive masking of transactivation domains *(23),* active inhibition of surrounding enhancer elements *(24–29),* or interference with transcription complex assembly *(30)*. Interactions with Rb are not always inhibitory, however, and a number of TFs are activated by Rb, although the mechanisms and biological significance of such interactions remain obscure. This chapter discusses recent

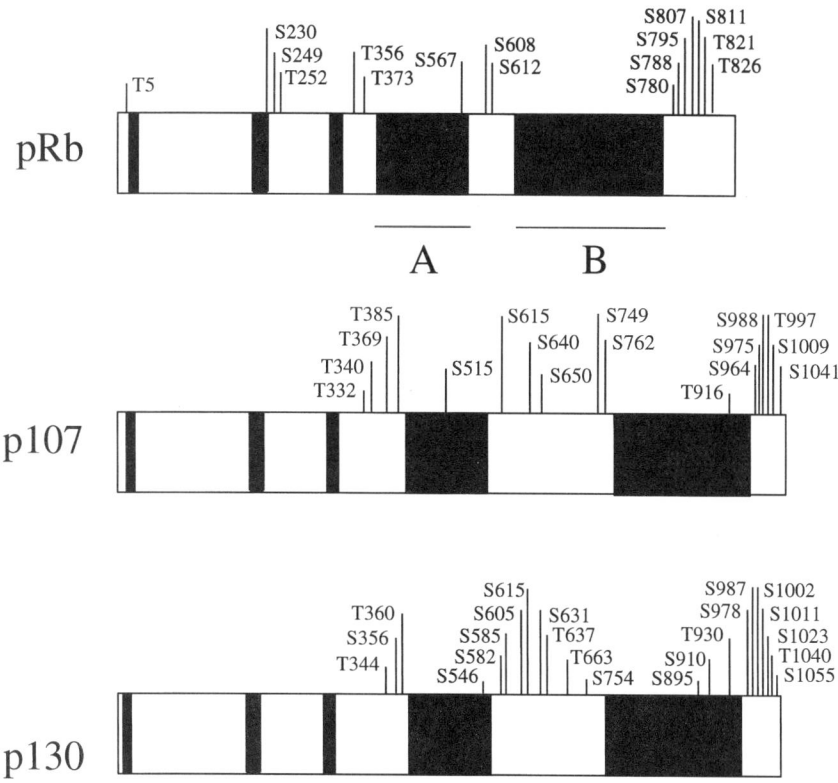

Fig. 2. Structural domains in pocket proteins. Conserved regions are shown as black boxes. Highly conserved A and B subdomains of A/B pocket domain are underlined. Consensus CDK phosphorylation sites are shown for each protein. Single letters refer to serine (S) and threonine (T) residues; numbers refer to the position of the amino acid in the protein.

advances in understanding how Rb controls cell cycle progression, differentiation, development, and apoptosis.

2. The Pocket Protein Family

In mammals, the *Rb* family (Fig. 2) contains three members, *Rb, p107,* and *p130,* with *p107* and *p130* being more closely related, based on sequence *(31–32).* These proteins are frequently referred to as "pocket proteins," because they all share a structural element called the A/B pocket, which serves as a protein–protein interaction domain. A number of proteins have been identified that can associated with the A/B pocket of all three family members, including DNA tumor virus oncoproteins, such as E1A and T-antigen (Table 1). The interaction with the A/B pocket is typically mediated by an LXCXE motif *(74)* located in the target protein. The structure of the LXCXE peptide from human papilloma virus E7 with the *Rb* pocket has recently been solved *(75),* which provides a model for how cellular proteins may interact with the A/B pocket. In addition, all three family members bind to members of the E2F family of TFs through sequences overlapping the A/B pocket and the C-terminus *(33,76–82).* The C pocket (located at the C-terminus of *Rb*) contains sequences that

Table 1
Viral and Cellular *Rb*-associated Proteins

Protein	Function	Ref
E2F	Transcription factor	*(23,26,33)*
SV40 large T-antigen	Simian virsus protein	*(12,16,17)*
E1A	Adenoviral protein	*(13,16,18)*
HPV E7	Human papilloma virus protein	*(14)*
BRLF1	Epstein-Barr virus protein	*(34)*
D1-3	Cyclins	*(35)*
MyoD	Muscle transcription factor	*(36)*
Myogenin	Muscle transcription factor	*(36)*
Mdm2	Oncoprotein	*(37)*
ATF-2	Transcription factor	*(38)*
Elf-1	Lymphoid transcription factor	*(39)*
HBP1	Transcriptional repressor	*(40)*
AP-1	Transcription factor	*(41)*
c-Jun	AP-1 family transcription factor	*(42)*
SP-1	Transcription factor	*(43)*
E4TF1	Transcription factor	*(44)*
AP-2	Transcription factor	*(45,46)*
Id-2	Transcription factor	*(47)*
UBF	Transcription factor (pol I)	*(30,48)*
TFIIIB	Transcriptional regulator (pol III)	*(49)*
TAF(II)250	TFIID subunit (pol II regulation)	*(50,51)*
HDAC1	Chromatin remodeling histone deacetylase	*(52–54)*
TFIID	TATA-box binding protein	*(55)*
c-Abl	Tyrosine kinase	*(56)*
Raf-1	Serine/threonine kinase	*(57)*
RbAp46	Component of mSin3 co-repressor complex	*(58)*
RbAp48	RbAp46-related protein	*(59,60)*
Trip230	Thyroid receptor co-activator	*(61)*
RIZ1	Zinc-finger protein	*(62,63)*
BOG	LXCXE-protein	*(64)*
RIM	Leucine zipper protein with LECEE motif	*(65)*
CDC25A	Protein phosphatase type 1 catalytic subunit	*(66)*
PU.1	Lymphoid transcription factor	*(55)*
c- and N-Myc	Transcription factor	*(67,68)*
HBRM	Disruption of nucleosome structure	*(69)*
Lamin A, C	Nuclear matrix protein	*(70,71)*
Hsc73	Heat shock protein	*(72)*
hBRG1	Transcriptional activator	*(73)*

can interact with c-Abl *(83,84)*. In addition to these interactions, which are relatively well understood, a large number of other *Rb*-binding proteins have been found (Table 1), although, in many cases, the biological relevance of these proteins to growth control has not been established. A feature that distinguishes *Rb* from *p107/p130* is that the latter contain a large spacer sequence between the A and B pockets, which allows *p107* and *p130* to form stable complexes with cyclin-dependent kinases (CDKs) *(85–88)*. Such tight interactions are not observed with *Rb,* although, as described below, *Rb* does contain cyclin-targeting sequences as well. Recently, genes encoding *Rb* homologs in *Caenorhabditis elegans* and *Drosophila* have been identified *(89,90).* These genes appear to be hybrids between the p107/p130 proteins and *Rb,* and, in *C. elegans,* there appears to be only a single *Rb* homolog, suggesting that there may be functional as well as structural overlap. The identification of *Rb* homologs and mutants of these genes, in genetically tractable organisms such as these, holds promise for defining higher-order signaling pathways that regulate development and proliferation.

The major growth-inhibitory activity of *Rb* is exerted during the G1 phase of the cell cycle, and this function is thought to reflect the interaction of *Rb* with one or more proteins. In its growth-inhibitory form, *Rb* binds to cellular proteins that normally function to promote cell division. This inhibitory function is then reversed in late G1, upon phosphorylation of *Rb* by cell-cycle-regulated protein kinases, the CDKs. Phosphorylation releases particular associated proteins from *Rb,* thereby reversing *Rb*-mediated inhibition of cell proliferation. However, currently it is not understood what *Rb* targets are most important for inhibition of proliferation, and whether there are particular phosphorylation events that regulate particular aspects of *Rb* function.

3. E2F TFS AS POCKET PROTEIN TARGETS

By far the best-understood targets of pocket proteins are the E2F family of TFs (Fig. 3). A detailed review of *Rb*–E2F interactions is beyond the scope of this review, and this topic has recently been covered in detail *(91).* E2F TFs coordinate the transcriptional program required for cell cycle progression, and are linked to the basic cell cycle machinery through pocket proteins *(23–26,92–96).* E2Fs contain a transcriptional activation domain, but associate with E2 sites in promoters through the DNA-binding proteins, DP-1 and DP-2. Five E2F TFs are known in mammalian cells, but it is not clear how these individual proteins may differentially regulate transcription of the more than two dozen E2F-dependent genes discovered so far. Pocket proteins associate with E2F complexes in a cell-cycle-regulated manner. Early models suggested that pocket proteins bind to E2F, and block access of E2Fs activation domain to relevant TFs. Although this may still be true in some instances, it is now known that *Rb* can function as a general transcriptional repressor that is brought to particular promoters through association with E2F (Fig. 4). *Rb*–E2F complexes are found mostly in the G1 phase of the cycle; p107–E2F complexes are found mostly in S phase, and p130–E2F complexes are prominent in G0 *(79,87,97–102).* Precisely how different pocket proteins regulate distinct E2F complexes is not well understood. Rb associates most avidly with E2F-1, E2F-2, E2F-3, and, to a lesser extent, with E2F-4; p107 preferentially binds E2F-4, and p130 can be detected in complexes with E2F-4 and E2F-5 *(76–78,103–107).* The available data indicate that removal of *Rb* or p107–p130 leads to derepression of different sets of genes *(106,108–110),* suggesting some level of specificity for repression of particular genes by pocket proteins. In addition to being regulated by pocket proteins, E2F1–3 are also regulated by association with and phosphorylation by CDKs *(111–113).* Unlike E2F1-5, E2F-6 does not contain transcriptional activation activity, but can associate with DPs in a process that represses transcription *(114–117).*

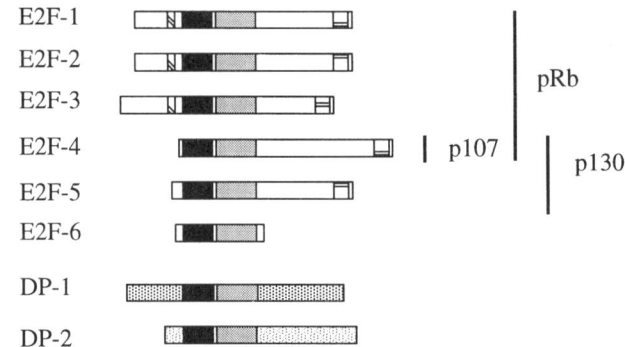

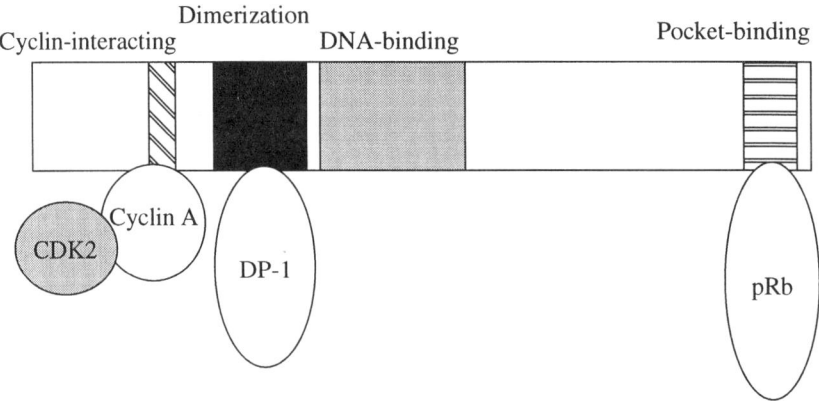

Fig. 3. Schematic diagram showing the structure of E2F and DP families members. E2F-6 does not have transcription activation domain, and is unable to bind pocket proteins. Domain structure of E2F-1 protein and its interacting proteins is depicted in more detail.

4. GROWTH CONTROL BY *RB* THROUGH REPRESSION OF POL I- AND POL III-MEDIATED TRANSCRIPTION

Genes regulated by E2F are transcribed by RNA polymerase II. Recent data also implicate Rb in the regulation of pol I- and pol III-mediated transcription (Fig. 5), which widens the potential roles of Rb in cellular function *(48,118–121)*. pol I is responsible for transcription of rRNAs, and pol III transcribes small nuclear RNA, small rRNA, and transfer tRNA. *Rb* associates with the dimeric pol I TF, UBF, thereby blocking its binding to DNA. In vivo, the extent of the UBF–*Rb* interaction changes in different cellular states, such as the process of cell cycle exit, suggesting that this function is regulated in some way *(30,48)*. Rb is also able to inhibit transcription of multiple pol III templates *(118–121,48)*. The level of pol III transcription in *Rb–/–* primary fibroblasts was 4–5-× higher than in *Rb+/+* fibroblasts, providing genetic evidence of a general role for *Rb* in pol III transcription. *Rb* physically associates with TFIIIB *(121–148)*, a component of the pol III complex, although it remains to be determined whether this interaction underlies inhibition of pol III by *Rb*. TFIIIB is responsible for

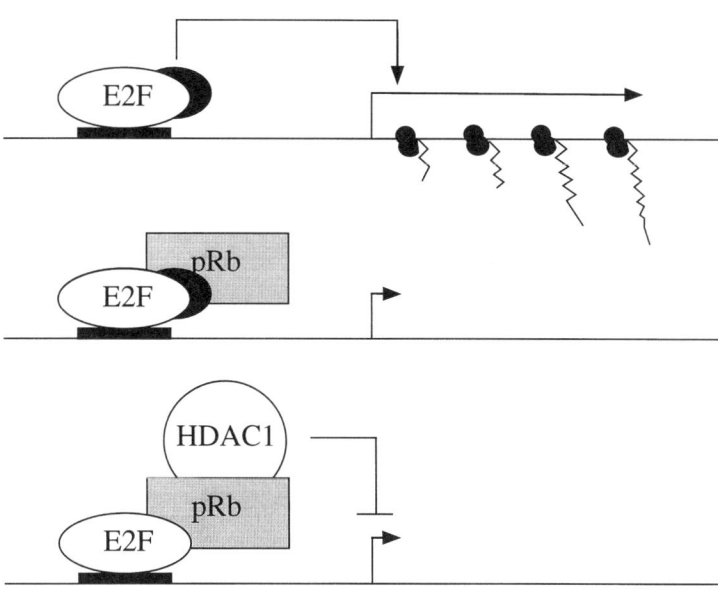

Fig. 4. Regulation of E2F-dependent promoters by *Rb*. (**A**) E2F activates transcription through its transcriptionactivating domain, indicated as a black domain on E2F. (**B**) *Rb* as a specific inhibitor of E2F transcription. Interaction of *Rb* with E2F blocks access of the activation domain with the transcriptional machinery, thereby blocking activation of target genes. (**C**) *Rb* as a transcriptional repressor that associates with a histone deacetylase, and which is brought to specific promoters through association with E2F. Recruitment of deacetylases to promoters by *Rb* may repress transcription in an E2F-dependent manner.

directing pol III to transcriptional start sites *(122)*. The activation of TFIIIB during the cell cycle is strikingly similar to that of E2F; TFIIIB is inactive during most of the G1, and its activity increases as cells progress into S phase. Both pol 1 and pol III activities are repressed in quiescent cells, and increase in parallel upon serum stimulation *(118,123,124)*.

The regulation of pol I and pol III transcription by Rb provides an interesting link between Rb and biosynthetic activity of the cell. By regulating the abundance of rRNA and tRNA, Rb may limit the rate of cell growth and protein accumulation *(30,48,49,125)*. Deregulation of this control may contribute to tumor development by supplying sufficient quantities of ribosomes and tRNAs for unrestrained cell growth. Indeed, some naturally occurring mutations in Rb affect pol I and pol III regulation, which correlates with loss of growth control.

5. MECHANISMS OF TRANSCRIPTIONAL INHIBITION BY RB

Recent data suggests that *Rb* may repress transcription at least in part by remodeling chromatin. *Rb* interacts with a histone deacetylase (HDAC1), and, when associated with E2F, *Rb* recruits HDAC1 to DNA (Fig. 4). The interaction of HDAC1 with Rb requires an intact pocket domain in Rb, and naturally occurring *Rb* mutations reduce this interaction *(52,54)*. A role for HDAC1 in Rb-mediated repression through E2F was demonstrated by experiments using specific inhibitor of HDAC1, trichostatin A (TSA).

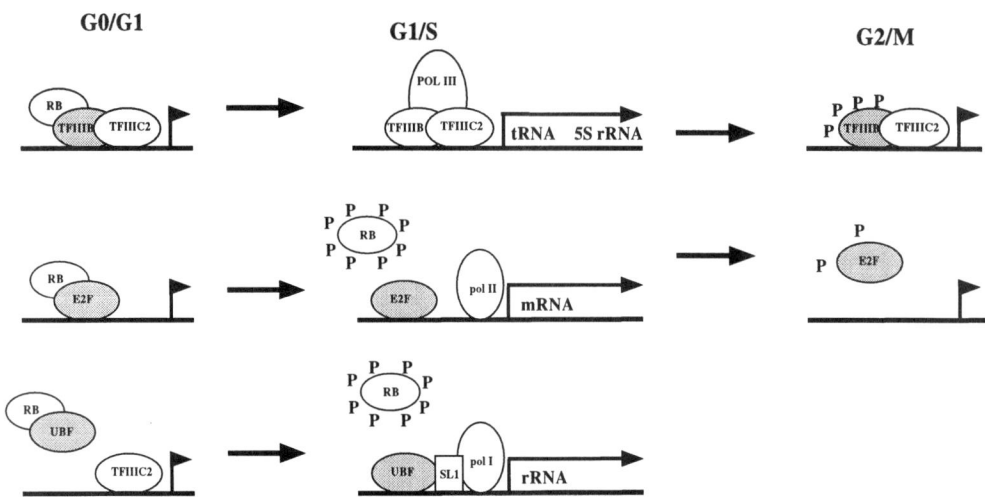

Fig. 5. Regulation of pol I, II, and III by Rb. *Rb* can regulate transcription by all three classes of polymerases. *Rb* represses pol I by preventing the general factor, UBF, from binding the complex. *Rb* represses pol II transcription through gene-specific TFs, such as E2F. *Rb* may repress pol III transcription through the general factor, TFIIB.

After treatment with TSA, *Rb*-mediated repression of an E2F-regulated promoter in a chromosomal context was abrogated *(52)*. Transected DNA templates, whose chromatin structure may not precisely mimic chromosomal chromatin, are also responsive to the TSA treatment, raising a possibility that HDAC1 may target other nonhistone proteins important for transcriptional regulation *(52–54)*. p107 and p130 were recently shown to interact in a similar manner with HDAC1 in repressing E2F4-mediated transcription *(126)*. Moreover, recent studies *(84)* indicate that E1A not only inactivates Rb-mediated recruitment of histone deacetylase, but also binds and activates CBP, increasing its histone acetyltransferase activity. Cumulatively, these two processes would be expected to significantly open chromatin structure, promoting gene expression *(128–132)*.

6. INVOLVEMENT OF RAS PATHWAY IN CDK ACTIVATION: LINKING INACTIVATION OF *RB* WITH EXTRACELLULAR SIGNALS

During the G1 phase of the cell cycle, cells integrate the various growth-stimulatory and -inhibitory signals in their environment, and make the decision to enter the division cycle or to exit the cell cycle. *Rb* appears to function both as an inhibitory barrier to S phase, which must be overcome through *Rb* inactivation to allow for S phase, and as a positive regulator of events which commit cells to a terminal differentiation pathway. A primary determinant of whether a cell will enter S phase appears to be the activity status of CDKs. These enzymes, composed of a catalytic CDK subunit and a regulatory cyclin subunit, control cell cycle transitions, and are the ultimate recipients of information intended to control proliferation. Two major classes of cyclins, D-type and E-type cyclins, have been implicated in the control of the G1–S transition, and in *Rb* inactivation *(133–141)*. D-type cyclins (D1, D2, and D3) serve as specific activators of CDK4 and CDK6; cyclin E activates CDK2. Because of the critical roles played by these two kinases in bringing about S phase, elaborate and diverse regulatory pathways have evolved to control the extent and timing of their activity. Ultimately, the activity status of CDKs is controled primarily by two parameters: the level of relevant cyclins and the levels of CDK inhibitors. Two classes of CDK inhibitors are used to control the G1–S transition. The p16^{INK4} class (p16, p15, p18, p19) specifically inhibits kinases activated by D-type cyclins; the p21^{CIP1} class (p21, p27, and p57) inhibit D-, E-, and A-type cyclin–CDK complexes. During the past few years, some inroads have been made in elucidating the hardware linking receptor tyrosine kinases with both the transcriptional program controlling cyclin expression and mechanisms that control the levels of CDK inhibitors. Activation of receptor tyrosine kinases lead to activation of the Ras/Raf(MEK1) mitogen-activated protein kinase pathway, which in turn leads to induction of cyclin D1 transcription *(142,143)*. Assembly of cyclin D1 with CDK4 is also regulated by mitogens in fibroblasts.

As described later in more detail, CDK4–cyclin D complexes are intimately involved in phosphorylation and inactivation of the negative growth control function of *Rb*. In keeping with this idea, microinjection of anti-Ras antibodies, or expression of dominant interfering alleles of Ras, block cyclin D1 accumulation, and cause cell cycle arrest in *Rb*-positive, but not *Rb*-negative, cells *(144,145)*. Moreover, expression of a constitutively active form of MEK1 induces cyclin D1 expression and assembly of active CDK4–cyclin D1 complexes *(146,147)*. Accumulation of cyclin D1–CDK4 complexes during G1 may have multiple consequences, depending on the status of CDK inhibitors, including *Rb* phosphorylation and titration of p27 to release active cyclin E–CDK2 complexes. In mitogen-deprived fibroblasts, p27 accumulates, and is thought to serve as a barrier to S-phase entry by blocking CDK-mediated *Rb* phosphorylation *(148–152)*, and one of the functions of mitogens is to facilitate destruction of p27 through a proteasome-dependent mechanism *(153)*. Cells engineered to overexpress cyclin D and CDK4 do not allow for S-phase entry in the absence of mitogens, apparently because of failure to assemble active kinase complexes. However, when these cells also express a constitutively active MEK1, still in the absence of mitogens, active cyclin D–CDK4 complexes are assembled and sequester p27, thereby releasing active CDK2/cyclin E complexes that further promote S-phase entry *(146,147)*. Biochemical studies suggest that, although p27 binds tightly to both cyclin D and cyclin E-dependent kinases, it is an effective inhibitor of only CDK2. Thus, accumulation of cyclin D–CDK4 complexes may play two roles in transit through G1 in fibroblast: *Rb* phosphorylation, and titration of p27 to activate cyclin E–CDK2. How pervasive this mechanism is in various cell types, and whether sequestration of p27 plays a role in progression from mitosis to S phase, a setting where p27 levels are generally low, remains to be determined. In addition to MEK1, Raf-1 has also been shown to overcome *Rb*-mediated growth arrest. Whether this represents a direct interaction of Raf-1 with *Rb*, as suggested by this paper *(57)*, or downstream effects of activation of cyclin D1 transcription and cyclin D–CDK4 assembly, remains to be determined.

7. *RB* PHOSPHORYLATION AND RELIEF OF GROWTH SUPPRESSION

Rb contains 16 possible CDK phosphorylation sites (Fig. 2), and undergoes cell-cycle-dependent phosphorylation. It is clear that cyclin D–CDK4 complexes are central to *Rb* inactivation *(35)*. There are several pieces of data that contribute to this idea, including the fact that overexpression of p16 can block *Rb*-positive, but not *Rb*-nega-tive, cells in G1 *(137,139,154–157)*. Because p16 is a specific inhibitor of CDK4/6 kinases, this data indicates that their ability to phosphorylate *Rb* is critical for cell cycle progression. In addition, cyclin D–CDK4 can efficiently phosphorylate *Rb* in vitro, and this phosphorylated *Rb* is unable to arrest *Rb*-negative cells in G1, upon microinjection *(96)*. The ability of *Rb* to be inactivated by cyclin D–CDK4 in vitro requires phospho-rylation of S795, a residue that is preferentially phosphorylated by CDK4–cyclin D in vitro. It is very likely that cyclin D–CDK4 is not sufficient to inactivate *Rb,* and, indeed, several studies suggest that cyclin E–CDK2 functions in combination with cyclin D–CDK4 to regulate *Rb (158–160)*. It has also been suggested that there is a temporal dependence in *Rb* phosphorylation, although this may reflect the timing of action of the two kinases, rather than a strict biochemical requirement. CDK2 and CDK4 display overlapping but distinct specificities toward *Rb,* and it is possible that different phosphorylation events are cumulatively responsible for *Rb* inactivation *(167–169)*. Recent studies indicate that some of the specificity in pocket protein phos-phorylation results from interactions of cyclin–CDK complexes with substrate recogni-tion motifs in pocket proteins. p107 and p130 associate tightly with cyclin A–CDK2 complexes through an RXL motif *(85–88),* a short sequence that is also found in the CIP/KIP family of CDK inhibitors. Tight association makes p107 and p130 obligate CDK substrates, and is required for efficient phosphorylation. In *Rb,* there are several RXL motifs located in the C-terminus of the protein *(165),* and one of these contributes significantly to the ability of cyclin A–CDK2 and cyclin D–CDK4 to phosphorylate *Rb* in the C-terminal region containing a cluster of phosphorylation sites. In contrast with p107, however, cyclins A and E do not bind tightly to Rb. In addition, cyclin D may interact with *Rb* through an N-terminal LXCXE motif in cyclin D *(137,138)*. However, this sequence is not required for *Rb* phosphorylation in vitro, or for cyclin D to over-come the growth suppressive effects of *Rb* in tissue culture cells.

Although cyclin E–CDK2 complexes probably contribute to *Rb* inactivation, they also have substrates other than *Rb.* This is based on the fact that cyclin E expression can bypass growth suppression by a nonphosphorylatable Rb protein, and can do so without activating E2F *(166–169)*. This activity probably reflects the ability of cyclin E–CDK2 to phosphorylate proteins whose activity is rate-limiting for S-phase entry. Although cyclin E can activate S-phase entry in this way, existing data indicate that this ectopic S phase is incomplete, and cells do not enter mitosis.

8. POCKET PROTEIN FUNCTION REVEALED
THROUGH KNOCKOUT MICE

8.1. Rb Deficiency

Knockout mice and fibroblast cell strains for all three known members of the family have been used extensively to analyze pocket protein function in vivo and in vitro. Mice lacking *Rb* die by embryonic day 16, and display multiple defects *(170–172)*. These defects are seen, starting at d 11.5, as swelling near the fourth ventricle, and poor blood distribution. Extensive p53-dependent and -independent cell death occurs throughout the central nervous system (CNS), but no defects were observed in periph-eral nervous system development. The likely cause of apoptosis in *Rb–/–* embryos is inappropriate cell divisions in the CNS *(173)*. In mutant mice, numerous dividing cells were observed in the nervous system, ocular lens, and liver, where wild-type (wt) embryos contain mostly differentiating and migrating cells. In the CNS of *Rb–/–* embryos, increased levels of cyclin E and free E2F were observed. Red blood cells (RBCs) of mutant embryos often have aberrant morphology, and peripheral red blood cells retain their nuclei, but peripheral blood of the wt embryos contain mostly enucleated erythrocytes *(170–172)*. Experiments with chimeric mice pinpointed the defect in RBC maturation in *Rb–/–* embryos to a defect in hepatic stromal cells, and not in erythrocytes themselves *(174–176)*. Because *Rb–/–* mice are inviable, analysis of transformation because of *Rb* loss has been analyzed in *Rb+/–* mice. *Rb* loss is associated with development of RB in humans, and, because of this, it was anticipated that mice deficient in *Rb* would also show retinal phenotypes. However, *Rb+/–* mice developed pituitary (PIT) tumors, and not RBs, indicating that mice and humans have distinct dependencies for particular tumor suppressor genes in specific tissues. Pituitary transformation was associated with loss of the wt *Rb* allele *(170–172,177–178)*.

A major question has been to what extent the functions of pocket proteins overlap. This has been investigated in vivo by crossing *Rb–/–* mice with mice deficient for p107 *(179)*. Where *Rb–/–* embryos die at 13.5–15.5 dpc, p107–/–*Rb–/–* embryos die at 10.5–12.5 dpc, and death is accompanied by increased rates of apoptosis in the CNS and liver. Thus, in the absence of *Rb,* p107 is required for some aspect of prenatal development. The elimination of p107 in Rb+/– mice leads to severe growth retardation and increased mortality, but these animals did not show any additional tumor phenotypes. However, these mice did develop focal lesions of the retina, with retinal displasia and elimination of the photoreceptor layer *(179,180)*. The stochastic appearance of these lesions suggests that additional somatic events, possibly including loss of the existing wt *Rb* allele, may be responsible for this weak phenotype. The enhanced phenotype observed in *Rb*/p107 knockouts may reflect shared functions in certain tissues, or functional compensation by p107 in the absence of the Rb function.

8.2. p107 and p130 Deficiency

p107 knockout mice have been created in two genetic backgrounds with different outcomes. p107 knockout mice, constructed on an 129/Sv genetic background, are fertile and viable, and do not show increased predisposition to tumor development. p107–/– embryos display subtle thickening of the long bones, compared with the wt embryos, implicating p107 in bone development. However, fibroblasts derived from p107–/– embryos did not reveal any significant defect in cell cycle regulation. These observations indicate the absence of a unique obligatory function for p107 in the embryonic development and cell cycle regulation in the embryonic fibroblasts *(179)*, at least in this strain background.

In the alternative p107 knockout mice model, strain-specific effects were observed *(181)*. Mice deficient for p107 on the Balb/cJ background were viable and fertile as adults, and embryos as well as newborn pups were indistinguishable from the wt and heterozygous littermates. In contrast with the situation in the 129/Sv background, pups in the Balb/cJ background display a severe postnatal growth deficiency, in which animals were approx 50% normal size at 3 wk of age. Postnatal development was not delayed, and animals could be weaned, and reached sexual maturity at normal times. As such, the basis of the growth rate decrease remains undefined. The immune response of mutant mice was compromised, with frequent development of a diathetic myeloproliferative disorder *(181)*.

Embryonic fibroblasts derived from these p107–/– Balb/cJ mice displayed a markedly increased growth rate, with the doubling time of mutant fibroblasts about one-half that of wt. The rate of progression through all cell cycle phases seemed to be enhanced to similar extents, since there was no obvious change in cell cycle parameters in these strains. The increased rate of cell cycle progression may reflect in part the fact that cyclin E is derepressed in these p107 mutant strains, in contrast to the situation found with the 129/Sv background. The basis of strain-specific effects are still not understood, but it appears that there are modifier genes that can compensate for p107 loss in particular background *(179,181)*.

p130–/– mice constructed on 129/Sv genetic background are viable, and are without overt phenotypes *(182)*. Which was surprising, given that p130 is a major component of the E2F complex in G0 *(79,99,107)*. Analysis of the E2F complexes in serum-starved p130–/– cells revealed that, although the population of the free E2F increased, the majority of E2F was now complexed with the p107, instead of p130. Thus, in this instance, p107 can functionally compensate for p130 deficiency. When p130 deficiency was generated on a Balb/cJ background, homozygous-null animals were inviable *(183)*. Embryonic death was observed at 11–13 dpc, and animals at this stage were 75% smaller than heterozygous littermates. Mutant hearts displayed an abnormal morphology caused by dilation, and resembled hearts of earlier-stage normal embryos with two chambers. At 10.5 dpc, mutant embryos fail to form and extend hind limb buds. Histological examination of mutant mice revealed defects in neurogenesis and myogenesis, a poorly formed notochord, and abnormal cardiac morphology. The levels of apoptosis were greatly increased in the CNS in the absence of p130. Moreover, tissue-specific effects on proliferation were observed *(183)*.

The absence of dramatic effects in p107- or p130-deficient mice in the 129/Sv background potentially reflects the ability of pocket proteins to compensate for one another, a view that is supported by the shared structural and functional similarities of pocket proteins. To address this question, mice lacking both p107 and p130 were constructed *(182)*. After crossing p107+/– and p130+/– animals, only seven possible genotypes were observed soon after birth. Double-knockout mice survive until birth, but die shortly thereafter. At 18.5 dpc, double-knockout embryos were somewhat smaller than littermates, and had shortened limbs. Skeletal preparations revealed significant abnormalities in bone structure and in the timing of the bone deposition in double knockouts. Effects were observed in both the developing ribs and long bones, which are formed through the process of endochondral ossification. In contrast, bones formed through intramembranous ossification, such as most of the cranial bones, were normal. These data indicate that p107 and p130 are required for endochondral, but not intramembranous, bone formation *(182)*. Limb bones were also shortened, thickened and abnormally shaped. The shortening and thickening of the bones can be attributed solely to p107 loss, because p107–/–p130+/– embryos had the same defect. The defects in bone formation were traced to alterations in chondrocyte proliferation. In the absence of the p130 and p107 function, chondrocytes in epiphyseal centers proliferate at an increased rate, and were delayed in cell cycle withdrawal and differentiation *(182)*.

Analysis of p130–/– mice has also demonstrated functional compensation by *Rb* family members in peripheral T-lymphocytes *(184)*. T-lymphocytes form a uniform population of quiescent cells, which rapidly proliferate and differentiate after an appropriate mitogenic stimulus *(99,106,185)*. Lymphocytes maintain the quiescent resting state, with high levels of p130–E2F-4 complexes, which are promptly disrupted upon entering S phase. The postulated role of p130 in G0, in maintaining the quiescent state, predicts that p130–/– mice will have some disturbance in this phase of the cell cycle, but analysis of quiescent T-lymphocytes failed to reveal a difference in either the ability to enter quiescence or the rate of cell cycle entry, between p130–/– and p130+/– mice. The level of free and E2F-complexed p107 was significantly higher in p130–/– cells, suggesting that it compensating for p130 loss, and compensation appears to be essentially complete. T-lymphocytes lacking both p130 and p107 have elevated levels of free E2F, and also contain a Rb–E2F-4 complex, providing an additional means for compensation. But, in this case, the compensation does not appear as complete; these T-lymphocytes are hypersensitive to mitogens, and have derepressed a subset of E2F-regulated genes *(184)*. Similar derepression was observed previously in embryonic fibroblasts, and in tissues of mice lacking both p107 and p130 *(108–182)*. p130–/–p107–/– cells contain a new E2F complex lacking *Rb,* suggesting the possibility that another family member exists. It is not known at the moment if this hypothetical *Rb* homolog is important for compensation of p130 and p107 functions, or plays a role in control of cell cycle progression *(108)*. Clearly, there may be functional overlap between pocket proteins in certain cell types, but whether this reflects a normal shared function or a gain of function, upon loss of another family member, is difficult to discern.

Analysis of pocket protein mutations in multiple-strain backgrounds has revealed major effects on phenotypes, and points to the importance of variable interacting genes that control the severity and penetrance of phenotypes. It is possible that the same modifiers are involved in the strain-specific phenotypes of both of these genes. Both p107- and p130-mutant mice do not display abnormal phenotypes on the 129/Sv or C57BL/6J genetic backgrounds, but their deletions on the Balb/cJ background lead to specific effects. Such differences could be explained by recessive loss-of-function mutations in the 129/Sv and C57BL/6J genetic background, or by dominant gain-of-function mutations in the BALB/cJ background. The mixture of both possibilities also cannot be excluded. The identification of these modifiers could provide insight into the specific and shared roles of pocket proteins in development and cell cycle control.

9. SUPPRESSION OF *RB* DELETION PHENOTYPES BY MUTATION OF E2F-1

Although heterozygosity for the *Rb* mutation leads to PIT and thyroid tumors in mice, homozygosity leads to midgestation lethality, because of defective erythroid and neuronal differentiation and apoptosis, and multiple cell cycle defects. The absence of Rb function leads to increased level of free, transcriptionally active E2F, and derepression of E2F target genes, such as cyclin E, which is important for cell cycle entry *(173)*. To test if the proliferative and apoptotic disorders in *Rb*-mutant mice are related to derepression or transactivation of a subset of E2F-regulated genes, mice mutant for both *Rb* and E2F were created. E2F–/– mice by themselves have a mild lymphoproliferative disorder, decreased level of apoptosis in thymocytes and a wide array of tumors at advanced age *(186,187)*. Loss of even one copy of E2F was enough to extend the life of Rb+/– mice by several months, and reduced the incidence of thyroid tumors. Deletion of both copies of E2F leads to an even more extended life-span for Rb+/– mice, and to a decrease PIT tumor incidence *(188,189,190)*. Rb–/– embryos die at E13–14, but Rb–/–E2F-1–/– embryos have a slightly prolonged life-span, dying at E17 with defects in blood, muscle, and lung development. Deletion of E2F-1 on an Rb–/– background leads to decreased levels of apoptosis in the lens, PNS, and CNS, and decreased incidence of abnormal S-phase entry that is characteristic of Rb–/– embryos. These observations indicate that the major consequence of Rb deletion is deregulation of E2F-1 function, although not all the effects of Rb deletion can be corrected by deletion of E2F-1 *(189)*.

10. *RB* AND DIFFERENTIATION

Differentiation is known to be tightly coupled to cell cycle arrest in G1, but how this occurs is still not well understood. Two model systems, muscle cell (MC) differentiation and lens differentiation, have been extensively used to examine the role of Rb in the differentiation process. MC differentiation is controlled by a set of four myogenic basic helix-loop-helix transcription factors: MyoD, myogenin, MRF4/Myf-6/herculin and Myf-5. Each of these factors can activate the program of MC differentiation, when expressed in non-MCs. MyoD, Myf5, myogenin, and MRF4 have overlapping but distinct expression patterns during development of the skeletal muscles. Analysis of knockout mice lacking one or a combination of these TFs, has revealed their roles in muscle differentiation. MyoD and Myf5 act redundantly early in embryogenesis to establish the muscle lineage. Myoblasts, established with the help of MyoD and Myf5, continue to proliferate until receiving an unspecified differentiation signal, upon which myoblasts exit the cell cycle and form myotubes, a process controlled by myogenin. MRF4 is involved later in the process of myotube maturation. Mice lacking myogenin or a combination of MyoD and MRF4 have severe defects in skeletal muscle development, because of a failure of myoblasts to differentiate *(191–193)*. Myoblasts are completely absent in mice lacking both MyoD and Myf5 *(194)*. Because inactivation of both MyoD and MRF4 is required to inhibit differentiation, and of both MyoD and Myf5 to disrupt myoblast specification, it is clear that these sets of proteins have redundant roles, and can be compensated by another member of the group *(193,195)*.

In order to terminally differentiate, cells must exit the cell cycle. In experiments involving ectopic expression of MyoD, differentiation was abolished if cyclin D, A, or E were co-expressed with MyoD *(196–198)*. The cyclins act with CDK4 or CDK2 to phosphorylate and inactivate Rb. It is possible that, by inactivating Rb, CDKs interfere with differentiation. This notion is strengthened by the observation that viral proteins, which associate with and inactivate Rb, also interfere with MC differentiation *(199,200)*. Cells lacking Rb fail to execute proper differentiation in vitro *(201,202)*, also indicating that Rb plays a role in differentiation. The analysis of Rb's role in differentiation in vivo is complicated by the fact that Rb–/– mice die prenatally, before executing a program for secondary muscle differentiation *(170–172)*. Myoblasts lacking Rb can differentiate into multinucleated myotubes in vitro, following serum withdrawal, in apparent contrast with observations in vivo. But these myotubes remain inappropriately sensitive to external signals, and can enter S phase and replicate their DNA in response to stimulation by serum growth factors *(201,203)*, resembling results of low level Rb expression in vivo *(204)*. Rb–/– fibroblasts retain an ability to activate an aberrant muscle differentiation program after ectopic expression of MyoD, probably through compensation by other family members that are upregulated *(202)*. This suggests potential functional compensation by p107 or p130, which can both regulate (HBP1)-mediated transcriptional inhibition of MyoD family members, and p107 is substantially upregulated in *Rb–/–* myoblasts in vitro *(201,203,40,205)*. HBP-1 is a transcriptional repressor and a cell cycle inhibitor induced during MC differentiation. The relative ratio of *Rb* to HBP1 appears to be an important factor in determining whether cell cycle exit or full differentiation will occur. When the Rb:HBP1 ratio is low, cell cycle exit occurs, but tissue-specific gene expression is not observed. With a high ratio, full MC differentiation occurs. Therefore, the relative ratio of Rb to HBP1 may be a signal for activation of MyoD, implicating Rb function in controlling more than one step in the MC differentiation *(40)*.

It is not clear which components of the myogenic regulatory pathway require Rb for function. It is known that Rb interacts with MyoD in vitro *(201)*, although MyoD activity seems unaffected by *Rb* loss. p21 is a potential MyoD target; it is induced by MyoD expression, and could therefore couple the differentiation program to cell cycle arrest in G1 by blocking CDK-mediated *Rb* phosphorylation. The absence of muscle differentiation defects in p21–/– mice raised a possibility of functional redundancy of p21 in MC differentiation *(206)*. Indeed, it was found that skeletal muscle expresses the p21 homolog p57^{KIP2}, and simultaneous deletion of both p21^{Cip1} and p57^{Kip2} leads to severe defects in skeletal muscle development, indicating that these two inhibitors cooperate to control cell cycle exit and muscle differentiation *(207)*. MCs lacking both p21 and p57 display increased proliferation and apoptosis, and a phenotype shared by MCs lacking myogenin, or expressing inadequate levels of Rb, indicate that p21 and p57 function upstream of *Rb*, to facilitate MC differentiation. Additional links are suggested by the fact that Rb can physically associate with MyoD and myogenin *(201)*, and that p21 may be a target of the myogenin, although its expression is not altered in myogenin-deficient cells *(208,209)*.

The lens has also been an important experimental system for studying the role of *Rb* in differentiation, primarily because of its simplicity. The lens contains an anterior layer of proliferating epithelial cells that migrate to the equatorial zone, where they differentiate into lens fiber cells, and elongate. In *Rb*-mutant mice, lens epithelial cells fail to appropriately exit the cell cycle, and many cells undergo inappropriate apoptosis. In a situation parallel to that found in muscle, loss of two CDK inhibitors (p57^{KIP2} and p27^{KIP1}) leads to a very similar phenotype, inappropriate S-phase entry, and apoptosis. One interpretation is that, in the absence of these inhibitors, *Rb* is inactivated by CDK-dependent phosphorylation; thus, the phenotype of the double-knockout mimics *Rb* loss, as with MCs. Additional studies are needed to understand whether *Rb* is required simply for cell cycle arrest, or is also directly involved in promoting the action of transcriptional pathways that promote lens epithelial cell differentiation.

11. OTHER ROLES FOR RB

11.1. Apoptosis

Cell proliferation, differentiation, and apoptosis are major pathways controling cell fate in the developing organism. The role of Rb in cell cycle control and differentiation well established. In the past few years, a role for Rb in apoptosis has been suggested, based on the fact that *Rb* deficiency leads to increased cell death in the developing lens and nervous system *(170–172)*. Moreover, Rb–/– fibroblasts show an increased tendency to undergo apoptosis in response to growth factor withdrawal and DNA damage *(210)*. p130-deficiency in certain genetic backgrounds also produces similar apoptotic effects in muscle, neural structures, and the developing eye *(183)*. These data indicate that both Rb and p130 may have a role in protecting cells against apoptosis.

Rb's apoptosis protection function may result from its regulation of E2F-1. E2F-1, when overexpressed, can induce apoptosis in certain cell types *(123,211–214)*, and, in the absence of *Rb*, E2F-1 may inappropriately activate apoptotic signals. That this is the case is suggested by reduced apoptosis when *Rb* deficiency is crossed onto an E2F-1-deficient background *(188,189)*. The ability of E2F-1 to promote apoptosis may reflect its ability to activate ARF *(215,216)*. The ARF protein is able to bind and inhibit the Mdm2 protein, which normally controls the levels of p53 through ubiquitin-mediated proteolysis *(217–220)*. Blocking p53 destruction may activate p53-dependent apoptosis in this context. In addition to possibly being required to suppress E2F-1-mediated apoptosis, *Rb* is also a target of caspases involved in cleaving proteins during apoptosis *(221,222,223,224)*. This area of research has recently been reviewed *(225)*. Whether cleaved *Rb* is required for apoptosis is still an open question.

11.2. Tumorigenesis and DNA Damage Arrest

The ability of a cell to arrest in response to DNA damage is essential for accurate transmission of genetic information. The failure of a cell to arrest in such circumstances can lead to propagation and accumulation of harmful mutations that facilitate transformation. The ability to confer cell cycle arrest in response to γ- or UV irradiation, or after treatment with chemotherapeutic agents, may be an important characteristic of certain classes of tumor suppressors. p53 is a tumor suppressor that is stabilized and activated upon DNA damage *(101,217–221)*. Activated p53 has increased transcriptional activity, and is required for both the G1–S DNA damage checkpoint and the G2–M spindle checkpoint. p21^{CIP1} is a major target of p53-mediated transcriptional activation following DNA damage, and is important for G1–S arrest *(212,226–230)*.

Rb was recently identified as a mediator of p53-dependent arrest imposed by a diverse range of DNA-damaging agents *(231)*. Rb–/– embryonic fibroblasts failed to respond to DNA damage by arresting the cell cycle, and continue to replicate damaged DNA *(170–172)*; fibroblasts with mutant p130, p107, or both p107 and p130, had a response similar to that of wt cells *(179–183)*. The rate of induction of p53 and p21 in response to damage was unchanged in Rb–/– and *Rb*+/– cells, indicating that Rb loss affects DNA damage pathways downstream of p21 *(170–172)*. These data are consistent with the idea that p21 blocks CDKs from inactivating *Rb*'s growth-suppressive function. The presence of functional Rb correlates with the ability of p21 to arrest cells *(230)*, although p21 can block *Rb*-negative cells when expressed at sufficient levels. Alternatively, the role for Rb in the G1 damage checkpoint may reflect an indirect role of Rb loss in gene regulation. The level of cyclin E, an important target of E2F–*Rb* regulation, is increased in *Rb*–/– cells *(108,232)*. Increased cyclin E levels may require higher levels of p21 for complete inhibition than are achievable through p53-dependent induction. A specific role for Rb in the G1 checkpoint may underlie the fact that, of the three pocket proteins, only Rb is a tumor suppressor. The exact biochemical mechanism of Rb-dependent cell cycle arrest remains unclear, and it would be important, for practical applications, to determine if the same pathway is at work in vivo.

12. THERAPEUTIC APPLICATIONS

Because the disruption of the Rb pathway (Fig. 1) is a common feature of tumors, the restoration of Rb function initially seemed to be an attractive way for correcting the deficiency. Indeed, introduction of wt *Rb* into transformed cells can suppress inappropriate growth. Despite the theoretical attractiveness of such an approach, there has been little progress in bringing Rb-based therapies to the clinic. Several problems of a conceptual and practical nature have yet to be solved. First, a highly efficient means of targeted drug delivery should be employed, because even rare cells with unrestored Rb function will be selected for growth, and any such cell can potentially give rise to tumors. Second, most human cancers are associated with multiple mutations, and restoration of just one mutated function may not lead to a fully functional restoration of the normal phenotype. Recently, it was shown that Rb is able to prevent replication of damaged DNA *(231),* with *Rb–/–* cells being able to replicate in the presence of the damage, despite the presence of functional p53. So, Rb loss may mediate propagation of mutations. This observation, made using Rb–/– mouse embryonic fibroblasts, may prove to be crucial to ideas on cancer treatment. Chemotherapy is a commonly prescribed procedure, but chemotherapeutic treatment of tumors with *Rb* mutations may actually propagate mutations, and so have deleterious effects. In addition, it is now clear that cyclin E overexpression can bypass the *Rb*–E2F pathway to initiate S phase *(166,167).* Thus, the use of nonphosphorylatable *Rb* as a growth inhibitor may not provide the expected outcome. As such, significant advances will need to be made before the *Rb* pathway becomes a viable target for cancer therapeutics.

13. CONCLUSION

The past 10 years has seen remarkable progress in the understanding of *Rb* function. However, there are several outstanding questions. Although a large number of *Rb*-interacting proteins have been identified, for most of these, very little is known about their roles in the *Rb* pathway or even whether they are important mediates of *Rb* function. In addition, although *Rb* is clearly required for differentiation of several cell types, it is unclear precisely what its role is in this process. The identification of *Rb* homologs in genetically tractable organisms will probably allow a better appreciation of these and other questions related to pocket protein function.

REFERENCES

1. Friend SH, Horowitz JM, Gerber MR, Wang X-F, Bogenmann E, Li FP, Weinberg RA. Deletions of a DNA sequence in retinoblastomas and mesenchymal tumors: organization of the sequence and its encoded protein. *Proc Natl Acad Sci USA* 1987; 84:9059–9063.
2. Lee EY-HP, To H, Shew J-Y, Bookstein R, Scully P, Lee W-H. Inactivation of the retinoblastoma susceptibility gene in human breast cancers. *Science* 1988; 241:218–221.
3. Harbour JW, Lai S-H, Whang-Peng J, Gazdar AF, Minna JD, Kaye FJ. Abnormalities in structure and expression of the human retinoblastoma gene in SCLC. *Science* 1988, 241:353–357.
4. Toguchida J, Ishizaki K, Sasaki MS, Ikenaga M, J, Ishizaki K, Sasaki MS, et al. Chromosomal reorganization for the expression of recessive mutation of retinoblastoma susceptibility gene in the development of osteosarcoma. *Cancer Res* 1988; 48:3939–3943.

5. Shew J-Y, Ling N, Yang X, Fodstad O, Lee W-H. Antibodies detecting abnormalities of the retinoblastoma susceptibility gene product (pp100RB) in osteosarcomas and synovial sarcomas. *Oncogene Res* 1989; 1:205–214.

6. Shew J-Y, Lin B, Chen P-L, Tseng BY, Yang-Feng TL, Lee W-H. C-terminal truncation of the RB protein leads to functional inactivation. *Proc Natl Acad Sci USA* 1990; 87:6–10.

7. Bookstein R, Lee EY, Peccei A, Lee WH. Human retinoblastoma gene: long-range mapping and analysis of its deletion in a breast cancer cell line. *Mol Cell Biol* 1989; 9:1628–1634.

8. Cheng J, Scully P, Shew J-Y, Lee W-H, Vila V, Haas M. Homozygous deletion of the retinoblastoma gene in an acute lymphoblastic leukemia (T) cell line. *Blood* 1990; 75:730–735.

9. Hensel CH, Hsieh CL, Gadzar AF, Johnson BE, Sakaguchi AY, Naylor SL, Lee W-H, Lee EY-HP. Altered structure and expression of the human retinoblastoma susceptibility gene in small cell lung cancer. *Cancer Res* 1990; 50:3067–3072.

10. Horwitz JM, Park S-H, Bogenmann E, Cheng J-C, Yandell DW, Kaye FJ, Minna JD, Dryja TP, Weinberg RA. Frequent inactivation of the retinoblastoma antioncogene is restricted to a subset of human tumors. *Proc Natl Acad Sci USA* 1990; 87:2775–2779.

11. Juge-Morineau N, Harousseau JL, Amiot M, Bataille R. The retinoblastoma susceptibility gene RB- 1 in multiple myeloma. *Leuk Lymphoma* 1997; 24:229–237.

12. DeCaprio JA, Ludlow JW, Figge J, Shew J-Y, Huang C-M, Lee W-H, et al. SV40 large tumor antigen forms a specific complex with the product of the retinoblastoma susceptibility gene. *Cell* 1988; 54:275–283.

13. Whyte P, Buchkovich KJ, Horowitz JM, Friend SH, Raybuck M, Weinberg RA, Harlow E. Association between an oncogene and an anti-oncogene: the adenovirus E1A proteins bind to the retinoblastoma gene product. *Nature* 1988; 334:124–129.

14. Munger K, Werness BA, Dyson N, Phelps WC, Harlow E, Howley PM. Complex formation of human papillomavirus E7 proteins with the retinoblastoma tumor suppressor gene product. *EMBO J* 1989; 8:4099–4105.

15. Whyte P, Williamson NM, Harlow E. Cellular targets for transformation by the adenovirus E1A proteins. *Cell* 1989; 56:67–75.

16. Hu QJ, Dyson N, Harlow E. The regions of the retinoblastoma protein needed for binding to adenovirus E1A or SV40 large T antigen are common sites for mutations. *EMBO J* 1990; 9:1147–1155.

17. Pilon AA, Desjardins P, Hassell JA, Mes-Masson AM. Functional implications of mutations within polyomavirus large T antigen Rb-binding domain: effects on pRb and p107 binding in vitro and immortalization activity in vivo. *J Virol* 1996; 70:4457–4465.

18. Kaelin WG Jr, Ewen ME, Livingston DM. Definition of the minimal simian virus 40 large T antigen- and adenovirus E1A-binding domain in the retinoblastoma gene product. *Mol Cell Biol* 1990; 10:3761–3769.

19. Moran E. A Region of SV40 large T antigen can substitute for a transforming domain of the adenovirus E1A products. *Nature* 1988; 334:168–170.

20. Moran E, Zarler B, Harrison TM, Mathews MB. Identification of separate domains in the adenovirus E1A gene for immortalization activity and the activation of virus early genes. *Mol Cell Biol* 1986; 6:3470–3480.

21. Souza RF, Yin J, Smolinski KN, Zou TT, Wang S, Shi YQ, et al. Frequent mutation of the E2F-4 cell cycle gene in primary human gastrointestinal tumors. *Cancer Res* 1997; 57:2350–2353.

22. Xu G, Livingston DM, Krek W. Multiple members of the E2F transcription factor family are the products of oncogenes. *Proc Natl Acad Sci USA* 1995; 92:1357–1361.

23. Flemington EK, Speck SH, and Kaelin WG. E2F1-mediated transactivation is inhibited by complex formation with the retinoblastoma susceptibility gene product. *Proc Natl Acad Sci USA* 1993; 90:6914–6918.

24. Weintraub SJ, Prater CA, Dean DC. Retinoblastoma protein switches the E2F site from positive to negative element. *Nature* 1992; 358:259–261.

25. Bremner R, Cohen BL, Sopta M, Hamel PA, Ingles CJ, Gallie BL, Phillips RA. Direct transcriptional repression by pRB and its reversal by specific cyclins. *Mol Cell Biol* 1995; 15:3256–3265.

26. Zacksenhaus E, Jiang Z, Phillips RA, Gallie BL. Dual mechanisms of repression of E2F1 activity by the retinoblastoma gene product. *EMBO J* 1996; 15:5917–5927.

27. Chow KN, Starostik P, Dean DC. The Rb family contains a conserved cyclin-dependent-kinase-regulated transcriptional repressor motif. *Mol Cell Biol* 1996; 16:7173–7181.

28. Starostik P, Chow KN, Dean DC. Transcriptional repression and growth suppression by the p107 pocket protein. *Mol Cell Biol* 1996; 16:3606–3614.

29. Lam EW, Watson RJ. An E2F-binding site mediates cell-cycle regulated repression of mouse B-myb transcription. *EMBO J* 1993; 12:2705–2713.

30. Voit R, Schafer K, Grummt I. Mechanism of repression of RNA polymerase I transcription by the retinoblastoma protein. *Mol Cell Biol* 1997; 17:4230–4237.

31. Chen G, Guy CT, Chen H-W, Hu N, Lee EY-HP, Lee W-H. Molecular cloning and developmental expression of mouse p130, a member of the retinoblastoma gene family. *J Biol Chem* 1996; 271:9567–9572.

32. Ewen ME, Xing Y, Lawrence JB, Livingston DM. Molecular cloning, chromosomal mapping, and expression of the cDNA for p107, a retinoblastoma gene product-related protein. *Cell* 1991; 66:1155–1164.

33. Chellappan SP, Hiebert S, Mudryi M, Horowitz JM, Nevins JR. The E2F transcription factor is a cellular target for the RB protein. *Cell* 1991; 65:1053–1061.

34. Zacny VL, Wilson J, Pagano JS. The Epstein-Barr virus immediate-early gene product, BRLF1, interacts with the retinoblastoma protein during the viral lytic cycle. *J Virol* 1998; 72:8043–8051.

35. Kato J-Y, Matsushime H, Hiebert SW, Ewen ME, Sherr CJ. Direct binding of cyclin D to the retinoblastoma gene product (Rb) and Rb phosphorylation by the cyclin D-dependent kinase Cdk4. *Genes Dev* 1993; 7:331–342.

36. Gu W, Schneider JW, Condorelli G, et al. Interaction of myogenic factors and the retinoblastoma protein mediates muscle cell commitment and differentiation. *Cell* 1993; 72:309–324.

37. Xiao Z-X, Chen J, Levine AJ, Modjtahedi N, Xing J, Sellers WR, Livingston DM, Interaction between the retinoblastoma protein and the oncoprotein MDM2. *Nature* 1995; 375:694–698.

38. Kim SJ, Wagner S, Liu F, O'Reilly MA, Robbins PD, Green MR. Retinoblastoma gene product activates expression of the human TGF-beta2 gene through transcription factor ATF-2. *Nature* 1992; 358:331–334.

39. Wang CY, Petryniak B, Thompson CB, kaelin WG, Leiden JM. Regulation of the Ets-related transcription factor ELF-1 by binding to the retinoblastoma protein. *Science* 1993; 260:1330–1335.

40. Shih HH, Tevosian SG, Yee AS. Regulation of differentiation by HBP1, a target of the retinoblastoma protein. *Mol Cell Biol* 1998; 18:4732–4743.

41. Robbins PD, Horowitz JM, Mulligan RC. Negative regulation of human c-fos expression by the retinoblastoma gene product. *Nature* 1990; 346:668–671.

42. Nead MA, Baglia LA, Antinore MJ, Ludlow JW, McCance DJ. Rb binds c-Jun activates transcription. *EMBO J* 1998; 17:2342–2352.

43. Park K, Choe J, Osifchin NE, Templeton DJ, Robbins PD, Kim SJ. The Human retinoblastoma susceptibility gene promoter is positively autoregulated by its own product. *J Biol Chem* 1994; 269:6083–6088.

44. Savoysky E, Mizuno T, Sowa Y, Watanabe H, Sawada J, Nomura H, et al. The retinoblastoma binding factor 1 (RBF-1) site in RB gene promoter binds preferentially E4TF1, a member of the Ets transcription factors family. *Oncogene* 1994; 9:1839–1846.

45. Batsche E, Muchardt C, Behrens J, Hurst HC, Cremisi C. RB and c-Myc activate expression of the E-cadherin gene in epithelial cells through interaction with transcription factor AP-2. *Mol Cell Biol* 1998; 18:3647–3658.

46. Wu F, Lee AS. Identification of AP-2 as an interactive target of Rb and a regulator of the G1/S control element of the hamster histone H3.2 promoter. *Nucleic Acids Res* 1998; 26:4837–4845.

47. Lasorella A, Iavarone A, Israel MA. Id2 specifically alters regulation of the cell cycle by tumor suppressor proteins. *Mol Cell Biol* 1996; 16:2570–2578.

48. Cavanaugh AH, Hempel WM, Taylor LJ, Rogalsky V, Todorov G, Rothblum LI. Activity of RNA polymerase I transcription factor UBF blocked by Rb gene product. *Nature* 1995; 374:177–180.

49. Larminie CGC, Cairns CA, Mital R, Martin K, Kouzarides T, Jackson SP, White RJ. Mechanistic analysis of RNA polymerase III regulation by the retinoblastoma protein. *EMBO J* 1997; 18:2061–2071.

50. Shao Z, Siegert JL, Ruppert S, Robbins PD. Rb interacts with TAF(II)250/TFIIID through multiple domains. *Oncogene* 1997; 15:385–392.

51. Siegert JL, Robbins PD. Rb inhibits the intrinsic kinase activity of TATA-binding protein-associated factor TAFII250. *Mol Cell Biol* 1999; 19:846–854.

52. Brehm A, Miska EA, McCance DJ, Reid JL, Bannister AJ, Kouzarides T. Retinoblastoma protein recruits histone deacetylase to repress transcription. *Nature* 1998; 391:597–601.

53. Magnaghi-Jaulin L, Groisman R, Naguibneve I, Robin P, Larain S, Le Villain JP, et al. Retinoblastoma protein represses transcription by recruiting a histone deacetylase. *Nature* 1998; 391:601–605.

54. Luo RX, Postigo AA, Dean DC. Rb interacts with histone deacetylase to repress transcription. *Cell* 1998; 92:463–473.

55. Hagemeier C, Bannister AJ, Cook A, Kouzarides T. The activation domain of transcription factor PU.1 binds the retinoblastoma (RB) protein and the transcription factor TFIID in vitro: RB shows sequence similarity to TFIID and TFIIB. *Proc Natl Acad Sci USA* 1993; 90:1580–1584.

56. Wen ST, Jackson PK, Van Etten RA. The cytostatic function of c-Abl is controlled by multiple nuclear localization signals and requires the p53 and Rb tumor suppressor gene products. *EMBO J* 1996; 15:1583–1595.

57. Wang S, Ghosh RN, Chellappan SP. Raf-1 physically interacts with Rb and regulates its function: a link between mitogenic signaling and cell cycle regulation. *Mol Cell Biol* 1998; 18:7487–7498.

58. Guan LS, Rauchman M, Wang ZY. Induction of Rb-associated protein (RbAp46) by Wilms' tumor suppressor WT1 mediates growth inhibition. *J Biol Chem* 1998; 273:27,047–27,050.

59. Ach RA, Taranto P, Gruissem W. A conserved family of WD-40 proteins binds to the retinoblastoma protein in both plants and animals. *Plant Cell* 1997; 9:1595–1606.

60. Qian YW, Wang YC, Hollingsworth RE Jr, Jones D, Ling N, Lee EY. A retinoblastoma-binding protein related to a negative regulator of Ras in yeast. *Nature* 1993; 364:648–652.

61. Chang KH, Chen Y, Chen TT, Chou WH, Chen PL, Ma YY, et al. Thyroid hormone receptor coactivator negatively regulated by the retinoblastoma protein. *Proc Natl Acad Sci USA* 1997; 94:9040–9045.

62. Huang S, Shao G, Liu L. The PR domain of the Rb-binding zinc finger protein R1Z1 is a protein binding interface and is related to the SET domain functioning in chromatin-mediated gene expression. *J Biol Chem* 1998; 273:15,933–15,939.

63. Buyse IM, Shao G, Huang S. The retinoblastoma protein binds to RIZ, a zinc-finger protein that shares an epitope with the adenovirus E1A protein. *Proc Natl Acad Sci USA* 1995; 92:4467–4471.

64. Woitach JT, Zhang M, Niu C-H, Thorgiersson SS. A retinoblastoma-binding protein that affects cell-cycle control and confers transforming ability. *Nat Genet* 1998; 19:371–374.

65. Fusco C, Reymond A, Zervos AS. Molecular cloning and characterization of a novel retinoblastoma-binding protein. *Genomics* 1998; 51:351–358.

66. Durfee T, Becherer K, Chen PL, Yeh SH, Yang Y, Kilburn AE, Lee WH, Elledge SJ. The retinoblastoma protein associates with the protein phosphatase type 1 catalytic subunit. *Genes Dev* 1993; 7:555–569.

67. Rustgi AK, Dyson N, Bernards R. Amino-terminal domains of c-myc and N-myc proteins mediate binding to the retinoblastoma gene product. *Nature* 1991; 352:541–544.

68. Adnane J, Robbins PD. The retinoblastoma susceptibility gene product regulates Myc-mediated transcription. *Oncogene* 1995; 10:381–387.

69. Trouche D, Le Chalony C, Muchardt C, Yaniv M, Kouzarides T. RB and hbrm cooperate to repress the activation functions of E2F1. *Proc Natl Acad Sci USA* 1997; 94:11,268–11,273.

70. Ozaki T, Saijo M, Murakami K, Enomoto H, Taya Y, Sakiyama S. Complex formation between lamin A and the retinoblastoma gene product: identification of the domain on lamin A required for its interaction. *Oncogene* 1994; 9:2649–2653.

71. Mancini MA, Shan B, Nickerson JA, Penman S, Lee WH. The retinoblastoma gene product is a cell cycle-dependent, nuclear matrix-associated protein. *Proc Natl Acad Sci USA* 1994; 91:418–422.

72. Inoue A, Torigoe T, Sogahata K, Kamiguchi K, Takahashi S, Sawada Y, et al. 70-kDa heat shock cognate protein interacts directly with the N-terminal region of the retinoblastoma gene product pRb. Identification of a novel region of pRb-mediating protein interaction. *J Biol Chem* 1995; 270:22,571–22,576.

73. Strober BE, Dunaief JL, Guna S, Goff SP. Functional interactions between the hBRM/hBRG1 transcriptional activators and the pRB family of proteins. *Mol Cell Biol* 1996; 16:1576–1583.

74. Radulescu RT, Bellitti MR, Ruvo M, Cassani G, Fassina G. Binding of the LXCXE insulin motif to a hexapeptide derived from retinoblastoma protein. *Biochem Biophys Res Commun* 1995; 206:97–102.

75. Lee JO, Russo AA, Pavletich NP. Structure of the retinoblastoma tumour-suppressor pocket domain bound to a peptide from HPV E7. *Nature* 1998; 391:859–865.

76. Bagchi S, Weinmann R, Raychaudhuri P. The retinoblastoma protein copurifies with E2F-1, an E1A-regulated inhibitor of the transcription factor E2F. *Cell* 1991; 65:1063–1072.

77. Lees JA, Saito M, Vidal M, Valentine M, Look T, Harlow E, Dyson N, Helin K. The retinoblastoma protein binds to a family of E2F transcription factors. *Mol Cell Biol* 1993; 13:7813–7825.

78. Li J-M, Hu PP-C, Shen X, Yu Y, Wang X-F. E2F4-Rb and E2F4-p107 complexes suppress gene expression by transforming growth factor β through E2F binding sites. *Proc Natl Acad Sci USA* 1997; 94:4948–4953.

79. Cobrinik D, Whyte P, Peeper DS, Jacks T, Weinberg RA. Cell cycle-specific association of E2F with the p130 E1A-binding domain. *Genes Dev* 1993; 7:2392–2404.

80. Shan B, Zhu X, Chen PL, Durfee T, Yang Y, Sharp D, Lee WH. Molecular cloning of cellular genes encoding retinoblastoma-associated proteins: identification of a gene with properties of the transcription factor E2F. *Mol Cell Biol* 1992; 12:5620–5631.

81. Whitaker LL, Su H, Baskaran R, Knudsen ES, Wang JYJ. Growth suppression by an E2F-binding-defective retinoblastoma protein (RB): Contribution from the RB C pocket. *Mol Cell Biol* 1998; 18:4032–4042.

82. Chittenden T, Livingston DM, Kaelin WG Jr. The T/E1A-binding domain of the retinoblastoma product can interact selectively with a sequence-specific DNA-binding protein. *Cell* 1991; 65:1073–1082.

83. Welch PJ, Wang JY. C-terminal protein-binding domain in the retinoblastoma protein regulates nuclear c-Abl tyrosine kinase in the cell cycle. *Cell* 1993; 75:779–790.

84. Whitaker LL, Su H, Baskaran R, Knudsen ES, Wang JY. Growth suppression by an E2F-binding-defective retinoblastoma protein (RB): contribution from the RB C pocket. *Mol Cell Biol* 1998; 18:4032–4042.

85. Faha B, Ewen ME, Tsai L-H, Livingston DM, Harlow E. Interaction between human cyclin A and adenovirus E1A-associated p107 protein. *Science* 1992; 255:87–90.

86. Ewen ME, Faha B, Harlow E, Livingston DM. Interaction of p107 with cyclin A independent of complex formation with viral oncoproteins. *Science* 1992; 255:85–87.

87. Hannon GJ, Demetrick D, Beach D. Isolation of the Rb-related p130 through its interaction with CDK2 and cyclins. *Genes Dev* 1993; 7:2378–2391.

88. Li Y, Graham C, Lacy S, Duncan AMV, Whyte P. The Adenovirus E1A-associated 130-kD protein is encoded by a member of the retinoblastoma gene family and physically interacts with cyclins A and E. *Genes Dev* 1993; 7:2366–2377.

89. Lu X, Horvitz HR. lin-35 and lin-53, two genes that antagonize a C. elegans Ras pathway, encode proteins similar to Rb and its binding protein RbAp48. *Cell* 1998; 95:981–991.

90. Du W, Vidal M, Xie J-E, Dyson N. RBF, a novel RB-related gene that regulates EF activity and interacts with cyclin E in *Drosophila*. *Genes Dev* 1996; 10:1206–1218.

91. Dyson N. The Regulation of E2F by pRb-family proteins. *Genes Dev* 1998; 12:2245–2262.

92. DeGregori J, Kowalik T, Nevins JR. Cellular targets for activation by the E2F1 transcription factor include DNA synthesis- and G1/S-regulatory genes. *Mol Cell Biol* 1995; 15:4215–4224.

93. Oswald F, Dobner T, Lipp M. The E2F transcription factor activates a replication-dependent human H2A gene in early S phase of the cell cycle. *Mol Cell Biol* 1996; 16:1889–1895.

94. Qin X-Q, Livingston DM, Ewen M, Sellers WR, Arany Z, Kaelin WG Jr. The Transcription factor E2F-1 is a downstream target of RB action. *Mol Cell Biol* 1995; 15:742–755.

95. Zwicker J, Liu N, Engeland K, Lucibello FC, Muller R. Cell cycle regulation of E2F site occupation in vivo. *Science* 1996; 271:1595–1597.

96. DeCaprio JA, Ludlow JW, Lynch D, Furukawa Y, Griffin J, Piwnica-Worms H, Huang C-M, Livingston DM. The Product of the retinoblastoma susceptibility gene has properties of a cell cycle regulatory element. *Cell* 1989; 58:1085–1095.

97. Schwarz JK, Devoto SH, Smith EJ, Chellappan SP, Jakoi L, Nevins JR. Interactions of the p107 and Rb proteins with E2F during the cell proliferation response. *EMBO J* 1993; 12:1013–1020.

98. Corbeil HB, Whyte P, Branton PE. Characterization of transcription factor E2F complexes during muscle and neuronal differentiation. *Oncogene* 1995; 11:909–920.

99. Smith EJ, Leone G, DeGregori J, Jakoi L, Nevins JR. The Accumulation of an E2F-p130 transcriptional repressor distinguishes a G0 cell state from a G1 cell state. *Mol Cell Biol* 1996; 16:6965–6976.

100. Knudsen ES, Buckmaster C, Chen TT, Feramisco JR, Wang JY. Inhibition of DNA synthesis by RB: effects on G1/S transition and S-phase progression. *Genes Dev* 1998; 12:2278–2292.

101. Niculescu III AB, Chen X, Smeets M, Hengst L, Prives C, Reed SI. Effects of p21Cip1/Waf1 at both the G1/S and the G2/M cell cycle transitions: pRb is a critical determinant in blocking DNA replication and in preventing endoreduplication. *Mol Cell Biol* 1998; 18:629–643.

102. Moberg K, Starz M, Lees JA. E2F-4 switches from p130 to p107 and pRb in response to cell cycle reentry. *Mol Cell Biol* 1996; 16:1436–1449.

103. Dyson N, Dembski M, Fattaey A, Ngwu C, Ewen M, Helin K. Analysis of p107-associated proteins: p107 associates with a form of E2F that differs from pRb-associated E2F-1. *J Vir.* 1993; 67:7641–7647.

104. Beijersbergen RL, Kerkhoven RM, Zhu L, Carlee L, Voorhoeve PM, Bernards R. E2F-4, a new member of the E2F gene family, has oncogenic activity and associates with p107 in vivo. *Genes Dev* 1994; 8:2680–2690.

105. Ginsberg D, Vairo G, Chittenden T, Xiao Z-X, Xu G, Wydner KL, et al. E2F-4, a new member of the E2F transcription factor family, interacts with p107. *Genes Dev* 1994; 8:2665–2679.

106. Vairo G, Livingston DM, Ginsberg D. Functional interaction between E2F-4 and p130: evidence for distinct mechanisms underlying growth suppression by different retinoblastoma protein family members. *Genes Dev* 1995; 9:869–881.

107. Hijmans EM, Voorhoeve PM, Beijersbergen RL, van't Veer LJ, Bernards R. E2F-5, a new E2F family member that interacts with p130 in vivo. *Mol Cell Biol* 1995; 15:3082–3089.

108. Hurford RK Jr, Cobrinik D, Lee MH, Dyson N. pRb and p107/p130 are required for the regulated expression of different sets of E2F responsive genes. *Genes Dev* 1997; 11:1447–1463.

109. Tao Y, Kassatly RF, Cress WD, Horowitz JM. Subunit composition determines E2F DNA-binding site specificity. *Mol Cell Biol* 1997; 17:6994–7007.

110. Morkel M, Wenkel J, Bannister AJ, Kouzarides T, Hagemeier C. An E2F-like repressor of transcription. *Nature* 1997; 390:567–568.

111. Dynlacht BD, Flores O, Lees JA, Harlow E. Differential regulation of E2F trans-activation by cyclin/cdk2 complexes *Genes Dev* 1994; 8:1772–1786.

112. Hofmann F, Livingston DM. Differential effects of cdk2 and cdk3 on the control of pRb and E2F function during G1 exit. *Genes Dev* 1996; 10:851–861.

113. Krek W, Xu G, Livingston DM. Cyclin A-kinase regulation of E2F-1 DNA binding function underlies suppression of an S phase checkpoint. *Cell* 1995; 83:1149–1158.

114. Bandara LR, Buck VM, Zamanian M, Johnston LH, La Thangue NB. Functional synergy between DP-1 and E2F-1 in the cell cycle-regulating transcription factor DRTF1/E2F. *EMBO J* 1993; 12:4317–4324.

115. Weintraub SJ, Prater CA, Dean DC. Retinoblastoma protein switches the E2F site from positive to negative element. *Nature* 1992; 358:259–261.

116. Cartwright P, Muller H, Wagener C, Holm K, Helin K. E2F-6: A novel member of the E2F family is an inhibitor of E2F-dependent transcription. *Oncogene* 1998; 17:611–623.

117. Trimarchi JM, Fairchild B, Verona R, Moberg K, Andon N, Lees JA. E2F-6, a member of the E2F family that can behave as a transcriptional repressor. *Proc Natl Acad Sci USA* 1998; 95:2850–2855.

118. White RJ, Trouche D, Martin K, Jackson SP, Kouzarides T. Repression of RNA polymerase III transcription by the retinoblastoma protein. *Nature* 1996; 382:88–90.

119. White RJ. Regulation of RNA polymerases I and III by the retinoblastoma protein: a mechanism for growth control? *TIBS* 1997; 22:77–80.

120. Cairns CA, White RJ. p53 is a general repressor of RNA polymerase III transcription. *EMBO J* 1998; 17:3112–3123.

121. White RJ. Transcription factor IIIB: An important determinant of biosynthetic capacity that is targeted by tumor suppressors and transforming proteins. *Int J Oncol* 1998; 12:741–748.

122. Geiduschek EP, Kassavetis GA. Comparing transcriptional initiation by RNA polymerases I and III. *Curr Opin Cell Biol* 1995; 7:344–351.

123. Johnson DG, Schwarx JK, Cress WD, Nevins JR. Expression of transcription factor E2F1 induces quiescent cells to enter S phase. *Nature* 1993; 365:349–352.

124. Mauck JC, Green H. Regulation of pre-transfer RNA synthesis during transition from resting to growing state. *Cell* 1974; 3:171–177.

125. Larminie CGC, Alzuherri HM, Cairns CA, McLees A, White RJ. Transcription by RNA polymerases I and III: a potential link between cell growth, protein synthesis and the retinoblastoma protein. *J Mol Med* 1998; 76:94–103.

126. Ferreira R, Magnaghi-Jaulin L, Robin P, Harel-Bellan A, Trouche D. The three members of the pocket proteins family share the ability to repress E2F activity through recruitment of a histone deacetylase. *Proc Natl Acad Sci USA* 1998; 95:10,493–10,498.

127. Ait-Si-Ali S, Ramirez S, Barre FX, Dkhissi F, Magnaghi-Jaulin L, Girault JA, et al. Histone acetyltransferase activity of CBP is controlled by cycle-dependent kinases and oncoprotein E1A. *Nature* 1998; 396:184–186.

128. Gu W, Roeder RG. Activation of p53 sequence-specific DNA binding by acetylation of the p53 C-terminal domain. *Cell* 1997; 90:595–606.

129. Wolffe AP. Transcriptional control. Sinful repression. *Nature* 1997; 387:16–17.

130. Struhl K. Histone acetylation and transcriptional regulatory mechanisms. *Genes Dev* 1998; 12:599–606.

131. Pazin MJ, Kadonaga JT. What's up and down with histone deacetylation and transcription? *Cell* 1997; 89:325–328.

132. DePinho RA. The cancer-chromatin connection. *Nature* 1998; 391:533–536.

133. LaTangue NB. DRTF1/E2F: an expanding family of heterodimeric transcription factors implicated in cell-cycle control. *Trends Biochem Sci* 1994; 19:108–114.

134. Matsushime H, Quelle DE, Shurtleff SA, Shibuya M, Sherr CJ, Kato J-Y. D-type cyclin-dependent kinase activity in mammalian cells. *Mol Cell Biol* 1994; 14:2066.

135. Sewing A, Burger C, Brusselbach S, Schalk C, Lucibello FC, Muller R. Human cyclin D1 encodes a labile nuclear protein wose synthesis is directly induced by growth factors and suppressed by cyclic AMP. *J Cell Sci* 1993; 104:545–554.

136. Zhu X, Ohsubo M, Bohmer RM, Roberts JM, Assoian RK. Adhesion-dependent cell cycle progression linked to the expression of cyclin D1, activation of cyclin E-cdk2, and phosphorylation of the retinoblastoma protein. *J Cell Biol* 1996; 133:391–403.

137. Dowdy SF, Hinds PW, Louie K, Reed SI, Arnold A, Weinberg RA. Physical interaction of the retinoblastoma protein with human D cyclins. *Cell* 1993; 73:499–511.

138. Ewen ME, Sluss HK, Sherr CJ, Matsushime H, Kato J-Y, Livingston DM. Functional interactions of the retinoblastoma protein with mammalian D-type cyclins. *Cell* 1993; 73:487–497.

139. Connell-Crowley L, Harper JW, Goodrich DW. Cyclin D1/Cdk4 regulates retinoblastoma protein-mediated cell cycle arrest by site-specific phosphorylation. *Mol Biol Cell* 1997; 8:287–301.

140. Ohtsubo M, Theodoras AM, Schumacher J, Roberts JM, Pagano M. Human cyclin E, a nuclear protein essential for the G1-to-S phase transition. *Mol Cell Biol* 1995; 15:2612–2624.

141. Knoblich JA, Sauer K, Jones L, Richardson H, Saint R, Lehner CF. Cyclin E controls S phase progression and its down-regulation during Drosophila embryogenesis is required for the arrest of cell proliferation. *Cell* 1994; 77:107–120.

142. Peeper DS, Upton TM, Ladha MH, Neuman E, Zalvide J, Bernards R, DeCaprio JA, Ewen ME. Ras signalling linked to the cell-cycle machinery by the retinoblastoma protein. *Nature* 1997; 386:177–181.

143. Arber N, Sutter T, Miyake M, Kahn SM, Venkatraj VS, Sobrino A, et al. Increased expression of cyclin D1 and the Rb tumor suppressor gene in c-K-ras transformed rat enterocytes. *Oncogene* 1996; 12:1903–1908.

144. Fan J, Bertino JR. K-ras modulates the cell cycle via both positive and negative regulatory pathways. *Oncogene* 1997; 14:2595–2607.

145. Mittnacht S, Paterson H, Olson MF, Marshall CJ. Ras signaling is required for inactivation of the tumour suppressor pRb cell-cycle control protein. *Curr Biol* 1997; 7:219–221.

146. Ladha MH, Lee KY, Upton TM, Reed MF, Ewen ME. Regulation of exit from quiescence by p27 and cyclin D1-CDK4. *Mol Cell Biol* 1998; 18:6605–6615.

147. Cheng M, Sexi V, Sherr CJ, Roussel MF. Assembly of cyclin D-dependent kinase and titration of p27Kip1 regulated by mitogen-activated protein kinase kinase (MEK1). *Proc Natl Acad Sci USA* 1998; 95:1091–1096.

148. Soos TJ, Kiyokawa H, Yan JS, Rubin MS, Giordano A, DeBlasio A, et al. Formation of p27-CDK complexes during the human mitotic cell cycle. *Cell Growth Differ* 1996; 7:135–146.

149. Durand B, Fero ML, Roberts JM, Raff MC. p27Kop1 alters the response of cells to mitogen and is part of a cell-intrinsic timer that arrests the cell cycle and initiates differentiation. *Curr Biol* 1998; 8:431–440.

150. Hengst L, Reed SI. Translational control of p27Kip1 accumulation during the cell cycle. *Science* 1996; 271:1861–1864.

151. Dietrich C, Wallenfang K, Oesch F, Wieser R. Differences in the mechanisms of growth control in contact-inhibited and serum-deprived human fibroblasts. *Oncogene* 1997; 15:2743–2747.

152. Rivard N, L'Allemain G, Bartek J, Pouyssegur J. Abrogation of p27Kip1 by cDNA antisense suppresses quiescence (G0 state) in fibroblasts. *J Biol Chem* 1996; 271:18,337–18,341.

153. Nguyen H, Gitig DM, Koff A. Cell-free degradation of p27(kip1), a G1 cyclin-dependent kinase inhibitor, is dependent on CDK2 activity and the proteasome. *Mol Cell Biol* 1999; 19:1190–1201.

154. Lukas J, Bartkova J, Rogde M, Strauss M, Bartek J. Cyclin D1 is dispensable for G1 control in retinoblastoma gene-deficient cells independently of cdk4 activity. *Mol Cell Biol* 1995; 15:2600–2611.

155. Lukas J, Parry D, Aagard L, Mann DJ, Bartkova J, Strauss M, Peters G, Bartek J. Retinoblastoma-protein-dependent cell-cycle inhibition by the tumor suppressor p16. *Nature* 1995; 375:503–506.

156. Guan K-L, Jenkins CW, Li Y, Nichols MA, Wu X, O'Keefe CL, Matera AG, Xiong Y. Growth suppression by p18, a p16INK4/MTS1 and p14INK4B/MS2-related Cdk6 inhibitor, correlates with wild-type Rb function. *Genes Dev* 1994; 8:2939–2952.

157. Medema RH, Herrera RE, Lamb F, Weinberg RA. Growth suppression by p16ink4 requires functional retinoblastoma protein. *Proc Natl Acad Sci USA* 1995; 92:6289–6293.

158. Horton LE, Qian Y, Templeton DJ. G1 cyclins control the retinoblastoma gene product growth regulation activity via upstream mechanisms. *Cell Growth Differ* 1995; 6:395–407.

159. Mittnacht S, Lees JA, Desai D, Harlow E, Morgan DO, Weinberg RA. Distinct sub-populations of the retinoblastoma protein show a distinct pattern of phosphorylation. *EMBO J* 1994; 13:118–127.

160. Akiyama T, Ohuchi T, Sumida S, Matsumoto K, Toyoshima K. Phosphorylation of the retinoblastoma protein by Cdk2. *Proc Natl Acad Sci USA* 1992; 89:7900–7904.

161. Knudsen ES, Wang JYJ. Differential regulation of retinoblastoma protein function by specific Cdk phosphorylation sites. *J Biol Chem* 1996; 14:8313–8320.

162. DeCaprio JA, Furukawa Y, Ajchenbaum F, Griffin JD, Livingston DM. The retinoblastoma-susceptibility gene product becomes phosphorylated in multiple stages during cell cycle entry and progression. *Proc Natl Acad Sci USA* 1992; 89:1795–1798.

163. Kitagawa M, Higashi H, Jung H-K, Suzuki-Takahashi I, Ikeda M, Tamai K, et al. The consensus motif for phosphorylation by cyclin D1-Cdk4 is different from that for phosphorylation by cyclin A/E-Cdk2. *EMBO J* 1996; 15:7060–7069.

164. Lundberg AS, Weinberg RA. Functional inactivation of the retinoblastoma protein requires sequential modification by at least two distinct cyclin-cdk complexes. *Mol Cell Biol* 1998; 18:753–761.

165. Adams PD, Li X, Sellers WR, Baker KB, Leng X, Harper JW, Taya Y, Kaelin WG Jr. The retinoblastoma protein contains a C-terminal motif that targets it for phosphorylation by cyclin/cdk2 complexes. *Mol Cell Biol,* in press.

166. Lukas J, Herzinger T, Hansen K, Moroni MC, Resnitzky D, Helin I, et al. Cyclin E-induced S phase without activation of the Rb/E2F pathway. *Genes Dev* 1997; 11:1479–1492.

167. Leng X, Connell-Crowley L, Goodrich D, Harper JW. S-phase entry upon ectopic expression of G1 cyclin-dependent kinases in the absence of retinoblastoma protein phosphorylation. *Curr Biol* 1997; 7:709–712.

168. Connell-Crowley L, Elledge SJ, Harper JW. G1 cyclin-dependent kinases are sufficient to initiate DNA synthesis in quiescent human fibroblasts. *Curr Biol* 1997; 8:65–68.

169. Leone G, DeGregori J, Jakoi L, et al. Collaborative role of E2F transcriptional activity and G1 cyclin dependent Kinase activity in the induction of S-phase. *Proc Natl Acad Sci USA* 1999; 96:6626–6631.

170. Lee EY-HP, Chang C-Y, Hu N, Wang Y-CJ, Lai C-C, et al. Mice deficient for Rb are nonviable and show defects in neurogenesis and haematopoiesis. *Nature* 1992; 359:288–295.

171. Jacks T, Fazeli A, Schmitt EM, Bronson RT, Goodell MA, Weinberg RA. Effects of an Rb mutation in the mouse. *Nature* 1992; 359:295–300.

172. Clarke AR, Maandag ER, van Roon M, van der Lugt NMT, van der Valk M, et al. Requirement for a functional Rb-1 gene in murine development. *Nature* 1993; 359:328–330.

173. Macleod KF, Hu Y, and Jacks T. Loss of Rb activates both p53-dependent and independent cell death pathways in the developing mouse nervous system. *EMBO J* 1996; 15:6176–6188.

174. Hu N, Gulley ML, Kung JT, Lee EY. Retinoblastoma gene deficiency has mitogenic but not tumorigenic effects on erythropoiesis. *Cancer Res* 1997; 57:4123–4129.

175. Williams BO, Schmitt EM, Remington L, Bronson RT, Albert DM, Weinberg RA, Jacks T. Extensive contribution of Rb-deficient cells to adult chimeric mice with limited histopathological consequences. *EMBO J* 1994; 13:4251–4259.

176. Maandag EC, van der Valk M, Vlaar M, Feltkamp C, O'Brien J, van Roon M, et al. Developmental rescue of an embryonic-lethal mutation in the retinoblastoma gene in chimeric mice. *EMBO J* 1994; 13:4260–4268.

177. Hu N, Gutsmann A, Herbert DC, Bradley A, Lee WH, Lee EY. Heterozygous Rb-1 delta 20/+ mice are predisposed to tumors of the pituitary gland with a nearly complete penetrance. *Oncogene* 1994; 9:1021–1027.

178. Nikitin AY, Lee WH. Early loss of the retinoblastoma gene is associated with impaired growth inhibitory innervation during melanotroph carcinogenesis in Rb+/– mice. *Genes Dev* 1996; 10:1870–1879.

179. Lee M-H, Williams BO, Mulligan G, Mukai S, Bronson RT, Dyson N, Harlow E, Jacks T. Targeted disruption of p107: functional overlap between p107 and Rb. *Genes Dev* 1996; 10:1621–1632.

180. Robanus-Mandag E, Dekker M, van der Valk M, Carrozza ML, Jeanny JC, Dannenberg JH, Berns A, Riele H. p107 is a suppressor of retinoblastoma development in pRb-deficient mice. *Genes Dev* 1998; 12:1599–1609.

181. LeCouter JE, Hardy WR, Ying C, Megeney LA, May LL, Rudnicki MA. Strain-dependent myeloid metaplasia, growth deficiency, and shortened cell-cycle in mice lacking p107. *Mol Cell Biol* 1998; 18:7455–7465.

182. Cobrinik D, Lee M-H, Hannon G, Mulligan G, Bronson RT, Dyson N, et al. Shared role of the pRB-related p130 and p107 proteins in limb development. *Genes Dev* 1996; 10:1633–1644.

183. LeCouter JE, Kablar B, Whyte PFM, Ying C, Rudnicki MA. Strain-dependent embryonic lethality in mice lacking the retinoblastoma-related p130 gene. *Development* 1998; 125:4669–4679.

184. Mulligan GJ, Wong J, Jacks T. p130 is dispensable in peripheral T lymphocytes: evidence for functional compensation by p107 and pRB. *Mol Cell Biol* 1998; 18:206–220.

185. Chittenden T, Livingston DM, DeCaprio JA. Cell cycle analysis of E2F in primary human T cells reveals novel E2F complexes and biochemically distinct forms of free E2F. *Mol Cell Biol* 1993; 13:3975–3983.

186. Field SJ, Tsai FY, Kuo F, Zubiaga AM, Kaelin WG Jr, Livingston DM, Orkin SH, Greenberg ME. E2F-1 functions in mice to promote apoptosis and suppress proliferation. *Cell* 1996; 85:549–561.

187. Yamasaki L, Jacks T, Bronson R, Goillot E, Harlow E, Dyson NJ. Tumor induction and tissue atrophy in mice lacking E2F-1. *Cell* 1996; 85:537–548.

188. Yamasaki L, Bronson R, Williams BO, Dyson NJ, Harlow E, Jacks T. Loss of E2F-1 reduces tumorigenesis and extends the lifespan of Rb1(+/–) mice. *Nat Genet* 1998; 18:360–364.

189. Tsai KY, Hu Y, Macleod KF, Crowley D, Yamasaki L, Jacks T. Mutation of E2F-1 suppresses apoptosis and inappropriate S phase entry and extends survival of Rb-deficient mouse embryos. *Mol Cell* 1998; 2:293–304.

190. Field SJ, Tsai F-Y, Kuo F, Zubiaga AM, Kaelin WG Jr, Livingston DM, Orkin SH, Greenberg ME. E2F-1 functions in mice to promote apoptosis and suppress proliferation. *Cell* 1996; 85:549–561.

191. Hasty P, Bradley A, Morris JH, Edmondson DG, Venuti JM, Olson EN, Klein WH. Muscle deficiency and neonatal death in mice with a targeted mutation in the myogenin gene. *Nature* 1993; 364:501–506.

192. Nabeshima Y, Hanaoka K, Hayasaka M, Esumi E, Li S, Nonaka I. Myogenin gene disruption results in perinatal lethality because of severe muscle defect. *Nature* 1993; 364:532–535.

193. Rawls A, Valdez MR, Zhang W, Richardson J, Klein WH, Olson EN. Overlapping functions of the myogenic bHLH genes MRF4 and MyoD revealed in double mutant mice. *Development* 1998; 125:2349–2358.

194. Rudnicki MA, Schnegelsberg PNJ, Stead RH, Braun T, Arnold H-H, Jaenisch R. MyoD or Myf-5 is required for the formation of skeletal muscle. *Cell* 1993; 75:1351–1359.

195. Rudnicki MA, Jaenisch R. The MyoD family of transcription factors and skeletal myogenesis. *BioEssays* 1995; 17:203–209.

196. Rao SS, Chu C, Kohtz DS. Ectopic expression of cyclin D1 prevents activation of gene transcription by myogenic basic helix-loop-helix regulators. *Mol Cell Biol* 1994; 14:5259–5267.

197. Skapek SX, Rhee J, Spicer DB, and Lassar AB. Inhibition of myogenic differentiation in proliferating myoblasts by cyclin D1-dependent kinase. *Science* 1995; 267:1022–1024.

198. Skapek SX, Rhee J, Kim PS, Novitch BG, and Lassar AB. Cyclin-mediated inhibition of muscle gene expression via a mechanism that is independent of pRB hyperphosphorylation. *Mol Cell Biol* 1996; 16:7043–7053.

199. Taylor DA, Kraus VB, Schwarz JJ, Olson EN, Kraus WE. E1A-mediated inhibition of myogenesis correlates with a direct physical interaction of E1A12S and basic helix-loop-helix proteins. *Mol Cell Biol* 1993; 13:4714–4727.

200. Crescenzi M, Soddu S, Sacchi A, Tato F. Adenovirus infection induces reentry into the cell cycle of terminally differentiated skeletal muscle cells. *Ann NY Acad Sci* 1995; 752:9–18.

201. Gu W, Schneider JW, Condoreili G, Kaushal S, Mahdave V, Nadal-Ginard B. Interaction of myogenic factors and the retinoblastoma protein mediates muscle cell commitment and differentiation. *Cell* 1993; 72:309–324.

202. Novitch BG, Mulligan GJ, Jacks T, Lassar AB. Skeletal muscle cells lacking the retinoblastoma protein display defects in muscle gene expression and accumulate in S and G2 phases of the cell cycle. *J Cell Biol* 1996; 135:441–456.

203. Schneider JW, Gu W, Zhu L, Mahdavi V, Nadal-Ginard B. Reversal of terminal differentiation mediated by p107 in Rb–/– muscle cells. *Science* 1994; 264:1467–1471.

204. Zacksenhaus E, Jiang Z, Chung D, Marth JD, Phillips RA, Gallie BL. pRb controls proliferation, differentiation, and death of skeletal muscle cells and other lineages during embryogenesis. *Genes Dev* 1996; 10:3051–3064.

205. Lavender P, Vandel L, Bannister AJ, Kouzarides T. The HMG-box transcription factor HBP1 is targeted by the pocket proteins and E1A. *Oncogene* 1997; 14:2721–2728.

206. Deng C, Zhang P, Harper JW, Elledge SJ, Leder P. Mice lacking p21CIP1/WAF1 undergo normal development, but are defective in G1 checkpoint control. *Cell* 1995; 82:675–684.

207. Zhang P, Wong C, Liu D, Finegold M, Harper JW, Elledge SJ. p21CIP1 and p57KIP2 control muscle differentiation at the myogenin step. *Genes Dev* 1999; 13:213–224.

208. Andres V, Walsh K. Myogenin expression, cell cycle withdrawal, and phenotypic differentiation are temporally separable events that precede cell fusion upon myogenesis. *J Cell Biol* 1996; 132:657–666.

209. Parker SB, Eichele G, Zhang P, Rawls A, Sands AT, Bradley A, Olson EN, Harper JW, Elledge SJ. p53 independent expression of p21Cip1 in muscle and other terminally differentiating cells. *Science* 1995; 267:1024–1027.

210. Almasan A, Yin Y, Kelly RE, Lee EY-HP, Bradley A, Li W, Bertino JR, Wahl GM. Deficiency of retinoblastoma protein leads to inappropriate S-phase entry, activation of E2F-responsive genes, and apoptosis. *Proc Natl Acad Sci USA* 1995; 92:5436–5440.

211. DeGregori J, Leone G, Miron A, Jakoi L, Nevins JR. Distinct roles for E2F proteins in cell growth control and apoptosis. *Proc Natl Acad Sci USA* 1997; 94:7245–7250.

212. Pan H, Yin C, Dyson NJ, Harlow E, Yamasaki L, Dyke TV. Key roles for E2F1 in signaling p53-dependent apoptosis and in cell division within developing tumors. *Mol Cell* 1998; 2:283–292.

213. Qin X-Q, Livingston DM, Kaelin WG Jr, Adams PD. Deregulated transcription factor E2F-1 expression leads to S-phase entry and p53-mediated apoptosis. *Proc Natl Acad Sci USA* 1994; 91:10,918–10,922.

214. Shan B, Lee W-H. Deregulated expression of E2F-1 induces S-phase entry and leads to apoptosis. *Mol Cell Biol* 1994; 14:8166–8173.

215. Bates S, Phillips AC, Clark PA, Stott F, Peters G, Ludwig RL, Vousden KH. p19ARF links the tumour suppressors RB and p53. *Nature* 1998; 395:124–125.

216. Palmero I, Pantoja C, Serrano M. p19ARF links the tumour suppressor p53 to Ras. *Nature* 1998; 395:125–126.

217. Pomerantz J, Schreiber-Agus N, Liegeois NJ, Silverman A, Alland L, Chin L, et al. The Ink4a tumor suppressor gene product, p19Arf, interacts with MDM2 and neutralizes MDM2's inhibition of p53. *Cell* 1998; 92:713–723.

218. Zhang Y, Xiong Y, Yarbrough WG. ARF promotes MDM2 degradation and stabilizes p53: ARF-INK4a locus deletion impairs both the Rb and p53 tumor suppression pathways. *Cell* 1998; 92:725–734.

219. Honda R, Yasuda H. Association of p19(ARF) with mdm2 inhibits ubiquitin ligase activity of mdm2 for tumor suppressor p53. *EMBO J* 1999; 18:22–27.

220. Kamijo T, Weber JD, Zambetti G, Zindy F, Roussel MF, Sherr CJ. Functional and physical interactions of the ARF tumor suppressor with p53 and Mdm2. *Proc Natl Acad Sci USA* 1998; 95:8292–8297.

221. Jones DL, Thompson DA, Munger K. Destabilization of the RB tumor suppressor protein and stabilization of p53 contribute to HPV type 16 E7-induced apoptosis. *Virology* 1997; 238:97–107.

222. Janicke RU, Walker PA, Lin XY, Porter AG. Specific cleavage of the retinoblastoma protein by an ICE-like protease in apoptosis. *EMBO J* 1996; 15:6969–6978.

223. An B, Dou QP. Cleavage of retinoblastoma protein during apoptosis: and interleukin 1 beta-converting enzyme-like protease as candidate. *Cancer Res* 1996; 56:438–442.

224. Bowen C, Spiegel S, Gelmann EP. Radiation-induced apoptosis mediated by retinoblastoma protein. *Cancer Res* 1998; 58:3275–3281.

225. Tan X, Wang JYJ. The caspase-RB connection in cell death. *Trends Cell Biol* 1998; 8:116–120.

226. Chen X, Ko LJ, Jayaraman L, Prives C. p53 levels, functional domains, and DNA damage determine the extent of the apoptotic response of tumor cells. *Genes Dev* 1996; 10:2438–2451.

227. Wyllie FS, Haughton MF, Bond JA, Rowson JM, Jones CJ, Wynford-Thomas D. S phase cell-cycle arrest following DNA damage is independent of the p53/p21(WAF1) signalling pathway. *Oncogene* 1996; 12:1077–1082.

228. Macleod KF, Sherry N, Hannon G, Beach D, Tokino T, Kinzler K, Vogelstein B, Jacks T. p53-dependent and independent expression of p21 during cell growth, differentiation, and DNA damage. *Genes Dev* 1995; 9:935–944.

229. Bunz F, Dutriaux A, Lengauer C, Waldman T, Zhou S, Brown JP, et al. Requirement for p53 and p21 to sustain G2 arrest after DNA damage. *Science* 1998; 282:1497–1501.

230. Niculescu AB III, Chen X, Smeets M, Hengst L, Prives C, Reed SI. Effects of p21(Cip1/Waf1) at both the G1/S and the G2/M cell cycle transitions: pRb is a critical determinant in blocking DNA replication and in preventing endoreduplication. *Mol Cell Biol* 1998; 18:629–643.

231. Harrington EA, Bruce JL, Harlow E, Dyson N. pRB plays an essential role in cell cycle arrest induced by DNA damage. *Proc Natl Acad Sci USA* 1998; 95:11,945–11,950.

232. Herrera RE, Sah VP, Williams BO, Makela TP, Weinberg RA, Jacks T. Altered cell cycle kinetics, gene expression, and G1 restriction point regulation in Rb-deficient fibroblasts. *Mol Cell Biol* 1996; 16:2402–2407.

7

p53 Tumor Suppressor Protein

Margaret Ashcroft, PhD and
Karen H. Vousden, PhD

CONTENTS

1. INTRODUCTION

The p53 protein was originally identified during the late 1970s, by several independent groups, as a novel cellular protein that was tightly associated with the large T antigen in cells transformed by simian virus-40 (SV40) *(1–3)*. Although originally thought to function as an oncogene, isolation of the wild-type (wt) gene encoding p53 led to the discovery that p53 functioned as a potent tumor suppressor *(4–8)*. A role for p53 in preventing malignant progression was subsequently demonstrated by the observations that transfection of p53 into cultured cells inhibited transformation by a number of oncogenes *(9)*, and that mice lacking the *p53* gene rapidly developed tumors with high incidence *(10–12)*. It is now known that *p53* is one of the most frequently mutated genes in human cancer, in which loss of function mutations contribute to the development of many major human malignancies. Approximately 50% of all human tumors carry a *p53* mutation, and at least 52 different types of tumor have *p53* mutations *(13–15)*. During the past decade, p53 has been brought to the forefront of cancer research, and intensive investigation has provided insight into how it mediates its tumor suppressor activities, and how these activities are regulated. Elucidation of the mechanisms that activate and regulate p53, and the identification of upstream and downstream effectors and targets involved in p53 function, should contribute to understanding how cancers arise, and to the development of new therapeutic tools for their treatment.

2. STRUCTURAL FEATURES OF THE P53 PROTEIN

The human *p53* gene encodes a protein of 393 amino acids, which can be divided into several well-characterized structural and functional domains, based chiefly on the

From: *Tumor Suppressor Genes in Human Cancer*
Edited by: D. E. Fisher © Humana Press Inc., Totowa, NJ

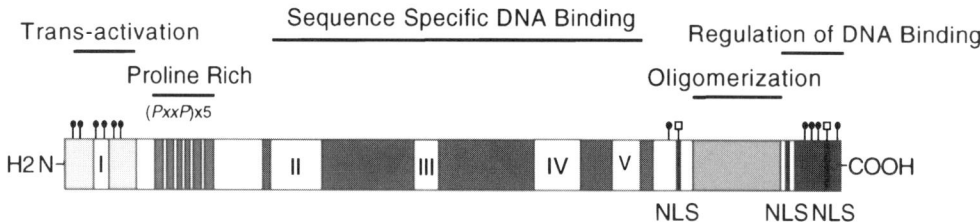

Fig. 1. Schematic representation of the human p53 protein. The important functional domains (transactivation, proline-rich, sequence-specific DNA-binding, oligomerization, regulation of DNA binding), conserved box regions I–V, sites of phosphorylation within the N- and C-terminal regions (filled circles), and sites of acetylation within the C-terminus (open squares), are indicated.

ability of p53 to function as a transcription factor transcription factors (Fig. 1). The *p53* gene has been identified in several species, from human to squid. Sequence comparison of the p53 protein from different species shows five highly conserved regions (Fig. 1). Four of these conserved regions lie within the central core domain of p53, which comprises a sequence-specific DNA-binding region. The N-terminal region of p53 is rich in acidic residues, and contains a transcriptional transactivation domain, which has been shown to form a direct contact with several basal transcription factors, such as the TATA-binding protein (TBP) *(16–20)*, TBP-associated factors in the TFIID complex *(21–23)*, and the transcriptional co-activators, p300 and CBP *(24,25)*. This region also contains the site for binding to Mdm2, the product of a p53-inducible gene that plays an important role in negatively regulating p53 function *(26,27)*. A proline-rich region links the transcriptional activation domain with the central core, and has been shown to be important for the growth-suppressor function of p53 *(28–31)*. The C-terminal region of p53 is rich in basic residues, and shows nonspecific DNA/RNA-binding, which can regulate the DNA-binding activity of the central core *(32,33)*. Modification of the extreme C-terminus of p53 by deletion, phosphorylation, or binding of a specific antibody activates the sequence-specific DNA-binding activity. The native p53 protein is a tetramer in solution *(34)*, and residues 324–355 within the C-terminal region are essential for oligomerization *(34–37)*. The C-terminal region also contains three nuclear localization sequences *(38–40)* and a nuclear export signal that regulates the subcellular localization of the p53 protein *(41)*. p53 is posttranslationally modified by phosphorylation at specific serine (Ser) residues within the N-terminus and C-terminus of the protein, and by acetylation at specific lysine residues within the C-terminal region *(42)*. Two common polymorphic variants of p53 exist, arising from a single base-pair substitution at codon 72, encoding either a proline or an arginine residue *(43,44)*. Although both polymorphic forms share similar growth-suppressive activities, recent studies suggesting subtle differences in their regulation and potency may be reflected in increased cancer susceptibility in some individuals *(45,46)*.

In recent years, two genes showing structural similarities to p53 have been identified *(47–53)*. The *p73* gene encodes at least four distinct isoforms; the full-length version, which gives rise to the protein called p73α, and three splice variants, which encode proteins referred to as p73β, γ, and δ. p73 has been shown to mediate at least some functions in common with p53, including apoptosis, transcriptional transactivation of *p21*[WAF1/CIP1], a known target of p53, and suppression of cell growth. The *p73* gene has been mapped to chromosome 1p36.3, a locus that is deleted in neuroblastoma and some other human cancers *(49,54)*. Studies to date indicate, however, that, in contrast to p53, p73 tumor-derived mutations appear infrequently *(54–56)*, although this may be a reflection of the observation that p73 is imprinted. Another p53 family member, referred to as *p40, p51, p63, p73L,* or the rat homolog, *KET,* encodes at least six isoforms that are expressed in a tissue-specific manner, and have different transactivation potentials *(53)*. A contribution of p63 to tumor suppression has not yet been established, although mice deleted of the *p63* gene suffer severe developmental defects *(57,58)*.

3. CELLULAR FUNCTIONS OF THE P53 PROTEIN

Although mice deleted of the *p53* gene show a high incidence of cancer, their viability and relatively normal development indicated that p53 function is not essential for cell growth and differentiation. In contrast, analysis of transgenic mice expressing a p53-responsive *lacZ* reporter gene demonstrated that tissue-specific overexpression of p53, following irradiation, is detrimental to normal embryonic development (59–61), indicating the importance of the tight negative regulation of p53 function during normal cell growth. It is now clear that p53 plays a major role in the cellular response to stress, and that activation of p53 following various insults prevents the growth of damaged and abnormal cells, and in some cases contributes to the ability of the cell to repair the damage.

Under normal circumstances, endogenous p53 protein is rapidly degraded and maintained at very low levels within the cell. However, a variety of stress signals, including DNA damage *(62,63)*, hypoxia *(64)*, oncogene activation *(65–68)*, heat shock, metabolic changes, viral infection, or cytokine treatment *(69)*, or withdrawal *(70)* lead to the rapid elevation of p53 levels, principally through stabilization of the protein. In most cell types, the activation of p53 results in either cell cycle arrest or cell death through apoptosis, thereby preventing the propagation of potentially abnormal cells. p53 can also contribute to cellular senescence *(71–73)* and differentiation *(74–77)*, and, in mouse cells, loss of p53 results in rapid loss of the mechanisms that normally limit the proliferative capacity of cells in culture. Further functions of p53 include regulation of adhesion, metastasis, and angiogenesis *(78,79)*. p53 activity is also important for normal centrosome duplication *(80)*, and a role for p53 in DNA repair, replication, and recombination has also been described *(81–87)*.

3.1. Transcriptional Activation by p53

As mentioned earlier, one of the most important functions of p53 in tumor suppression is its ability to inhibit cell growth by inducing cell cycle arrest or apoptosis. Although these are independent activities of p53, the ability of p53 to function as a transcription factor contributes to both responses *(88)*. Many cellular genes have been shown to be transcriptional targets of p53 *(89)*, although only a few of these have been verified as direct mediators of the p53 response. These genes can be broadly divided into those likely to contribute to activation of a cell cycle arrest, and those that are more likely to play a role in mediating the apoptotic response *(88)*.

Activation of p53 in many cell types leads to an arrest at G1 and G2–M phases of the cell cycle, which under some circumstances, is reversible *(90)*. One of the principal mediators of both G1 and G2/M arrests is the *p21$^{WAF1/CIP1}$* gene, a direct target of p53 transcriptional activation *(91)*. p21$^{WAF1/CIP1}$ is a broad inhibitor of cyclin-dependent kinase (CDK) complexes, which are essential for cell cycle progression *(92,93)*. p21$^{WAF1/CIP1}$ also binds to proliferating cell nuclear antigen (PCNA), a subunit of DNA polymerase, and directly inhibits DNA replication *(94)*. Overexpression of p21$^{WAF1/CIP1}$ in a number of cellular backgrounds leads to G1 and G2 arrests indistinguishable from those mediated by p53 *(95)*. Deletion of *p21$^{WAF1/CIP1}$* from human cells dramatically reduces the ability of the cell to undergo cell cycle arrest in response to p53 activation *(96)*, although the retention of some activity in p21$^{WAF1/CIP1}$-null mouse cells suggests that other p53-inducible genes also contribute to this response *(97,98)*. Mice deficient in the *p21$^{WAF1/CIP1}$* gene develop normally, but do not show the high incidence of tumor development seen in p53-deficient animals, suggesting that this response is not the primary mechanism by which p53 functions as a tumor suppressor. A number of other p53-responsive target genes have been described that may contribute to the cell cycle arrest. The *GADD45* gene is a member of a group of growth arrest and DNA damage-inducible genes (GADD) *(99)*, and is induced by ionizing radiation in many cell types containing wt p53 *(100)*. Expression of GADD45 has been associated with the activation of both the G1 and G2/M arrest *(101,102)*. GADD45 binds PCNA, to stimulate excision repair of damaged DNA *(103)*, and has recently been shown to modify DNA accessibility on damaged chromatin *(104)*. p53-dependent activation of *GADD45* can be mediated via direct p53/DNA interactions, as described above, but also through interaction of p53 with the product of the Wilm's tumor gene, *WT1 (105)*, indicating that further p53-responsive genes, which lack p53-binding sites within their promoter region, may remain to be identified. Recently, *PA26*, another novel p53 target gene that belongs to the GADD family, was identified *(106)*. Although no functional clues have been revealed by its sequence, *PA26* appears to be negatively regulated by serum factors, and so, like *GADD45*, may also play a role in growth regulation *(106)*. Another p53-target gene, which could play a role in regulating proliferation, is *IGFBP-3*, which encodes a secreted inhibitor of insulin-like growth factors *(107)*. The human homolog of the *Drosophila sina* gene, *SIAH-1*, has also been shown to play a role in p53-dependent cell-cycle arrest *(108,109)* and the 14-3-3 σ protein has been shown to be a potent p53-mediated regulator of G2–M progression *(110)*. 14-3-3 proteins have also been shown to specifically bind to a DNA-damage-inducible 14-3-3 consensus binding site within the C-terminal region of p53, which enhances sequence-specific DNA-binding function in vitro *(111)*. However, the precise physiological role for this interaction, in terms of p53 transcriptional regulation, remains unclear.

Activation of transcription also plays a role in the ability of p53 to activate apoptotic cell death *(112)*. The apoptotic targets of p53 appear to be independent of the cell cycle arrest targets, and activation of genes, such as *p21$^{WAF1/CIP1}$*, play no role in the induction of cell death *(112)*. Apoptosis is a form of programmed cell death, and is characterized by morphological changes that include cell shrinkage, plasma membrane blebbing, nuclear condensation, and DNA fragmentation *(113)*. These cellular changes have been shown to involve a family of cysteine proteases called caspases (ICE/CED-3 proteases) *(114)*, which can be activated through two main pathways, one involving the activation of death receptors, such as Fas/APO1 and DR5, at the cell surface *(115)*, and the other involving cytochrome-*c* dependent activation of the adaptor protein, Apaf-1 *(116)*. Although the mechanisms by which p53 initiates apoptosis are not fully understood, both the death-receptor-associated pathways and the Apaf-1-dependent apoptotic pathway have been shown to be involved in mediating p53-dependent cell death *(117)*. Several transcriptional targets of p53 have been identified, which directly link p53 to caspase activation, including the proapoptotic Bcl-2 family member, Bax *(118)*, which functions by releasing cytochrome-*c (119)*, and the cell surface death receptors, DR5 and Fas *(120,121)*. None of these appear to represent the only component of the p53-mediated death mechanism, however, since mice deficient in *bax* or *Fas* are still able to mediate apoptosis in response to activation of p53 in some cell types *(122–125)*.

Other potential apoptotic transcriptional targets of p53 include IGFBP-3 *(107)*, which inhibits IGF survival function; *PAG608,* which encodes a nuclear zinc finger protein with apoptotic activity *(126);* and the gene that encodes the cathepsin-D protease, which contributes to cytokine-mediated apoptosis *(127)*. Reactive oxygen species participate in cell death mechanisms mediated by many different stimuli, and it has been proposed that p53 may transcriptionally regulate genes *(128)* or cooperate with cellular targets that are sensitive to reactive oxygen species *(129)*. Several further novel p53-induced genes, which are highly expressed in the colorectal cancer line, DLD-1, before the onset of apoptosis, have been identified using sequence analysis gene expression *(128)*, although their contribution to the p53 response has not yet been fully characterized.

Finally, p53 has also been shown to activate the expression of a number of metastasis- or angiogenesis-related genes, including epidermal growth factor, matrix metalloprotease (MMP-2, also called human type IV collagenase), cathepsin-D, thrombospondin-1 *(130)* and BAI1, an inhibitor of angiogenesis *(131)*. Drugs specifically targeting some of these genes (MMP inhibitors), have been used, in combination with radiotherapy, to combat the induction of metastases in the treatment of some tumors that express wt p53 *(130)*.

3.2. Transcriptional Repression by p53

In addition to activating genes with p53-binding sites, p53 can also repress promoters that lack the p53-binding element, and this may be of considerable biological importance in mediating the apoptotic response of p53 *(112,132)*. Indeed, a number of genes, including interleukin-6 *(133)*, nuclear factor-κB RelA (NF κB) *(134,135)*, cyclin A *(136,137)*, PCNA *(138)*, and a number of metastasis-related genes *(130)*, have been shown to be transcriptionally repressed by p53 in this way. Additionally, both RNA polymerase II and III transcription can be repressed by p53 *(139,140)*. Despite genetic evidence showing a strong correlation between the ability of p53 to induce apoptosis and the retention of transcriptional repression, identifying the important targets that are transcriptionally repressed by p53 has been difficult. Relatively few of the potential target genes for repression by p53 have been verified by showing downregulation of endogenous gene expression following p53 expression, and the mechanisms of function of those that have been authenticated in this way, such as mitogen-activated protein 4 *(141,142)* remain unclear. Furthermore, the observation that transcriptional repression of some genes may be the consequence, rather than the cause, of apoptosis may further complicate the identification of *bona fide* targets of p53 *(143)*. The difficulty in examining this activity of p53 is clearly demonstrated by the c-*fos* gene promoter, which had for many years represented the archetypal target for transcriptional repression by p53 *(144)*, until it was recently shown that the endogenous gene is transcriptionally activated by p53 *(145)*.

3.3. Transcriptionally Independent Activities of p53

Despite the clear contribution of transcriptional activation to p53-mediated apoptosis in many systems, considerable evidence has accumulated to support the existance of a transcriptionally independent function of p53 in the activation of cell death. p53 can mediate cell death when RNA and protein synthesis are blocked *(146)*, and mutational studies have shown that, although loss of transcriptional activity invariably results in loss of the ability to induce cell cycle arrest, the ability to induce apoptosis is not necessarily impaired *(147)*. Conversely, C-terminal mutants of p53 that retain transcriptional activity show defects in apoptotic activity *(148,149)*, despite retaining cell cycle arrest functions. The proline-rich region of p53 has also been shown to be important for apoptosis in several systems *(28–31)*, although this region may in fact play a role in differential regulation of p53 transcriptional function *(150)*. The observation that the requirements for p53-induced apoptosis are dependent on the cell type under examination has further complicated the identification of these activities, although several functions of p53 that are not dependent on transcription have been described. One activity of p53 with clear implications for the induction of apoptosis is the ability to traffic the death receptor, Fas, to the cell surface, thereby sensitizing cells to Fas-mediated apoptosis *(151)*. Whether p53 shows a general ability to redistribute death receptors to the cell surface remains unknown. The direct interaction of the extreme C-terminal region of p53 with the XPB and XPD DNA helicases has also been shown to play a role in p53-mediated apoptosis *(152,153)*, although this activity is clearly not required under all circumstances *(154)*. The N-terminal proline-rich domain within p53 shows some similarities to an SH3-binding domain, and binds to the SH3 domain of c-Abl tyrosine kinase. This interaction has recently been shown to enhance p53 expression by neutralizing the inhibitory effects of Mdm2 *(155)*, and so far it is unclear whether the interaction of c-Abl with p53 via the proline-rich domain is important for p53-mediated apoptosis *(156,157)*.

3.4. Cell Cycle Arrest or Apoptosis

The identification of distinct p53 target genes and mechanisms that are involved in p53's ability to mediate cell cycle arrest or apoptosis indicates that these are separable and independent physiological functions of p53, and p53 mutants, which retain one, but not the other, function, have been described in several studies *(28,30,31,143,147,158)*. The cellular response to p53 in undergoing cell cycle arrest or apoptosis represents a critical choice between a potentially transient delay in cell growth and the clearly irreversible process of cell death. Although the choice of response to p53 is strongly influenced by the cell type, the cellular environment (including presence of survival factors) and other genetic abnormalities sustained by the cell, it is clear that, to some extent, the choice can be mediated by p53 itself. Overall p53 levels can determine the choice of response *(149)*, and it has been suggested that this in part determines the response to DNA damage. In this model, low levels of damage, which may be repairable, induce low levels of p53, and lead to a cell cycle arrest; extensive, irreparable damage, which induces high levels of p53, leads to apoptosis.

Because distinct transcriptional targets of p53 mediate the cell cycle arrest and apoptotic responses, it is possible that activation of the two sets of target genes may be dissociated under some circumstances. The observation that apoptotic cell death in response to p53 obscures an underlying cell cycle arrest *(159,160)* implies that, in these cells, the choice of response is not either cell cycle arrest or apoptosis, but rather cell cycle arrest, or cell cycle arrest with apoptosis. These observations suggest that the transcriptional targets that mediate cell cycle arrest are more sensitive to p53 than the apoptotic targets, with some evidence that the promoters of cell-cycle-arrest genes contain p53-binding sites with higher affinities for p53 than the apoptotic target gene promoters *(161,162)*. Several mutants of p53, which contain amino acid substitutions in the DNA binding domain, show a temperature-sensitive loss of the ability to activate apoptotic targets, while retaining activation of cell-cycle-arrest target genes *(143)*. This may reflect a reduction in DNA binding activity of these mutants, so that they fall below a threshold necessary to activate the apoptotic targets. Similarly, mutation within the proline-rich domain of p53, which regulates sequence-specific DNA binding *(30,163)*, also results in loss of the ability to activate a specific subset of p53-responsive genes.

4. REGULATION OF P53 FUNCTION

4.1. Regulation of p53 Protein Stability

The principal mechanism by which p53 function is regulated is through the stability of the protein, although transcriptional and translational control of p53 expression has also been reported *(164–166)*. Endogenous p53 protein is maintained at very low levels within the cell, because of its rapid degradation by ubiquitin-dependent proteolysis. It is now clear that a direct transcriptional target of p53, Mdm2, is an essential component of this process *(26,27)*. Mdm2, unlike the other transcriptional targets described above, has not been shown to directly contribute to cell cycle arrest or apoptotic functions of p53 *(167,168)*. However, Mdm2 directly binds to residues within the N-terminal transactivation domain of p53, repressing transcriptional activation *(169)* and targeting p53 for degradation *(26,27,170)*. In this way, *Mdm2* negatively regulates p53 stability and transcriptional function. The importance of this regulation was clearly illustrated by the development of mice deficient in Mdm2, which resulted in lethality at the implantation stage of development *(171,172)*. This lethality was rescued by breeding onto a *p53*-null background, demonstrating that the ability of Mdm2 to regulate p53 protein levels is essential during development.

The abundance of cellular Mdm2 protein depends mostly on its transcriptional activation by p53, and, in normal cells, p53 and Mdm2 exist in a tightly regulated feedback loop, through the direct binding of Mdm2 to p53. This mutual regulation of p53 and Mdm2 is lost in cells expressing mutant forms of p53, which have lost the ability to activate transcription, such as those expressed in many cancers. These p53 mutants are unable to activate the expression of Mdm2, and are therefore usually stable and expressed at high levels *(170,173,174)*. The observation that mutant p53 is usually expressed at high levels has been used as a prognostic indicator in some tumor types *(175)*.

Two activities of Mdm2 have recently been described that are likely to contribute to the regulation of p53 stability: the ability to function as a ubiquitin ligase, and the regulation of subcellular localization. Mdm2 has specific ubiquitin ligase E3 activity, and will readily ubiquitinate p53, upon association *(176)*, thus targeting p53 for degradation *(177)*. This degradation depends on the direct interaction between the two proteins, although regions outside the binding sites on both p53 and Mdm2 are also essential *(170,174,178)*. For example, deletion of the last 30 amino acids from the C-terminus of p53 renders the protein partially resistant to Mdm2-mediated degradation *(174,179)*, and deletion of the p53-conserved box II sequences in the DNA-binding region also results in slight resistance to Mdm2-mediated degradation *(173,174)*. It is possible that the latter observation reflects a contribution of binding between p53, Mdm2, and the transcriptional co-activator p300, in allowing efficient degradation of p53 *(180)*. Sequences in Mdm2 apart from the p53-binding region are also necessary for degradation, with evidence that the RING finger domain at the C-terminus of Mdm2 is critical for ubiquitin ligase activity *(177)*. Mdm2 also appears to regulate its own stability. Mutants of Mdm2 that fail to mediate ubiquitin ligase activity are themselves stable *(170,178)*, and Mdm2 can target its own ubiqutination *(176,177)*. Mdm2 can also target the degradation of other proteins to which it can bind *(181,182)*, and the stability of other important components of the cell cycle machinery, such as the transcription factor E2F, may be regulated by Mdm2 *(183)*.

In addition to acting as a ubiquitin ligase, Mdm2 also plays a role in regulating the subcellular localization of p53. The Mdm2 protein contains both nuclear import and export sequences, and the importance of the nuclear export function of Mdm2 in the degradation of p53 suggests that Mdm2 plays a role in relocating p53 to the cytoplasm, where degradation takes place *(184)*. This interpretation is complicated by the identification of nuclear export sequences in p53, which also result in transport of p53 to the cytoplasm, without a requirement for Mdm2 *(41)*, although this activity may depend on the oligomerization status of p53. It seems likely that the subcellular location of these proteins will be as important as their expression levels and activities in determining their ability to function.

Although Mdm2 is a major player in regulating p53 stability, other proteins also contribute to the degradation of p53. The jun-N-terminal kinase (JNK) plays a role in targeting p53 for ubiquitination and degradation, in an Mdm2-independent manner *(185,186)*, although the mechanism is unknown, and does not depend on the kinase activity of JNK. Viral proteins, such as the papillomavirus E6 and E1B, and E4orf from adenovirus, can also target p53 for degradation *(187–189)*. p53 stability may also be regulated by other proteases, such as calpain *(187,190)*, although the pathways involved are not yet understood. The p53-related protein, p73, retains the ability to bind Mdm2, but is not degraded following this interaction *(191,192)*, although p73 protein levels do appear to be regulated in a proteasome-dependent manner *(191)*. It is possible that other Mdm2-related proteins, such as MdmX *(193–195)*, play a role in regulating the stability of p53 family members, although there is at present no evidence for such a mechanism.

Since the Mdm2 protein plays a central role in negatively regulating p53 stability, it seems highly likely that inhibition of this regulatory mechanism would be essential for stabilization of the protein under conditions of cellular stress. It is becoming clear that many different pathways can lead to the inhibition of Mdm2-mediated degradation of p53, and different stress signals are likely to utilize different mechanisms to stabilize p53 (Fig. 2). One way to render p53 resistant to Mdm2 is to prevent interaction

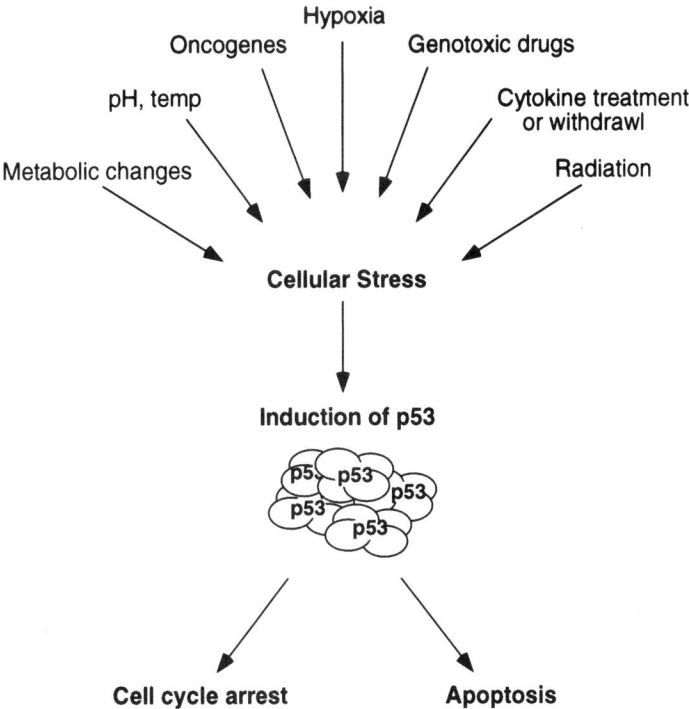

Fig. 2. Multiple pathways lead to the stabilization of p53.

between the two proteins, and phosphorylation of p53 or Mdm2 has been shown to reduce the p53–Mdm2 interaction *(196,197),* and to inhibit Mdm2-mediated ubiquitination in vitro *(177).*

Several kinases have been shown to phosphorylate the N-terminus of p53 in vitro, close to or within the Mdm2-binding region. One candidate is the ATM kinase, which is encoded by the gene mutated in the human genetic disorder, ataxia telangiectasia, and is activated in response to ionizing radiation, but not UV radiation *(198).* ATM can directly associate with p53, and phosphorylate it at Ser15 in vitro *(198–200),* and this residue is inducibly phosphorylated upon DNA damage, simultaneously with stabilization of the protein *(201,202).* Cells deficient for ATM show a delayed ability to stabilize p53 in response to ionizing radiation, although the UV response remains intact, supporting a role for ATM in governing p53 stability *(202).* The ATR kinase, a protein closely related to ATM, also phosphorylates Ser15 and Ser37 in vitro, and overexpression of an inactive ATR kinase prevents phosphorylation of Ser15 in response to UV radiation *(203).* Recently, phosphorylation of p53 at Ser20, within the Mdm2-binding site, has also been shown to occur in response to some forms of DNA damage *(204),* and has been proposed to play a role in regulating p53 stability *(205),* although the kinase responsible for this phosphorylation has not been identified.

DNA damage-induced phosphorylation of p53 provides an attractive mechanism for mediating the stabilization of the protein. However, analyses of p53 mutants that cannot be phosphorylated, either at individual sites or at multiple sites within N- or C-terminal regions of the protein, have shown that phosphorylation is not essential for stabilization in response to several forms of DNA damage in vivo *(173,206)*. Indeed, the p53 protein is stabilized in response to ionizing radiation, even under conditions when Ser15 phosphorylation has been abolished by the expression of an antisense ATM construct *(200)*. Furthermore, some DNA-damaging agents, such as actinomycin D, induce rapid stabilization of p53 in the absence of detectable Ser15 phosphorylation *(207)*. These studies indicate that phosphorylation is not essential for DNA-damage-induced stabilization of p53, and other mechanisms may also play a role. One mechanism employed by topoisomerase inhibitors seems to be the specific downregulation of Mdm2 expression at the transcriptional levels, thereby releasing p53 from degradation by reducing Mdm2 protein levels in the cell *(207,208)*. Other proteins that bind Mdm2 or p53, such as the retinoblastoma tumor suppressor protein (Rb) *(209)* or c-Abl *(155)*, have also been shown to regulate the degradation of p53.

In addition to DNA damage, several other forms of stress stabilize and activate p53. Many studies have implicated a role for p53 in responding to abnormal proliferative signals generated by oncogene activation, and stabilization of p53 in response to such oncogenic stress represents a powerful fail-safe mechanism to protect cells from malignant progression. Stabilization of p53 in response to signals generated by oncoproteins, such as E2F1, Myc, Ras, and adenovirus E1A, is mediated though the p14^{ARF} (mouse, p19^{ARF}) protein *(65,66,210)*, which binds directly to Mdm2 *(211–213)*, and inhibits the ability of Mdm2 to function as a ubiquitin ligase *(177)*. Expression of p14^{ARF} therefore results in the stabilization of both p53 and Mdm2. p14^{ARF} is a direct transcriptional target of E2F1; therefore, most forms of deregulated proliferation, which lead to E2F1 activation, are likely to result in p14^{ARF} activation. Complex feedback mechanisms, through which p53 can inhibit p14^{ARF} expression, exist *(68,212,214)*, probably to prevent activation of a p53 response during each normal cell cycle.

Although p14^{ARF} very efficiently stabilizes p53, it does not appear to be involved in the DNA damage response pathway, because cells deficient in p14^{ARF} mediate p53 induction in response DNA damage *(215)*. Conversely, ATM, which plays a role in the activation of p53 in response to some DNA-damaging agents, is dispensable for the activation of p53 by proliferative abnormalities *(216)*. The observation that many tumor cell lines that retain wt p53 suffer loss of p14^{ARF} suggests that it is loss of the p14^{ARF} pathway that is of critical importance during tumor development. Therefore, the failure to respond to proliferative abnormalities following loss of either p14^{ARF} or p53 is a key step for tumor development.

4.2. Regulation of p53 Activity

Although activation of a p53 response is clearly dependent on stabilization of the protein, in many cases, other mechanisms that regulate p53 activity have also been described. Evidence from in vitro studies suggests that the conformation of the p53 protein is crucially important for its ability to bind DNA, and there is evidence that this activity can be regulated by the C-terminus of the p53 protein. One model suggests that this C-terminal region of p53 allosterically regulates activation of the latent non-DNA-binding form of p53 to the active wt conformation *(35,217)*. Others have suggested that the C-terminus reciprocally regulates the ability of p53 to bind DNA via the sequence-specific DNA-binding core *(218,219)*. Many different mechanisms, which modify the C-terminus of p53, have been shown to activate DNA binding. These include phosphorylation *(220,221)*, acetylation *(25,222,223)*, O-glycosylation *(224)*, binding to the C-terminal antibody, PAb421 *(33)*, binding to short strands of DNA *(225)*, C-terminal deletion, or point mutations *(226)*. Dephosphorylation within this region, following IR, has also been shown to correlate with binding of the 14-3-3 proteins to the p53 C-terminus, and subsequent activation of transcription *(111)*. A number of other proteins that bind to p53 have also been shown to regulate its function *(221)*. The redox/repair protein, Ref-1, is a potent activator of p53 DNA-binding and transcriptional activation functions *(227)*, and a recently identified activator of p53, HMG-1, has also been shown to enhance p53 DNA-binding function *(228)*.

Despite the clear ability to activate p53 protein produced in vitro, the contribution of these regulatory modifications to controling p53 activity and stability is much less clear in a physiological context. Several studies have shown a dissociation between activation of p53 function and stabilization of p53 protein *(220,229)*, and, conversely, stabilization of inactive p53 has been described in teratocarcinoma cells *(230)*. These results suggest that regulation of latent and active conformational states could play an important role in controling p53 function in cells.

Regulation of protein binding to the N-terminus of p53 also contributes to the modulation of p53 activity. Phosphorylation at the N-terminus of p53 decreases its interaction with TFIID in virus-transformed cells *(231);* by contrast, phosphorylation of Ser15 within this region of p53 increases the recruitment of transcriptional co-activators CBP/p300 *(232)*. Both CBP and p300 show histone acetyl transferase activity, and have also been shown to acetylate the C-terminus of p53 *(233,234)*. This has led to the hypothesis that phosphorylation at the N-terminus of p53 mediates acetylation at the C-terminus, and thus enhances DNA-binding and transcriptional activation function *(223)*. In an interesting elaboration of this system, p300 was recently shown to be more important for p53 activation of Mdm2 than other transcriptional targets that mediate the cell cycle arrest and apoptotic responses *(235)*. Inhibition of p300 activity, following adenovirus E1A expression, for example, can therefore stabilize p53 by inhibiting Mdm2 expression *(235)*, the growth-inhibitory activities of p53 may be regulated, under physiological conditions, by activation of other transcription factors, such as NF-κB, which compete with p53 for binding to p300 or CBP *(236)*.

4.3. p53 Cellular Localization

In addition to regulation of protein levels and protein activity, p53 function can also be modulated by control of subcellular localization. p53 is subject to both nuclear import and export mechanisms *(38);* the C-terminal region of p53 contains three nuclear localization sequences *(38–40)* and a nuclear export signal *(41)*. Translocation to the nucleus is essential for p53 function *(237)*, and inactive wt p53, which fails to localize to the nucleus has been described in some tumor cells *(238,239)*. The anti-apoptotic protein, Bcl-2, has also been shown to regulate p53 nuclear import *(240)*. As described earlier, degradation of p53 by Mdm2 protein has been shown to be dependent on export from the nucleus *(184,241)*, although, in this case, nuclear export appears to depend on signals in Mdm2, rather than p53 itself.

5. p53 IN TUMOR DEVELOPMENT

Increased understanding of the regulation and functions of p53 has allowed a reassessment of the frequency of p53 functional loss during tumorigenesis. Mutations

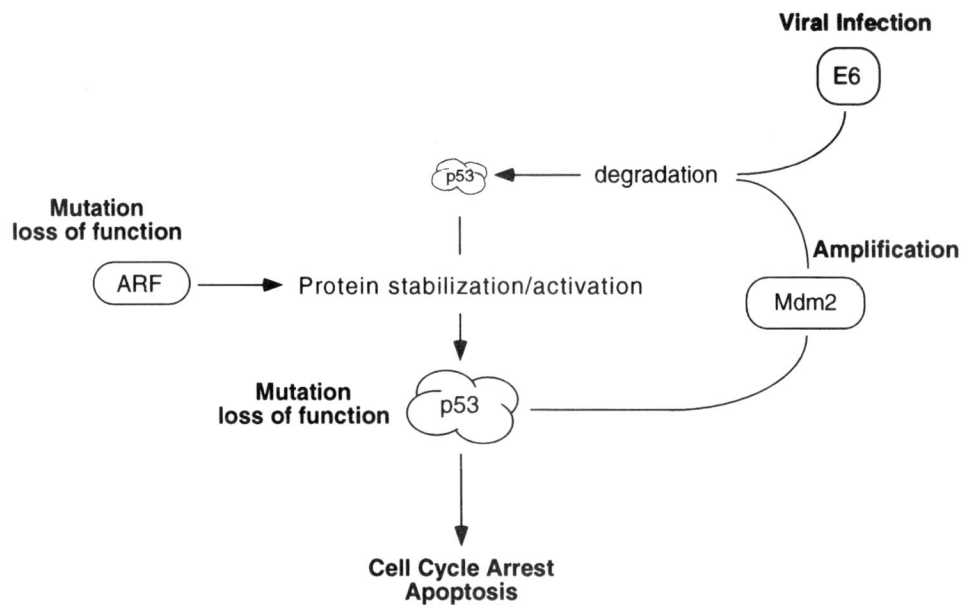

Fig. 3. Multiple mechanisms for loss of p53 function during tumor development.

in *p53* are common event in many major human malignancies, yet around one-half of all human cancers arise despite retention of wt p53 sequences. It is now apparent that many of these cancers have lost p53 function through other mechanisms, such as over-expression of Mdm2 or the human papillomavirus E6 protein, which target p53 for degradation, thus preventing stabilization in response to stress (Fig. 3). Another common defect that prevents p53 activation in response to abnormal proliferation is loss of the p14^{ARF} protein, which normally inhibits Mdm2 degradation of p53 following onco-gene activation. Other proteins that function like p14^{ARF} may also play a role in medi-ating tumor-suppressor activities of p53. Tumors may also show loss of the downstream effectors of p53 function, and mutation of Bax, for example, may protect tumor cells from apoptosis following activation of p53 *(242)*. Taken together, it seems likely that loss of the normal p53 response pathway, through one mechanism or another, is a common and possibly even obligatory step in cancer development.

   Despite the identification of several mechanisms by which p53 activity can be lost, mutation within the *p53* gene itself is likely to be the most efficient mechanism to release the cell from the negative growth control mediated by p53. Mutant forms of p53 that are expressed in many cancer cells can show dominant-negative activity in inhibit-ing co-expressed wt p53 *(243)*, and even show an oncogenic gain of function indepen-dent of the expression of wt p53 *(244–246)*. Tumor cells in which evasion of p53-mediated tumor suppression is the result of loss of upstream activators of p53, such as p14^{ARF}, show only partial protection from p53 activation. The observation that p14^{ARF}-negative cells retain the ability to activate p53 in response to DNA damage suggests that tumors of this type may be more sensitive to DNA-damaging chemother-apeutic treatments than tumors with mutations in p53 itself. Furthermore, in addition to loss of function, mutant forms of p53 show dominant transforming activities that may reflect their ability to oligomerize with, and inhibit the activity of, other p53 family members, such as p63 and p73 *(247)*.

The observation that p53 responds to many signals appears to be mirrored in the ability of p53 to function at several stages during tumor development. The p53 response to DNA damaging events, such as UV radiation through excessive exposure to sunlight *(248,249),* could inhibit the accumulation of potentially oncogenic mutations and the initial stages of tumor development, and loss of p53 may also be necessary to tolerate the loss of DNA-repair-associated genes, such as *BRCA-1 (250,251).* Abnormal proliferation, a hallmark of cancer cells, also activates p53, and the ability of E2F-1, both to drive cell proliferation and to induce p53, embodies part of the complex protective mechanism that make cells exquisitely sensitive to abnormal growth. Finally, p53 may also be involved in suppression of later stages of tumor development, responding to hypoxic conditions that occur as tumors grow in size, and inhibiting angiogenesis necessary for progressive tumor growth.

The high frequency of loss of p53 function in a broad range of human tumors has encouraged many attempts to restore p53 pathways in tumor therapy. The prospects for these approaches have been further strengthened by indications that activation of p53 might preferentially kill tumor cells, sparing normal cells by inducing a reversible cell cycle arrest. The basis for this selectivity appears to reside in the deregulation of E2F-1 activity, which occurs in almost all tumors. In addition to stabilizing p53 through p14^{ARF}, E2F-1 also sensitizes cells to apoptotic signals in a p53-independent manner *(252),* and activation of E2F-1 and p53 potently induces cell death under circumstances in which activation of p53 alone would lead to only cell cycle arrest *(68).* The concept that tumor cells, with deregulated E2F-1, are more likely to undergo apoptosis in response to p53 than normal cells, has generated a great deal of enthusiasm for the use of p53 in tumor therapy, and several mechanisms for the activation of p53 function are under investigation. Depending on the situation, these could involve inhibition of Mdm2 activity, restoration of p14^{ARF} expression, or, in cells expressing high levels of mutant p53, reactivation of the DNA-binding function of p53 itself *(253,254).* Whatever the mechanism, restoration of p53 function to selectively kill tumor cells in a broad range of major human cancers may be a distant, but attainable goal.

ACKNOWLEDGMENTS

The authors are grateful to members of the Vousden lab for advice and helpful criticisms of this review. This work was sponsored by the National Cancer Institute, Department of Health and Human Services, under contract with ABL.

REFERENCES

1. DeLeo AB, Jay G, Appella E, Dubois GC, Law LW, Old LJ. Detection of transformation-related antigen in chemically induced sarcomas and other transformed cells of the mouse. *Proc Natl Acad Sci* 1979; 76:2420–2424.
2. Lane DP, Crawford LV. T antigen is bound to a host protein in SV40-transformed cells. *Nature* 1979; 278:261–263.
3. Linzer DIH, Levine AJ. Characterization of a 54K dalton cellular SV40 tumor antigen present in SV40-transformed cells and uninfected embryonal carcinoma cells. *Cell* 1979; 17:43–52.
4. Baker SJ, Markowitz S, Fearon ER, Willson JKV, Vogelstein B. Suppression of human colorectal carcinoma cell growth by wild-type p53. *Science* 1990; 249:912–915.

5. Chen PL, Chen YM, Bookstein R, Lee WH. Genetic mechanisms of tumor suppression by the human p53 gene. *Science* 1990; 250:1576–1580.

6. Eliyahu D, Michalovitz D, Eliyahu S, Pinashi-Kimhi O, Oren M. Wild-type p53 can inhibit onco-gene-mediated focus formation. *Proc Natl Acad Sci USA* 1989; 86:8763–8767.

7. Finlay CA, Hinds PW, Levine AJ. The p53 proto-oncogene can act as a suppressor of transformation. *Cell* 1989; 57:1083–1093.

8. Mercer WE, Shields MT, Amin M, Sauve GJ, Appella E, Romano JW, Ullrich SJ. Negative growth regulation in a glioblastoma tumor cell line that conditionally expresses human wild-type p53. *Proc Natl Acad Sci USA* 1990; 87:6166–6170.

9. Michalovitz D, Halevy O, Oren M. Conditional inhibition of transformation and of cell proliferation by a temperature-sensitive mutant of p53. *Cell* 1990; 62:671–680.

10. Donehower LA, Harvey M, Slagle BL, McArthur MJ, Montgomery CA Jr, Butel JS, Bradley A. Mice deficient for p53 are developmentally normal but susceptible to spontaneous tumors. *Nature* 1992; 356:215–221.

11. Jacks T, Remington L, Williams BO, Schmitt EM, Halachmi S, Bronson RT, Weinberg RA. Tumor spectrum analysis in p53-mutant mice. *Curr Biol* 1994; 4:1–7.

12. Purdie CA, Harrison DJ, Peter A, Dobbie L, White S, Howie SE, et al. Tumour incidence, spectrum and ploidy in mice with a large deletion in the p53 gene. *Oncogene* 1994; 9:603–609.

13. Greenblatt MS, Bennett WP, Hollstein M, Harris CC. Mutations in the *p53* tumor suppressor gene: clues to cancer etiology and molecular pathogenesis. *Cancer Res* 1994; 54:4855–4878.

14. Hollstein M, Sidransky D, Vogelstein B, Harris CC. p53 mutations in human cancers. *Science* 1991; 253:49–53.

15. Nigro JM, Baker SJ, Preisinger AC, Jessup JM, Hostetter R, Cleary K, et al. Mutations in the p53 gene occur in diverse human tumour types. *Nature* 1989; 342:705–708.

16. Horikoshi N, Usheva A, Chen JD, Levine AJ, Weinmann R, Shenk T. Two domains of p53 interact with the TATA-binding protein, and the adenovirus 13S E1A protein disrupts the association, reliev-ing p53-mediated transcriptional repression. *Mol Cell Biol* 1995; 15:227–234.

17. Liu X, Miller CW, Koeffler PH, Berk AJ. The p53 Activation domain binds the TATA box-binding polypeptide in holo-TFIID, and a neighboring p53 domain inhibits transcription. *Mol Cell Biol* 1993; 13:3291–3300.

18. Martin DW, Munoz RM, Subler MA, Deb S. p53 Binds to the TATA-binding protein–TATA complex. *J Biol Chem* 1993; 268:13,062–13,067.

19. Seto E, Usheva A, Zambetti GP, Momand J, Horikoshi N, Weinmann R, Levine AJ, Shenk T. Wild-type p53 binds to the TATA-binding protein and represses transcription. *Proc Natl Acad Sci USA* 1992; 89:12,028–12,032.

20. Truant R, Xiao H, Ingles CJ, Greenblatt J. Direct interaction between the transcriptional activation domain of human p53 and the TATA box-binding protein. *J Biol Chem* 1993; 268:2284–2287.

21. Farmer G, Colgan J, Nakatani Y, Manley JL, Prives C. Functional interaction between p53, the TATA-binding potein (TBP), and TBP-associated factors in vivo. *Mol Cell Biol* 1996; 16:4295–4304.

22. Lu H, Levine AJ. Human $TAF_{II}31$ protein is a transcriptional coactivator of the p53 protein. *Proc Natl Acad Sci USA* 1995; 92:5154–5158.

23. Thut CJ, Chen J-L, Klemm R, Tjian R. p53 transcriptional activation mediated by coactivators $TAF_{II}40$ and $TAF_{II}60$. *Science* 1995; 267:100–104.

24. Avantaggiati ML, Ogryzko V, Gardner K, Giordano A, Levine AS, Kelly K. Recruitment of p300/CBP in p53-dependent signal pathways. *Cell* 1997; 89:1175–1184.

25. Lill NL, Grossman SR, Ginsberg D, DeCaprio J, Livingston DM. Binding and modulation of p53 by p300/CBP coactivators. *Nature* 1997; 387:823–827.

26. Haupt Y, Maya R, Kazaz A, Oren M. Mdm2 promotes the rapid degradation of p53. *Nature* 1997; 387:296–299.

27. Kubbutat MHG, Jones SN, Vousden KH. Regulation of p53 stability by Mdm2. *Nature* 1997; 387:299–303.

28. Ruaro EM, Collavin L, Del Sal G, Haffner R, Oren M, Levine AJ, Schneider C. A proline-rich motif in p53 is required for transactivation-independent growth arrest as induced by Gas1. *Proc Natl Acad Sci USA* 1997; 94:4675–4680.

29. Sakamuro D, Sabbatini P, White E, Prendergast GC. The polyproline region of p53 is required to acti-vate apoptosis but not growth arrest. *Oncogene* 1997; 15:887–898.

30. Venot C, Maratrat M, Dureuil C, Conseiller E, Bracco L, Debussche L. The requirement for the p53 proline-rich functional domain for mediation of apoptosis is correlated with specific PIG3 gene transactivation and with transcriptional repression. *EMBO J* 1998; 17:4668–4679.
31. Walker KK, Levine AJ. Identification of a novel p53 functional domain that is necessary for efficient growth suppression. *Proc Natl Acad Sci USA* 1996; 93:15,335–15,340.
32. Halazonetis TD, Davis LJ, Kandil AN. Wild-type p53 adopts a 'mutant'-like conformation when bound to DNA. *EMBO J* 1993; 12:1021–1028.
33. Hupp TR, Meek DW, Midgley CA, Lane DP. Regulation of the specific DNA binding function of p53. *Cell* 1992; 71:875–886.
34. Sakamoto H, Lewis MS. Specific sequences from the carboxyl terminus of human p53 gene product form anti-parallel tetramers in solution. *Proc Natl Acad Sci* 1994; 91:8974–8978.
35. Hupp TR, Lane DP. Allosteric activation of latent p53 tetramers. *Curr Biol* 1994; 4:865–75.
36. Jeffrey PD, Gorina S, Pavletich NP. Crystal structure of the tetramerization domain of the p53 tumor suppressor at 1.7 angstroms. *Science* 1995; 267:1498–1502.
37. Waterman JL, Shenk JL, Halazonetis TD. The dihedral symmetry of the p53 tetramerization domain mandates a conformational switch upon DNA binding. *EMBO J* 1995; 14:512–519.
38. Middeler G, Zerf K, Jenovai S, Thulig A, Tschodrich-Rotter M, Kubitscheck U, Peters R. The tumor suppressor p53 is subject to both nuclear import and export, and both are fast, energy-dependent and lectin-inhibited. *Oncogene* 1997; 14:1407–1417.
39. Shaulsky G, Goldfinger N, Ben-Ze'ev A, Rotter V. Nuclear accumulation of p53 protein is mediated by several nuclear localization signals and plays a role in tumorigenesis. *Mol Cell Biol* 1990; 10:6565–6577.
40. Shaulsky G, Goldfinger N, Tosky MS, Levine A, Rotter V. Nuclear localization is essential for the activity of p53 protein. *Oncogene* 1991; 6:2055–2065.
41. Stommel JM, Marchenko ND, Jimenez GS, Moll UM, Hope TJ, Wahl GM. A leucine-rich nuclear export signal in the p53 tetramerization domain: regulation of subcellular localization and p53 activity by NES masking. *EMBO J* 1999; 18:1660–1672.
42. Prives C. Signaling to p53: Breaking the MDM2-p53 circuit. *Cell* 1998; 95:5–8.
43. Matlashewski G, Pim D, Banks L, Crawford L. Alternative splicing of human p53 transcripts. *Oncogene Res* 1987; 1:77–85.
44. Matlashewski GJ, Tuck S, Pim D, Lamb P, Schneider J, Crawford LV. Primary structure polymorphism at amino acid residue 72 of human p53. *Mol Cell Biol* 1987; 7:961–963.
45. Storey A, Thomas M, Kalita A, Harwood C, Gardiol D, Mantovani F, et al. Role of a p53 polymorphism in the development of human papillomavirus-associated cancer. *Nature* 1998; 393:229–234.
46. Thomas M, Kalita A, Labrecque S, Pim D, Banks L, Matlashewski G. Two polymorphic variants of wild-type p53 differ biochemically and biologically. *Mol Cell Biol* 1999; 19:1092–1100.
47. De Laurenzi V, Costanzo A, Barcaroli D, Terrinoni M, Falco M, Annicchiarico-Petruzzelli M, Levrero M, Melino G. Two new p73 splice variants, γ and δ, with different transcriptional activity. *J Exp Med* 1998; 188:1763–1768.
48. Jost CA, Marin MC, Kaelin Jr WG. p73 is a human p53-related protein that can induce apoptosis. *Nature* 1997; 389:191–194.
49. Kaghad M, Bonnet H, Yang A, Creancier L, Biscan J-C, Valent A, et al. Monoallelically expressed gene related to p53 at 1p36, a region frequently deleted in neuroblastoma and other human cancers. *Cell* 1997; 90:809–819.
50. Osada M, Ohba M, Kawahara C, Ishioka C, Kanamaru R, Katoh I, et al. Cloning and functional analysis of human p51, which structurally and functionally resembles p53. *Nat Med* 1998; 4:839–843.
51. Schmale H, Bamberger C. A novel protein with strong homolgy to the tumor suppressor p53. *Oncogene* 1997; 15:1363–1367.
52. Senoo M, Seki N, Ohira M, Sugano S, Watanabe M, Inuzuka S, et al. A second p53-related protein, p73L, with high homology to p73. *Biochem Biophys Res Commun* 1998; 248:603–607.
53. Yang A, Kaghad M, Wang Y, Gillett E, Fleming MD, Dotsch V, et al. p63, a p53 homolog at 3q27–29, encodes multiple products with transactivating, death-inducing, and dominant-negative activities. *Mol Cell* 1998; 3:305–316.
54. Ichimiya S, Nimura Y, Kageyama H, Takada N, Sunahara M, Shishikura T, et al. p73 at chromosome 1p36.3 is lost in advanced stage neuroblastoma but its mutation is infrequent. *Oncogene* 1999; 18:1061–1066.

55. Kroiss MM, Bosserhoff AK, Vogt T, Buettner R, Bogenrieder T, Landthaler M, Stolz W. Loss of expression or mutations in the p73 tumour suppressor gene are not involved in the pathogenesis of malignant melanomas. *Melanoma Res* 1998; 8:504–509.

56. Yokomizo A, Mai M, Tindall DJ, Cheng L, Bostwick DG, Naito S, Smith DI, Liu W. Overexpression of the wild type p73 gene in human bladder cancer. *Oncogene* 1999; 18:1629–1633.

57. Mills AA, Zheng B, Wang XJ, Vogel H, Roop DR, Bradley A. p63 is a p53 homologue required for limb and epidermal morphogenesis. *Nature* 1999; 398:708–713.

58. Yang A, Schweitzer R, Sun D, Kaghad M, Walker N, Bronson RT, et al. p63 is essential for regenerative proliferation in limb, craniofacial and epithelial development. *Nature* 1999; 398:714–717.

59. Godley LA, Kopp JB, Eckhaus M, Paglino JJ, Owens J, Varmus HE. Wild-type p53 transgenic mice exhibit altered differentiation of the ureteric bud and possess small kidneys. *Genes Dev* 1996; 10:836–850.

60. Gottlieb E, Haffner R, King A, Asher G, Gruss P, Lonai P, Oren M. Transgenic mouse model for studying the transcriptional activity of the p53 protein: age-and tissue-dependent changes in radiation-induced activation during embryogenesis. *EMBO J* 1997; 16:1381–1390.

61. Komarova EA, Chernov MV, Franks R, Wang K, Armin G, Zelnick CR, et al. Transgenic mice with p53-responsive *lacZ:* p53 activity varies dramatically during normal development and determines radiation and drug sensitivity *in vivo*. *EMBO J* 1997; 16:1391–1400.

62. Kastan MB, Onyekwere O, Sidransky D, Vogelstein B, Craig RW. Participation of p53 protein in the cellular response to DNA damage. *Cancer Res* 1991; 51:6304–6311.

63. Lu X, Lane DP. Differential induction of transcriptionally active p53 following UV or ionizing radiation: defects in chromosome instability syndromes? *Cell* 1993; 75:765–778.

64. Graeber TG, Osmanian C, Jacks T, Housman DE, Koch CJ, Lowe SW, Giaccia AJ. Hypoxia-mediated selection of cells with diminished apoptotic potential in solid tumours. *Nature* 1996; 379:88–91.

65. Bates S, Phillips AC, Clarke PA, Stott F, Peters G, Ludwig RL, Vousden KH. p14[ARF] links the tumour suppressors RB and p53. *Nature* 1998; 395:124–125.

66. de Stanchina E, McCurrach ME, Zindy F, Shieh SY, Ferbeyre G, Samuelson AV, et al. E1A signaling to p53 involves the p19(ARF) tumor suppressor. *Genes Dev* 1998; 12:2434–2442.

67. Palmero I, Pantoja C, Serrano M. p19ARF links the tumour suppressor p53 to Ras. *Nature* 1998; 395:125–126.

68. Sherr CJ. Tumor surveillance via the ARF=p53 pathway. *Genes Dev* 1998; 12:2984–2991.

69. Eizenberg O, Faber-Elman A, Gottlieb E, Oren M, Rotter V, Schwartz M. Direct involvement of p53 in programmed cell death of oligodendrocytes. *EMBO J* 1995; 14:1136–1144.

70. Canman CE, Gilmer TM, Coutts SB, Kastan MB. Growth factor modulation of p53-mediated growth arrest versus apoptosis. *Genes Dev* 1995; 9:600–611.

71. Atadja P, Wong H, Garkavtsev I, Veillette C, Riabowol K. Increased activity of p53 in senescing fibroblasts. *Proc Natl Acad Sci USA* 1995; 92:8348–8352.

72. Bond J, Haughton M, Blaydes J, Gire V, Wynford-Thomas D, Wyllie F. Evidence that transcriptional activation by p53 plays a direct role in the induction of cellular senescence. *Oncogene* 1996; 13:2097–2104.

73. Gire V, Wynford-Thomas D. Reinitiation of DNA synthesis and cell division in senescent human fibroblasts by microinjection of anti-p53 antibodies. *Mol Cell Biol* 1998; 18:1611–1621.

74. Coffman FD, Studzinski GP. Differentiation-related mechanisms which suppress DNA replication. *Exp Cell Res* 1999; 248:58–73.

75. Feinstein E, Gale RP, Reed J, Canaani E. Expression of the normal p53 gene induces differentiation of K562 cells. *Oncogene* 1992; 7:1853–1857.

76. Mahdi T, Alcalay D, Cognard C, Tanzer J, Kitzis A. Rescue of K562 cells from MDM2-modulated p53-dependent apoptosis by growth factor-induced differentiation. *Biol Cell* 1998; 90:615–627.

77. Masuda T, Ohmi K, Yamaguchi H, Hasegawa K, Sugiyama T, Matsuda Y, Iino M, Nonomura Y. Growing and differentiating characterization of aortic smooth muscle cell line, p53LMAC01 obtained from p53 knock out mice. *Mol Cell Biochem* 1999; 190:99–104.

78. Nigro JM, Aldape KD, Hess SM, Tlsty TD. Cellular adhesion regulates p53 protein levels in primary human keratinocytes. *Cancer Res* 1997; 57:3635–3639.

79. Nikiforov MA, Hagen K, Ossovskaya VS, Connor TM, Lowe SW, Deichman GI, Gudkov AV. p53 modulation of anchorage independent growth and experimental metastasis. *Oncogene* 1996; 13:1709–1719.

80. Fukasawa K, Choi T, Kuriyama R, Rulong S, Vande Woude GF. Abnormal centrosome amplification in the absence of p53. *Science* 1996; 271:1744–1747.

81. Cox LS, Hupp T, Midgley CA, Lane DP. A direct effect of activated human p53 on nuclear DNA replication. *EMBO J* 1995; 14:2099–2105.
82. Dudenhoffer C, Rohaly G, Will K, Deppert W, Wiesmuller L. Specific mismatch recognition in heteroduplex intermediates by p53 suggests a role in fidelity control of homologous recombination. *Mol Cell Biol* 1998; 18:5332–42.
83. Dutta A, Ruppert JM, Aster JC, Winchester E. Inhibition of DNA replication factor RPA by p53. *Nature* 1993; 365:79–82.
84. Janus F, Albrechtsen N, Dornreiter I, Wiesmuller L, Grosse F, Deppert W. The dual role model for p53 in maintaining genomic integrity. *Cell Mol Life Sci* 1999; 55:12–17.
85. Miller SD, Farmer G, Prives C. p53 inhibits DNA replication in vitro in a DNA-binding-dependent manner. *Mol Cell Biol* 1995; 15:6554–6560.
86. Notterman D, Young S, Wainger B, Levine AJ. Prevention of mammalian DNA reduplication, following the release from the mitotic spindle checkpoint, requires p53 protein, but not p53-mediated transcriptional activity. *Oncogene* 1998; 17:2743–2751.
87. Smith ML, Chen IT, Zhan Q, O'Connor PM, Fornace AJJ. Involvement of the p53 tumor suppressor in repair of u.v.-type DNA damage. *Oncogene* 1995; 10:1053–1059.
88. Bates S, Vousden KH. p53 in signalling checkpoint arrest or apoptosis. *Current Opin Genet Dev* 1996; 6:1–7.
89. Ko LJ, Prives C. p53: puzzle and paradigm. *Genes Dev* 1996; 10:1054–1072.
90. Bates S, Hickman ES, Vousden KH. Reversal of p53-induced cell-cycle arrest. *Mol Carcinog* 1999; 24:7–14.
91. El-Deiry W, Tokino T, Velculescu VE, Levy DB, Parson VE, Trent JM, et al. WAF1, a potential mediator of p53 tumour suppression. *Cell* 1993; 75:817–825.
92. Gu Y, Turck CW, Morgan DO. Inhibition of CDK2 activity in vivo by an associated 20K regulatory subunit. *Nature* 1993; 366:707–710.
93. Harper JW, Adami GR, Wei N, Keyomarsi K, Elledge SJ. The p21 CDK-interacting protein cip1 is a potent inhibitor of G1 cyclin-dependent kinases. *Cell* 1993; 75:805–816.
94. Luo Y, Hurwitz J, Massagué J. Cell-cycle inhibition by independent CDK and PCNA binding domains in p21. *Nature* 1995; 375:159–161.
95. Bates S, Ryan KM, Phillips AC, Vousden KH. Cell cycle arrest and DNA endoreduplication following p21$^{Waf1/Cip1}$ expression. *Oncogene* 1998; 17:1691–1703.
96. Waldman T, Kinzler KW, Vogelstein B. p21 is necessary for the p53-mediated G1 arrest in human cancer cells. *Cancer Res* 1995; 55:5187–5190.
97. Brugarolas J, Chandrasekaran C, Gordon JI, Beach D, Jacks T, Hannon GJ. Radiation-induced cell cycle arrest compromised by p21 deficiency. *Nature* 1995; 377:552–556.
98. Deng C, Zhang P, Harper JW, Elledge SJ, Leder P. Mice lacking p21$^{CIP1/WAF1}$ undergo normal development, but are defective in G1 checkpoint control. *Cell* 1995; 82:675–684.
99. Fornace AJ, Alamo I, Hollander MC. DNA damage inducible transcripts in mammalian cells. *Proc Nat. Acad Sci USA* 1988; 85:8800–8804.
100. Papathanasiou MA, Kerr NC, Robbins JH, McBride OW, Alamo IJ, Barrett SF, Hickson ID, Fornace AJJ. Induction by ionizing radiation of the gadd45 gene in cultured human cells: lack of mediation by protein kinase C. *Mol Cell Biol* 1991; 11:1009–1016.
101. Chin PL, Momand J, Pfeifer GP. *In vivo* evidence for binding of p53 to consensus binding sites in the *p21* and *GADD45* genes in response to ionizing radiation. *Oncogene* 1997; 15:87–99.
102. Wang XW, Zhan Q, Coursen JD, Khan MA, Kontny HU, Yu L, et al. GADD45 induction of a G2/M cell cycle checkpoint. *Proc Natl Acad Sci USA* 1999; 96:3706–3711.
103. Smith ML, Chen I-T, Zhan Q, Bae I, Chen C-Y, Gilmer TM, et al. Interaction of the p53-regulated protein Gadd45 with proliferating cell nuclear antigen. *Science* 1994; 266:1376–1380.
104. Carrier F, Georgel PT, Pourquier P, Blake M, Kontny HU, Antinore MJ, et al. Gadd45, a p53-responsive stress protein, modifies DNA accessibility on damaged chromatin. *Mol Cell Biol* 1999; 19:1673–1685.
105. Zhan Q, Chen IT, Antimore MJ, Fornace AJ. Tumor suppressor p53 can participate in transcriptional induction of the GADD45 promoter in the absence of direct DNA binding. *Mol Cell Biol* 1998; 18:2768–2778.
106. Velasco-Miguel S, Buckbinder L, Jean P, Gelbert L, Talbott R, Laidlaw J, Seizinger B, Kley N. PA26, a novel target of the p53 tumor suppressor and member of the GADD family of DNA damage and growth arrest inducible genes. *Oncogene* 1999; 18:127–137.
107. Buckbinder L, Talbott R, Velasco-Miguel S, Takenaka I, Faha B, Seizinger BR, Kley N. Induction of the growth inhibitor IGF-binding protein 3 by p53. *Nature* 1995; 377:646–649.

108. Matsuzawa S, Takayama S, Froesch BA, Zapata JM, Reed JC. p53-inducible human homogue of Drosophila seven in absentia (Siah) inhibits cells growth: suppression by BAG-1. *EMBO J* 1998; 17:2736–2747.

109. Nemani M, Linares-Cruz G, Bruzzoni-Giovanelli H, Roperch JP, Tuynder M, Bougueleret L. Activation of the human homologue of the Drosophila sina gene in apoptosis and tumor suppression. *Proc Natl Acad Sci USA* 1996; 93:9039–9042.

110. Hermeking H, Lengauer C, Polyak K, He T-C, Zhang L, Thiagalingam S, Kinzler KW, Vogelstein B. *14-3-3σ* is a p53-regulated inhibitor of G2/M progression. *Mol Cell* 1997; 1:3–11.

111. Waterman MJ, Stavridi ES, Waterman JL, Halazonetis TD. ATM-dependent activation of p53 involves dephosphorylation and association with 14-3-3 proteins. *Nat Genet* 1998; 19:175–178.

112. Bates S, Vousden KH. Mechanisms of p53-mediated apoptosis. *Cell Mol Life Sci* 1999; 55:28–37.

113. Kerr JF, Wyllie AH, Currie AR. Apoptosis: a basic biological phenomenon with wide-ranging implications in tissue kinetics. *Br J Cancer* 1972; 26:239–257.

114. Salvesen GS, Dixit VM. Caspases: intracellular signaling by proteolysis. *Cell* 1997; 91:443–446.

115. Ashkenazi A, Dixit VM. Death receptors: Signaling and modulation. *Science* 1998; 281:1305–1308.

116. Green DR, Reed JC. Mitochondria and apoptosis. *Science* 1998; 281:1309–1312.

117. Soengas MS, Alarcon RM, Yoshida H, Giaccia AJ, Hakem R, Mak TW, Lowe SW. Apaf-1 and caspase-9 in p53-dependent apoptosis and tumor inhibition. *Science* 1999; 284:156–159.

118. Miyashita T, Reed JC. Tumor suppressor p53 is a direct transcriptional activator of the human bax gene. *Cell* 1995; 80:293–299.

119. Rossé T, Olivier R, Monney L, Rager M, Conus S, Fellay I, Jansen B, Borner C. Bcl-2 prolongs cell surivial after Bax-induced release of cytochrome *c*. *Nature* 1998; 391:496–499.

120. Kastan M. On the TRAIL from p53 to apoptosis? *Nat Genet* 1997; 17:130–131.

121. Wu G, Burns TF, McDonald ER, Jiang W, Meng R, Krantz ID, et al. KILLER/DR5 is a DNA damage-inducible p53-regulated death receptor gene. *Nat Genet* 1997; 17:141–143.

122. Knudson MC, Tung KSK, Tourtellotte WG, Brown GAJ, Korsmeyer SJ. Bax-deficient mice with lymphoid hyperplasia and male germ cell death. *Science* 1995; 270:96–98.

123. McCurrach ME, Connor TMF, Knudson MC, Korsmeyer SJ, Lowe SW. *bax-* deficiency promotes drug resistance and oncogenic transformation by attenuating p53-dependent apoptosis. *Proc Natl Acad Sci USA* 1997; 94:2345–2349.

124. Reinke V, Lozano G. The p53 targets *mdm2* and *Fas* are not required as mediators of apoptosis *in vivo*. *Oncogene* 1997; 15:1527–1534.

125. Yin C, Knudson CM, Korsmeyer SJ, Van Dyke T. Bax suppresses tumorigenesis and stimulates apoptosis *in vivo*. *Nature* 1997; 385:637–640.

126. Israeli D, Tessler E, Haupt Y, Elkeles A, Wilder S, Amson R, Telerman A, Oren M. A novel p53-inducible gene, *PAG608,* encodes a nuclear zinc finger protein whose expression promotes apoptosis. *EMBO J* 1997; 16:4384–4392.

127. Wu GS, Saftig P, Peters C, El-Deiry WS. Potential role for cathepsin D in p53-dependent tumor suppression and chemosensitivity. *Oncogene* 1998; 16:2177–2183.

128. Polyak K, Xia Y, Zweier JL, Kinzler KW, Vogelstein B. A model for p53-induced apoptosis. *Nature* 1997; 389:300–305.

129. Dumont A, Hehner SP, Hofmann TG, Ueffing M, Droge W, Schmitz ML. Hydrogen peroxide-induced apoptosis is CD95-independent, requires the release of mitochondria-derived reactive oxygen species and the activation of NF-kappaB. *Oncogene* 1999; 18:747–757.

130. Sun Y, Wicha M, Leopold WR. Regulation of metastasis-related gene expression by p53: a potential clinical implication. *Mol Carcinog* 1999; 24:25–28.

131. Nishimori H, Shiratsuchi T, Urano T, Kimura Y, Kiyono K, Tatsumi K, et al. Novel brain-specific p53-target gene, BAI1, containing thrombospondin type 1 repeats inhibits experimental angiogenesis. *Oncogene* 1997; 15:2145–2150.

132. White E. Life, death, and the pursuit of apoptosis. *Genes Dev* 1996; 10:1–15.

133. Santhanam U, Ray A, Sehgal PB. Repression of the interleukin 6 gene promoter by p53 and the retinoblastoma susceptibility gene product. *Proc Natl Acad Sci USA* 1991; 88:7605–7609.

134. Ravi R, Mookerjee B, van Hensbergen Y, Bedi GC, Giordano A, El-Deiry WS, Fuchs EJ, Bedi A. p53-mediated repression of nuclear factor-kappaB RelA via the transcriptional integrator p300. *Cancer Res* 1998; 58:4531–4536.

135. Wadgaonkar R, Phelps KM, Haque Z, Williams AJ, Silverman ES, Collins T. CREB-binding protein is a nuclear integrator of nuclear factor-kappaB and p53 signaling. *J Biol Chem* 1999; 274:1879–1882.

136. Desdouets C, Ory C, Matesic G, Soussi T, Brechot C, Sobczak-Thepot J. ATF/CREB site mediated tran-
     scriptional activation and p53 dependent repression of the cyclin A promoter. *FEBS Lett* 1996: 385.
137. Sugrue MM, Shin DY, Lee SW, Aaronson SA. Wild-type p53 triggers a rapid senescence program in
     human tumor cells lacking functional p53. *Proc Natl Acad Sci USA* 1997; 94:9648–9653.
138. Yamaguchi M, Hayashi Y, Matsuoka S, Takahashi T, Matsukage A. Differential effect of p53 on the
     promoters of mouse DNA polymerase beta gene and proliferating-cell-nuclear-antigen gene. *Eur J
     Biochem* 1994; 221:227–237.
139. Cairns CA, White RJ. p53 is a general repressor of RNA polymerase III transcription. *EMBO J* 1998;
     17:3112–3123.
140. Chesnokov I, Chu WM, Botchan MR, Schmid CW. p53 inhibits RNA polymerase III-directed tran-
     scription in a promoter-dependent manner. *Mol Cell Biol* 1996; 16:7084–7088.
141. Murphy M, Hinman A, Levine AJ. Wild-type p53 negatively regulates the expression of a micro-
     tubule-associated protein. *Genes Dev* 1996; 10:2971–2980.
142. Zhang CC, Yang JM, White E, Murphy M, Levine A, Hait WN. The role of MAP4 expression in the
     sensitivity to paclitaxel and resistance to vinca alkaloids in p53 mutant cells. *Oncogene* 1998;
     16:1617–1624.
143. Ryan KM, Vousden KH. Characterization of structural p53 mutants which show selective defects in
     apoptosis, but not cell cycle arrest. *Mol Cell Biol* 1998, 18:3692–3698.
144. Kley N, Chung RY, Fay S, Loeffler JP, Seizinger BR. Repression of the basal c-fos promoter by wild-
     type p53. *Nucleic Acids Res* 1992; 20:4083–4087.
145. Elkeles A, Juven-Gershon T, Israeli D, Wilder S, Zalcenstein A, Oren M. c-fos proto-oncogene is a
     target for transactivation by the p53 tumor suppressor. *Mol Cell Biol* 1999; 19:2594–2600.
146. Caelles C, Helmberg A, Karin M. p53-dependent apoptosis in the absence of transcriptional activa-
     tion of p53-target genes. *Nature* 1994; 370:220–223.
147. Haupt Y, Rowan S, Shaulian E, Vousden KH, Oren M. Induction of apoptosis in HeLa cells by trans-
     activation deficient p53. *Genes Dev* 1995; 9:2170–2183.
148. Almog N, Li R, Peled A, Schwartz D, Wolkowicz R, Goldfinger N, Pei H, Rotter V. The murine C'-
     terminally alternatively spliced form of p53 induces attenuated apoptosis in myeloid cells. *Mol Cell
     Biol* 1997; 17:713–722.
149. Chen X, Ko LJ, Jayaraman L, Prives C. p53 levels, functional domains, and DNA damage determine
     the extent of the apoptotic response of tumor cells. *Genes Dev.* 1996; 10:2438–2451.
150. Zhu J, Jiang J, Zhou W, Zhu K, Chen X. Differential regulation of cellular target genes devoid of the
     PXXP motifs with impaired apoptotic function. *Oncogene* 1999; 18:2149–2155.
151. Bennett M, Macdonald K, Chan S, Luzio JP, Simari R, Weissberg. Cell surface trafficking of Fas: a
     rapid mechanism of p53 mediated apoptosis. *Science* 1998; 282:290–293.
152. Wang XW, Vermeulen W, Coursen JD, Gibson M, Lupold SE, Forrester K, et al. XPB and XPD
     DNA helicases are components of the p53-mediated apoptosis pathway. *Genes Dev* 1996;
     10:1219–1232.
153. Wang XW, Yeh H, Schaeffer L, Roy R, Moncollin V, Egly JM, et al. p53 modulation of TFIIH-associ-
     ated nucleotide excision repair activity. *Nat Genet* 1995; 10:188–195.
154. Haupt Y, Barak Y, Oren M. Cell type-specific inhibition of p53-mediated apoptosis by mdm2. *EMBO
     J* 1996; 15:1596–1606.
155. Sionov RV, Moallem E, Berger M, Kazaz A, Gerlitz O, Ben-Neriah Y, Oren M, Haupt Y. c-Abl neu-
     tralizes the inhibitory effect of Mdm2 on p53. *J Biol Chem* 1999; 274:8371–8374.
156. Theis S, Roemer K. c-Abl tyrosine kinase can mediate tumor cell apoptosis independently of the Rb
     and p53 tumor suppressors. *Oncogene* 1998; 17:557–564.
157. Yuan ZM, Huang Y, Ishiko T, Kharbanda S, Weichselbaum R, Kufe D. Regulation of DNA damage-
     induced apoptosis by the c-Abl tyrosine kinase. *Proc Natl Acad Sci USA* 1997; 94:1437–1440.
158. Rowan S, Ludwig RL, Haupt Y, Bates S, Lu X, Oren M, Vousden KH. Specific loss of apoptotic but
     not cell cycle arrest funtion in a human tumour derived p53 mutant. *EMBO J* 1996; 15:827–838.
159. Chiou SK, Rao L, White E. Bcl-2 blocks p53-dependent apoptosis. *Mol Cell Biol* 1994;
     14:2556–2563.
160. Guillouf C, Grana X, Selvakumaran M, De Luca A, Giordano A, Hoffman B, Liebermann DA. Dis-
     section of the genetic programs of p53-mediated G1 growth arrest and apoptosis: blocking p53-
     induced apoptosis unmasks G1 arrest. *Blood* 1995; 85:2691–2698.
161. Friedlander P, Haupt Y, Prives C, Oren M. A mutant p53 that discriminated between p53 responsive
     genes cannot induce apoptosis. *Mol Cell Biol* 1996; 16:4961–4971.

162. Ludwig RL, Bates S, Vousden KH. Differential transcriptional activation of target cellular promoters by p53 mutants with impaired apoptotic function. *Mol Cell Biol* 1996; 16:4952–4960.

163. Muller-Tiemann BF, Halazonetis TD, Elting JJ. Identification of an additional negative regulatory region for p53 sequence-specific DNA binding. *Proc Natl Acad Sci USA* 1998; 95:6079–6084.

164. Fu L, Benchimol S. Participation of the human p53 3′UTR in translational repression and activation following gamma-irradiation. *EMBO J* 1997; 16:4117–4127.

165. Fu L, Minden MD, Benchimol S. Translational regulation of human p53 gene expression. *EMBO J* 1996; 15:4392–4401.

166. Reisman D, Loging WT. Transcriptional regulation of the p53 tumor suppressor gene. *Semin Cancer Biol* 1998; 8:317–324.

167. Lozano G, Montes de Oca Luna R. MDM2 function. Biochim Biophys Acta 1998; 1377:M55–59.

168. Marston NJ, Crook T, Vousden KH. Interaction of p53 with MDM2 is independent of E6 and does not mediate wild type transformation suppressor function. *Oncogene* 1994; 9:2707–2716.

169. Thut CJ, Goodrich JA, Tjian R. Repression of p53-mediated transcription by MDM2: a dual mechanism. *Genes Dev* 1997; 11:1974–1986.

170. Kubbutat MHG, Vousden KH. Keeping an old friend under control: regulation of p53 stability. *Mol Med Today* 1998; 4:250–256.

171. Jones SN, Roe AE, Donehower LA, Bradley A. Rescue of embyonic lethality in Mdm2-deficient mice by absence of p53. *Nature* 1995; 378:206–208.

172. Montes de Oca Luna R, Wagner DS, Lozano G. Rescue of early embryonic lethality in *mdm2*-deficient mice by deletion of *p53*. *Nature* 1995; 378:203–206.

173. Ashcroft M, Kubbutat MH, Vousden KH. Regulation of p53 function and stability by phosphorylation. *Mol Cell Biol* 1999; 19:1751–1758.

174. Kubbutat MHG, Ludwig RL, Ashcroft M, Vousden KH. Regulation of Mdm2 directed degradation by the C-terminus of p53. *Mol Cell Biol* 1998; 18:5690–5698.

175. Elledge RM, Allred DC. Prognostic and predictive value of p53 and p21 in breast cancer. *Breast Cancer Res Treat* 1998; 52:79–98.

176. Honda R, Tanaka H, Yasuda H. Oncoprotein MDM2 is a ubiquitin ligase E3 for tumor suppressor p53. *FEBS Lett* 1997; 420:25–27.

177. Honda R, Yasuda H. Association of p19[ARF] with Mdm2 inhibits ubiquitin ligase activity of Mdm2 for tumor suppressor p53. *EMBO J* 1999; 18:22–27.

178. Kubbutat MHG, Ludwig RL, Levine AJ, Vousden KH. Analysis of the degradation function of Mdm2. *Cell Growth Differ* 1999, in press.

179. Midgley CA, Lane DP. p53 protein stability in tumour cells is not determined by mutation but is dependent on Mdm2 binding. *Oncogene* 1997; 15:1179–1189.

180. Grossman SR, Perez M, Kung AL, Joseph M, Mansur C, Xiao ZX, et al. p300/MDM2 complexes participate in MDM2-mediated p53 degradation. *Mol Cell* 1998; 2:405–415.

181. Juven-Gershon T, Shifman O, Unger T, Elkeles A, Haupt Y, Oren M. The Mdm2 oncoprotein interacts with the cell fate regulator Numb. *Mol Cell Biol* 1998; 18:3974–3982.

182. Oren M, Rotter V. Introduction: p53—the first twenty years. *Cell Mol Life Sci* 1999; 55:9–11.

183. Blattner C, Sparks A, Lane D. Transcription factor E2F-1 is upregulated in response to DNA damage in a manner analogous to that of p53. *Mol Cell Biol* 1999; 19:3704–3713.

184. Freedman DA, Levine AJ. Nuclear export is required for degradation of endogenous p53 by MDM2 and human papillomavirus E6. *Mol Cell Biol* 1998; 18:7288–7293.

185. Fuchs SY, Adler V, Buschmann T, Yin Z, Wu X, Jones SN, Ronai Z. JNK targets p53 ubiquitination and degradation in nonstressed cells. *Genes Dev* 1998; 12:2658–2663.

186. Fuchs SY, Adler V, Pincus MR, Ronai Z. MEKK1/JNK signaling stabilizes and activates p53. *Proc Natl Acad Sci USA* 1998; 95:10541–10546.

187. Kubbutat MHG, Vousden KH. New HPV E6 binding proteins: dangerous liaisons? *Trends Microbiol* 1998; 6:173–175.

188. Querido E, Marcellus RC, Lai A, Charbonneau R, Teodoro JG, Ketner G, and Branton PE. Regulation of p53 levels by the E1B 55-kilodalton protein and E4orf6 in adenovirus-infected cells. *J Virol* 1997; 71:3788–3798.

189. Roth J, Dobbelstein M, Freedman DA, Shenk T, Levine AJ. Nucleocytoplasmic shuttling of the hdm2 oncoprotein regulates the levels of the p53 protein via a pathway used by the human immunodeficiency virus rev protein. *EMBO J* 1998; 17:554–564.

190. Kubbutat MHG, Vousden KH. Proteolytic cleavage of human p53 by calpain: a potential regulator of protein stability. *Mol Cell Biol* 1997; 17:460–468.

191. Balint, S, Vousden KH. *Oncogene* 1999; 18:3923–3929.

192. Zeng X, Chen L, Jost CA, Maya R, Keller D, Wang X, et al. MDM2 suppresses p73 function without promoting p73 degradation. *Mol Cell Biol* 1999; 19:3257–3266.

193. Bottger V, Bottger A, Garcia-Echeverria C, Ramos YF, van der Eb AJ, Jochemsen AG, Lane DP. Comparative study of the p53-mdm2 and p53-MDMX interfaces. *Oncogene* 1999; 18:189–199.

194. Shvarts A, Steegenga WT, Riteco N, van Laar T, Dekker P, Bazuine M, et al. MDMX: a novel p53-binding protein with some functional properties of MDM2. *EMBO J* 1996; 15:5349–5357.

195. Tanimura S, Ohtsuka S, Mitsui K, Shirouzu K, Yoshimura A, Ohtsubo M. MDM2 interacts with MDMX through their RING finger domains. *FEBS Lett* 1999; 447:5–9.

196. Mayo LD, Turchi JJ, Berberich SJ. Mdm-2 phosphorylation by DNA-dependent protein kinase prevents interaction with p53. *Cancer Res* 1997; 57:5013–5016.

197. Shieh S-Y, Ikeda M, Taya Y, Prives C. DNA damage-induced phosphorylation of p53 alleviates inhibition by MDM2. *Cell* 1997; 91:325–334.

198. Canman CE, Lim D-S, Cimprich KA, Taya Y, Tamai K, Sakaguchi K, et al. Activation of the ATM kinase by ionizing radiation and phosphorylation of p53. *Science* 1998; 281:1677–1679.

199. Banin S, Moyal L, Shieh S-Y, Taya Y, Anderson CW, Chessa L, et al. Enhanced phosphorylation of p53 by ATM in response to DNA damage. *Science* 1998; 281:1674–1677.

200. Khanna KK, Keating KE, Kozlov S, Scott S, Gatei M, Hobson K, et al. ATM associates with and phosphorylates p53: mapping the region of interaction. *Nat Genet* 1998; 20:398–400.

201. Nakagawa K, Taya Y, Tamai K, Yamaizumi M. Requirement of ATM in phosphorylation of the human p53 protein at serine 15 following DNA double-strand breaks. *Mol Cell Biol* 1999; 19:2828–2834.

202. Siliciano JD, Canman CE, Taya Y, Sakaguchi K, Appella E, Kastan MB. DNA damage induces phosphorylation of the amino terminus of p53. *Genes Dev* 1997; 11:3471–3481.

203. Tibbetts RS, Brumbaugh KM, Williams JM, Sarkaria JN, Cliby WA, Shieh SY, et al. A role for ATR in the DNA damage-induced phosphorylation of p53. *Genes Dev* 1999; 13:152–157.

204. Shieh SY, Taya Y, Prives C. DNA damage-inducible phosphorylation of p53 at N-terminal sites including a novel site, Ser20, requires tetramerization. *EMBO J* 1999; 18:1815–1823.

205. Unger T, Juven-Gershon T, Moallem E, Berger M, Vogt Sionov R, Lozano G, Oren M, Haupt Y. Critical role for Ser20 of human p53 in the negative regulation of p53 by Mdm2. *EMBO J* 1999; 18:1805–1814.

206. Blattner C, Tobiasch E, Litfen M, Rahmsdorf HJ, Herrlich P. DNA damage induced p53 stabilization: no indication for an involvement of p53 phosphorylation. *Oncogene* 1999; 18:1723–1732.

207. Ashcroft M, Taya Y, Vousden KH. Stress signals utilize multiple pathways to stabilize p53 Mol. Cell. Biol. 2000, 20:3224–3233.

208. Arriola EL, Rodriguez Lopez A, Chresta CM. Differential regulation of p21[waf-1/cip-1] and Mdm2 by etoposide: etoposide inhibits the p53-Mdm2 autoregulatory loop. *Oncogene* 1999; 18:1081–1091.

209. Hsieh JK, Chan FS, O'Connor DJ, Mittnacht S, Zhong S, Lu X. RB regulates the stability and the apoptotic function of p53 via MDM2. *Mol Cell* 1999; 3:181–193.

210. Zindy F, Eischen CM, Randle DH, Kamijo T, Cleveland JL, Sherr CJ, Roussel MF. Myc signaling via the ARF tumor suppressor regulates p53-dependent apoptosis and immortalization. *Genes Dev* 1998; 12:2424–2433.

211. Pomerantz J, Schreiber-Agus N, Liégeois NJ, Silverman A, Alland L, Chin L, et al. *Ink4a* tumor suppressor gene product, p19[Arf], interacts with MDM2 and neutralizes MDM2's inhibition of p53. *Cell* 1998; 92:713–723.

212. Stott F, Bates SA, James M, McConnell BB, Starborg M, Brookes S, et al. The alternative product from the human *CDKN2A* locus, p14[ARF], participates in a regulatory feedback loop with p53 and MDM2. *EMBO J* 1998; 17:5001–5014.

213. Zhang Y, Xiong Y, Yarbrough WG. ARF promotes MDM2 degradation and stabilizes p53: *ARF-INK4a* locus deletion impairs both the Rb and p53 tumor suppression pathways. *Cell* 1998; 92:725–734.

214. Robertson KD, Jones PA. The human ARF cell cycle regulatory gene promoter is a CpG island which can be silenced by DNA methylation and down-regulated by wild-type p53. *Mol Cell Biol* 1998; 18:6457–6473.

215. Kamijo T, Weber JD, Zambetti G, Zindy F, Roussel MF, Sherr CJ. Functional and physical interactions of the ARF tumor suppressor with p53 and Mdm2. *Proc Natl Acad Sci USA* 1998; 95:8292–8297.

216. Liao MJ, Yin C, Barlow C, Wynshaw-Boris A, van Dyke T. Atm is dispensable for p53 apoptosis and tumor suppression triggered by cell cycle dysfunction. *Mol Cell Biol* 1999; 19:3095–3102.

217. Halazonetis TD, Kandil AN. Conformational shifts propagate from the oligomerization domain of p53 to its tetrameric DNA binding domain and restore DNA binding to select p53 mutants. *EMBO J* 1993; 12:5057–5064.

218. Anderson ME, Woelker B, Reed M, Wang P, Tegtmeyer P. Reciprocal interference between the sequence-specific core and nonspecific C-terminal DNA binding domains of p53: implications for regulation. *Mol Cell Biol* 1997; 17:6255–6264.

219. Bayle JH, Elenbaas B, Levine AJ. The carboxyl-terminal domain of the p53 protein regulates sequence-specific DNA binding through its nonspecific nucleic acid-binding activity. *Proc Natl Acad Sci USA* 1995; 92:5729–5733.

220. Hupp TR, Lane DP. Two distinct signaling pathways activate the latent DNA binding function of p53 in a casein kinase II-independent manner. *J Biol Chem* 1995; 270:18,165–18,174.

221. Prives C, Hall PH. The p53 pathway. *J Pathol* 1999; 187:112–126.

222. Gu W, Roeder RG. Activation of p53 sequence-specific DNA binding by acetylation of the C-terminal domain. *Cell* 1997; 90:595–606.

223. Sakaguchi K, Herrera JE, Saito S, Miki T, Bustin M, Vassilev A, Anderson CW, Appella E. DNA damage activates p53 through a phosphorylation-acetylation cascade. *Genes Dev* 1998; 12:2831–2841.

224. Shaw P, Freeman J, Bovey R, Iggo R. Regulation of specific DNA binding by p53: evidence for a role of O-glycosylation and charged residues at the carboxy-terminus. *Oncogene* 1996; 12:921–930.

225. Jayaraman L, Prives C. Activation of p53 sequence-specific DNA binding by short single strands of DNA requires the p53 C-terminus. *Cell* 1995; 81:1021–1029.

226. Marston NJ, Ludwig RL, Vousden KH. Activation of p53 DNA binding activity by point mutation. *Oncogene* 1998; 24:3123–3131.

227. Jayaraman L, Murthy KGK, Zhu C, Curran T, Xanthoudakis S, Prives C. Identification of redox/repair protein Ref-1 as a potent activator of p53. *Genes Dev* 1997; 11:558–570.

228. Jayaraman L, Moorthy NC, Murthy KG, Manley JL, Bustin M, Prives C. High mobility group protein-1 (HMG-1) is a unique activator of p53. *Genes Dev* 1998; 12:462–472.

229. Lu X, Burbridge SA, Griffin S, Smith HM. Discordance between accumulated p53 protein levels and its transcriptional activity in response to U.V. radiation. *Oncogene* 1997; 13:413–418.

230. Lutzker SG, Levine AJ. A functionally inactive p53 protein in teratocarcinoma cells is activated by either DNA damage or cellular differentiation. *Nat Med* 1996; 2:804–810.

231. Pise-Masison C, Radonovich M, Sakaguchi K, Appella E, Brady JN. Phosphorylation of p53: a novel pathway for p53 inactivation in human T-cell lymphotropic virus type 1-transformed cells. *J Virol* 1998; 72:6348–6355.

232. Lambert PF, Kashanchi F, Radonovich MF, Shiekhattar R, Brady JN. Phosphorylation of p53 serine 15 increases interaction with CBP. *J Biol Chem* 1998; 273:33,048–33,053.

233. Chakravarti D, Ogryzko V, Kao HY, Nash A, Chen H, Nakatani Y, Evans RM. A viral mechanism for inhibition of p300 and PCAF acetyltransferase activity. *Cell* 1999; 96:393–403.

234. Liu L, Scolnick DM, Trievel RC, Zhang HB, Marmorstein R, Halazonetis TD, Berger SL. p53 sites acetylated in vitro by PCAF and p300 are acetylated in vivo in response to DNA damage. *Mol Cell Biol* 1999; 19:1202–1209.

235. Thomas A, White E. Suppression of the p300-dependent *mdm2* negative-feedback loop induced the p53 apoptotic function. *Genes Dev* 1998; 12:1975–1985.

236. Webster GA, Perkins ND. Transcriptional cross talk between NF-kappaB and p53. *Mol Cell Biol* 1999; 19:3485–3495.

237. Martinez JD, Craven MT, Joseloff E, Milczarek G, Bowden GT. Regulation of DNA binding and transactivation in p53 by nuclear localization and phosphorylation. *Oncogene* 1997; 14:2511–2520.

238. Moll UM, Riou G, Levine AJ. Two distinct mechanisms alter p53 in breast cancer: mutation and nuclear exclusion. *Proc Natl Acad Sci USA* 1992; 89:7262–7266.

239. Takahashi K, Suzuki K. DNA synthesis-associated nuclear exclusion of p53 in normal human breast epithelial cells in culture. *Oncogene* 1994; 9:183–188.

240. Beham A, Marin MC, Fernandez A, Herrmann J, Brisbay S, Tari AM, et al. Bcl-2 inhibits p53 nuclear import following DNA damage. *Oncogene* 1997; 15:2767–2772.

241. Roth J, König C, Wienzek S, Weigel S, Ristea S, Dobbelstein M. Inactivation of p53 but not p73 by adenovirus type 5 E1B 55-kilodalton and E4 34-kilodalton oncoproteins. *J Virol* 1998; 72:8510–8516.

242. Rampino N, Yamamoto H, Ionov Y, Li Y, Sawai H, Reed JC, Perucho M. Somatic frameshift mutations in the BAX gene in colon cancers of the microsatellite mutator phenotype. *Science* 1997; 275:967–969.

243. Hansen R, Oren M. p53; from inductive signal to cellular effect. *Curr Opin Genet Dev* 1997; 7:46–51.
244. Blandino G, Levine AJ, Oren M. Mutant p53 gain of function: differential effects of different p53 mutants on resistance of cultures cells to chemotherapy. *Oncogene* 1999; 18:477–485.
245. Dittmer D, Pati S, Zambetti G, Chu S, Teresky AK, Moore M, Finlay C, Levine AJ. Gain of function mutations in p53. *Nat Genet* 1993; 4:42–46.
246. Li R, Sutphin PD, Schwartz D, Matas D, Almog N, Wolkowicz R, et al. Mutant p53 protein expression interferes with p53-independent apoptotic pathways. *Oncogene* 1998; 16:3269–3277.
247. Di Como CJ, Gaiddon C, Prives C. p73 function is inhibited by tumor-derived p53 mutants in mammalian cells. *Mol Cell Biol* 1999; 19:1438–1449.
248. Hall PA, McKee PH, Menage HD, Dover R, Lane DP. High levels of p53 protein in UV irradiated human skin. *Oncogene* 1993; 8:203–207.
249. Liang SB, Ohtsuki Y, Furihata M, Takeuchi T, Iwata J, Chen BK, Sonobe H. Sun-exposure- and aging-dependent p53 protein accumulation results in growth advantage for tumour cells in carcinogenesis of nonmelanocytic skin cancer. *Virchows Arch* 1999; 434:193–199.
250. Crook T, Brooks LA, Crossland S, Osin P, Barker KT, Waller J, et al. p53 mutation with frequent novel condons but not a mutator phenotype in BRCA1- and BRCA2-associated breast tumours. *Oncogene* 1998; 17:1681–1689.
251. Crook T, Crossland S, Crompton MR, Osin P, Gusterson BA. p53 mutations in BRCA1-associated familial breast cancer. *Lancet* 1997; 350:638–639.
252. Phillips AC, Bates S, Ryan KM, Helin K, Vousden KH. Induction of DNA synthesis and apoptosis are separable functions of E2F-1. *Genes Dev* 1997; 11:1853–1863.
253. Selivanova G, Iotsova V, Okan I, Fritsche M, Ström M, Groner B, Grafström RC, Wiman KG. Restoration of the growth suppression function of mutant *p53* by a synthetic peptide derived from the p53 C-terminal domain. *Nat Med* 1997; 3:632–638.
254. Selivanova G, Ryabchenko L, Jansson E, Iotsova V, Wiman KG. Reactivation of mutant p53 through interaction of a C-terminal peptide with the core domain. *Mol Cell Biol* 1999; 19:3395–3402.

8

p16 Tumor Suppressor

Alexander Kamb, PhD and Ken McCormack, PhD

CONTENTS

1. INTRODUCTION

p16, a protein named for its migration rate on denaturing gels, is a member of a class of functionally similar proteins called cyclin-dependent kinase inhibitors (CDKI). p16 and its relatives bind to CDKs and inhibit their kinase activity. Because cyclins and CDKs form the core of the cell cycle apparatus, p16 is positioned to directly regulate some of the most basic cell cycle decisions. In eukaryotic cells, the cell cycle is typically described in terms of four component phases: G1, S, G2, and M. Most regulation occurs at the G1–S and G2–M transitions. In higher eukaryotes, the G1 checkpoint is especially important. This is precisely where p16 acts, and its activity is biochemically and functionally linked to other regulators of the G1–S transition, including the well-known cell cycle regulator (and tumor suppressor protein), Rb.

Over the past 5 years, a variety of evidence has accumulated that ties p16 dysfunction to the development of cancer. As an inhibitor of CDKs, p16 is a compelling candidate for a tumor suppressor gene (TSG). Consistent with this view, the *p16* gene is a target for inactivation in many types of tumor, and germ line mutations in *p16* predispose to melanoma. In addition, as the complexities of the *p16* locus have been teased apart, it is becoming apparent that this locus sits astride a second pathway of growth control that involves the notorious tumor suppressor, p53.

Here is reviewed current knowledge about p16, giving special attention to its role in cancer. The chapter addresses the biochemical and physiological functions of p16, the extraordinary molecular biology of the locus, and some of the outstanding questions related to the *p16* gene and its neighboring sequences.

From: *Tumor Suppressor Genes in Human Cancer*
Edited by: D. E. Fisher © Humana Press Inc., Totowa, NJ

2. P16'S INVOLVEMENT IN HUMAN CANCER

Genes involved in growth control and other aspects of biology relevant to cancer may be identified by mutations in sporadic tumor cells or in the germ line. Certain cancer-related genes are found to be altered only in sporadic tumors; other mutant cancer genes are only seen in the germ line. Most commonly, however, genes involved in cancer predisposition syndromes contain both germ line and somatic mutations. *p16* is a constituent of this latter class, with mutations that occur in sporadic tumors and in the germ line of cancer-prone individuals. Although p16 was initially identified biochemically, based on its interaction with CDK4 *(1)*, its role in cancer was first demonstrated by analysis of lesions in tumor cells and cell lines *(2,3)*.

2.1. p16 Somatic Mutations

Loss of heterozygosity (LOH), deletions of chromosomal regions or entire homologs, can be found in nearly all tumors and cell lines. These karyotypic abnormalities are not only characteristic of cancer, but in certain instances can also be related causally to tumor progression *(4)*. It has been presumed that regions frequently deleted in cancer cells harbor TSGs; inactivation of the resident genes by deletion and/or mutation results in a growth advantage conferred upon the evolving tumor cell *(5)*. Often, one member of an allelic pair of genes is inactivated by a detectable deletion, providing the means to localize the underlying mutant gene.

LOH on the short arm of chromosome 9 (9p21) is one of the most common chromosomal aberrations observed in human cancer *(6)*. Furthermore, homozygous deletions that remove tens, hundreds, or even thousands of kilobase pairs in 9p21 are relatively common, allowing an especially straightforward route to the presumptive TSG. At the center of these deletions lies the *p16* locus. In one study *(2)*, this locus was shown to be deleted in nearly one-half of all cell lines derived from tumors (Table 1). An impressive range of tumor cell lines exhibits such homozygous *p16* deletions, including melanoma, glioma, breast, and bladder malignancies. However, not all types of tumor cell line display *p16* loss; for instance, cell lines from colon cancers and neuroblastomas rarely contain 9p21 homozygous deletions. *p16* deletions are also observed in a variety of primary tumors, although several studies have reported detection of homozygous deletions at lower rates than those observed in cognate cell lines *(8)*. This disparity may relate to heterogeneity of primary tumors, growth advantages in tissue culture enjoyed by cells lacking p16, and/or the difficulty of detecting homozygous deletions in primary tumors that contain stromal tissue *(9)*.

In contrast to other TSGs, inactivation by mutation is relatively rare in *p16*, compared to homozygous deletion. But, in a fraction of cells derived from tumors that do not harbor homozygous deletions of *p16*, mutant *p16* genes are found. Almost invariably, the mutations occur in tumors that exhibit 9p21 LOH. The smaller lesions include nonsense, missense, and frameshift mutations, and microdeletions. In many cases that have been analyzed, missense mutations have been shown to affect p16's ability to inhibit CDK4 enzymatic activity in vitro *(10–13)*. Yet homozygous deletion of the locus is by far the most common mechanism of inactivation. For example, in a survey of melanoma cell lines, 60% contained homozygous deletions; only 15% had mutant *p16* *(2)*.

Methylation leading to decreased transcription may be yet another mechanism by which *p16* is rendered inactive in tumors. There have been numerous reports of methy-

Table 1
p16 Mutation/Deletion Frequencies

Germ line	<1/10,000
Somatic Homozygous Deletions in Tumor Lines[a]	
	(%)
Astrocytoma	82
Bladder	33
Breast	60
Colon	0
Glioma	88
Leukemia	64
Lung	36
Melanoma	62
Neuroblastoma	0

[a] From ref. 2.

lation-silenced *p16* alleles in cell lines and primary malignancies. For example, in cancers of the colon, bladder, and lung (nonsmall cell), methylation of C residues in presumptive gene control regions near *p16* has been proposed to interfere with p16 expression *(14–17)*. The occurrence of methylated, nontranscribed *p16* genes in tumors is indisputable, but it has been difficult to link methylation causally to tumor progression. Nevertheless, *p16* methylation remains a tenable hypothesis for epigenetic alteration of growth control during tumor development.

2.2. p16 Germ-line Mutations

Germ-line mutations in *p16* were identified through analysis of melanoma-prone kindreds *(7,18)*. These kindreds were ascertained and studied because they display unusually high incidence of melanoma, as well as a preponderance of individuals with skin covered by moles, or nevi. Initially, the relationship to moles, thought to be precursors to melanoma, was strongly emphasized. A syndrome called "dysplastic nevus syndrome" (DNS) was defined, characterized by a large number of abnormal (dysplastic) moles and a tendency to develop melanoma *(19)*. DNS has been a focus for controversy concerning its diagnosis, genetic basis, and relation to melanoma *(20)*. The major melanoma-predisposing gene, *MLM,* was mapped by ignoring DNS in one kindred set, and concentrating on melanoma as a dominant trait *(21)*. *MLM* maps within 9p21, with some kindreds showing high-confidence statistical linkage to markers in 9p21 logarithm of the odds of linkage (LOD scores > 3).

The identification of *p16* as the 9p21-linked target for deletion and mutation in tumor cell lines led to its analysis as a candidate for *MLM (7,18)*. To date, a large number of melanoma-prone kindreds have been examined, and numerous germ-line mutations in *p16* (missense, nonsense, frameshift, microdeletion) have been documented *(22,61)*. However, in some kindreds with high LOD scores, mutations in the *p16*-coding sequences have not been detected *(7)*. This has fostered a belief that another cancer gene located in 9p21 plays a part in familial melanoma *(23)*. However, there is no solid evidence for such a gene, and it remains possible that some kindreds may carry *p16* genes with mutations that lie in noncoding sequences that have not been systematically screened. Indeed, at least one such regulatory mutation has been defined *(24)*.

The overall incidence of germ-line *p16* mutations in the population is probably very low. Estimates of the percentage of familial melanoma, as opposed to nonfamilial or sporadic melanoma, cluster between 5 and 10% *(21)*. Yet, in several studies in which there is some evidence of familiality, but in which linkage analysis is impossible, very few mutations in *p16* have been observed *(7)*. In a few other studies, however, high rates of *p16* mutation have been detected *(25)*. Persons afflicted by sporadic melanoma also rarely carry *p16* mutations. On balance, therefore, it appears that the majority of melanomas typically described as familial, as well as sporadic melanoma, cannot be explained by germ-line mutations in *p16*. Because melanoma afflicts roughly 1% of the U.S. population, the incidence of *p16* germ-line mutations in the population at large is probably very low, perhaps on the order of 1/10,000. Nevertheless, *p16* remains the most significant cancer gene discovered to date that controls melanoma susceptibility.

Germ-line *p16* mutations increase melanoma risk by roughly 50-fold in certain populations, depending on the level of sun exposure and skin pigmentation *(26)*. The penetrance of mutant *p16* alleles, i.e., the probability of incurring disease over a lifetime, is approx 50%. Thus, inheritance of a defective *p16* gene does not guarantee melanoma: It merely increases risk. Because risk can be significantly modified by sun exposure, it is possible to prescribe a lifestyle that, even in the presence of mutant *p16* alleles, should substantially reduce the likelihood of melanoma.

The relationship between *p16* and other malignancies is unclear. The prevalence of *p16* mutations and deletions in many cancer types suggests a role for *p16* germ-line mutations in cancers other than melanoma. Some studies have linked *p16* mutations to increased pancreatic cancer risk, but the effect appears to be substantially less than for melanoma risk *(27)*. Thus, despite *p16*'s involvement in numerous sporadic cancer types, germ-line mutations of *p16* predominantly, if not entirely, alter melanoma risk. This apparent paradox may be resolved by consideration of rate-limiting genetic changes in cancer progression. Perhaps only in melanoma development is p16 inactivation rate-limiting.

The connection between *p16* and DNS is also obscure. Some studies have found a correlation between *p16* mutations and mole number and size *(26)*. If heterozygous *p16* mutations influence moles, they must be dominant or co-dominant, or the nevi must display loss of the wild-type *p16* allele. In support of this view, some reports of 9p21 LOH in dysplastic nevi have appeared *(28)*.

At least one other familial melanoma gene has been discovered, probably affecting less than one-tenth as many people as *p16 (29)*. This second melanoma gene is CDK4, one target of p16's inhibitory action. A specific missense substitution in CDK4 produces a CDK4 protein that is insensitive to p16 inhibition *(30)*. Mutant CDK4 acts as an oncogene, stimulating cell growth even in the presence of wild-type components of the p16–Rb pathway. The elucidation of the role of mutant CDK4 in familial melanoma further underscores the importance of the p16 pathway in cancer (Table 2).

3. BIOCHEMICAL FUNCTION OF P16

p16, as a CDKI, binds to specific CDKs. CDKs comprise a variety of types, denoted CDK1–6 in mammals *(31)*. They and their ancillary factors are responsible for transi-

Table 2
Role of p16 in Human Cancer

Germ-line mutation	Melanoma predisposition; pancreatic cancer?
Somatic mutation	Immortalization?
Somatic overexpression	G1 cell cycle arrest
Biochemical activity	CDK4/6 inhibitor
Physiological activity	Cell cycle arrest, life-span control
Other genes mutated in pathway	*CDK4, Rb1*

tion through different control points in the cell cycle, the so-called "checkpoints". CDKs are convergence sites for signals emanating from the cell's environment, which affect decisions related to the cell cycle. Thus, CDKs form part of a cellular switch that controls the all-or-none commitments to replicate DNA and undergo cytokinesis.

Cyclins are protein co-factors that form complexes with CDKs. Like the CDKs, cyclins comprise a family of proteins, termed "cyclins A–E" *(31)*. Certain cyclins preferentially associate with particular CDKs, and specific complexes regulate the G1–S or G2–M transitions. CDK4(6)–cyclin D and CDK2–cyclin E complexes influence the G1–S choice to proliferate or not. These complexes, when appropriately phosphorylated by the action of additional regulatory kinases and phosphatases, phosphorylate a variety of substrates. The fundamental features of this control mechanism are conserved in all eukaryotes studied, including fungi, plants, and animals. *(31)*.

To date, several CDKIs have been identified, including CDKIs from single-cell eukaryotes, such as yeast *(32)*. In mammals, these molecules fall into two general classes: the p16 family (p15, p16, p18, p19), which contains sequence motifs present in the cytoskeletal protein, ankyrin, and the p21 family (p21, p27, p57), which shares other homology elements. Although all CDKIs interfere with CDK activity, the mechanism of action of p16 family inhibitors is different from p21 family members *(32)*. p16 and its homologs bind to free CDKs, and competitively inhibit binding of the activating cyclin co-factors. In contrast, p21-like inhibitors bind to the CDK–cyclin complex, and block its interaction with substrates. The end result is the same: All CDKIs prevent phosphorylation of CDK substrates.

Certain CDKIs, such as p21, bind to all CDK–cyclin complexes; p16 is more selective *(32)*: It specifically binds to CDK4 and CDK6 in the free form (i.e., not in complex with cyclin D). The crystal structure of a complex between p16 and CDK6 has been solved *(33)*, providing a detailed, atomic-resolution view of the biochemical interaction. p16 resides near the adenosine triphosphate-binding region of the catalytic site, opposite the cyclin-binding site. p16 binding appears to prevent cyclin interaction by altering the conformation of the cyclin-binding site.

p16 exerts its biochemical effects at the G1–S checkpoint (Fig. 1). CDK4 and CDK6, in association with D-type cyclins, promote transition through the checkpoint. A variety of proteins, including histones, HMG proteins, E2F proteins, and Rb, are substrates for CDKs *(34)*. From the perspective of growth control, Rb and E2Fs are especially significant substrates. Rb in its hypophosphorylated form binds several proteins. Among these are members of the E2F family of transcription factors, which play important roles in downstream aspects of growth regulation. E2F1, for example, is

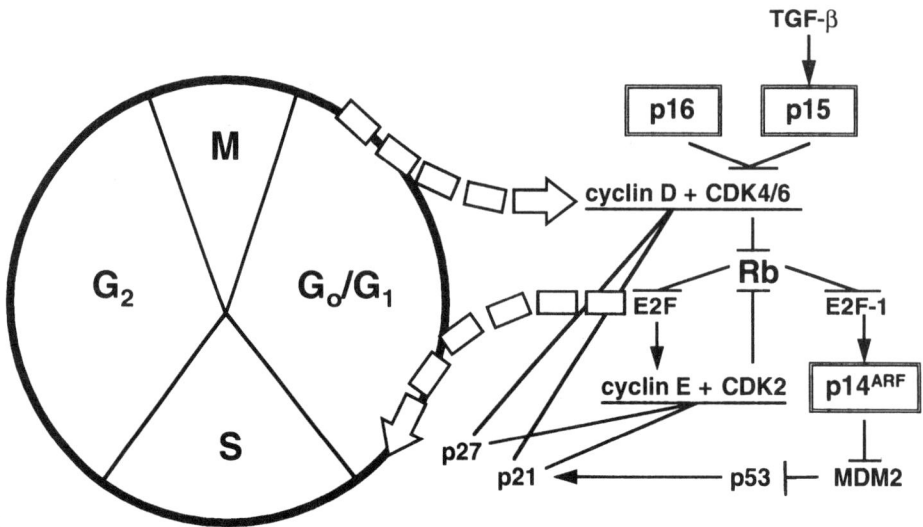

Fig. 1. Cell cycle and components of the p16, p15, and p14^{ARF} pathways. Arrows indicate activation; blunt-ended lines indicate inhibition.

implicated in transcription of genes in early S phase, such as thymidylate synthase and dihydrofolate reductase, enzymes involved in synthesis of DNA precursors *(35)*. In addition, cyclin E expression is induced by E2Fs, thus propelling cells across the G1–S boundary. The entire set of biochemical interactions involving p16 and its downstream pathway components supports a model whereby p16 inhibits CDK4/6 by interfering with cyclin D binding; in the absence of phosphorylation by CDK4, Rb binds E2Fs; E2Fs sequestered by Rb are unable to activate transcription of cyclin E to enter S phase, thus trapping cells in G1.

4. PHYSIOLOGICAL FUNCTION OF P16

As described above, p16's biochemical function is to inhibit activators of cell division (i.e., specific CDKs). Thus, it has a principal physiological role in regulating cell growth. In this general sense, it differs little from other CDK inhibitors, such as Pho81 and FAR1 in yeast, or p21 and p27. All these molecules participate in control of cell growth. What distinguishes p16 from other CDK inhibitors is its physiological role.

Growth control is important in biology, and there are multiple layers of growth regulation in each cell, as well as a plethora of tissue-specific mechanisms for regulating cell division. Cell growth is obviously critical during development, but it is also important in tissue regeneration and maintenance. Apart from nonmitotic, terminally differentiated cells, all cells retain the ability to divide, and must do so in an appropriate context. In general, a go/no-go decision is made after receiving positive, growth-stimulatory signals from outside and/or a threshold-crossing decrease in growth-inhibitory stimuli. Normal cells not only integrate signals from their environment, but also assess their internal condition. Cellular machinery surveys the integrity of the genomic DNA, the availability of DNA biosynthetic precursors, and the state of the mitotic spindle. If problems are detected, cell cycle progression may be aborted at one of the checkpoints.

This may cause a delay in division, followed by resumption of the orderly cell cycle program, or it may result in apoptosis. The feedback loops which regulate the decisions to proceed, temporarily arrest, or commit suicide are increasingly apparent *(36)*. Many components and several independent pathways participate. A number of these components are targets for mutation in cancer.

A second aspect of growth control involves genetically programmed limits on cellular life-span. With the exception of germ-line cells (i.e., embryonic stem cells), normal cells are not capable of indefinite division, and eventually senesce. The genetic determinants responsible for the timing and the mechanism of senescence are beginning to emerge *(37)*. Again, some of these components play a role in tumorigenesis.

Cancers progress from relatively benign types, which differ little from normal tissue, to malignant, highly abnormal cell masses. As they evolve, cancer cells gradually lose their dependence on growth signals in their environment. Checkpoint controls go awry, genomes become unstable, and cells avoid senescence to become immortal. Many CDKIs are implicated in specific aspects of tumor progression. For example, p21 is related to the apoptosis decision, and p27 functions in transforming growth factor β (TGF-β)-mediated cell cycle arrest *(38,39)*.

p16, however, is unique, because it is involved intimately with tumor progression. p16, in contrast to other CDKIs, is a principal target for mutation during tumor development. Overexpression of p16 (like some other CDKIs) causes G1-phase cell cycle arrest in normal cells, and in many tumor cells. However, in Rb cells that have lost *Rb1* gene function, p16 overexpression does not cause arrest *(40,41)*. This result suggests that p16 functions in the same pathway upstream of Rb, an observation consistent with the biochemical roles of the proteins. In addition, *p16* mutations and *Rb1* mutations, though commonly found in many tumor types, are seldom found in the same tumors, suggesting that inactivation of one of the components is sufficient to compromise the pathway *(42)*.

A role of p16 in cellular senescence has been proposed *(43,44)*. Evidence supporting this proposition comes from studies of p16 levels in aging cells. As cells approach the crisis point at which cellular life-span controls come into play, p16 mRNA and protein levels rise. In cells that traverse this point, p16 levels fall dramatically. This senescent state can be mimicked in younger cells by overexpression of the *ras* oncogene, causing G1 arrest and buildup of p16 and p53 *(43)*. Inactivation of p16 permits escape from this condition, at least in certain cells. These studies support the view that p16 participates in the physiology of cellular life-span.

One observation that does not fit neatly with the senescence model for p16's physiological role involves the timing of p16 loss in tumor development. In certain tumor types that have been studied carefully, *p16* gene inactivation appears to occur early during progression *(6,62)*. Because immortalization is generally considered to be a late-stage tumor phenotype, these findings imply that p16 has a different activity regarding tumor formation. It is possible that p16 has multiple physiological roles, at least one of which is relevant in early phases of tumor growth, and another in the later stage of immortalization.

Genetic analysis of p16 mutations in mice prove the hypothesis that p16 functions as a tumor suppressor in vivo. Gene knockouts (i.e., homozygous, loss-of-function mutations) are revealing with respect to p16's part in physiology. Homozygous-null p16 mice are viable, demonstrating that p16 is neither essential for the survival of normal

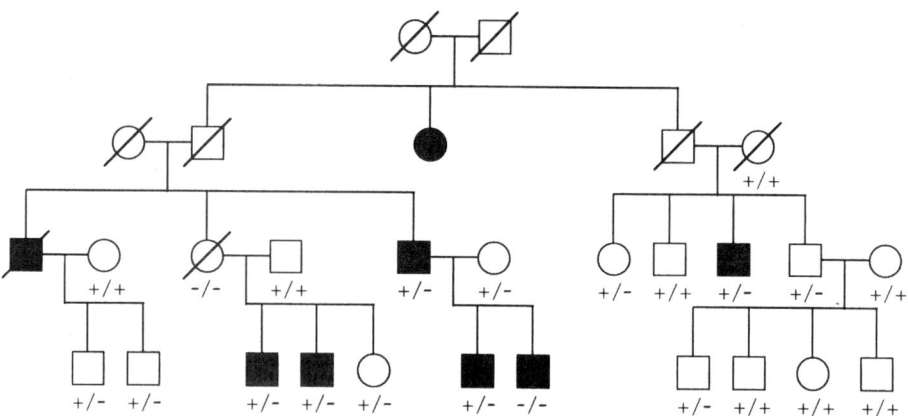

Fig. 2. Pedigree of Dutch melanoma-prone family segregating a *p16* microdeletion. Filled circles/squares indicate individuals affected by melanoma. The *p16* genotype is shown below each family member. Note the presence of two *p16* homozygous nulls (–/–) in the pedigree, one affected, one unaffected *(46)*.

cells, nor for the proper development of the mouse *(45)*. Homozygous-null mutants develop tumors more readily than control animals, confirming a cancer-predisposing role for p16 in mammals. Furthermore, fibroblasts isolated from the homozygous-null mice are demonstrably different from normal mouse fibroblasts. These fibroblasts are exceptionally sensitive to transformation by introduced *ras* oncogenes, suggesting that they have sustained at least one hit in their growth-control pathways. Heterozygous *p16*-mutant mice are not cancer-prone, however, unless they are stressed with carcinogens *(45)*.

Homozygous p16 human knockouts also exist *(46)*. Two individuals have been identified who inherited the same defective *p16* allele from each parent (Fig. 2). Both individuals were normal, although one developed two primary melanomas by age 15 yr. The other individual, however, was melanoma-free, dying at age 55 yr of an adenocarcinoma. These observations confirm that, as in mice, *p16* is not an essential gene. Furthermore, they illustrate that the same cancer-predisposing mutation can produce different consequences depending on genetic background, environmental factors, or random biological processes.

5. MOLECULAR BIOLOGY OF *P16* LOCUS

The *p16* gene and its neighboring sequences comprise a locus of exceptional interest and complexity. Two genes other than *p16* contribute to this complexity: *p15* and *p14*[ARF] (*p16E1β*). Together, they endow *p16* with some of the most fascinating and surprising molecular biology of any human locus.

p16 consists of three exons: E1α, E2, and E3. E3 contains only three codons, with the remainder of the coding sequence roughly split between the first two exons *(48)*; (Fig. 3). All the exons are remarkably GC-rich, averaging 70% G or C residues. The p16 mRNA gives rise to the 156-amino-acid-long p16 protein *(1,48)*.

Located about 12 kilobase pairs upstream from E1α is another exon, E1β, which is spliced to E2 and E3, forming a second transcript type, the β transcript, originating

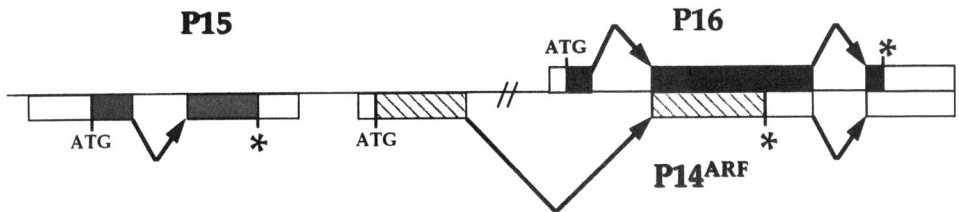

Fig. 3. *p16* locus. Asterisks indicate translation termination sites; ATG indicates translational initiation. Distances between genes and exons not drawn to scale.

from the *p16* locus. This transcript is not alternatively spliced from the p16-encoding transcript; rather, it is generated from an independent promotor upstream from E1β *(48)*. If translated in the p16 reading frame, the β transcript could initiate translation at a methionine codon located in E1β, eight codons from the E1β–E2 junction. This hypothetical protein would be approx 10 kDa. A second alternative open reading frame exists, which could produce a 14 kDa protein (19 kDa in mouse), if translated. This protein has been detected in vivo, and shown to produce cell cycle arrest when expressed at high levels in cultured cells *(49)*. Moreover, knockout constructs that specifically target E1β give rise to mice that resemble *p16*-null mice in phenotype *(50)*. These findings have produced a flurry of inquiries directed at p14[ARF].

Despite intensive efforts to find mutations that specifically affect the p14[ARF] protein, none has been identified. To date, all *p16* germ-line and somatic mutations that have been described affect the p16 protein, though some also alter p19[ARF] (Table 3). The only lesions that affect the β transcript, leaving *p16α* intact, are deletions. Several examples of homozygous deletions that disrupt the β transcript, but leave the *p16* mRNA intact, have been defined *(47,63)*.

Further investigation of the p14[ARF] protein has uncovered several other features of special relevance to its role in cancer. p14[ARF] appears to activate p53 by blocking its degradation *(51–55)*. Moreover, one of Rb's targets, E2F1, directly induces p14[ARF] transcription. Thus, when Rb levels fall, E2F levels rise, leading to increased p14[ARF] expression. This in turn elevates p53 levels, potentially providing a feedback circuit that compensates for failure of the Rb pathway. It has also been observed that activated Ras protein induces p14[ARF] expression, causing apoptosis via the p53 pathway *(56)*. Again, the implication is that the *p16* locus and p53 are functionally intertwined, connected through p14[ARF]. p14[ARF]'s involvement with the p53 pathway may have deep biological significance, bridging the p16–Rb pathway and the p53 pathway.

Immediately upstream of E1β is a third gene, *p15,* that encodes a protein with substantial similarity to p16. *p15* comprises two exons that are, like *p16* exons, extremely GC-rich. The *p15* and *p16* genes are similar over their entire length, but especially so near their 3′ ends. Comparative studies of these genes in humans and mice indicates that they were involved in a gene conversion event that occurred about the time of the establishment of the ape lineage *(57)*. The p15 protein is essentially indistinguishable from p16 in its binding and inhibition of CDK4 and 6, but, despite this similarity, *p15* does not appear to be a solitary target for mutation in cancer (Table 3). No germ-line mutations in *p15* have been found, although the gene has been subjected to intense scrutiny, particularly in 9p21-linked melanoma kindreds for which *p16* mutations have

Table 3
Genetics of *p16* Locus

	p16	*p15*	*p14^{ARF}*
Somatic (human and mouse)			
Homozygous deletions	Yes	Yes	Yes*^a
Microlesions	Yes	No	No
Methylation	Yes	Yes	–
Germ line (human)			
Heterozygous phenotype	Yes	–	–
Homozygous phenotype	Yes (viable)	–	–
Germ line (mouse)			
Heterozygous phenotype	No	–	No
Homozygous phenotype (knockout)	Yes (viable)	–	Yes

^{a} All homozygous deletions that affect *p14^{ARF}*, but not *p16,* also affect *p15*.

not been detected. In addition, virtually no somatic mutations have been seen in *p15* in melanomas or other tumor types *(47)*. The majority of homozygous deletions that remove *p16* also delete *p15*. However, apart from the few that also affect E1β, no deletions remove *p15* alone, leaving *p16* intact. Thus, there is no evidence that *p15* is involved in cancer as a mutation target.

All three genes of the *p16* locus are regulated independently: *p15* is upregulated in response to TGF-β in certain cell lines; *p16* E1α and E1β are not *(58)*. There is no appreciable cell cycle regulation of E1α transcription in lymphocytes; E1β levels oscillate, depending on cell cycle phase *(47)*. E1α expression is substantially higher in Rb cells, suggesting a role in a feedback loop with Rb; the expression of the other transcripts is unchanged.

6. OUTSTANDING QUESTIONS

Since the discovery of p16 in 1993, progress has been made in understanding its role in normal tissues and in tumorigenesis. However, several aspects of p16 biology remain to be clarified. The central question pertains to the difference between p16 and other CDKIs. Why is the *p16* gene an attractive target for mutation, but other CDKI genes are never, or infrequently, mutated? This question applies especially to *p15* and *p14^{ARF}*, which do not suffer mutations in the manner of *p16*. It is of course possible that the prevalence of *p16* homozygous deletions reflects the presence of neighboring genes, *p15* and *p14^{ARF}*, which can be inactivated jointly by a single deletional event. However, the puzzle remains as to why p16 has point mutations but other genes such as p15 do not.

Although several components of the p16 pathway have been defined, the upstream molecules that regulate *p16* expression have not been elucidated. These factors are especially important, because it is likely that they are the key to p16 physiology. The biochemical similarity between p15 and p16 implies that their different physiological roles may result from idiosyncrasies of gene regulation. If p16 plays a major role in controlling cellular life-span, it will be interesting to dissect the circuitry that links growth limits to p16 expression.

A further intriguing aspect of p16 biology involves the peculiar structure of the *p16* locus. A teleological explanation for *p16* and *p14^{ARF}* is not obvious. What, if anything, is the value of such densely packed genetic information? In the case of certain bacteriophages, it is possible to rationalize overlapping genes, based on maximizing protein-coding sequence in the face of DNA packaging constraints. But, in light of the profligate use of DNA among mammals, it is not clear what evolutionary pressures underpin the compressed features of the *p16* locus.

Even more uncertain is the potential of p16 as a genetic diagnostic or prognostic indicator of disease. In the realm of risk analysis, it seems likely that germ-line screening for *p16* mutations will be a useful strategy for preventing melanoma. Sun exposure is a great risk factor for melanoma that can be conveniently reduced. Thus, genetic information could be used to encourage lifestyle modification. But the low frequency of *p16* germ-line mutations in the population may produce economic barriers to testing that must be overcome by, e.g., technical advances in screening. Somatic *p16* mutations may also ultimately prove useful in directing cancer therapy *(59)*. However, clinical correlation studies must be completed before this type of molecular pathology has a significant impact on therapy and outcome.

As always, therapeutic implications of a cancer gene, such as *p16,* are most difficult to predict. Several companies are presently developing and testing drugs that mimic CDKIs. The obvious shortcoming of this approach is the likelihood of side effects. p16 overexpression in normal cells causes growth arrest. Thus, dividing cells, including hematopoietic stem cells and intestinal epithelial precursors, would probably also experience arrest. Gene therapy, with functional *p16* sequences to complement loss in tumors, remains an appealing prospect. However, tumor targeting and escape are formidable problems. An alternative strategy is to use *p16* to arrest and protect normal cells prior to exposure to therapeutic doses of cancer drugs *(60)*. A lesson learned from the study of p16 is that even when a cancer gene is relatively well understood, such as *p16,* the inherent complexities of physiology provide daunting obstacles to therapy that must be overcome.

REFERENCES

1. Serrano M, Hannon G, Beach D. A new regulatory motif in cell-cycle control causing specific inhibition of cyclin D/CDK4. *Nature* 1993; 366:740–747.
2. Kamb A, et al. A cell cycle regulator potentially involved in genesis of many tumor types. *Science* 1994; 264:436–440.
3. Nobori T, et al. Deletions of the cyclin-dependent kinase-4 inhibitor gene in multiple human cancers. *Nature* 1994; 368:753–756.
4. Fearon E, Vogelstein B. A genetic model for colorectal tumorigenesis. *Cell* 1990; 61(5):759–767.
5. Knudson AJ. Mutation and cancer: statistical study of retinoblastoma. *Proc Natl Acad Sci USA* 1971; 68:820–823.
6. Liggett WJ, Sidransky D. Role of the p16 tumor suppressor gene in cancer. *J Clin Oncol* 1998; 16:1197–1206.
7. Kamb A, et al. Analysis of the p16 gene (CDKN2) as a candidate for the chromosome 9p melanoma susceptibility locus. *Nat Genet* 1994; 8:23–26.
8. Spruck CR, et al. p16 gene in uncultured tumours. *Nature* 1994; 370:183–184.
9. Kamb A, et al. Rates of p16 (MTS1) mutations in primary tumors with 9p loss (response). *Science* 1994; 265:416.
10. Ranade K, et al. Mutations associated with familial melanoma impair p16INK4 function. *Nat Genet* 1995; 10:114–116.
11. Koh J, et al. Tumour-derived p16 alleles encoding proteins defective in cell-cycle inhibition. *Nature* 1995; 375:506–510.

12. Lilischkis R, et al. Cancer-associated mis-sense and deletion mutations impair p16INK4 CDK inhibitory activity. 66, 1996; 2:249–254.

13. Parry D, Peters G. Temperature-sensitive mutants of p16CDKN2 associated with familial melanoma. *Mol Cell Biol* 1996; 16:3844–3852.

14. Merlo A, et al. 5′ CpG island methylation is associated with transcriptional silencing of the tumour suppressor p16/CDKN2/MTS1 in human cancers. *Nat Med* 1995; 1:686–692.

15. Otterson G, et al. CDKN2 gene silencing in lung cancer by DNA hypermethylation and kinetics of p16INK4 protein induction by 5-aza 2′deoxycytidine. *Oncogene* 1995; 11:1211–1216.

16. Herman J, et al. Inactivation of the CDKN2/p16/MTS1 gene is frequently associated with aberrant DNA methylation in all common human cancers. *Cancer Res* 1995; 55:4525–4530.

17. Belinsky S, et al. Aberrant methylation of p16(INK4a) is an early event in lung cancer and a potential biomarker for early diagnosis. *Proc Natl Acad Sci USA* 1998; 95:11,891–11,896.

18. Hussussian C, et al. Germline p16 mutations in familial melanoma. *Nat Genet* 1994; 8:15–21.

19. Elder D, et al. Dysplastic nevus syndrome: a phenotypic association of sporadic cutaneous melanoma. *Cancer* 1980; 46:1787–1794.

20. Piepkorn M. A perspective on the dysplastic nevus controversy. *Pathology* (Phila) 1994; 2:259–279.

21. Cannon-Albright L, et al. Assignment of a locus for familial melanoma, MLM, to chromosome 9p13-p22. *Science* 1992; 258:1148–1152.

22. Hogg D, et al. Role of the cyclin-dependent kinase inhibitor CDKN2A in familial melanoma. *J Cutan Med Surg* 1998; 2:172–179.

23. Flores J, et al. Loss of the p16INK4a and p15INK4b genes, as well as neighboring 9p21 markers, in sporadic melanoma. *Cancer Res* 1996; 56:5023–5032.

24. Liu L, et al. Mutation of the CDKN2A 5′ UTR creates an aberrant initiation codon and predisposes to melanoma. *Nat Genet* 1999; 21:128–132.

25. Soufir N, et al. Prevalence of p16 and CDK4 germline mutations in 48 melanoma-prone families in France. The French Familial Melanoma Study Group. *Hum Mol Genet* 1998; 7:209–216.

26. Cannon-Albright L, et al. Penetrance and expressivity of the chromosome 9p melanoma susceptibility locus (MLM). *Cancer Res* 1994; 54:6041–6044.

27. Flanders T, Foulkes W. Pancreatic adenocarcinoma: epidemiology and genetics. *J Med Genet* 1996; 33:889–898.

28. Park W, et al. Allelic detection at chromosome 9p21(p16) and 17p13(p53) in microdissected sporadic dysplastic nevus. *Hum Pathol* 1998; 29:127–130.

29. Zuo L, et al. Germline mutations in the p16INK4a binding domain of CDK4 in familial melanoma. *Nat Genet* 1996; 12:97–99.

30. Wolfel T, et al. A p16INK4a-insensitive CDK4 mutant targeted by cytolytic T lymphocytes in a human melanoma. *Science* 1995; 269:1281–1284.

31. Nurse P, Masui Y, Hartwell L. Understanding the cell cycle. *Nat Med* 1998; 4:1103–1106.

32. Vogt P, Reed S. In: Vogt P, Reed S, eds., Cyclin Dependent Kinase (CDK) Inhibitors, Berlin, Heidelberg: Springer-Verlag. 1998:139–148.

33. Russo A, et al. Structural basis for inhibition of the cyclin-dependent kinase Cdk6 by the tumour suppressor p16INK4a. *Nature* 1998; 395:237–243.

34. Dynlacht B, et al. Specific regulation of E2F family members by cyclin-dependent kinases. *Mol Cell Biol* 1997; 17:3867–3875.

35. DeGregori J, Kowalik T, Nevins J. Cellular targets for activation by the E2F1 transcription factor include DNA synthesis- and G1/S-regulatory genes. *Mol Cell Biol* 1995; 15:4215–4224.

36. Reed J. Mechanisms of apoptosis avoidance in cancer. *Curr Opin Oncol* 1999; 11:68–75.

37. Johnson F, Marciniak R, Guarente L. Telomeres, the nucleolus and aging. *Curr Opin Cell Biol* 1998; 10:332–338.

38. el-Deiry W, et al. WAF1/CIP1 is induced in p53-mediated G1 arrest and apoptosis. *Cancer Res* 1994; 1:1169–1174.

39. Polyak K, et al. p27Kip1, a cyclin-Cdk inhibitor, links transforming growth factor-beta and contact inhibition in cell cycle arrest. *Genes Dev* 1994; 8:9–22.

40. Lukas J, et al. Retinoblastoma-protein-dependent cell-cycle inhibition by the tumour suppressor p16. *Nature* 1995; 375:503–506.

41. Medema R, et al. Growth suppression by p16ink4 requires functional retinoblastoma protein. *Proc Natl Acad Sci USA* 1995; 92:6289–6293.

42. Ueki K, et al. CDKN2/p16 or RB alterations occur in the majority of glioblastomas and are inversely correlated. *Cancer Res* 1996; 56:150–153.

43. Serrano M, et al. Oncogenic ras provokes premature cell senescence associated with accumulation of p53 and p16INK4a. *Cell* 1997; 88:593–602.

44. McConnell B, et al. Inhibitors of cyclin-dependent kinases induce features of replicative senescence in early passage human diploid fibroblasts. *Curr Biol* 1998; 8:351–354.

45. Serrano M, et al. Role of the INK4a locus in tumor suppression and cell mortality. *Cell* 1996; 85:27–37.

46. Gruis N, et al. Homozygotes for CDKN2 (p16) germline mutation in Dutch familial melanoma kindreds. *Nat Genet* 1995; 10:351–353.

47. Stone S, et al. Genomic structure, expression and mutational analysis of the P15 (MTS2) gene. *Oncogene* 1995; 11:987–991.

48. Stone S, et al. Complex structure and regulation of the P16 (MTS1) locus. *Cancer Res* 1995; 55:2988–2994.

49. Quelle D, et al. Alternative reading frames of the INK4a tumor suppressor gene encode two unrelated proteins capable of inducing cell cycle. *Cell* 1995; 83:993–1000.

50. Kamijo T, et al. Tumor suppression at the mouse INK4a locus mediated by the alternative reading frame product p19ARF. *Cell* 1997; 91:649–659.

51. Pomerantz J, et al. The Ink4a tumor suppressor gene product, p19Arf, interacts with MDM2 and neutralizes MDM2's inhibition of p53. *Cell* 1998; 92:713–723.

52. Zhang Y, Xiong Y, Yarbrough W. ARF promotes MDM2 degradation and stabilizes p53: ARF-INK4a locus deletion impairs both the Rb and p53 tumor suppression pathways. *Cell* 1998; 92:725–734.

53. Kamijo T, et al. Functional and physical interactions of the ARF tumor suppressor with p53 and Mdm2. *Proc Natl Acad Sci USA* 1998; 95:8292–8297.

54. Stott F, et al. The alternative product from the human CDKN2A locus, p14(ARF), participates in a regulatory feedback loop with p53 and MDM2. *EMBO J* 1998; 17:5001–5014.

55. Bates S, et al. p14ARF links the tumour suppressors RB and p53. *Nature* 1998; 395:124–125.

56. Palmero I, C Pantoja M. Serrano p19ARF links the tumour suppressor p53 to Ras. *Nature* 1998; 395:125–126.

57. Jiang P, et al. Comparative analysis of Homo sapiens and Mus musculus cyclin-dependent kinase (CDK) inhibitor genes p16 (MTS1) and p15 (MTS2). *J Mol Evol* 1995; 41:795–802.

58. Hannon G, Beach D. p15INK4B is a potential effector of TGF-beta-induced cell cycle arrest. *Nature* 1995; 371:257–261.

59. Orlow I, et al. Alterations of INK4A and INK4B genes in adult soft tissue sarcomas: effect on survival. *J Natl Cancer Inst* 1999; 91:73–79.

60. Stone S, Dayananth P, Kamb A. Reversible, p16-mediated cell cycle arrest as protection from chemotherapy. *Cancer Res* 1996; 56:3199–3202.

61. Foulkes W, et al. The CDKN2A (p16) gene and human cancer. *Mol Med* 1997; 3:5–20.

62. Barrett M, et al. Allelic loss of 9p21 and mutation of the CDKN2/p16 gene develop as early lesions during neoplastic progresion in Barrett's esophagus. *Oncogene* 1996. 13:1867–1873.

63. Glendening J, et al. Homozygous loss of the p16INK4B gene (and not the p16INK4 gene) during tumor progression in a sporadic melanoma patient. *Cancer Res* 1995; 55:5531–5535.

9 DNA Mismatch Repair in Tumor Suppression

Guo-Min Li, PHD, Scott McCulloch, PHD, and Liya Gu, PHD

CONTENTS

INTRODUCTION
E. COLI METHYL-DIRECTED MISMATCH REPAIR (MMR)
MMR IN HUMAN CELLS
MMR DEFICIENCY AND COLORECTAL CANCER (CC)
MMR DEFICIENCY AND NON-CC
MMR AS SENSOR OF GENOMIC DAMAGE

1. INTRODUCTION

It is believed that cancer is caused by mutations *(1)*. In addition to the mutations induced by DNA damage, mutations can arise from mismatched base pairs (bps) generated during DNA replication and recombination. However, to avoid mutagenesis, cells possess mutation avoidance systems, one of which is the DNA mismatch repair (MMR) pathway. MMR is the primary cellular pathway that is responsible for correcting mispairs that arise during normal DNA metabolism. In bacteria, the importance of MMR in maintaining genomic stability was demonstrated with the observation, made more than 25 years ago, that defects in this pathway lead to elevated spontaneous mutability *(2,3)*. Inactivation of MMR in eukaryotic cells results in a mutator phenotype, and thus MMR is also crucial in maintaining genomic stability in eukaryotes. A dramatic example of this fact is the direct association of MMR deficiency with human hereditary and sporadic cancer.

There are multiple MMR pathways in both bacteria and eukaryotic cells. In the past several years, a rapid increase in knowledge of MMR in yeast and humans has led to the current understanding of the molecular mechanism(s) of MMR in eukaryotic cells. The focus of this chapter is human MMR and its role in cancer avoidance. Because the human MMR pathway and understanding of it are based on its homology to the *Escherichia coli* methyl-directed and MutS/MutL-dependent pathway, the *E. coli* pathway is discussed briefly. Readers interested in other MMR pathways and MMR in other organisms are referred to a number of excellent reviews *(4–13)*.

From: *Tumor Suppressor Genes in Human Cancer*
Edited by: D. E. Fisher © Humana Press Inc., Totowa, NJ

2. *E. COLI* METHYL-DIRECTED MMR

The best-characterized MMR pathway is the *E. coli* methyl-directed and MutHLS-dependent repair system, which has been reconstituted using 10 purified proteins *(14)*. Among these proteins, MutS, MutL, and MutH are the three key components specifically required for this system. The other proteins involved in the process are UvrD (helicase II), single-stranded DNA-binding protein (SSB), three exonucleases (ExoI, RecJ, and ExoVII), DNA polymerase III holoenzyme, and DNA ligase. The MutS protein (97 kDa) is the mismatch recognition protein that binds to both base–base mismatches *(15)* and small nucleotide (1–3) insertion/deletion (ID) mispairs *(16)*. The function of MutL is not clear, although it interacts with MutS, and is required for activation of MutH *(17,18)*. Recently, MutL has been shown to load UvrD to the repair initiation site *(19)*, and to possess an adenosine triphosphatase (ATPase) activity *(20)*. Since the interaction between MutS, MutL, and MutH is an ATP-dependent reaction *(17,18)*, the ATPase activity of MutL may drive this process. MutH is a latent endonuclease, which nicks hemimethylated d(GATC) sequences on the unmethylated strand, when in its activated form *(21)*. This strand-specific nicking activity of MutH overcomes the dilemma that exists when both bases in a mismatch are the normal components of a Watson-Crick bp. Normally, DNA in *E. coli* is methylated at the N^6 position of adenine residues in d(GATC) sequences. But newly replicated DNA is transiently unmethylated in these sequences, which provides a signal for MutH to identify and target repair to the newly synthesized daughter strand. The requirement for d(GATC) hemimethylation and MutH is bypassed, if mismatched DNA contains a pre-existing strand break *(14)*.

A model illustrating current understanding of the *E. coli* methyl-directed MMR pathway is shown in Fig. 1. Mismatch correction is initiated by the binding of a MutS homodimer to the mismatch. The MutS–DNA complex recruits MutL and MutH proteins to the repair complex, in a reaction dependent on ATP. The repair complex then bidirectionally draws flanking DNA through itself, to form an α-loop structure *(22)*. When the protein complex encounters a hemimethylated d(GATC) site, activated MutH binds to the sequence, and cleaves the unmethylated strand in the DNA substrate. With the help of MutL, UvrD loads at the nick *(19,23)*. UvrD unwinds the duplex from the nick toward the mismatch, revealing a ssDNA region of the unnicked strand to which SSB binds, preventing attack on it by nucleases. Depending on the position of the strand break relative to the mismatch, exonuclease I (3′–5′ exonuclease) or exonuclease VII or the RecJ exonuclease (both 5′ and 3′ exonucleases), degrades the nicked strand from the nicked site up to and slightly past the mismatch. The resulting single-stranded region is filled in by DNA polymerase III holoenzyme, and the nick is sealed by DNA ligase.

The *E. coli* methyl-directed and MutHLS-dependent MMR pathway possesses the following features: First, it is strand-specific, i.e., the system can recognize which strand is correct and only replace the incorrect base; second, the repair is bidirectional, being able to process mispairs (base–base and ID) that lie either 5′ or 3′ to the strand discrimination signal *(24,25)*; finally, the system has a defined, broad specificity as to what type of substrates it can process. These characteristics of *E. coli* MMR depend on MutS, MutL, and MutH proteins: For example, the bidirectional capability of the system, which is achieved through the action of exonucleases, requires functional MutS

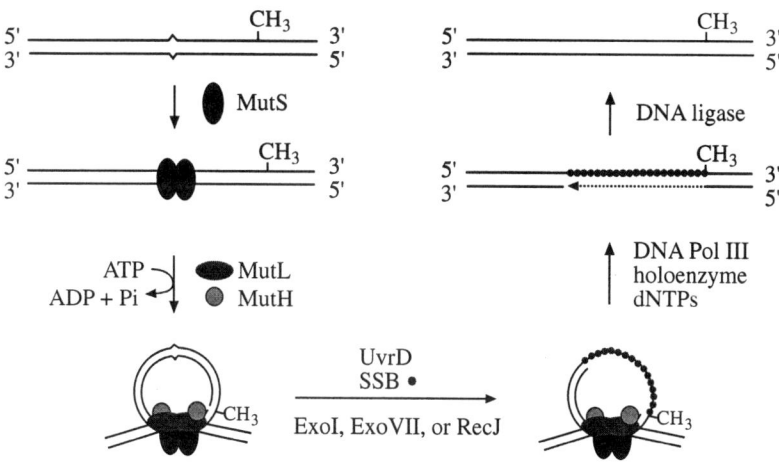

Fig. 1. Mechanism of methyl-directed MMR in *E. coli.*

and MutL. Although *E. coli* methyl-directed MMR is the best-studied MMR system, it is far from being perfectly understood. Further studies are required to address many unanswered questions, such as characterizing the interaction between MutS, MutL, and MutH, and precisely defining the function of MutL.

3. MMR IN HUMAN CELLS

Human cells possess a MMR pathway homologous to the *E. coli* methyl-directed and MutHLS-dependent system *(26,27)*. As shown in Table 1, these two pathways share many similarities. Like the *E. coli* pathway, the human system efficiently repairs both base–base mismatches and small ID mispairs *(26–30)*. However, human cells have a broader substrate specificity than *E. coli,* as evidenced by their ability to efficiently repair C–C mismatches and ID mispairs larger than four nucleotides *(26,28–33)* which cannot be corrected by the *E. coli* system. The strand discrimination signal in human cells does not appear to involve methylation *(26),* but it does involve recognition of and targeting to a strand containing a pre-existing strand break *(31)*. This suggests that, like the *E. coli* reaction, the strand-specific repair in human cells is also nick-directed. As in *E. coli,* the human MMR reaction is capable of bidirectional processing of a mispaired base. Regardless of the polarity (5′ or 3′) of the strand break, relative to the mismatch, mismatch-provoked excision always occurs between the nick and the mismatch *(31)*.

A search has been undertaken to find related protein components that are involved in the *E. coli* and human MMR systems. This approach has provided convincing evidence that the human strand-specific MMR reaction is a homolog of the *E. coli* methyl-directed and MutHLS-dependent MMR reaction. Human homologs of *E. coli* MutS and MutL, two of the three known MMR-specific *E. coli* functions, have been identified and shown to be required in the human MMR pathway (*see* Subheading 3.2.).

The DNA resynthesis step in human nuclear extracts is sensitive to aphidicolin, an inhibitor for eukaryotic DNA polymerases α, δ, and ε, suggesting that MMR in human cells may require replicative DNA polymerases *(26,27,31)*. The requirement for prolif-

Table 1
Similarities Between MMR in *E. coli* and Human Cells

	E. coli	*Human*
Substrate	Base–base and ID mispairs	Base–base and ID mispairs
Strand discrimination signal	Methyl group-directed strand break	Single-strand nick
Bidirectionality	Yes	Yes
Protein components		
MMR-specific	MutS homodimer	hMutSα, hMutSβ
	MutL homodimer	hMutLα, hMutLβ
	MutH	?
Helicase	UvrD	?
Exonuclease	ExoI, RecJ, ExoVII	ExoI?
Replicative polymerase	PolIII holoenzyme	Polδ, PCNA
ssDNA binding	SSB	RPA
Ligase	DNA ligase	?

erating cellular nuclear antigen (PCNA), an accessory factor for DNA polymerases δ and ε during MMR-associated DNA resynthesis, supports the involvement of one or both polymerases in MMR *(34)*. The identification of DNA polymerase δ in the human strand-specific MMR reaction *(35)* has assigned a role for the polymerase in MMR, although it remains possible that the other two aphidicolin-sensitive polymerases are also involved in MMR in human cells.

Activities involved in the unwinding and excision steps of the human reaction have not been specified. A human homolog of yeast EXO1 *(36)* has been proposed to participate in human MMR, based on the interaction of the yeast protein with MSH2 *(37)*. However, given the participation of EXO1 in a number of DNA transactions, its actual role in MMR awaits further characterization. Recently, human ssDNA-binding protein RPA has also been implicated in human MMR *(38,39)*. The discovery of similarities in substrate specificity, component activities, and repair mechanism(s) in prokaryotes and eukaryotes has greatly advanced current understanding of the human MMR pathway.

3.1. Function of Human MutS Homologs

The first identified human gene homologous to *E. coli MutS* was *hMSH3,* initially called *DUG-1 (40)*. In fact, the *hMSH3* gene was identified fortuitously during study of the *DHFR* gene promoter, which is a divergent promoter that also regulates the expression of *hMSH3*. Even though the gene was identified 10 yr ago, the actual function of *hMSH3* was only determined recently. By virtue of their homology to bacterial MutS, additional human MutS homologs were identified, including *hMSH2 (41,42), hMSH4 (43),* and *hMSH5 (44,45),* thanks to the pioneering work in yeast by Kolodner et al. *(46,47)*. *hMSH6* was cloned based on partial protein sequence information obtained from peptides of the 160 kDa G–T mismatch-binding protein, also called GTBP *(48–50)*. All MutS homologs identified to date share a highly conserved carboxy-terminal region, which includes the consensus ATP-binding site. Among these human MutS homologs, only hMSH2, hMSH3, and hMSH6 are known to participate in strand-specific MMR in human nuclear extracts.

The human MSH2 protein was initially thought to be the human mismatch recognition protein, because it binds with high specificity to both base–base mismatches and ID mispairs *(51,52)*. However, later experiments revealed that functional human mismatch recognition activity involves an hMSH2-containing heterodimer. This was illustrated by co-purification of the 105-kDa hMSH2 with a 160-kDa polypeptide *(30)*. This heterodimer, designated as hMutSα *(30)*, was subsequently reconstituted using recombinant hMSH2 and hMSH6 proteins *(53,54)*. Although the identification of hMutSα as a mismatch recognition complex was a step forward, the existence of this complex is not sufficient to adequately explain the distinct mutator phenotypes of *hMSH2-* and *hMSH6*-mutant cells. Tumor cells with mutations in *hMSH2* are defective in repair of both base–base mismatches and small ID mispairs *(29,30,55)*. Accordingly, these cells exhibit an elevated number of mutations in both the *HPRT* gene and in simple repeated sequences *(56)*. hMSH6 mutant cells are deficient in repair of base–base and single-nucleotide ID mispairs, but partially proficient in the processing of ID mispairs that are two nucleotides or larger *(30,57)*. This observation is consistent with the fact that *hMSH6*-deficient cells have no detectable instability in di- and trinucleotide repeated sequences *(58)*, while displaying hypermutability in single-base ID at the *HPRT* locus *(57,58)*. As expected, purified hMutSα restores full MMR activity to both types of mutants *(30)*. These findings strongly suggest that base–base mismatches and ID mispairs are recognized by different human MutS homologs, with the former recognized by hMutSα and the latter recognized by an activity involving hMSH2, but not hMSH6 *(30)*.

This hypothesis has been confirmed by experiments in both yeast and humans. Marsischky et al. *(59)* demonstrated that yeast *MSH3* mutants show a low but increased rate of frameshift mutations, but essentially no increase in the rate of base substitutions. However, the *MSH3* and *MSH6* double mutants display the same phenotype as *MSH2* single mutants. It was concluded that there are two mismatch recognition heterodimers: one that consists of MSH2 and MSH6, and the other that consists of MSH2 and MSH3. Indeed, there is evidence for physical interaction between MSH2 and MSH3 *(59,60)* and a MSH2–MSH3 heterodimer, which is referred to as MutSβ *(33,61)*, has been isolated from human and yeast cells in native form *(33)* and recombinant form *(53,61,62)*. Recently, protein interaction domains mediating the interface between MSH2 and MSH6, and between MSH2 and MSH3, have been identified *(63,64)*. MSH2 interacts with MSH3 and MSH6 through the same domain *(64)*.

The function of MutS-like proteins is to recognize mispairs. The mismatch binding capability of hMutSα and hMutSβ heterodimers has been demonstrated, indicating that these protein complexes are indeed mismatch recognition proteins. hMutSα is capable of binding to both base–base mismatches and ID mispairs *(30,53,54)*, but hMutSβ only recognizes ID mispairs, which is consistent with the in vitro repair data *(30,33)*. This result substantiates the earlier proposal by Marsischky et al. *(59)* regarding the substrate specificity of these two MutS heterodimers.

In addition to mismatch recognition, MutS-like proteins also possess a weak intrinsic ATPase activity *(65–67)*, which has been associated with hMutSα *(54,68)*. A clear function of the ATPase activity has not yet been established, but it is likely to be important, because a single mutation in the MutS consensus ATP-binding domain leads to a mutator phenotype *(65,67)*. ATP affects several key steps of MMR process: ATP negatively regulates the binding of MutS homologs to mismatches *(30,54,68)*; ATP is required for interactions between MutS and MutL homologs *(34,69)*; and it drives translocation of MMR proteins along the DNA helix *(22,33)* or the molecular switch model proposed by Fishel et al. *(68,70)*. Thus, ATPase may play an important role as a motor for MutS homologs in mediating these transactions.

3.2. Function of Human MutL Homologs

Three human MutL homologs (hMLH1, hPMS1, and hPMS2) have been identified *(71–73)*, based on their homology to the *E. coli* and yeast MutL proteins. Early genetic studies on *Saccharomyces cerevisiae* mutants with increased postmeiotic segregation (PMS) identified the yeast MutL homolog, *PMS1 (74)*. The homology between yPMS1 and bacterial MutL protein has led to identification of *yMLH1 (75)* and *yMLH3 (11,76)*, and the three mammalian *MutL* homologs *(71–73,77,78)*.

Similar to the eukaryotic MutS homologs, functional human MutL homologs are also heterodimers. In fact, the concept of Mut homologs as heterodimers originated from the studies of yeast MutL homologs. Based on the indistinguishable mutator phenotypes of *yMLH1–yPMS1* double mutants and either *yMLH1* or *yPMS1* single mutants, Prolla et al. *(75)* proposed that these two MutL homologs exist as a heterodimer that functions in the MMR pathway. This idea is supported by the demonstration that yMLH1 and yPMS2 physically interact with each other *(75,79,80)*. However, more direct evidence supporting the heterodimeric form of the MutL homolog was the co-purification of hMLH1 and hPMS2 (human homolog of yPMS1) from HeLa nuclear extracts, which was designated hMutLα *(81)*. Although efforts to determine the biochemical function of hMutLα have not yet succeeded *(81)*, it was established that individual components of hMutLα are absolutely required for strand-specific MMR. Cells defective in either *hMLH1* or *hPMS2* are hypermutable, and are deficient in the repair of both base–base mismatches and ID mispairs *(28,82,83)*. hMutLα is capable of fully restoring MMR to mutants of *hMLH1 (81)* and *hPMS2* (Li, unpublished results).

The MMR pathway in both *E. coli* and humans possesses a bidirectional processing capability *(25,31)*, i.e., it is capable of removing mismatches either from a 5' to 3' orientation or from a 3' to 5' orientation, depending on the relative position of the strand break and the mismatch. MMR components that have a polar activity are helicases and exonucleases. The authors have found an *hMLH1*-deficient cell line that is only defective for repair in one orientation, suggesting that human MutL homologs are involved in selecting MMR orientation. This cell line possesses a wild-type (wt) *hMLH1* gene, but the expression of the gene is inhibited by hypermethylation of its promoter region. As a result, this cell line is defective in the repair of heteroduplexes with a strand break 3' to the mismatch, but proficient in the processing of heteroduplexes with a strand break 5' to the heterology (Li and Modrich P, unpublished results). A similar result has also been found in a cisplatin-resistant cell line with defective *hMLH1 (83)*. Also, though cells defective in *hMLH1* fail to carry out repair in a 3' to 5' orientation, a cell line defective in *hPMS2* has been shown to be defective in repair in the 5' to 3' orientation *(84)*. It was recently observed that *E. coli* MutL plays a role in loading UvrD at the site of the strand break *(19,23)*. Thus, the association of human MutL homologs with repair in a specific orientation suggests that hMLH1 may be involved in delivering a helicase or exonuclease to the 3′ nick accordingly hPMS2 may be responsible for directing these enzymes to the 5′ nick. Further biochemical studies are required to address and resolve these issues.

The role of hPMS1 in MMR has not been determined. However, a recent study in yeast has demonstrated that deficiency in *MLH3*, a yeast homolog of *hPMS1*, causes a small but significant increase in the frameshift mutation rate. This phenotype is similar to the phenotype of *yMSH3* single and *yMSH3–yMLH3* double mutants, suggesting that MSH3 and MLH3 are involved in the same ID mispair correction pathway *(76)*. In addition, this study showed that yMLH3 interacts with yMLH1 in a two-hybrid system, implying the existence of a second heterodimeric MutL complex consisting of yMLH1 and yMLH3 *(76)*. Given the close homology between MMR systems in all species, it would not be surprising if a second human MutL heterodimer of hMLH1–hPMS1 were identified. Indeed, Jiricny et al. have recently found that hMLH1 and hPMS1 proteins co-immunoprecipitate with each other, both in HeLa nuclear extracts and in baculovirus recombinant proteins, although the biochemical function of this complex has not been determined *(84a)*.

The bacterial MutL protein has been described as a molecular matchmaker *(85)* or a molecular chaperon *(12)* that facilitates the interactions between MutS and MutH. Although direct evidence for human MutL homologs as molecular matchmakers or chaperons is lacking, hMutLα has been shown to be crucial for formation of the initiation complex. Gu et al. *(34)* have demonstrated that hMSH2, hMLH1, hPMS2, and PCNA can be co-immunoprecipitated in HeLa nuclear extracts, but this co-precipitation is not observed in the extracts derived from *hMLH1*-deficient cells. In addition to enhancing protein–protein interactions, MutL homologs have been shown to enhance the binding of MutS homologs to mismatches in DNA *(79,86)*.

Taken together, a model can be proposed for the role of MutS and MutL homologs in human MMR (Fig. 2). Mismatch recognition involves two heterodimers of human MutS homologs, hMutSα and hMutSβ: hMutSα recognizes both base–base and ID mispairs; hMutSβ binds only to ID mispairs. The substrate specificity of these two complexes implies that human cells may use hMutSα in routine repair processes, and hMutSβ as a partial backup system, although both complexes function redundantly on ID mispairs. A number of lines of evidence support this view. First, cells with genetic deficiency in *MSH6* display a much more severe mutator phenotype than *MSH3*-defective cells *(87,88)*, because the former cells are defective in base–base mismatch correction, but retain partial repair activity on ID mispairs *(30)*. Second, the overexpression of MSH3 (which competes with MSH6 for MSH2) leads to a strong mutator phenotype, despite the fact that these cells are proficient in ID mismatch correction *(89,90)*. Third, under normal circumstances, human cells possess much more hMutSα than hMutSβ *(33,89)*. The two hMutL heterodimers, hMutLα and hMutLβ (hMLH1 and hPMS1), may have a similar relationship. hMutLα seems to interact with both hMutSα and hMutSβ; hMutLβ only interacts with hMutSβ. The proposed roles for hMutL homologs are consistent with genetic studies in yeast *(76)*, mice *(91)*, and human tumors, in which only a limited number of germ-line mutations are documented in *hPMS1* and *hPMS2;* a relatively large number are reported in *hMLH1* (see Table 2).

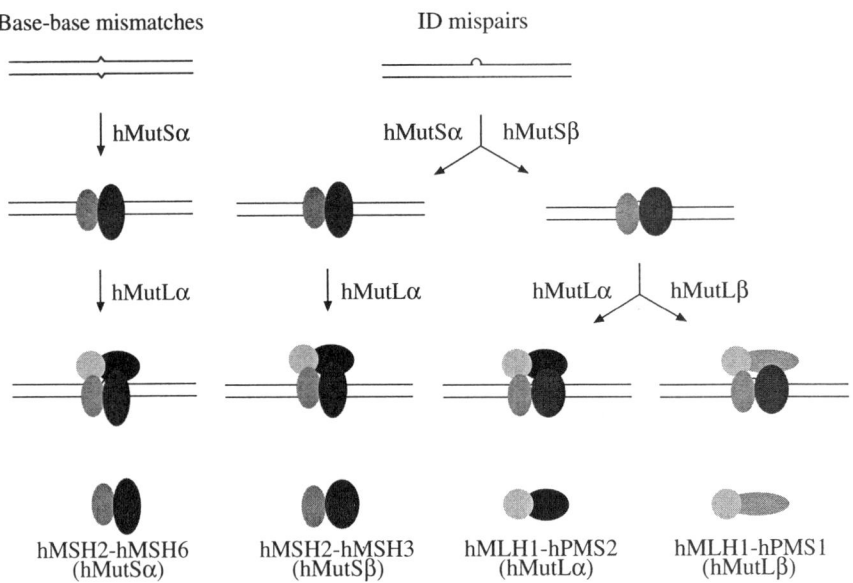

Fig. 2. Proposed function for MutS and MutL homologs in human MMR.

4. MMR DEFICIENCY AND COLORECTAL CANCER

MMR has long been postulated to be an important cellular mechanism to prevent genomic instability. However, it was not implicated as a cause of cancer susceptibility until 1993, when MMR defects were associated with hereditary nonpolyposis colorectal cancer (HNPCC) and a subset of sporadic colon cancers (CCs). HNPCC is a heritable autosomal dominant disease that is defined by the presence of at least three family members with CC in two successive generations, with one affected member having been diagnosed at less than 50 yr of age *(92)*. HNPCC accounts for 4–13% of all CC *(93)*, one of the most common forms of neoplasia in Western populations *(94)*. In addition to CC, patients with HNPCC are at increased risk to develop cancers of the endometrium, ovary, stomach, urinary tract, brain, and other epithelial organs *(95,96)*. Although the HNPCC syndrome was suspected to be a heritable disease more than 85 yr ago *(97)*, the molecular pathogenesis of this disease was not established until 1993.

4.1. Microsatellite Instability in HNPCC and Sporadic CC

The discovery of the molecular basis of HNPCC started with the observation of frequent alterations in simple repeated sequences (also called microsatellite instability [MSI]) in HNPCC tumors. Tumor susceptibility genes are often subject to loss of heterozygosity in tumors. To explore the hypothesis that the genetic basis of HNPCC involved a tumor suppressor gene, Aaltonen et al. *(98)* employed microsatellite markers to determine allelic losses in two large HNPCC kindreds in chromosome (chr) 2p15–16, an HNPCC locus identified previously by genetic linkage analysis *(99)*. However, allelic losses were not found in this HNPCC locus; instead, they observed insertion or deletion mutations at these repetitive sequences in 11/14 tumors examined *(98)*. These unexpected mutations were evident in each di- and trinucleotide repeat tested, and were referred to as replication error positive *(98)*. In addition, those authors also found a subset of sporadic CCs with a similar phenotype, but occurring at a much lower incidence (6/46) *(98)*.

At the same time, two other groups, Ionov et al. *(100)* and Thibodeau et al. *(101)*, independently reported microsatellite alterations in 12–15% of sporadic CCs. These observations were made using random microsatellite markers throughout the genome, rather than the markers within the 2p15–16 region, used by the first group. Taken together, these findings suggest that MSI in HNPCC, and a subset of sporadic CCs, is a genome-wide phenomenon, and may be caused by a common mechanism. Although the genetic basis of HNPCC remained unidentified at that time, these studies provided an important clue as to the mechanism of its action.

4.2. MMR Deficiency Is the Genetic Basis of HNPCC

4.2.1. BREAKTHROUGH

The identification of MSI in CCs received a great deal of attention from cancer investigators, as well as from geneticists and biochemists working on DNA MMR, because the mutational fingerprint is similar in HNPCC and MMR-deficient cells. At that time, the following points had been established: Loss of MMR function leads to genome-wide base–base substitutions, as well as to frameshift mutations *(4,57)*; MMR proteins recognize and process ID mispairs *(16)*; and repetitive dinucleotide sequences undergo frequent contractions or expansions in MMR-deficient bacterial cells *(102,103)*. Therefore, the hypothesis was made that the genetic defects in HNPCC involve the loss of MMR function.

Several groups tested this hypothesis using different approaches. First, Strand et al. *(104)* examined the stability of poly(GT) tracts in yeast strains with either a single or double knockout of *MSH2, MLH1*, or *PMS1*. All mutants (both single and double mutants) exhibited 100–700-fold elevated levels of tract instability involving insertion or deletion of 2–4 repeated units *(104)*. That study strongly supports an association of MMR defects with the HNPCC syndrome.

Second, two teams, Fishel et al. *(41)* and Leach et al. *(105)*, independently searched for human MMR homolog genes, and determined their association with HNPCC kindreds. Both teams reported in succession the cloning of the *hMSH2* gene, using polymerase chain reaction products of degenerate primers derived from two highly conserved regions of the known bacterial MutS and yeast MSH proteins, and located the gene on chr 2p (there was a subtle conflict on the exact position of the gene on the chr) *(41,105)*. Germ-line mutations of *hMSH2* were indeed identified in HNPCC families *(105)*.

Third, two laboratories *(28,29)* took a biochemical approach, and examined the MMR proficiency of tumor cells derived from HNPCC and sporadic CCs with MSI. Both laboratories demonstrated that cell extracts of these tumor cells are completely defective in repair of base–base and ID mispairs *(28,29)*. These in vitro biochemical studies provided definitive evidence supporting the hypothesis that MMR defects are the genetic basis of HNPCC.

4.2.2. GENETIC EVIDENCE

Immediately after the first HNPCC-linked locus was mapped to chr 2p15–16 *(99)*, Lindblom et al. *(106)* identified a second locus linked to HNPCC predisposition at chr

3p21–23. Since the MMR pathway involves multiple components (*see* Subheading 3.) and characterization of the first HNPCC-associated locus led to the cloning of *hMSH2* *(41,105)*, investigators looked for additional HNPCC-associated human Mut homologs. In a remarkably short period of time, three human *MutL* homolog genes *(hMLH1, hPMS1,* and *hPMS2)* were cloned *(71–73)*. Bronner et al. *(71)* identified the *hMLH1* gene, using an approach similar to that used to identify *hMSH2 (41,105)*. The gene was mapped to chr 3p, and a missense germ-line mutation was characterized in a family with a history of HNPCC *(71)*. At the same time, Papadopoulos et al. *(73)* searched a large database of human cDNAs and also identified the *hMLH1* gene. In addition, they reported two more human *MutL* homologs, *hPMS1* and *hPMS2,* located on chr 2q and 7p, respectively *(72)*. Germ-line mutations of each human *MutL* homolog were found in HNPCC kindreds, with defects in *hMLH1* present in the majority of the HNPCC kindreds *(72,73)*.

Since the initial identification of HNPCC-linked genes, HNPCC kindreds have been extensively screened for mutations in each of these genes *(107–119)*. It is now clear that mutations of *hMSH2* and *hMLH1* are very common, and are found in more than 90% of HNPCC kindreds, but mutations of *hMPS1* and *hMPS2* are rare (*see* Table 2). Less is known about germ-line mutations in *hPMS1* and *hPMS2,* two MMR genes not identified through HNPCC linkage analysis. There is one report of a germ-line mutation in *hMSH6 (120),* and no reports of *hMSH3* mutations in HNPCC. However, somatic mutations in each of the MMR genes in sporadic cancer and cell lines with MSI, as well as in HNPCC, have been documented *(87,121–130),* although mutations of *hMSH2* and *hMLH1* are also predominant. The protein products of these two genes play a central role in MMR, through their essential functions in hMutS and hMutL heterodimers. In contrast, overlapping roles exist for the hMSH3 and hMSH6 proteins, and for the hPMS1 and hPMS2 proteins (*see* above). Thus, the observed distribution of mutations in these genes, in HNPCC and sporadic cancer, is consistent with the relative importance of their functional roles in MMR.

Mice with a knockout mutation in each *MMR* gene are available *(77,78,91,131–134b)*. Although most of them display a typical hypermutable phenotype and a predisposition to develop cancer, it is surprising that none of these MMR-deficient mice develop CC as in HNPCC (Table 3). Mice deficient in *MSH2* and *MLH1* demonstrate a similar pattern of cancer development. Both of these mice genotypes usually develop lymphomas *(78,91,131,132),* intestinal adenomas and adenocarcinomas, and, to a lesser extent, skin neoplasms and sarcomas *(78,91,132)*. However, the *MLH1*-deficient mice (both male and female) are infertile *(78,133)*. In contrast to *MSH2-* or *MLH1*-deficient mice, *PMS2*-deficient animals do not develop intestinal adenomas and adenocarcinomas, although they do appear prone to sarcomas and lymphomas *(77,91)*. In *PMS2* knockouts, only male mice are infertile *(77)*. *MSH6*-deficient animals develop a spectrum of tumors, but the most predominant are gastrointestinal tumors and lymphomas *(134),* suggesting that germ-line mutations of *MSH6* may be associated with a cancer predisposition. Unlike all the MMR-deficient mice described above, mice deficient in *PMS1* gene show no instability in repeat sequences, except for a small mutation rate in mononucleotide repeats *(91)*. The *PMS1*-knockout mice do not develop any tumors *(91)*. The same seems to be true for the *MSH3*-knockout mice (134a, 134b), although MMR-deficient mice do not develop CC, these studies certainly

Table 2
MMR Genes in HNPCC

Gene	Chromosomal location	Mutations in HNPCC
MutS Homologs		
MSH2	2p	45%[a]
MSH3	5q11–13	None
MSH6	2p16	One family
MutL homologs		
MLH1	3p21–23	49%[a]
PMS1	2q31–33	One family
PMS2	7p22	6%[a]

[a]From ref. 120.
[b]From ref. 72.

Table 3
Characteristics of MMR-deficient Knockout Mice

Gene	Tumor	MSI	Other
MSH2	Lymphoma Intestinal adenoma	Yes	
MSH3	None	Not determined	
MSH6	Lymphoma Gastrointestinal	No instability in dinucleotide repeats	
MLH1	Lymphoma Intestinal adenoma Skin tumor and sarcoma	Yes	Male and female infertility
PMS1	None	Mononucleotide repeats only	
PMS2	Lymphoma Sarcoma	Yes	Male infertility

support the view that MMR defects lead to genomic instability, and eventually to cancer, as originally proposed, based on studies of the HNPCC syndrome.

Strong evidence supporting the concept that MMR genes are crucial to genome stability was provided by chr or gene-transfer experiments *(135–139)*. Koi et al. *(135)* reported that the transfer of human chr 3, carrying wt *hMLH1* gene to an *hMLH1*-deficient colorectal tumor cell line, restores MMR to the cell line *(135)*. Similarly, human chr 2, containing both the *hMSH2* and *hMSH6* genes, can complement MMR defects in *hMSH2*- and *hMSH6*-deficient tumor cell lines *(136,137,140)*. Recently, restoration of MMR repair to a *PMS2*-deficient human cell line and a *MLH1*-knockout mouse cell line has been achieved by introduction of *hPMS2* or *hMLH1* genes into these lines, respectively *(138,139)*. These studies further confirm that the MMR system plays an essential role in the maintenance of genomic stability, and provides molecular insights with potential for application in HNPCC gene therapy.

4.2.3. BIOCHEMICAL EVIDENCE

The most convincing evidence that the HNPCC syndrome is caused by MMR defects may be the biochemical studies of this disease. Functional biochemical assays of extracts prepared from a number of cell lines, which were derived from HNPCC and sporadic tumors with MSI, have clearly demonstrated that these cells are deficient in nick-directed MMR *(28–30,55,82,128,141–144)*. Further characterization of these cell lines has defined at least two in vitro complementation groups *(29,30,55)*, which led to the isolation of hMutLα and hMutSα *(30,81)*. Using an in vitro complementation analysis, Li and Modrich *(81)* purified a heterodimer from HeLa nuclear extracts that restores strand-specific *MMR* to an *hMLH1*-deficient colorectal tumor cell line. This heterodimer has two components, 85 kDa hMLH1 and 110 kDa hPMS2, and is called hMutLα *(81)*. Using a similar approach, Drummond et al. *(30)* isolated hMutSα (105 kDa hMSH2 and 160 kDa hMSH6), based on its ability to restore MMR to tumor cells defective in either *hMSH2* or *hMSH6*.

4.2.4. EPIGENETIC EVIDENCE

As discussed previously, mutations in *MMR* genes, which account for the hyper-mutable phenotype, are associated with the HNPCC syndrome and a subset of sporadic CCs with MSI. However, in a significant fraction of sporadic MSI colon tumors, no mutations have been identified in *MMR* genes *(124,125,145)*, suggesting that a novel mechanism may be involved in causing MSI in these cases. The search for the novel mechanism has linked these tumors again to MMR defects. This time, it is an epige-netic factor, methylation, responsible for suppressing the expression of *MMR* genes *(121,123,143,146–149)*.

Kane et al. *(123)* demonstrated that hypermethylation of the *hMLH1* promoter is corre-lated with a lack of *hMLH1* expression in several sporadic colon tumors and cell lines that are free of mutations in the *hMLH1* gene. Those authors suggested that hypermethylation is probably a common mode of *MMR* gene inactivation in sporadic cancer *(123)*. Since then, hypermethylation of *MMR* genes, especially *hMSH2* and *hMLH1*, has been exten-sively studied *(143,146–149)*. According to the Bethesda guidelines *(150)*, sporadic tumors can be classified into three types, based on their MSI status: microsatellite stable, low-frequency MSI (MSI-L), and high-frequency MSI (MSI-H). It has been reported that more than 95% of MSI-H tumors are caused by loss of expression of *hMLH1* *(151)*. Almost all MSI-H tumors that do not have a detectable mutation within the *hMLH1* gene demonstrate hypermethylation in the *hMLH1* promoter *(148,149)*. In contrast, hyperme-thylation of the *hMSH2* gene is not observed in sporadic tumors *(148)*.

To determine the nature of the hypermethylation of the *hMLH1* promoter in these MSI tumors, two independent research groups, Veigl et al. *(143)* and Herman et al. *(149)*, treated several tumor cell lines deficient in *hMLH1* expression because of hypermethylation in the *hMLH1* promoter with the demethylating agent 5-aza-deoxy-cytidine. This treatment successfully restores hMLH1 protein expression in all tumor cells that lack *hMLH1* expression because of a methylated *hMLH1* promoter. The expression of hMLH1 is associated with the presence of unmethylated *hMLH1* alleles *(143,147)*. More importantly, extracts derived from the drug-treated cells are capable of performing strand-specific MMR *(147)*. These experiments indicate that both genetic defects and epigenetic modification in MMR genes can result in a mutator phenotype.

5. MMR DEFICIENCY AND NON-CC

5.1. Microsatellite Instability in Non-CC

The identification of MSI in HNPCC in 1993 led to the dramatic elucidation of the molecular pathogenesis of this disease. Since then, a great body of work has been published demonstrating that MSI is also associated with a wide variety of non-HNPCC and noncolonic tumors (for detailed reviews, *see* refs. *150,152,* and *153).* These tumors include endometrial *(154–166),* ovarian *(167–173),* gastric *(174–190),* cervical *(163,191–195),* breast *(196–208),* skin *(209–220),* lung *(221–226),* glioma *(227–229),* prostate *(230–238),* bladder *(239–242),* and lymphoma *(243,244)* (Table 4). These studies were carried out using various numbers of different microsatellite markers, and employing different numbers of samples; thus, it is not surprising that the observed mutation rates vary from study to study, and in some cases are not in agreement with one another *(195,210,245–256).* Therefore, it is difficult to draw a firm conclusion based on the MSI work published to date. To address this problem, an international workshop was recently held in Bethesda, Md, to develop international criteria for MSI studies, referred to as the "Bethesda guidelines" *(150).* The workshop resulted in the suggestion that at least five loci should be used in future MSI studies, with instability in one of five loci scored as MSI-L, and instability in two or more loci scored as MSI-H *(150).*

Table 4 shows results of some of the MSI studies in non-HNPCC and noncolonic tumors that can be judged by the Bethesda guidelines *(150).* As observed in sporadic CCs, noncolorectal tumors exhibit the MSI-H and MSI-L phenotypes. Most of the sporadic endometrial and gastric tumors, lung cancers, and lymphomas display a high level of MSI in many markers (Table 4). Some tumors demonstrate greater instability in one marker than another. Tumors with MSI can be divided into two groups: one that displays elevated instability at mono- and dinucleotide markers and, to a lesser degree, at larger-repeat markers, and a second group that displays elevated instability only at specific larger-repeat markers, such as tri- and tetranucleotide repeats. Endometrial and gastric tumors usually belong to the first group; bladder, lung, head, and neck cancers belong to the second group *(150,257).*

The presence of MSI in these sporadic noncolonic tumors stimulated a search for somatic mutations in the *hMSH2* and *hMLH1* genes in these tumors. Initially, no mutations of these two HNPCC genes were detected *(183,258–261).* However, several laboratories have recently demonstrated *(149,162,262)* that the *hMLH1* promoter is hypermethylated in sporadic endometrial and gastric tumors. This result suggests that hypermethylation of *hMLH1,* and its subsequent transcriptional silencing, may be a major cause of sporadic noncolonic tumors with MSI. In addition, somatic frameshift mutations in *hMSH2, hMSH3,* and *hMSH6* have been reported in gastric cancers *(130,263,264).* Biochemical studies have demonstrated that cell lines derived from sporadic endometrial, ovarian, and prostate cancers are defective in strand-specific MMR *(29,55,128).* These findings suggest that MMR defects are a likely cause of noncolonic sporadic cancer with MSI, although other mechanisms may also be involved in causing the MSI mutator phenotype.

5.2. MMR Deficiency and Inactivation of Genes Critical for Cellular Growth

The work described above conclusively demonstrates that loss of function in MMR genes can cause HNPCC. This result implies that MMR genes are tumor suppressor

Table 4
MSI in Noncolonic Tumors[a]

Tumor	Loci	MSI	Frequency no. (%)	Ref.
Endometrial	Dinucleotide	≥2/8 loci	10/109 (9)	(159)
Endometrial	D1S126	≥2/5 loci	18/77 (22)	(259)
	D2S393			
	D3S1067			
	D5S644			
	TP53			
Ovarian		≥2/5 loci	2/68 (3)	
Endometrial	Dinucleotide	≥2/8 loci	12/68 (18.5)	(156)
Gastric	D2S123	≥2/5 loci	15/24 (62.5)	(333)
	D2S126			
	D3S1067			
	D11S922			
	TP53			
Gastric	$AP(\delta)3\ (A)_{18}$	≥2/5 loci	25/167 (15)	(130)
	$hMSH3\ (A)_{26}$			
	D1S158			
	D5S421			
	D8S199			
	$BAX\ (G)_8$		16/25 (64)	
	$hMSH3\ (A)_8$		16/25 (64)	
	$hMSH6\ (G)_8$		13/25 (52)	
Gastric	Dinucleotide	≥2/6 loci	5/98 (2)	(177)
Cervix	Dinucleotide	≥2/30 loci	5/98 (3.3)	(244)
	Trinucleotide			
	Tetranucleotide			
Breast	Dinucleotide	≥2/12 loci	2/100 (2)	(207)
Skin cancers	47 loci[b]	≥2 of loci tested		(213)
Basal			1/47 (2)	
Squamous			1/49 (2)	
Melanoma			0/41 (0)	
Lymphoma				
MALT[c]	D3S1262	≥2/5 loci	21/40 (52.5)	(334)
	D3S1265			
	D3S11			
	D3S1261			
	D6S262			
HIV-positive	9 loci	≥2/5 loci	4/6 (66)	(335)
Lung	Dinucleotide	≥2/8 loci	12/35 (34)	(336)
Gliomas		≥1/2 loci		(228)
Glioblastoma	D2S123	≥1/2 loci	5/24 (21)	
	D3S1067			
Astrocytoma	D2S123	≥1/2 loci	2/16 (12.5)	
	D3S1067			
Glioblastoma	$TGF\beta\ RII\ (A)_{10}$		4/24 (17)	
Astrocytoma	$TGF\beta\ RII\ (A)_{10}$		2/16 (12.5)	
Prostate	23 loci	>2/23 loci	4/47 (8.5)	(237)

[a]Used with permission from ref. (150)

[b]Not all tumors investigated at all loci.

[c]MALT, mucosa-associated lymphoid tissue.

genes, at least with respect to certain cancer phenotypes. It has also established that the MMR pathway is a mutation avoidance system or a caretaker system *(265)*. Therefore, loss of MMR function will have a great impact on genome stability, especially on the stability of genes critical for cellular growth, such as other tumor suppressor genes and oncogenes. Because of technical limitations, it is impossible to assess the impact of MMR-deficiency on a genome-wide basis, and to identify all mutations that accumulate as a result of MMR deficiency. However, using MSI analysis, it is possible to readily detect frameshift mutations in genes that contain simple repeat sequences within their coding regions, which in most cases lead to truncated proteins. Therefore, repeat tracts within coding regions have been studied in MMR-deficient cells, to determine if they are frequent mutational targets.

Markowitz et al. *(266)* reported that mutations in the type II transforming growth factor-β receptor *(TGFβ RII)* gene is associated with sporadic CC cells defective in MMR. These mutations are all frameshift mutations, and occur either in a 6-bp GTGTGT repeat or in an $(A)_{10}$ mononucleotide repeat *(266)*. In each case, the frameshift mutation results in a mutant protein form of TGFβ RII. Subsequent studies *(267–273)* have demonstrated that frameshift mutations of simple repeat tracts in the *TGFβ RII* gene are common in colorectal tumors with MSI. Similar mutations of *TGFβ RII* have also been observed in MSI gastric cancer *(274–277)*, glioma *(228)*, uterine cervical cancer *(278)*, squamous carcinoma of the head and neck *(279)*, ulcerative colitis-associated neoplasm *(280)*, and sporadic cecum cancer *(281)*. It is known that *TGFβ RII* is required for transduction of the TGFβ growth inhibitory signal to suppress epithelial cell growth. The loss of TGFβ RII function in tumors with MSI represents a crucial mechanism for escape from growth control. The targeted mutations in simple repeated sequences in *TGFβ RII* are characteristic of what is expected from a MMR defective system.

In addition to *TGFβ RII*, somatic frameshift mutations of mononucleotide runs have been documented in several genes critical for cellular growth in tumors with MSI. These genes include the apoptosis gene *bax (160,282–288)*, insulin-like growth factor 2 receptor *IGF2R (285,289–291)*, transcription factor *E2F-4 (276,292)*, MMR genes *hMSH3* and *hMSH6 (129,130,263,264)*, and tumor suppressor genes *APC (293–295)* and *PTEN (162,289,296–298)*. All of these genes are crucial for cellular growth control, and the inactivation of any of these genes would be a key mechanism by which tumors with MSI become neoplastic. Therefore, the potential impact of loss of the tumor suppressor function of MMR is not only relevant to HNPCC, but to virtually all types of cancer.

It is worth noting that not all genes that contain simple repeat sequences undergo frameshift mutations in MSI tumors. For example, this type of mutation is not associated with the *p53* gene *(275,299–301)*. It appears that there is a large variation in the propensity for mutation in repeat sequences at different sites in the genome in MMR-deficient cells, which is not yet completely understood. It is also important to point out that defects in MMR cause base–base substitutions and frameshift mutations in nonrepeat sequences, and these mutations can also inactivate essential genes. This kind of mutation has not been extensively examined in MMR-deficient cells, because it can be tedious and difficult to detect mutations in nonrepeat sequences. Nevertheless, it has been documented that MSI tumors and tumor cell lines that are defective in MMR exhibit an elevated rate of base–base substitutions and frameshift mutations in the

HPRT locus *(56,57,302)*. Therefore, further studies are required to determine the impact of MMR defects on mutagenesis and cancer.

6. MMR as a Sensor of Genomic Damage

As discussed in subheading 5.2., MMR acts as a tumor suppressor, through its ability to maintain genomic stability by correcting base–base and ID mispairs. However, recent studies *(303,304)* suggest that MMR participates in other cellular functions, such as transcription-coupled nucleotide excision repair. DNA recombination *(305,306)*, and chr synapsis *(77,78)*. In addition, human MMR plays an important role in cellular response to DNA damage induced by many physical and chemical agents, a concept supported by the fact that MMR-deficient cells are much more resistant to killing by these agents than MMR-proficient cells.

The first evidence for this concept came from studies of *E. coli dam⁻* mutants *(307–309)*. These cells lack the methylase activity required for strand discrimination during MMR, and are sensitive to killing by *N*-methyl-*N'*-nitro-*N*-nitrosoguanidine (MNNG), a methylating agent that modifies guanine residues in DNA to produce O^6-methylguanine (O^6-meG). However, double mutants that are also defective in *mutS* or *mutL* are highly resistant to killing by MNNG. The mutagenic capability of the drug is similar regardless of the genotype (wt, *dam⁻*, *dam⁻/mutS⁻*, *dam⁻/mutL⁻*), leading to the conclusion that the wt MMR system is involved in the MNNG-induced cytotoxicity of *dam⁻* cells. A similar phenotype was also observed in MMR-deficient human cells.

The best-characterized MNNG-resistant human cell line is MT1. MT1 was derived from a lymphoblastoid cell line by selection with a high dose of MNNG *(310)*. Compared with the parental TK6 cells, MT1 cells not only exhibit more than a 500-fold increased resistance to killing by MNNG, but also display an elevated mutability at the *HPRT* locus in the absence of the drug *(57,310)*. The MT1 cell line was found to harbor a mutation in *hMSH6 (58)*, and to be defective in strand-specific MMR *(57)*. While MMR deficiency can be acquired following treatment with the alkylating agent MNNG, human tumor cells also demonstrate a correlation between MMR-deficiency and resistance to alkylating agents. For example, an *hMLH1*-defective colorectal tumor cell line is resistant to killing by MNNG; however, when it receives a wt copy of *hMLH1* by chr 3 transfer, it becomes sensitive to MNNG-induced cytotoxicity and killing *(135,311)*. Similarly, cellular deficiency in other MMR genes also confers resistance to alkylating agents *(82,131,312–315)*.

Tumor cell lines resistant to other chemotherapeutic drugs, including cisplatin, procarbazine, and temozolomide, are also known to be deficient in MMR. Cisplatin damages DNA by forming inter- and intrastrand crosslinks between two guanine residues *(316,317)*. It has been reported that several ovarian tumor cell lines lose MMR function and gain resistance to cisplatin by a single-step selection process *(318,319)*. Accordingly, MMR-deficient cells are resistant to the cytotoxic effects of cisplatin *(320)*. Like MNNG, temozolomide and procarbazine are also DNA-methylating agents, and exert their cytotoxic effects primarily through methylation at the O^6 position of guanine. Similar results have been obtained with cisplatin and these two drugs. Serial in vivo exposure of a glioblastoma multiforme xenograft in nude mice, with procarbazine, led to a resistant tumor that lost expression of *hMSH2 (321)*. It is also documented that MMR-deficient cells are resistant to temozolomide *(315,321–323)*. Recently, the list of agents to which MMR-deficient cells gain resistance has been extended to include ionizing radiation *(324,325)* and environmental chemical carcinogens, such as polycyclic aromatic hydrocarbons and aromatic amines, which react covalently with DNA (preferentially at guanine residues) to form bulky DNA adducts. These findings suggest that MMR components may have a function in sensing DNA damage induced by drugs or carcinogens.

Although the mechanism for drug resistance in MMR-deficient cells is unclear, several models have been proposed. The first model suggests that the death of wt MMR cells, in response to drug or chemical treatments, results from an attempt by the MMR pathway to correct a lesion in the template strand, by repairing the newly synthesized strand, which leads to a reiterative cycle of excision and resynthesis of the daughter strand, and leaves the adducts in the template strand unaffected. Such a futile repair cycle may lead to cell death by an uncharacterized pathway (for reviews, *see* refs. *326* and *327*). Alternatively, cell death may result from the binding of MMR proteins to DNA adducts, which may block DNA transactions, such as replication, transcription, and proper damage repair. A third model proposes that a carcinogen-induced dsDNA break acts as the signal for cell death *(12)*. In this model, DNA adducts may dissociate the DNA replication machinery at the adduct site, and terminate chain elongation. DNA synthesis can be reinitiated behind the lesion by a newly formed replication complex, which can lead to a single-strand gap opposite the DNA adduct. A double-stranded break is formed during the subsequent cycle of DNA replication, when the gapped strand is used as a template. This model includes no proposed role for MMR components.

Strong support for the first two models has recently been provided by both in vitro and in vivo experiments. First, gel shift analyses demonstrated that human MutS homologs specifically recognize DNA adducts induced by MNNG *(328)*, cisplatin *(328–331)*, aromatic amines *(329)*, and polycyclic aromatic hydrocarbons *(325a)*. Second, D'Atri et al. *(323)* have shown that cell death induced by temozolomide is associated with increased expression of p53. The carcinogen-induced p53-dependent apoptosis was also shown to be dependent on functional hMutS and hMutL homologs *(325a)*. Finally, O^6-msG can provoke strand-specific MMR in vitro *(329a)*. These observations strongly support the first hypothesis.

Taken together, the mechanism by which MMR triggers carcinogen-induced apoptosis can be proposed as follows (Fig. 3). When a DNA replication fork encounters a carcinogen–DNA adduct in the template strand, DNA polymerase incorporates a base opposite the adduct, which can be recognized by hMutS homologs as a mismatch. The binding of hMutS homologs to the damaged bp initiates a strand-specific MMR reaction. Because repair is restricted to the daughter DNA strand, the adduct persists in the template strand. Thus, repair resynthesis actually restores the original adduct-containing bp, which will retrigger a strand-specific MMR process. This futile repair cycle is a signal to induce p53 expression, which leads to programmed cell death. Recently, a p73-dependent apoptotic response has also been proposed *(329b)*.

The recognition and processing of carcinogen–DNA adducts by MMR may play an important role in ensuring genomic integrity. Normally, DNA damage induced by physical and chemical carcinogens is repaired by other repair systems, e.g., O^6-meG by O^6-meG methyltransferase, and bulky DNA adducts by base and/or nucleotide excision repair. However, when these systems are not available and genomic integrity is in dan-

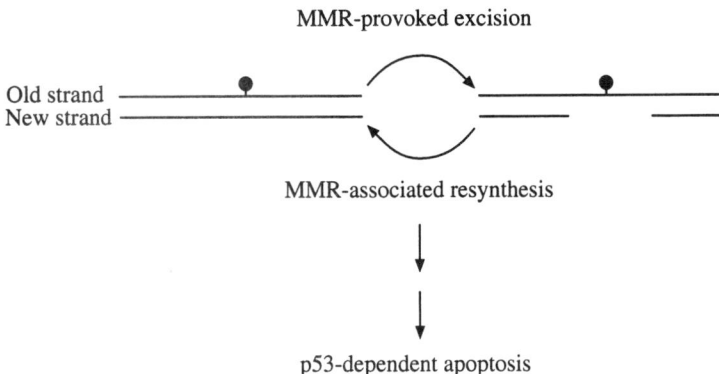

Fig. 3. Proposed mechanism for MMR triggering apoptosis. The filled circle represents the carcinogen adduct.

ger, cells will rely on MMR to commit suicide. Therefore, the authors propose that the MMR pathway possesses at least two functions that preserve genomic stability: MMR corrects mismatches, and it also provides a signal for cell death when DNA damage cannot be repaired.

Although MMR maintains genomic stability in two different ways, the differential response of MMR-deficient and MMR-proficient cells to a number of chemotherapeutic drugs may provide a significant basis for understanding the effects of cancer chemotherapy. First, certain widely used clinical drugs e.g., temozolomide, procarbazine, and cisplatin, are expected to be harmful for patients with tumors caused by MMR defects, because these drugs actually kill the patient's normal tissue, rather than the tissue of the tumor. Second, since some tumor cells can acquire MMR deficiency upon exposure to certain drugs *(301,318,319,321)*, the use of these drugs in clinical practice may lead to a secondary cancer characterized by MMR defects.

The rapid advances in the field of eukaryotic MMR, described in this review, demonstrate important new connections between DNA repair and the fields of carcinogenesis and toxicology. The implications of these findings are only beginning to be fully appreciated. Areas of importance for future exploration in this field may include screening, diagnosis, and gene therapy for cancer with MMR defects.

ACKNOWLEDGMENTS

The authors thank Joe Jiricny for provision of unpublished data, and Miriam Sander for critical reading of the manuscript. This work was supported by grants (to GML) CA72956 from the National Cancer Institute and 4755 from the Council for Tobacco Research. SM was supported in part by an National Institutes of Environmental Health Sciences Training Grant (ES07266).

REFERENCE

1. Loeb LA. Mutator phenotype may be required for multistage carcinogenesis. *Cancer Res* 1991; 51:3075–3079.
2. Tiraby JG, Fox MS. Marker discrimination in transformation and mutation of pneumococcus. *Proc Natl Acad Sci USA* 1973; 70:3541–3545.

3. Nevers P, Spatz H. Escherichia coli mutants *uvrD uvrE* deficient in gene conversion of lambda heteroduplexes. *Mol Gen Genet* 1975; 139:233–243.
4. Modrich P. Mechanisms and biological effects of mismatch repair. *Ann Rev Genet* 1991; 25:229–253.
5. Kolodner RD. Mismatch repair: mechanisms and relationship to cancer susceptibility. *Trends Biochem Sci* 1995; 20:397–401.
6. Friedberg EC, Walker GC, and Siede W. *DNA Repair and Mutagenesis.* Washington DC: ASM 1995.
7. Fishel R, Kolodner RD. Identification of mismatch repair genes and their role in the development of cancer. *Curr Opin Genet Dev* 1995; 5:382–395.
8. Modrich P, Lahue R. Mismatch repair in replication fidelity, genetic recombination, and cancer biology. *Annu Rev Biochem* 1996; 65:101–133.
9. Kolodner R. Biochemistry and genetics of eukaryotic mismatch repair. *Genes Dev* 1996; 10:1433–1442.
10. Umar A, Kunkel TA. DNA-replication fidelity, mismatch repair and genome instability in cancer cells. *Eur J Biochem* 1996; 238:297–307.
11. Crose G. Mismatch repair systems in *Saccharomyces cerevisiae.* Totowa, NJ: Humana. 1997; 411–448.
12. Jiricny J. Eukaryotic mismatch repair: an update. *Mutt Res* 1998; 409:107–121.
13. Jiricny J. Replication errors: cha(lle)nging the genome. *EMBO J* 1998; 17:6427–6436.
14. Lahue RS, Au KG, Modrich P. DNA mismatch correction in a defined system. *Science* 1989; 245:160–164.
15. Su S-S, Modrich P. *Escherichia coli mutS*-encoded protein binds to mismatched DNA base pairs. *Proc Natl Acad Sci USA* 1986; 83:5057–5061.
16. Parker BO, Marinus MG. Repair of DNA heteroduplexes containing small heterologous sequences in *Escherichia coli., Proc Natl Acad Sci USA* 1992; 89:1730–1734.
17. Grilley M, Welsh KM, Su S-S, Modrich P. Isolation and characterization of the *Escherichia coli mutL* gene product. *J Biol Chem* 1989; 264:1000–1004.
18. Au KG, Welsh K, Modrich P. Initiation of methyl-directed mismatch repair. *J Biol Chem* 1992; 267:12,142–12,148.
19. Dao V, Modrich P. Mismatch-, MutS-, MutL-, and helicase II-dependent unwinding from the single-strand break of an incised heteroduplex. *J Biol Chem* 1998; 273:9202–9207.
20. Ban C, Yang W. Crystal structure and ATPase activity of MutL: implications for DNA repair and mutagenesis. *Cell* 1998; 95:541–552.
21. Welsh KM, Lu A-L, Clark S, Modrich P. Isolation and characterization of the *Escherichia coli mutH* gene product. *J Biol Chem* 1987; 262:15,624–15,629.
22. Allen DJ, Makhov A, Grilley M, Taylor J, Thresher R, Modrich P. Griffith JD. MutS mediates heteroduplex loop formation by a translocation mechanism. *EMBO J* 1997; 16:4467–4476.
23. Yamaguchi M, Dao V, Modrich P. MutS and MutL activate DNA helicase II in a mismatch-dependent manner. *J Biol Chem* 1998; 273:9197–9201.
24. Grilley M, Griffith J, Modrich P. Bidirectional excision in methyl-directed mismatch repair. *J Biol Chem* 1993; 268:11,830–11837.
25. Cooper DL, Lahue RS, Modrich P. Methyl-directed mismatch repair is bidirectional. *J Biol Chem* 1993; 268:11,823–11,829.
26. Holmes J, Clark S, Modrich P. Strand-specific mismatch correction in nuclear extracts of human and *Drosophila melanogaster* cell lines. *Proc Natl Acad Sci USA* 1990; 87:5837–5841.
27. Thomas DC, Roberts JD, Kunkel TA. Heteroduplex repair in extracts of human HeLa cells. *J Biol Chem* 1991; 266:3744–3751.
28. Parsons R, Li GM, Longley MJ, Fang WH, Papadopoulos N, Jen J, et al. Hypermutability and mismatch repair deficiency in RER$^+$ tumor cells. *Cell* 1993; 75:1227–1236.
29. Umar A, Boyer JC, Thomas DC, Nguyen DC, Risinger JI, Boyd J, et al. Defective mismatch repair in extracts of colorectal and endometrial cancer cell lines exhibiting microsatellite instability. *J Biol Chem* 1994; 269:14,367–14,370.
30. Drummond JT, Li GM, Longley MJ, Modrich P. Isolation of an hMSH2-p160 heterodimer that restores DNA mismatch repair to tumor cells. *Science* 1995; 268:1909–1912.
31. Fang WH, Modrich P. Human strand-specific mismatch repair occurs by a bidirectional mechanism similar to that of the bacterial reaction. *J Biol Chem* 1993; 268:11,838–11,844.
32. Umar A, Boyer JC, Kunkel TA. DNA loop repair by human cell extracts. *Science* 1994; 266:814–816.
33. Genschel J, Littman SJ, Drummond JT, Modrich P. Isolation of MutSβ from human cells and comparison of the mismatch repair specificities of MutSβ and MutSα. *J Biol Chem* 1998; 273:19,895–19,901.

34. Gu L, Hong Y, McCulloch S, Watanabe H, Li G-M. ATP-dependent interaction of human mismatch repair proteins and dual role of PCNA in mismatch repair. *Nucleic Acids Res* 1998; 26:1173–1178.

35. Longley MJ, Pierce AJ, Modrich P. DNA polymerase δ is required for human mismatch repair in vitro. *J Biol Chem* 1997; 272:10,917–10,921.

36. Tishkoff DX, Amin NS, Viars CS, Arden KC, Kolodner RD. Identification of a human gene encoding a homologue of Saccharomyces cerevisiae EXO1, an exonuclease implicated in mismatch repair and recombination. *Cancer Res* 1998; 58:5027–5031.

37. Tishkoff DX, Boerger AL, Bertrand P, Filosi N, Gaida GM, Kane MF, Kolodner RD. Identification and characterization of Saccharomyces cerevisiae EXO1, a gene encoding an exonuclease that interacts with MSH2. *Proc Natl Acad Sci USA* 1997; 94:7487–7492.

38. Umezu K, Sugawara N, Chen C, Haber JE, Kolodner RD. Genetic analysis of yeast RPA1 reveals its multiple functions in DNA metabolism. *Genetics* 1998; 148:989–1005.

39. Lin YL, Shivji MK, Chen C, Kolodner R, Wood RD, Dutta A. The evolutionarily conserved zinc finger motif in the largest subunit of human replication protein A is required for DNA replication and mismatch repair but not for nucleotide excision repair. *J Biol Chem* 1998; 273:1453–1461.

40. Fujii H, Shimada T. Isolation and characterization of cDNA clones derived from the divergently transcribed gene in the region upstream from the human dihydrofolate reductase gene. *J Biol Chem* 1989; 264:10,057–10,064.

41. Fishel R, Lescoe MK, Rao MR, Copeland NG, Jenkins NA, Garber J, Kane M, Kolodner R. The human mutator gene homolog *MSH2* and its association with hereditary nonpolyposis colon cancer. *Cell* 1993; 75:1027–1038.

42. Leach FS, Nicolaides NC, Papadopoulos N, Liu B, Jen J, Parsons R, et al. Mutations of a mutS homolog in hereditary nonpolyposis colorectal cancer. *Cell* 1993; 75:1215–1225.

43. Paquis-Flucklinger V, Santucci-Darmanin S, Paul R, Saunieres A, Turc-Carel C, Desnuelle C. Cloning and expression analysis of a meiosis-specific MutS homolog: the human MSH4 gene. *Genomics* 1997; 44:188–194.

44. Her C, Doggett NA. Cloning, structural characterization, and chromosomal localization of the human orthologue of saccharomyces cerevisiae MSH5 gene. *Genomics* 1998; 52:50–61.

45. Bocker T, Barusevicius A, Snowden T, Rasio D, Guerrette S, Robbins D, et al. hMSH5: a human MutS homologue that forms a novel heterodimer with hMSH4 and is expressed during spermatogenesis. *Cancer Res* 1999; 59:816–822.

46. Reenan RA, Kolodner RD. Characterization of insertion mutations in the *Saccharomyces cerevisiae MSH1* and *MSH2* genes: evidence for separate mitochondrial and nuclear functions. *Genetics* 1992; 132:975–985.

47. Reenan RA, Kolodner RD. Isolation and characterization of two *Saccharomyces cerevisiae* genes encoding homologs of the bacterial HexA and MutS mismatch repair proteins. *Genetics* 1992; 132:963–973.

48. Hughes MJ, Jiricny J. The purification of a human mismatch-binding protein and identification of its associated ATPase and helicase activities. *J Biol Chem* 1992; 267:23,876–23,882.

49. Palombo F, Gallinari P, Iaccarino I, Lettieri T, Hughes M, D'Arrigo A, et al. GTBP, a 160-kilodalton protein essential for mismatch-binding activity in human cells. *Science* 1995; 268:1912–1914.

50. Nicolaides NC, Palombo F, Kinzler KW, Vogelstein B, Jiricny J. Molecular cloning of the N-terminus of GTBP. *Genomics* 1996; 31:395–397.

51. Fishel R, Ewel A, Lee S, Lescoe MK, Griffith J. Binding of mismatched microsatellite DNA sequences by the human MSH2 protein. *Science* 1994; 266:1403–1405.

52. Fishel R, Ewel A, Lescoe MK. Purified human MSH2 protein binds to DNA containing mismatched nucleotides. *Cancer Res* 1994; 54:5539–5542.

53. Acharaya S, Wilson T, Gradia S, Kane MF, Guerrette S, Marsischky GT, Kolodner R, Fishel R. hMSH2 forms specific mispair-binding complexes with hMSH3 and hMSH6. *Proc Natl Acad Sci USA* 1996; 93:13,629–13,634.

54. Iaccarino I, Marra G, Palombo F, Jiricny J. hMSH2 and hMSH6 play distinct roles in mismatch binding and contribute differently to the ATPase activity of hMutSalpha. *EMBO J* 1998; 17:2677–2686.

55. Boyer JC, Umar A, Risinger JI, Lipford JR, Kane M, Yin S, et al. Microsatellite instability, mismatch repair deficiency, and genetic defects in human cancer cell lines. *Cancer Res* 1995; 55:6063–6070.

56. Bhattacharyya NP, Skandalis A, Ganesh A, Groden J, Meuth M. Mutator phenotypes in human colorectal carcinoma cell lines. *Proc Natl Acad Sci USA* 1994; 91:6319–6323.

57. Kat A, Thilly WG, Fang WH, Longley MJ, Li GM, Modrich P. An alkylation-tolerant, mutator human cell line is deficient in strand-specific mismatch repair. *Proc Natl Acad Sci USA* 1993; 90:6424–6428.

58. Papadopoulos N, Nicolaides NC, Liu B, Parsons RE, Palombo F, D'Arrigo A, et al. Mutations in *GTBP* in genetically unstable cells. *Science* 1995; 268:1914–1917.

59. Marsischky GT, Filosi N, Kane MF, Kolodner R. Redundancy of Saccharomyces cerevisiae MSH3 and MSH6 in MSH2-dependent mismatch repair. *Genes Dev* 1996; 10:407–420.

60. Johnson RE, Kovvali GK, Prakash L, Prakash S. Requirement of the yeast MSH3 and MSH6 genes for MSH2-dependent genomic stability. *J Biol Chem* 1996; 271:7285–7288.

61. Palombo F, Iaccarino I, Nakajima E, Ikejima M, Shimada T, Jiricny J. hMutSbeta, a heterodimer of hMSH2 and hMSH3, binds to insertion/deletion loops in DNA. *Curr Biol* 1996; 6:1181–1184.

62. Habraken Y, Sung P, Prakash L, Prakash S. Binding of insertion/deletion DNA mismatches by the heterodimer of yeast mismatch repair proteins MSH2 and MSH3. *Curr Biol* 1996; 6:1185–1187.

63. Alani E. The Saccharomyces cerevisiae Msh2 and Msh6 proteins form a complex that specifically binds to duplex oligonucleotides containing mismatched DNA base pairs. *Mol Cell Biol* 1996; 16:5604–5615.

64. Guerrette S, Wilson T, Gradia S, Fishel R. Interactions of human hMSH2 with hMSH3 and hMSH2 with hMSH6: examination of mutations found in hereditary nonpolyposis colorectal cancer. *Mol Cell Biol* 1998; 18:6616–6623.

65. Haber LT, Walker GC. Altering the conserved nucleotide binding motif in the *Salmonella typhimurium* MutS mismatch repair protein affects both its ATPase and mismatch binding activities. *EMBO J* 1991; 10:2707–2715.

66. Chi NW, Kolodner RD. Purification and characterization of MSH1, a yeast mitochondrial protein that binds to DNA mismatches. *J Biol Chem* 1994; 269:29,984–29,992.

67. Alani E, Sokolsky T, Studamire B, Miret JJ, Lahue RS. Genetic and biochemical analysis of Msh2p-Msh6p: role of ATP hydrolysis and Msh2p-Msh6p subunit interactions in mismatch base pair recognition. *Mol Cell Biol* 1997; 17:2436–2447.

68. Gradia S, Acharya S, Fishel R. The human mismatch recognition complex hMSH2-hMSH6 functions as a novel molecular switch. *Cell* 1997; 91:995–1005.

69. Habraken Y, Sung P, Prakash L, Prakash S. ATP-dependent assembly of a ternary complex consisting of a DNA mismatch and the yeast MSH2-MSH6 and MLH1-PMS1 protein complexes. *J Biol Chem* 1998; 273:9837–9841.

70. Fishel R. Mismatch repair, molecular switches, and signal transduction. *Genes Dev* 1998; 12:2096–2101.

71. Bronner CE, Baker S, Morrison PT, Warren G, Smith LG, Lescoe MK, et al. Mutation in the DNA mismatch repair gene homologue hMLH1 is associated with hereditary non-polyposis colon cancer. *Nature* 1994; 368:258–261.

72. Nicolaides NC, Papadopoulos N, Liu B, Wei Y-F, Carter KC, Ruben SM, et al. Mutations of two PMS homologues in hereditary nonpolposis colon cancer. *Nature* 1994; 371:75–80.

73. Papadopoulos N, Nicolaides NC, Wei Y-F, Ruben SM, Carter KC, Rosen CA, et al. Mutation of a *mutL* homolog in hereditary colon cancer. *Science* 1994; 263:1625–1629.

74. Kramer W, Kramer B, Williamson MS, Fogel S. Cloning and nucleotide sequence of DNA mismatch repair gene *PMS1* from *Saccharomyces cerevisiae*: homology of PMS1 to procaryotic MutL and HexB. *J Bacteriol* 1989; 171:5339–5346.

75. Prolla TA, Christie DM, Liskay RM. Dual requirement in yeast DNA mismatch repair for *MLH1* and *PMS1*, two homologs of the bacterial mutL gene. *Mol Cell Biol* 1994; 14:407–415.

76. Flores-Rozas H, Kolodner RD. The Saccharomyces cerevisiae MLH3 gene functions in MSH3-dependent suppression of frameshift mutations. *Proc Natl Acad Sci USA* 1998; 95:12,404–12,409.

77. Baker SM, Bronner CE, Zhang L, Plug AW, Robatzek M, Warren G, et al. Male mice defective in the DNA mismatch repair gene PMS2 exhibit abnormal chromosome synapsis in meiosis. *Cell* 1995; 82:309–319.

78. Baker SM, Plug AW, Prolla TA, Bronner CE, Harris AC, Yao X, et al. Involvement of mouse Mlh1 in DNA mismatch repair and meiotic crossing over. *Nat Genet* 1996; 13:336–342.

79. Prolla TA, Pang Q, Alani E, Kolodner RD, Liskay RM. MLH1, PMS1, and MSH2 interactions during the initiation of DNA mismatch repair in yeast. *Science* 1994; 265:1091–1093.

80. Pang Q, Prolla TA, Liskay RM. Functional domains of the Saccharomyces cerevisiae Mlh1p and Pms1p DNA mismatch repair proteins and their relevance to human hereditary nonpolyposis colorectal cancer-associated mutations. *Mol Cell Biol* 1997; 17:4465–4473.

81. Li GM, Modrich P. Restoration of mismatch repair to nuclear extracts of H6 colorectal tumor cells by a heterodimer of human MutL homologs. *Proc Natl Acad Sci USA* 1995; 92:1950–1954.

82. Risinger JI, Umar A, Barrett JC, Kunkel TA. A hPMS2 mutant cell line is defective in strand-specific mismatch repair. *J Biol Chem* 1995; 270:18,183–18,186.

83. Drummond JT, Anthoney A, Brown R, Modrich P. Cisplatin and Adriamycin resistance are associated with MutLα and mismatch repair deficiency in an ovarian tumor cell line. *J Biol Chem* 1996; 271:19,645–19,648.

84. Nicolaides NC, Littman SJ, Modrich P, Kinzler KW, Vogelstein B. A naturally occurring hPMS2 mutation can confer a dominant negative mutator phenotype. *Mol Cell Biol* 1998; 18:1635–1641.

84a. Raschle M, Marra G, Nystrom-Lahti M, et al. Identification of hMutLbeta, a Heterodimer of hMLH1 and hPMS1. *J Biol Chem* 1999; 274:32368–32375.

85. Sancar A, Hearst JE. Molecular matchmakers. *Science* 1993; 259:1415–1420.

86. Habraken Y, Sung P, Prakash L, Prakash S. Enhancement of MSH2–MSH3-mediated mismatch recognition by the yeast MLH1-PMS1 complex. *Curr Biol* 1997; 7:790–793.

87. Inokuchi K, Ikejima M, Watanabe A, Nakajima E, Orimo H, Nomura T, Shimada T. Loss of expression of the human MSH3 gene in hematological malignancies. *Biochem Biophys Res Commun* 1995; 214:171–179.

88. Strand M, Earley MC, Crouse GF, Petes TD. Mutations in the MSH3 gene preferentially lead to deletions within tracts of simple repetitive DNA in Saccharomyces cerevisiae. *Proc Natl Acad Sci USA* 1995; 92:10,418–10,421.

89. Drummond JT, Genschel J, Wolf E, Modrich P. DHFR/MSH3 amplification in methotrexate-resistant cells alters the hMutSalpha/hMutSbeta ratio and reduces the efficiency of base-base mismatch repair. *Proc Natl Acad Sci USA* 1997; 94:10,144–10,149.

90. Marra G, Iaccarino I, Lettieri T, Roscilli G, Delmastro P, Jiricny J. Mismatch repair deficiency associated with overexpression of the MSH3 gene. *Proc Natl Acad Sci USA* 1998; 95:8568–8573.

91. Prolla TA, Baker SM, Harris AC, Tsao JL, Yao X, Bronner CE, et al. Tumour susceptibility and spontaneous mutation in mice deficient in Mlh1, Pms1 and Pms2 DNA mismatch repair. *Nat Genet* 1998; 18:276–279.

92. Lynch HT, Smyrk TC, Watson P, Lanspa SJ, Lynch JF, Lynch PM, Cavalieri RJ, Boland CR. Genetics, natural history, tumor spectrum, pathology of hereditary nonpolyposis colorectal cancer: an updated review. *Gastroenterology* 1993; 104:1535–1549.

93. Lynch HT, Schuelke GS, Kimberling WJ, Albano WA, Lynch JF, Biscone KA, et al. Hereditary nonpolyposis colorectal cancer (Lynch syndromes I and II). II. Biomarker studies. *Cancer* 1985; 56:939–951.

94. Dunlop MG. Screening for large bowel neoplasms in individuals with a family history of colorectal cancer. *Br J Surg* 1992; 79:488–494.

95. Lynch HT, Lanspa S, Smyrk T, Boman B, Watson P, Lynch J. Hereditary nonpolyposis colorectal cancer (Lynch syndromes I & II). Genetics, pathology, natural history, and cancer control, Part I. *Cancer Genet Cytogenet* 1991; 53:143–160.

96. Watson P, Lynch HT. Extracolonic cancer in hereditary nonpolyposis colorectal cancer. *Cancer* 1993; 71:677–685.

97. Warthin AS. Heredity with reference to carcinoma. *Arch Intern Med* 1913; 12:546–555.

98. Aaltonen LA, Peltomäki P, Leach FS, Sistonen P, Pylkkänen L, Mecklin J-P, et al. Clues to the pathogenesis of familial colorectal cancer. *Science* 1993; 260:812–816.

99. Peltomäki P, Aaltonen LA, Sistonen P, Pylkkänen L, Mecklin J-P, Järvinen H, et al. Genetic mapping of a locus predisposing to human colorectal cancer. *Science* 1993; 260:810–812.

100. Ionov Y, Peinado MA, Malkhosyan S, Shibata D, Perucho M. Ubiquitous somatic mutations in simple repeated sequences reveal a new mechanism for colonic carcinogenesis. *Nature* 1993; 363:558–561.

101. Thibodeau SN, Bren G, Schaid D. Microsatellite instability in cancer of the proximal colon. *Science* 1993; 260:816–819.

102. Levinson G, Gutman GA. High frequencies of short frameshifts in poly-CA/TG tandem repeats borne by bacteriophage M13 in *Escherichia coli* K-12. *Nucleic Acids Res* 1987; 15:5323–5338.

103. Levinson G, Gutman GA. Slipped-strand mispairing: a major mechanism for DNA sequence evolution. *Mol Biol Evol* 1987; 4:203–221.

104. Strand M, Prolla TA, Liskay RM, Petes TD. Destabilization of tracts of simple repetitive DNA in yeast by mutations affecting DNA mismatch repair. *Nature* 1993; 365:274–276.

105. Leach FS, Nicolaides NC, Papadopoulos N, Liu B, Jen J, Parsons R, et al. Mutations of a mutS homolog in hereditary nonpolyposis colorectal cancer. *Cell* 1993; 75:1215–1225.

106. Lindblom A, Tannergard P, Werelius B, Nordenskjold M. Genetic mapping of a second locus predisposing to hereditary non-polyposis colon cancer. *Nat Genet* 1993; 5:279–282.

107. Liu B, Parsons RE, Hamilton SR, Peterson GM, Lynch HT, Watson P, et al. hMSH2 mutations in hereditary nonpolyposis colorectal cancer kindreds. *Cancer Res* 1994; 54:4590–4594.

108. Lynch HT, Drouhard T, Lanspa S, Smyrk T, Lynch P, Lynch J, et al. Mutation of an mutL homologue in a Navajo family with hereditary nonpolyposis colorectal cancer. *J Natl Cancer Inst* 1994; 86:1417–1419.

109. Mary JL, Bishop T, Kolodner R, Lipford JR, Kane M, Weber W, et al. Mutational analysis of the hMSH2 gene reveals a three base pair deletion in a family predisposed to colorectal cancer development. *Hum Mol Genet* 1994; 3:2067–2069.

110. Hemminki A, Peltomaki P, Mecklin JP, Jarvinen H, Salovaara R, Nystrom LM, de la, Chapelle A, Aaltonen LA. Loss of the wild type MLH1 gene is a feature of hereditary nonpolyposis colorectal cancer. *Nat Genet* 1994; 8:405–410.

111. Wijnen J, Vasen H, Khan PM, Menko FH, van der, Klift H, van Leeuven C, et al. Seven new mutations in hMSH2, an HNPCC gene, identified by denaturing gradient-gel electrophoresis. *Am J Hum Genet* 1995; 56:1060–1066.

112. Liu B, Parsons R, Papadopoulos N, Nicolaides NC, Lynch HT, Watson P, et al. Analysis of mismatch repair genes in hereditary non-polyposis colorectal cancer patients. *Nat Med* 1996; 2:169–174.

113. Han HJ, Maruyama M, Baba S, Park JG, Nakamura Y. Genomic structure of human mismatch repair gene, hMLH1, and its mutation analysis in patients with hereditary non-polyposis colorectal cancer (HNPCC). *Hum Mol Genet* 1995; 4:237–242.

114. Kolodner RD, Hall NR, Lipford J, Kane MF, Morrison PT, Finan PJ, et al. Structure of the human MLH1 locus and analysis of a large hereditary nonpolyposis colorectal carcinoma kindred for *mlh1* mutations. *Cancer Res* 1995; 55:242–248.

115. Luce MC, Marra G, Chauhan DP, Laghi L, Carethers JM, Cherian SP, et al. In vitro transcription/translation assay for the screening of hMLH1 and hMSH2 mutations in familial colon cancer. *Gastroenterology* 1995; 109:1368–1374.

116. Tannergard P, Lipford JR, Kolodner R, Frodin JE, Nordenskjold M, Lindblom A. Mutation screening in the hMLH1 gene in Swedish hereditary nonpolyposis colon cancer families. *Cancer Res* 1995; 55:6092–6096.

117. Maliaka YK, Chudina AP, Belev NF, Alday P, Bochkov NP, Buerstedde JM. CpG dinucleotides in the hMSH2 and hMLH1 genes are hotspots for HNPCC mutations. *Hum Genet* 1996; 97:251–255.

118. Wijnen J, Khan PM, Vasen H, Menko F, van der Klift H, van den Broek M, et al. Majority of hMLH1 mutations responsible for hereditary nonpolyposis colorectal cancer cluster at the exonic region 15–16. *Am J Hum Genet* 1996; 58:300–307.

119. Liu T, Wahlberg S, Rubio C, Holmberg E, Gronberg H, Lindblom A. DGGE screening of mutations in mismatch repair genes (hMSH2 and hMLH1) in 34 Swedish families with colorectal cancer. *Clin Genet* 1998; 53:131–135.

120. Miyaki M, Konishi M, Tanaka K, Kikuchi YR, Muraoka M, Yasuno M, et al. Germline mutation of MSH6 as the cause of hereditary nonpolyposis colorectal cancer. *Nat Genet* 1997; 17:271–272.

121. Benachenhou N, Guiral S, Gorska-Flipot I, Michalski R, Labuda D, Sinnett D. Allelic losses and DNA methylation at DNA mismatch repair loci in sporadic colorectal cancer. *Carcinogenesis* 1998; 19:1925–1929.

122. Homfray TF, Cottrell SE, Ilyas M, Rowan A, Talbot IC, Bodmer WF, Tomlinson IP. Defects in mismatch repair occur after APC mutations in the pathogenesis of sporadic colorectal tumours. *Hum Mutat* 1998; 11:114–120.

123. Kane MF, Loda M, Gaida GM, Lipman J, Mishra R, Goldman H, Jessup JM, Kolodner R. Methylation of the hMLH1 promoter correlates with lack of expression of hMLH1 in sporadic colon tumors and mismatch repair-defective human tumor cell lines. *Cancer Res* 1997; 57:808–811.

124. Liu B, Nicolaides NC, Markowitz S, Willson JKV, Parsons RE, Jen J, et al. Mismatch repair gene defects in sporadic colorectal cancers with microsatellite instability. *Nat Genet* 1995; 9:48–55.

125. Borresen AL, Lothe RA, Meling GI, Lystad S, Morrison P, Lipford J, et al. Somatic mutations in the hMSH2 gene in microsatellite unstable colorectal carcinomas. *Hum Mol Genet* 1995; 4:2065–2072.

126. Benachenhou N, Guiral S, Gorska-Flipot I, Labuda D, Sinnett D. High resolution deletion mapping reveals frequent allelic losses at the DNA mismatch repair loci hMLH1 and hMSH3 in non-small cell lung cancer. *Int J Cancer* 1998; 77:173–180.

127. Swisher EM, Mutch DG, Herzog TJ, Rader JS, Kowalski LD, Elbendary A, Goodfellow PJ. Analysis of MSH3 in endometrial cancers with defective DNA mismatch repair. *J Soc Gynecol Invest* 1998; 5:210–216.

128. Risinger J, Umar A, Boyd J, Berchuck A, Kunkel TA, Barrett JC. Mutation of MSH3 in endometrial cancer and evidence for its functional role in heteroduplex repair. *Nat Genet* 1996; 14:102–105.
129. Yamamoto H, Sawai H, Weber TK, Rodriguez BM, Perucho M. Somatic frameshift mutations in DNA mismatch repair and proapoptosis genes in hereditary nonpolyposis colorectal cancer. *Cancer Res* 1998; 58:997–1003.
130. Yamamoto H, Sawai H, Perucho M. Frameshift somatic mutations in gastrointestinal cancer of the microsatellite mutator phenotype. *Cancer Res* 1997; 57:4420–4426.
131. de Wind N, Dekker M, Berns A, Radman M, te Riele H. Inactivation of the mouse Msh2 gene results in mismatch repair deficiency, methylation tolerance, hyperrecombination, and predisposition to cancer. *Cell* 1995; 82:321–330.
132. Reitmair AH, Schmits R, Ewel A, Bapat B, Redston M, Mitri A, et al. MSH2 deficient mice are viable and susceptible to lymphoid tumours. *Nat Genet* 1995; 11:64–70.
133. Edelmann W, Cohen PE, Kane M, Lau K, Morrow B, Bennett S, et al. Meiotic pachytene arrest in MLH1-deficient mice. *Cell* 1996; 85:1125–1134.
134. Edelmann W, Yang K, Umar A, Heyer J, Lau K, Fan K, et al. Mutation in the mismatch repair gene Msh6 causes cancer susceptibility. *Cell* 1997; 91:467–477.
134a. de Wind N, Dekker M, Claij, N, et al. HNPCC-like cancer predisposition in mice through simultaneous loss of Msh3 and Msh6 mismatch-repair protein functions. *Nat Genet* 1999; 23:359–62.
134b. Edelmann W, Umar A, Yang K, et al. The DNA mismatch repair genes Msh3 and Msh6 cooperate in intestinal tumor suppression. *Cancer Res* 2000; 60:803–7.
135. Koi M, Umar A, Chauhan DP, Cherian SP, Carethers JM, Kunkel TA, Boland CR. Human chromosome 3 corrects mismatch repair deficiency and microsatellite instability and reduces N-methyl-N′-nitro-N-nitrosoguanidine tolerance in colon tumor cells with homozygous hMLH1 mutation. *Cancer Res* 1994; 54:4308–4312.
136. Umar A, Koi M, Risinger JI, Glaab WE, Tindall KR, Kolodner RD, et al. Correction of hypermutability, N-methyl-N′-nitro-N-nitrosoguanidine resistance, and defective DNA mismatch repair by introducing chromosome 2 into human tumor cells with mutations in MSH2 and MSH6. *Cancer Res* 1997; 57:3949–3955.
137. Tindall KR, Glaab WE, Umar A, Risinger JI, Koi M, Barrett JC, Kunkel TA. Complementation of mismatch repair gene defects by chromosome transfer. *Mutat Res* 1998; 402:15–22.
138. Risinger JI, Umar A, Glaab WE, Tindall KR, Kunkel TA, Barrett JC. Single gene complementation of the hPMS2 defect in HEC-1-A endometrial carcinoma cells. *Cancer Res* 1998; 58:2978–2981.
139. Buermeyer AB, Wilson-Van Patten C, Baker SM, Liskay RM. The human MLH1 cDNA complements DNA mismatch repair defects in Mlh1-deficient mouse embryonic fibroblasts. *Cancer Res* 1999; 59:538–541.
140. Umar A, Risinger JI, Glaab WE, Tindall KR, Barrett JC, Kunkel TA. Functional overlap in mismatch repair by human MSH3 and MSH6. *Genetics* 1998; 148:1637–1646.
141. Parsons R, Li GM, Longley M, Modrich P, Liu B, Berk T, et al. Mismatch repair deficiency in phenotypically normal human cells. *Science* 1995; 268:738–740.
142. Eshleman JR, Markowitz SD, Donover PS, Lang EZ, Lutterbaugh JD, Li G-M, et al. Diverse hypermutability of multiple expressed sequence motifs present in a cancer with microsatellite instability. *Oncogene* 1996; 12:1425–1432.
143. Veigl ML, Kasturi L, Olechnowicz J, Ma AH, Lutterbaugh JD, Periyasamy S, et al. Biallelic inactivation of hMLH1 by epigenetic gene silencing, a novel mechanism causing human MSI cancers. *Proc Natl Acad Sci USA* 1998; 95:8698–8702.
144. Eshleman JR, Donover PS, Littman SJ, Swinler SE, Li GM, Lutterbaugh JD, et al. Increased transversions in a novel mutator colon cancer cell line. *Oncogene* 1998; 16:1125–1130.
145. Thibodeau SN, French AJ, Roche PC, Cunningham JM, Tester DJ, Lindor NM, et al. Altered expression of hMSH2 and hMLH1 in tumors with microsatellite instability and genetic alterations in mismatch repair genes. *Cancer Res* 1996; 56:4836–4840.
146. Larsen CJ. Methylation is a frequent mechanism of inactivation of the hMLH1 gene copies in sporadic colonic cancer. *Bull Cancer* 1998; 85:741–742.
147. Herman JG, Umar A, Polyak K, Graff JR, Ahuja N, Issa JP, et al. Incidence and functional consequences of hMLH1 promoter hypermethylation in colorectal carcinoma. *Proc Natl Acad Sci USA* 1998; 95:6870–6875.
148. Cunningham JM, Christensen ER, Tester DJ, Kim CY, Roche PC, Burgart LJ, Thibodeau SN. Hypermethylation of the hMLH1 promoter in colon cancer with microsatellite instability. *Cancer Res* 1998; 58:3455–3460.

149. Leung SY, Yuen ST, Chung LP, Chu KM, Chan AS, Ho JC. hMLH1 promoter methylation and lack of hMLH1 expression in sporadic gastric carcinomas with high-frequency microsatellite instability. *Cancer Res* 1999; 59:159–164.

150. Boland CR, Thibodeau SN, Hamilton SR, Sidransky D, Eshleman JR, Burt RW, et al. National Cancer Institute Workshop on Microsatellite Instability for cancer detection and familial predisposition: development of international criteria for the determination of microsatellite instability in colorectal cancer. *Cancer Res* 1998; 58:5248–5257.

151. Thibodeau SN, French AJ, Cunningham JM, Tester D, Burgart LJ, Roche PC. et al. Microsatellite instability in colorectal cancer: different mutator phenotypes and the principal involvement of hMLH1. *Cancer Res* 1998; 58:1713–1718.

152. Eshleman JR, Markowitz SD. Microsatellite instability in inherited and sporadic neoplasms. *Curr Opin Oncol* 1995; 7:83–89.

153. Arzimanoglou II, Gilbert F, Barber HR. Microsatellite instability in human solid tumors. *Cancer* 1998; 82:1808–1820.

154. Burks RT, Kessis TD, Cho KR, Hedrick L. Microsatellite instability in endometrial carcinoma. *Oncogene* 1994; 9:1163–1166.

155. Duggan BD, Felix JC, Muderspach LI, Tourgeman D, Zheng J, Shibata D. Microsatellite instability in sporadic endometrial carcinoma. *J Natl Cancer Inst* 1994; 86:1216–1221.

156. Katabuchi H, van Rees B, Lambers AR, Ronnett BM, Blazes MS, Leach FS, Cho KR, Hedrick L. Mutations in DNA mismatch repair genes are not responsible for microsatellite instability in most sporadic endometrial carcinomas. *Cancer Res* 1995; 55:5556–5560.

157. Kobayashi K, Sagae S, Kudo R, Saito H, Koi S, Nakamura Y. Microsatellite instability in endometrial carcinomas: frequent replication errors in tumors of early onset and/or of poorly differentiated type. *Genes Chromosomes Cancer* 1995; 14:128–132.

158. Risinger JI, Berchuck A, Kohler MF, Watson P, Lynch HT, Boyd J. Genetic instability of microsatellites in endometrial carcinoma. *Cancer Res* 1993; 53:5100–5103.

159. Caduff RF, Johnston CM, Svoboda-Newman SM, Poy EL, Merajver SD, Frank TS. Clinical and pathological significance of microsatellite instability in sporadic endometrial carcinoma. *Am J Pathol* 1996; 148:1671–1678.

160. Catasus L, Matias-Guiu X, Machin P, Munoz J, Prat J. BAX somatic frameshift mutations in endometrioid adenocarcinomas of the endometrium: evidence for a tumor progression role in endometrial carcinomas with microsatellite instability. *Lab Invest* 1998; 78:1439–1444.

161. Catasus L, Machin P, Matias-Guiu X, Prat J. Microsatellite instability in endometrial carcinomas: clinicopathologic correlations in a series of 42 cases. *Hum Pathol* 1998; 29:1160–1164.

162. Gurin CC, Federici MG, Kang L, Boyd J. Causes and consequences of microsatellite instability in endometrial carcinoma. *Cancer Res* 1999; 59:462–466.

163. Helland A, Borresen-Dale AL, Peltomaki P, Hektoen M, Kristensen GB, Nesland JM, de la Chapelle A, Lothe RA. Microsatellite instability in cervical and endometrial carcinomas. *Int J Cancer* 1997; 70:499–501.

164. Jovanovic AS, Boynton KA, Mutter GL. Uteri of women with endometrial carcinoma contain a histopathological spectrum of monoclonal putative precancers, some with microsatellite instability. *Cancer Res* 1996; 56:1917–1921.

165. Sakamoto T, Murase T, Urushibata H, Kato K, Takada H, Imamura T, Mori H, Wake N. Microsatellite instability and somatic mutations in endometrial carcinomas. *Gynecol Oncol* 1998; 71:53–58.

166. Sirchia SM, Pariani S, Rossella F, Garagiola I, De Andreis C, Bulfamante G, et al. Cytogenetic abnormalities and microsatellite instability in endometrial adenocarcinoma. *Cancer Genet Cytogenet* 1997; 94:113–119.

167. Orth K, Hung J, Gazdar A, Mathis M, Bowcock A, Sambrook J. Ovarian tumors display persistent microsatellite instability caused by mutation in the mismatch repair gene hMSH-2. *Cold Spring Harbor Symp Quant Biol* 1994; 59:349–356.

168. Arzimanoglou II, Lallas T, Osborne M, Barber H, Gilbert F. Microsatellite instability differences between familial and sporadic ovarian cancers. *Carcinogenesis* 1996; 17:1799–1804.

169. Fujita M, Enomoto T, Yoshino K, Nomura T, Buzard GS, Inoue M, Okudaira Y. Microsatellite instability and alterations in the hMSH2 gene in human ovarian cancer. *Int J Cancer* 1995; 64:361–366.

170. King BL, Carcangiu ML, Carter D, Kiechle M, Pfisterer J, Pfleiderer A, Kacinski BM. Microsatellite instability in ovarian neoplasms. *Br J Cancer* 1995; 72:376–382.

171. Park TW, Felix JC, Wright TC, Jr. X chromosome inactivation and microsatellite instability in early and advanced bilateral ovarian carcinomas. *Cancer Res* 1995; 55:4793–4796.

172. Shih YC, Kerr J, Hurst TG, Khoo SK, Ward BG, Chenevix-Trench G. No evidence for microsatellite instability from allelotype analysis of benign and low malignant potential ovarian neoplasms. *Gynecol Oncol* 1998; 69:210–213.
173. Tangir J, Loughridge NS, Berkowitz RS, Muto MG, Bell DA, Welch WR, Mok SC. Frequent microsatellite instability in epithelial borderline ovarian tumors. *Cancer Res* 1996; 56:2501–2505.
174. Chong J-M, Fukayama M, Hayashi Y, Takizawa T, Koike M, Konishsi M, Kikuchi-Yanoshita R, Miyaki M. Microsatellite instability in the progression of gastric carcinoma. *Cancer Res* 1994; 54:4595–4597.
175. Rhyu MG, Park WS, Meltzer SJ. Microsatellite instability occurs frequently in human gastric carcinoma. *Oncogene* 1994; 9:29–32.
176. Zelada HM, Iselius L, Gunven P, Weger A, Nordenskjold M, Skoog L, Lindblom A. Genetic rearrangements in sporadic and familial gastric carcinomas detected with microsatellite markers. *Eur J Surg Oncol* 1994; 20:667–673.
177. Buonsanti G, Calistri D, Padovan L, Luinetti O, Fiocca R, Solcia E, Ranzani GN. Microsatellite instability in intestinal- and diffuse-type gastric carcinoma. *J Pathol* 1997; 182:167–173.
178. Chong JM, Fukayama M, Hayashi Y, Takizawa T, Koike M, Konishi M, Kikuchi-Yanoshita R, Miyaki M. Microsatellite instability in the progression of gastric carcinoma. *Cancer Res* 1994; 54:4595–4597.
179. Chung YJ, Song JM, Lee JY, Jung YT, Seo EJ, Choi SW, Rhyu MG. Microsatellite instability-associated mutations associate preferentially with the interstinal type of primary gastric carcinomas in a high-risk population. *Cancer Res* 1996; 56:4662–4665.
180. dos Santos NR, Seruca R, Constancia M, Seixas M, Sobrinho-Simoes M. Microsatellite instability at multiple loci in gastric carcinoma: clinicopathologic implications and prognosis. *Gastroenterology* 1996; 110:38–44.
181. Hayden JD, Cawkwell L, Quirke P, Dixon MF, Goldstone AR, Sue-Ling H, Johnston D, Martin IG. Prognostic significance of microsatellite instability in patients with gastric carcinoma. *Eur J Cancer* 1997; 33:2342–2346.
182. Jee MS, Koo C, Kim MH, Choi C, Lee KM, Choi SK, Rew JS, Yoon CM. Microsatellite instability in Korean patients with gastric adenocarcinoma. *Korean J Intern Med* 1997; 12:144–154.
183. Keller G, Grimm V, Vogelsang H, Bischoff P, Mueller J, Siewert JR, Hofler H. Analysis for microsatellite instability and mutations of the DNA mismatch repair gene hMLH1 in familial gastric cancer. *Int J Cancer* 1996; 68:571–576.
184. Lin JT, Wu MS, Shun CT, Lee WJ, Wang JT, Wang TH, Sheu JC. Microsatellite instability in gastric carcinoma with special references to histopathology and cancer stages. *Eur J Cancer* 1995; 31A:1879–1882.
185. Lin JT, Wu MS, Shun CT, Lee WJ, Sheu JC, Wang TH. Occurrence of microsatellite instability in gastric carcinoma is associated with enhanced expression of erbB-2 oncoprotein. *Cancer Res* 1995; 55:1428–1430.
186. Nakashima H, Honda M, Inoue H, Shibuta K, Arinaga S, Era S, et al. Microsatellite instability in multiple gastric cancers. *Int J Cancer* 1995; 64:239–242.
187. Nakashima H, Inoue H, Mori M, Ueo H, Ikeda M, Akiyoshi T. Microsatellite instability in Japanese gastric cancer. *Cancer* 1995; 75:1503–1507.
188. Ottini L, Palli D, Falchetti M, D'Amico C, Amorosi A, Saieva C, et al. Microsatellite instability in gastric cancer is associated with tumor location and family history in a high-risk population from Tuscany. *Cancer Res* 1997; 57:4523–4529.
189. Sasaki A, Nagashima M, Shiseki M, Katai H, Maruyama K, Iwanaga R, et al. Microsatellite instability in gastric cancer prone families. *Cancer Lett* 1996; 99:169–175.
190. Seruca R, Santos NR, David L, Constancia M, Barroca H, Carneiro F, et al. Sporadic gastric carcinomas with microsatellite instability display a particular clinicopathologic profile. *Int J Cancer* 1995; 64:32–36.
191. Hampton GM, Penny LA, Baergen RN, Larson A, Brewer C, Liao S, et al. Loss of heterozygosity in cervical carcinoma: subchromosomal localization of a putative tumor-suppressor gene to chromosome 11q22–q24. *Proc Natl Acad Sci USA* 1994; 91:6953–6957.
192. Chu TY, Shen CY, Lee HS, Liu HS. Monoclonality and surface lesion-specific microsatellite alterations in premalignant and malignant neoplasia of uterine cervix: a local field effect of genomic instability and clonal evolution. *Genes Chromosomes Cancer* 1999; 24:127–134.
193. Jimenez P, Canton J, Concha A, Torres LM, Abril E, Real LM, et al. Microsatellite instability in cervical intraepithelial neoplasia. *J Exp Clin Cancer Res* 1998; 17:361–366.

194. Kisseljov F, Semionova L, Samoylova E, Mazurenko N, Komissarova E, Zourbitskaya V, et al. Instability of chromosome 6 microsatellite repeats in human cervical tumors carrying papillomavirus sequences. *Int J Cancer* 1996; 69:484–487.

195. Rodriguez JA, Barros F, Carracedo A, Herckenrode CM. Low incidence of microsatellite instability in patients with cervical carcinomas. *Diagn Mol Pathol* 1998; 7:276–282.

196. Yee CJ, Roodi N, Verrier CS, Parl FF. Microsatellite instability and loss of heterozygosity in breast cancer. *Cancer Res* 1994; 54:1641–1644.

197. Contegiacomo A, Palmirotta R, De Marchis L, Pizzi C, Mastranzo P, Delrio P, Microsatellite instability and pathological aspects of breast cancer. *Int J Cancer* 1995; 64:264–268.

198. Aldaz CM, Chen T, Sahin A, Cunningham J, Bondy M. Comparative allelotype of in situ and invasive human breast cancer: high frequency of microsatellite instability in lobular breast carcinomas. *Cancer Res* 1995; 55:3976–3981.

199. Shaw JA, Walsh T, Chappell SA, Carey N, Johnson K, Walker RA. Microsatellite instability in early sporadic breast cancer. *Br J Cancer* 1996; 73:1393–1397.

200. Patel U, Grundfest-Broniatowski S, Gupta M, Banerjee S. Microsatellite instabilities at five chromosomes in primary breast tumors. *Oncogene* 1994; 9:3695–3700.

201. Bergthorsson JT, Egilsson V, Gudmundsson J, Arason A, Ingvarsson S. Identification of a breast tumor with microsatellite instability in a potential carrier of the hereditary non-polyposis colon cancer trait. *Clin Genet* 1995; 47:305–310.

202. Contegiacomo A, Palmirotta R, De Marchis L, Pizzi C, Mastranzo P, Delrio P, et al. Microsatellite instability and pathological aspects of breast cancer. *Int J Cancer* 1995; 64:264–268.

203. Rush EB, Calvano JE, van Zee KJ, Zelenetz AD, Borgen PI. Microsatellite instability in breast cancer. *Ann Surg Oncol* 1997; 4:310–315.

204. Toyama T, Iwase H, Yamashita H, Iwata H, Yamashita T, Ito K, et al. Microsatellite instability in sporadic human breast cancers. *Int J Cancer* 1996; 68:447–451.

205. Walsh T, Chappell SA, Shaw JA, Walker RA. Microsatellite instability in ductal carcinoma in situ of the breast. *J Pathol* 1998; 185:18–24.

206. Wu Y, Barnabas N, Russo IH, Yang X, Russo J. Microsatellite instability and loss of heterozygosity in chromosomes 9 and 16 in human breast epithelial cells transformed by chemical carcinogens. *Carcinogenesis* 1997; 18:1069–1074.

207. Toyama T, Iwase H, Iwata H, Hara Y, Omoto Y, Suchi M, et al. Microsatellite instability in in situ and invasive sporadic breast cancers of Japanese women. *Cancer Lett* 1996; 108:205–209.

208. Sourvinos G, Kiaris H, Tsikkinis A, Vassilaros S, Spandidos DA. Microsatellite instability and loss of heterozygosity in primary breast tumours. *Tumour Biol* 1997; 18:157–166.

209. Field JK, Kiaris H, Howard P, Vaughan ED, Spandidos DA, Jones AS. Microsatellite instability in squamous cell carcinoma of the head and neck. *Br J Cancer* 1995; 71:1065–1069.

210. Muzeau F, Flejou JF, Belghiti J, Thomas G, Hamelin R. Infrequent microsatellite instability in oesophageal cancers. *Br J Cancer* 1997; 75:1336–1339.

211. Nakashima H, Mori M, Mimori K, Inoue H, Shibuta K, Baba K, Mafune K, Akiyoshi T. Microsatellite instability in Japanese esophageal carcinoma. *Int J Cancer* 1995; 64:286–289.

212. Piccinin S, Gasparotto D, Vukosavljevic T, Barzan L, Sulfaro S, Maestro R, Boiocchi M. Microsatellite instability in squamous cell carcinomas of the head and neck related to field cancerization phenomena. *Br J Cancer* 1998; 78:1147–1151.

213. Quinn AG, Healy E, Rehman I, Sikkink S, Rees JL. Microsatellite instability in human non-melanoma and melanoma skin cancer. *J Invest Dermatol* 1995; 104:309–312.

214. Zaphiropoulos PG, Soderkvist P, Hedblad MA, Toftgard R. Genetic instability of microsatellite markers in region q22.3–q31 of chromosome 9 in skin squamous cell carcinomas. *Biochem Biophys Res Commun* 1994; 201:1495–1501.

215. El-Naggar AK, Hurr K, Huff V, Clayman GL, Luna MA, Batsakis JG. Microsatellite instability in preinvasive and invasive head and neck squamous carcinoma. *Am J Pathol* 1996; 148:2067–2072.

216. Walker GJ, Palmer JM, Walters MK, Nancarrow DJ, Hayward NK. Microsatellite instability in melanoma. *Melanoma Res* 1994; 4:267–268.

217. Tomlinson IP, Beck NE, Bodmer WF. Allele loss on chromosome 11q and microsatellite instability in malignant melanoma. *Eur J Cancer* 1996; 32A:1797–1802.

218. Talwalkar VR, Scheiner M, Hedges LK, Butler MG, Schwartz HS. Microsatellite instability in malignant melanoma. *Cancer Genet Cytogenet* 1998; 104:111–114.

219. Richetta A, Silipo V, Calvieri S, Frati L, Ottini L, Cama A, Mariani-Costantini R. Microsatellite instability in primary and metastatic melanoma. *J Invest Dermatol* 1997; 109:119–120.
220. Peris K, Keller G, Chimenti S, Amantea A, Kerl H, Hofler H. Microsatellite instability and loss of heterozygosity in melanoma. *J Invest Dermatol* 1995; 105:625–628.
221. Merlo A, Mabry M, Gabrielson E, Vollmer R, Baylin SB, Sidransky D. Frequent microsatellite instability in primary small cell lung cancer. *Cancer Res* 1994; 54:2098–2101.
222. Shridhar V, Siegfried J, Hunt J, del Mar Alonso M, Smith DI. Genetic instability of microsatellite sequences in many non-small cell lung carcinomas. *Cancer Res* 1994; 54:2084–2087.
223. Adachi J, Shiseki M, Okazaki T, Ishimaru G, Noguchi M, Hirohashi S, Yokota J. Microsatellite instability in primary and metastatic lung carcinomas. *Genes Chromosomes Cancer* 1995; 14:301–306.
224. Fong KM, Zimmerman PV, Smith PJ. Microsatellite instebility and other molecular abnormalities in non-small cell lung cancer. *Cancer Res* 1995; 55:28–30.
225. Kim CH, Yoo CG, Han SK, Shim YS, Kim YW. Genetic instability of microsatellite sequences in non-small cell lung cancers. *Lung Cancer* 1998; 21:21–25.
226. Ryberg D, Lindstedt BA, Zienolddiny S, Haugen A. A hereditary genetic marker closely associated with microsatellite instability in lung cancer. *Cancer Res* 1995; 55:3996–3999.
227. Leung SY, Chan TL, Chung LP, Chan AS, Fan YW, Hung KN, et al. Microsatellite instability and mutation of DNA mismatch repair genes in gliomas. *Am J Pathol* 1998; 153:1181–1188.
228. Izumoto S, Arita N, Ohinisi T, Hiraga S, Taki T, Tomita N, Ohue M, Hayakawa T. Microsatellite instability and mutated type II transforming growth factor-beta receptor gene in gliomas. *Cancer Lett* 1997; 112:251–256.
229. Zhu JJ, Santarius T, Wu X, Tsong J, Guha A, Wu JK, Hudson TJ, Black PM. Screening for loss of heterozygosity and microsatellite instability in oligodendrogliomas. *Genes Chromosomes Cancer* 1998; 21:207–216.
230. Schoenberg MP, Hakimi JM, Wang S, Bova GS, Epstein JI, Fischbeck KH, et al. Microsatellite mutation ($CAG_{24\rightarrow18}$) in the androgen receptor gene in human prostate cancer. Biochem. *Biophys Res Comm* 1994; 198:74–80.
231. Suzuki H, Komiya A, Aida S, Akimoto S, Shiraishi T, Yatani R, Igarashi T, Shimazaki J. Microsatellite instability and other molecular abnormalities in human prostate cancer. *Jpn J Cancer Res* 1995; 86:956–961.
232. Uchida T, Wada C, Wang C, Ishida H, Egawa S, Yokoyama E, Ohtani H, Koshiba K. Microsatellite instability in prostate cancer. *Oncogene* 1995; 10:1019–1022.
233. Egawa S, Uchida T, Suyama K, Wang C, Ohori M, Irie S, Iwamura M, Koshiba K. Genomic instability of microsatellite repeats in prostate cancer: relationship to clinicopathological variables. *Cancer Res* 1995; 55:2418–2421.
234. Lacombe L, Orlow I, Reuter VE, Fair WR, Dalbagni G, Zhang ZF, Cordon-Cardo C. Microsatellite instability and deletion analysis of chromosome 10 in human prostate cancer. *Int J Cancer* 1996; 69:110–113.
235. Terrell RB, Wille AH, Cheville JC, Nystuen AM, Cohen MB, Sheffield VC. Microsatellite instability in adenocarcinoma of the prostate. *Am J Pathol* 1995; 147:799–805.
236. Watanabe M, Imai H, Shiraishi T, Shimazaki J, Kotake T, Yatani R. Microsatellite instability in human prostate cancer. *Br J Cancer* 1995; 72:562–564.
237. Watanabe M, Imai H, Kato H, Shiraishi T, Ushijima T, Nagao M, Yatani R. Microsatellite instability in latent prostate cancers. *Int J Cancer* 1996; 69:394–397.
238. Watanabe M, Shiraishi T, Muneyuki T, Nagai M, Fukutome K, Murata M, Kawamura J, Yatani R. Allelic loss and microsatellite instability in prostate cancers in Japan. *Oncology* 1998; 55:569–574.
239. Gonzalez-Zulueta M, Ruppert JM, Tokino K, Tsai YC, Spruck CH, Nichols PW, et al. Microsatellite instability in bladder cancer. *Cancer Res* 1993; 53:5620–5623.
240. Mao L, Schoenberg MP, Scicchitano M, Erozan YS, Merlo A, Schwab D, Sidransky D. Molecular detection of primary bladder cancer by microsatellite analysis. *Science* 1996; 271:659–662.
241. Christensen M, Jensen MA, Wolf H, Orntoft TF. Pronounced microsatellite instability in transitional cell carcinomas from young patients with bladder cancer. *Int J Cancer* 1998; 79:396–401.
242. Mourah S, Cussenot O, Vimont V, Desgrandchamps F, Teillac P, Cochant-Priollet B, et al. Assessment of microsatellite instability in urine in the detection of transitional-cell carcinoma of the bladder. *Int J Cancer* 1998; 79:629–633.
243. Bedi GC, Westra WH, Farzadegan H, Pitha PM, Sidransky D. Microsatellite instability in primary neoplasms from HIV+ patients. *Nat Med* 1995; 1:65–68.

244. Larson RS, Manning S, Macon WR, Vnencak-Jones C. Microsatellite instability in natural killer cell-like T-cell lymphomas in immunocompromised and immunocompetent individuals. *Blood* 1997; 89:1114–1115.

245. Berg PE, Liu J, Yin J, Rhyu MG, Frantz CN, Meltzer SJ. Microsatellite instability is infrequent in neuroblastoma. *Cancer Epidemiol Biomarkers Prev* 1995; 4:907–909.

246. Gasparian AV, Laktionov KK, Belialova MS, Pirogova NA, Tatosyan AG, Zborovskaya IB. Allelic imbalance and instability of microsatellite loci on chromosome 1p in human non-small-cell lung cancer. *Br J Cancer* 1998; 77:1604–1611.

247. Gleeson CM, Sloan JM, McGuigan JA, Ritchie AJ, Weber JL, Russell SE. Ubiquitous somatic alterations at microsatellite alleles occur infrequently in Barrett's-associated esophageal adenocarcinoma. *Cancer Res* 1996; 56:259–263.

248. Gleeson CM, Sloan JM, McGuigan JA, Ritchie AJ, Weber JL, Russell SE. Widespread microsatellite instability occurs infrequently in adenocarcinoma of the gastric cardia. *Oncogene* 1996; 12:1653–1662.

249. Jonsson M, Johannsson O, Borg A. Infrequent occurrence of microsatellite instability in sporadic and familial breast cancer. *Eur J Cancer* 1995.

250. Kaneko H, Horiike S, Taniwaki M, Misawa S. Microsatellite instability is an early genetic event in myelodysplastic syndrome but is infrequent and not associated with TGF-beta receptor type II gene mutation. *Leukemia* 1996; 10:1696–1699.

251. Shimada M, Horii A, Sasaki S, Yanagisawa A, Kato Y, Yamashita K, et al. Infrequent replication errors at microsatellite loci in tumors of patients with multiple primary cancers of the esophagus and various other tissues. *Jpn J Cancer Res* 1995; 86:511–515.

252. Tasak T, Lee S, Spira S, Takeuchi S, Hatta Y, Nagai M, Takahara J, Koeffler HP. Infrequent microsatellite instability during the evolution of myelodysplastic syndrome to acute myelocytic leukemia. *Leuk Res* 1996; 20:113–117.

253. Teng DH, Bogden R, Mitchell J, Baumgard M, Bell R, Berry S, et al. Low incidence of BRCA2 mutations in breast carcinoma and other cancers. *Nat Genet* 1996; 13:241–244.

254. Tashiro H, Lax SF, Gaudin PB, Isacson C, Cho KR, Hedrick L. Microsatellite instability is uncommon in uterine serous carcinoma. *Am J Pathol* 1997; 150:75–79.

255. Takenoshita S, Hagiwara K, Gemma A, Nagashima M, Ryberg D, Lindstedt BA, et al. Absence of mutations in the transforming growth factor-beta type II receptor in sporadic lung cancers with microsatellite instability and rare H-ras1 alleles. *Carcinogenesis* 1997; 18:1427–1429.

256. Chen T, Yamamoto S, Gen H, Murai T, Mori S, Oohara T, et al. Infrequent involvement of microsatellite instability in urinary bladder carcinomas of the NON/Shi mouse treated with N-butyl-N-(4-hydroxybutyl)nitrosamine. *Cancer Lett* 1998; 123:41–45.

257. Mao L, Lee DJ, Tockman MS, Erozan YS, Askin F, Sidransky D. Microsatellite alterations as clonal markers for the detection of human cancer. *Proc Natl Acad Sci USA* 1994; 91:9871–9875.

258. Katabuchi H, van Rees B, Lambers AR, Ronnett BM, Blazes MS, Leach FS, Cho KR, Hedrick L. Mutations in DNA mismatch repair genes are not responsible for microsatellite instability in most sporadic endometrial carcinomas. *Cancer Res* 1995; 55:5556–5560.

259. Kobayashi K, Matsushima M, Koi S, Saito H, Sagae S, Kudo R, Nakamura Y. Mutational analysis of mismatch repair genes, hMLH1 and hMSH2, in sporadic endometrial carcinomas with microsatellite instability. *Jpn J Cancer Res* 1996; 87:141–145.

260. Lim PC, Tester D, Cliby W, Ziesmer SC, Roche PC, Hartmann L, et al. Absence of mutations in DNA mismatch repair genes in sporadic endometrial tumors with microsatellite instability. *Clin Cancer Res* 1996; 2:1907–1911.

261. Wu MS, Sheu JC, Shun CT, Lee WJ, Wang JT, Wang TH, Cheng AL, Lin JT. Infrequent hMSH2 mutations in sporadic gastric adenocarcinoma with microsatellite instability. *Cancer Lett* 1997; 112:161–166.

262. Esteller M, Levine R, Baylin SB, Ellenson LH, Herman JG. MLH1 promoter hypermethylation is associated with the microsatellite instability phenotype in sporadic endometrial carcinomas. *Oncogene* 1998; 17:2413–2417.

263. Malkhosyan S, Rampino N, Yamamoto H, Perucho M. Frameshift mutator mutations. *Nature* 1996; 382:499–500.

264. Yin J, Kong D, Wang S, Zou TT, Souza RF, Smolinski KN, et al. Mutation of hMSH3 and hMSH6 mismatch repair genes in genetically unstable human colorectal and gastric carcinomas. *Hum Mutat* 1997; 10:474–478.

265. Kinzler KW, Vogelstein B. Cancer-susceptibility genes. Gatekeepers and caretakers. *Nature* 1997; 386:761–763.

266. Markowitz S, Wang J, Myeroff L, Parsons R, Sun L, Lutterbaugh J, et al. Inactivation of the type II TGF-beta receptor in colon cancer cells with microsatellite instability. *Science* 1995; 268:1336–1338.
267. Yagi OK, Akiyama Y, Ohkura Y, Ban S, Endo M, Saitoh K, Yuasa Y. Analyses of the APC and TGF-beta type II receptor genes, and microsatellite instability in mucosal colorectal carcinomas. *Jpn J Cancer Res* 1997; 88:718–724.
268. Samowitz WS, Slattery ML. Transforming growth factor-beta receptor type 2 mutations and microsatellite instability in sporadic colorectal adenomas and carcinomas. *Am J Pathol* 1997; 151:33–35.
269. Percesepe A, Kristo P, Aaltonen LA, Ponz de Leon M, de la Chapelle A, Peltomaki P. Mismatch repair genes and mononucleotide tracts as mutation targets in colorectal tumors with different degrees of microsatellite instability. *Oncogene* 1998; 17:157–163.
270. Parsons R, Myeroff LL, Liu B, Willson JK, Markowitz SD, Kinzler KW, Vogelstein B. Microsatellite instability and mutations of the transforming growth factor beta type II receptor gene in colorectal cancer. *Cancer Res* 1995; 55:5548–5550.
271. Wang J, Sun L, Myeroff L, Wang X, Gentry LE, Yang J, et al. Demonstration that mutation of the type II transforming growth factor beta receptor inactivates its tumor suppressor activity in replication error-positive colon carcinoma cells. *J Biol Chem* 1995; 270:22,044–22,049.
272. Grady WM, Rajput A, Myeroff L, Liu DF, Kwon K, Willis J, Markowitz S. Mutation of the type II transforming growth factor-beta receptor is coincident with the transformation of human colon adenomas to malignant carcinomas. *Cancer Res* 1998; 58:3101–3104.
273. Myeroff LL, Parsons R, Kim SJ, Hedrick L, Cho KR, Orth K, et al. A transforming growth factor beta receptor type II gene mutation common in colon and gastric but rare in endometrial cancers with microsatellite instability. *Cancer Res* 1995; 55:5545–5547.
274. Akiyama Y, Iwanaga R, Ishikawa T, Sakamoto K, Nishi N, Nihei Z, et al. Mutations of the transforming growth factor-beta type II receptor gene are strongly related to sporadic proximal colon carcinomas with microsatellite instability. *Cancer* 1996; 78:2478–2484.
275. Renault B, Calistri D, Buonsanti G, Nanni O, Amadori D, Ranzani G. N. Microsatellite instability and mutations of p53 and TGF-beta RII genes in gastric cancer. *Hum Genet* 1996; 98:601–607.
276. Souza RF, Yin J, Smolinski KN, Zou TT, Wang S, Shi YQ, et al. Frequent mutation of the E2F-4 cell cycle gene in primary human gastrointestinal tumors. *Cancer Res* 1997; 57:2350–2353.
277. Ohue M, Tomita N, Monden T, Miyoshi Y, Ohnishi T, Izawa H, et al. Mutations of the transforming growth factor beta type II receptor gene and microsatellite instability in gastric cancer. *Int J Cancer* 1996; 68:203–206.
278. Chu TY, Lai JS, Shen CY, Liu HS, Chao CF. Frequent aberration of the transforming growth factor-beta receptor II gene in cell lines but no apparent mutation in pre-invasive and invasive carcinomas of the uterine cervix. *Int J Cancer* 1999; 80:506–510.
279. Wang D, Song H, Evans JA, Lang JC, Schuller DE, Weghorst CM. Mutation and downregulation of the transforming growth factor beta type II receptor gene in primary squamous cell carcinomas of the head and neck. *Carcinogenesis* 1997; 18:2285–2290.
280. Souza RF, Lei J, Yin J, Appel R, Zou TT, Zhou X, et al. Transforming growth factor beta 1 receptor type II mutation in ulcerative colitis-associated neoplasms. *Gastroenterology* 1997; 112:40–45.
281. Togo G, Toda N, Kanai F, Kato N, Shiratori Y, Kishi K, et al. A transforming growth factor beta type II receptor gene mutation common in sporadic cecum cancer with microsatellite instability. *Cancer Res* 1996; 56:5620–5623.
282. Rampino N, Yamamoto H, Ionov Y, Li Y, Sawai H, Reed JC, Perucho M. Somatic frameshift mutations in the BAX gene in colon cancers of the microsatellite mutator phenotype. *Science* 1997; 275:967–969.
283. Brimmell M, Mendiola R, Mangion J, Packham G. BAX frameshit mutations in cell lines derived from human haemopoietic malignancies are associated with resistance to apoptosis and microsatellite instability. *Oncogene* 1998; 16:1803–12.
284. Molenaar JJ, Gerard B, Chambon-Pautas C, Cave H, Duval M, Vilmer E, Grandchamp B. Microsatellite instability and frameshift mutations in BAX and transforming growth factor-beta RII genes are very uncommon in acute lymphoblastic leukemia in vivo but not in cell lines. *Blood* 1998; 92:230–233.
285. Oliveira C, Seruca R, Seixas M, Sobrinho-Simoes M. The clinicopathological features of gastric carcinomas with microsatellite instability may be mediated by mutations of different "target genes": a study of the TGFbeta RII, IGFII R, and BAX genes. *Am J Pathol* 1998; 153:1211–1219.
286. Sakakibara T, Nakamura T, Yamamoto M, Matsuo M. Microsatellite instability in Japanese hereditary non-polyposis colorectal cancer does not induce mutation of a simple repeat sequence of the bax gene. *Cancer Lett* 1998; 124:193–197.

287. Sakao Y, Noro M, Sekine S, Nozue M, Hirohashi S, Itoh T, Noguchi M. Microsatellite instability and frameshift mutations in the Bax gene in hereditary nonpolyposis colorectal carcinoma. *Jpn J Cancer Res* 1998; 89:1020–1027.

288. Leung SY, Yuen ST, Chung LP, Chu KM, Wong MP, Branicki FJ, Ho JC. Microsatellite instability, Epstein-Barr virus, mutation of type II transforming growth factor beta receptor and BAX in gastric carcinomas in Hong Kong Chinese. *Br J Cancer* 1999; 79:582–588.

289. Yoshinaga K, Sasano H, Furukawa T, Yamakawa H, Yuki M, Sato S, Yajima A, Horii A. The PTEN, BAX and IGFIIR genes are mutated in endometrial atypical hyperplasia. *Jpn J Cancer Res* 1998; 89:985–990.

290. Souza RF, Appel R, Yin J, Wang S, Smolinski KN, Abraham JM, et al. Microsatellite instability in the insulin-like growth factor II receptor gene in gastrointestinal tumours. *Nat Genet* 1996; 14:255–257.

291. Ouyang H, Shiwaku HO, Hagiwara H, Miura K, Abe T, Kato Y, et al. The insulin-like growth factor II receptor gene is mutated in genetically unstable cancers of the endometrium, stomach, and colorectum. *Cancer Res* 1997; 57:1851–1854.

292. Yoshitaka T, Matsubara N, Ikeda M, Tanino M, Hanafusa H, Tanaka N, Shimizu K. Mutations of E2F-4 trinucleotide repeats in colorectal cancer with microsatellite instability. *Biochem Biophys Res Commun* 1996; 227:553–557.

293. Huang J, Papadopoulos N, McKinley AJ, Farrington SM, Curtis LJ, Wyllie AH, et al. APC mutations in colorectal tumors with mismatch repair deficiency. *Proc Natl Acad Sci USA* 1996; 93:9049–9054.

294. Laken SJ, Petersen GM, Gruber SB, Oddoux C, Ostrer H, Giardiello FM, et al. Familial colorectal cancer in Ashkenazim due to a hypermutable tract in APC. *Nat Genet* 1997; 17:79–83.

295. Reitmair AH, Cai JC, Bjerknes M, Redston M, Cheng H, Pind MT, et al. MSH2 deficiency contributes to accelerated APC-mediated intestinal tumorigenesis. *Cancer Res* 1996; 56:2922–2926.

296. Kong D, Suzuki A, Zou TT, Sakurada A, Kemp LW, Wakatsuki S, et al. PTEN1 is frequently mutated in primary endometrial carcinomas. *Nat Genet* 1997; 17:143–144.

297. Levine RL, Cargile CB, Blazes MS, van Rees B, Kurman RJ, Ellenson LH. PTEN mutations and microsatellite instability in complex atypical hyperplasia, a precursor lesion to uterine endometrioid carcinoma. *Cancer Res* 1998; 58:3254–3258.

298. Tashiro H, Blazes MS, Wu R, Cho KR, Bose S, Wang SI, et al. Mutations in PTEN are frequent in endometrial carcinoma but rare in other common gynecological malignancies. *Cancer Res* 1997; 57:3935–3940.

299. Strickler JG, Zheng J, Shu Q, Burgart LJ, Alberts SR, Shibata D. p53 mutations and microsatellite instability in sporadic gastric cancer: when guardians fail. *Cancer Res* 1994; 54:4750–4755.

300. Lleonart ME, Garcia-Foncillas J, Sanchez-Prieto R, Martin P, Moreno A, Salas C, Ramon y Cajal S. Microsatellite instability and p53 mutations in sporadic right and left colon carcinoma: Different clinical and molecular implications. *Cancer* 1998; 83:889–895.

301. Ben-Yehuda D, Krichevsky S, Caspi O, Rund D, Polliack A, Abeliovich D, et al. Microsatellite instability and p53 mutations in therapy-related leukemia suggest mutator phenotype. *Blood* 1996; 88:4296–4303.

302. Eshleman JR, Lang EZ, Bowerfind GK, Parsons R, Vogelstein B, Willson JK, et al. Increased mutation rate at the *hprt* locus accompanies microsatellite instability in colon cancer. *Oncogene* 1995; 10:33–37.

303. Mellon I, Champe GN. Products of DNA mismatch repair genes mutS and mutL are required for transcription-coupled nucleotide-excision repair of the lactose operon in *Escherichia coli*. *Proc Natl Acad Sci USA* 1996; 93:1292–1297.

304. Mellon I, Rajpal DK, Koi M, Boland CR, Champe GN. Transcription-coupled repair deficiency and mutations in human mismatch repair genes. *Science* 1996; 272:557–560.

305. Feng WY, Lee EH, Hays JB. Recombinagenic processing of UV-light photoproducts in nonreplicating phage DNA by the *Escherichia coli* methyl-directed mismatch repair system. *Genetics* 1991; 129:1007–1020.

306. Worth L, Jr, Clark S, Radman M, Modrich P. Mismatch repair proteins MutS and MutL inhibit RecA-catalyzed strand transfer between diverged DNAs. *Proc Natl Acad Sci USA* 1994; 91:3238–3241.

307. Fram RJ, Cusick PS, Wilson JM, Marinus MG. Mismatch repair of *cis*-diamminedichloroplatinum(II)-induced DNA damage. *Mol Pharmacol* 1985; 28:51–55.

308. Jones M, Wagner R. *N*-Methyl-*N'*-nitro-*N*-nitrosoguanidine sensitivity of *E. coli* mutants deficient in DNA methylation and mismatch repair. *Mol Gen Genet* 1981; 184:562–563.

309. Karran P, Marinus M. Mismatch correction at O^6-methylguanine residues in *E. coli* DNA. *Nature* 1982; 296:868–869.

310. Goldmacher VS, Cuzick RA, Thilly WG. Isolation and partial characterization of human cell mutants differing in sensitivity to killing and mutation by methylnitrosourea and *N*-methyl-*N'*-nitro-nitrosoguanidine. *J Biol Chem* 1986; 261:12462–12471.

311. Hawn MT, Umar A, Carethers JM, Marra G, Kunkel TA, Boland CR, Koi M. Evidence for a connection between the mismatch repair system and the G2 cell cycle checkpoint. *Cancer Res* 1995; 55:3721–3725.

312. Branch P, Aquilina G, Bignami M, Karran P. Defective mismatch binding and a mutator phenotype in cells tolerant to DNA damage. *Nature* 1993; 362:652–654.

313. Aquilina G, Hess P, Branch P, MacGeoch C, Casciano I, Karran P, Bignami M. Mismatch recognition defect in colon carcinoma confers DNA microsatellite instability and a mutator phenotype. *Proc Natl Acad Sci USA* 1994; 91:8905–8909.

314. Aquilina G, Hess P, Fiumicino S, Ceccotti S, Bignami M. A mutator phenotype characterizes one of two complementation groups in human cells tolerant to methylation damage. *Cancer Res* 1995; 55:2569–2575.

315. Levati L, Marra G, Lettieri T, D'Atri S, Vernole P, Tentori L, et al. Mutation of the mismatch repair gene hMSH2 and hMSH6 in a human T-cell leukemia line tolerant to methylating agents. *Genes Chromosomes Cancer* 1998; 23:159–166.

316. Eastman A. Characterization of the adduct produced in DNA by *cis*-diamminedichloroplatinum(II) and *cis*-dichloro(ethylenediamine)-platinum(II). *Biochemistry* 1983; 22:3927–3933.

317. Fichtinger-Schepman AMJ, van Veer JL, den Hartog JH, Lohman PH, Reedijk J. Adducts of the antitumor drug *cis*-diamminedichloroplatinum(II) with DNA: formation, identification, and quantitation. *Biochemistry* 1985; 24:707–713.

318. Anthoney DA, McIlwrath AJ, Gallagher WM, Edlin ARM, Brown R. Microsatellite instability, apoptosis, and loss of p53 function in drug-resistant tumor cells. *Cancer Res* 1996; 56:1374–1381.

319. Aebi S, Kurdi-Haidar B, Gordon R, Cenni B, Zheng H, Fink D, et al. Loss of DNA mismatch repair in acquired resistance to cisplatin. *Cancer Res* 1996; 56:3087–3090.

320. Fink D, Zheng H, Nebel S, Norris PS, Aebi S, Lin TP, et al. In vitro and in vivo resistance to cisplatin in cells that have lost DNA mismatch repair. *Cancer Res* 1997; 57:1841–1845.

321. Friedman HS, Johnson SP, Dong Q, Schold SC, Rasheed BK, Bigner SH. et al. Methylator resistance mediated by mismatch repair deficiency in a glioblastoma multiforme xenograft. *Cancer Res* 1997; 57:2933–2936.

322. Fink D, Aebi S, Howell SB. Role of DNA mismatch repair in drug resistance. *Clin Cancer Res* 1998; 4:1–6.

323. D'Atri S, Tentori L, Lacal PM, Graziani G, Pagani E, Benincasa E, et al. Involvement of the mismatch repair system in temozolomide-induced apoptosis. *Mol Pharmacol* 1998; 54:334–341.

324. Fritzell JA, Narayanan L, Baker SM, Bronner CE, Andrew SE, Prolla TA, et al. Role of DNA mismatch repair in the cytotoxicity of ionizing radiation. *Cancer Res* 1997; 57:5143–5147.

325. Reitmair AH, Risley R, Bristow RG, Wilson T, Ganesh A, Jang A, et al. Mutator phenotype in Msh2-deficient murine embryonic fibroblasts. *Cancer Res* 1997; 57:3765–3771.

325a. Wu J, Gu L, Geacintov NE, and Li G-M. Mismatch repair processing of carcinogen-DNA adducts triggers apoptosis. *Mol Cell Biol* 1999; 19:8292–9301.

326. Karran P. Bignami M. DNA damage tolerance, mismatch repair and genome instability. *Bioessays* 1994; 16:833–839.

327. Modrich P. Strand-specific mismatch repair in mammalian cells. *J Biol Chem* 1997; 272:24727–24730.

328. Duckett DR, Drummond JT, Murchie AIH, Reardon JT, Sancar A, Lilley DM, Modrich P. Human MutSα recognizes damaged DNA base pairs containing O^6-methylguanine, O^4-methylthymine, or the cisplatin-d(GpG) adduct. *Proc Natl Acad Sci USA* 1996; 93:6443–6447.

329. Li G-M, Wang H, Romano LJ. Human MutSα specifically binds to DNA containing aminofluorene and acetylaminofluorene adducts. *J Biol Chem* 1996; 271:24084–24088.

329a. Duckett DR, Bronstein SM, Taya Y, Modrich P. hMutSa- and hMutLa-dependent phosphorylation of p53 in response to DNA methylator damage. *Proc Natl Acad Sci USA* 1999; 96:12384–12388.

329b. Gong, JG, Costanzo A, Yang HQ, et al. The tyrosine kinase c-Abl regulates p73 in apoptotic response to cisplatin-induced DNA damage [In Process Citation]. *Nature* 1999; 399:806–809.

330. Mello JA, Acharya S, Fishel R, Essigmann JM. The mismatch-repair protein hMSH2 binds selectively to DNA adducts of the anticancer drug cisplatin. *Chem Biol* 1996; 3:579–589.

331. Mu D, Tursun M, Duckett DR, Drummond JT, Modrich P, Sancar A. Recognition and repair of compound DNA lesions (base damage and mismatch) by human mismatch repair and excision repair systems. *Mol Cell Biol* 1997; 17:760–769.

332. Kinzler KW, Vogelstein B. Lessons from hereditary colorectal cancer. *Cell* 1996; 87:159–170.
333. Horii A, Han HJ, Shimada M, Yanagisawa A, Kato Y, Ohta H, et al. Frequent replication errors at microsatellite loci in tumors of patients with multiple primary cancers. *Cancer Res* 1994; 54:3373–3375.
334. Peng H, Chen G, Du M, Singh N, Isaacson PG, Pan L. Replication error phenotype and p53 gene mutation in lymphomas of mucosa-associated lymphoid tissue. *Am J Pathol* 1996; 148:643–648.
335. Bedi GC, Westra WH, Farzadegan H, Pitha PM, Sidransky D. Microsatellite instability in primary neoplasms from HIV + patients. *Nat Med* 1995; 1:65–68.
336. Rosell R, Pifarre A, Monzo M, Astudillo J, Lopez-Cabrerizo MP, Calvo R, et al. Reduced survival in patients with stage-I non-small-cell lung cancer associated with DNA-replication errors. *Int J Cancer* 1997; 74:330–334.

10 PTEN

Regulator of Phosphoinositide 3-Kinase Signal Transduction

Jen Jen Yeh, MD, and William R. Sellers, MD

CONTENTS

1. INTRODUCTION

The chromosome (chr) region spanning 10q22–10q25 exhibits loss of heterozygosity (LOH) in multiple tumor types, including carcinoma of the prostate, endometrium, breast, kidney, and thyroid, as well as glioblastoma, melanoma, and meningioma *(1–14)*. In addition, the gene for Cowden disease (CD), an autosomal dominant familial cancer syndrome that is characterized by multiple hamartomas of the skin, breast, thyroid, and intestines, and an increased risk of breast and thyroid malignancies, has been localized to chr bands 10q22–23 *(15)* by linkage analysis. Microcell-mediated transfer of chr 10q into glioblastoma cells inhibits soft-agar colony formation and tumor formation in nude mice *(16)*. Likewise, chr transfers of 10q into rat prostate cancer (PC) cells suppresses the metastatic ability of the highly metastatic parental cells *(17)*. Taken together, these findings suggest the presence of one or more tumor suppressor genes (TSGs) in the chr 10q22–25 region. In 1997, the TSG phosphate and tensin homolog deleted on chr 10 *(PTEN)*/mutated in multiple advanced cancers 1 *(MMAC1)*/transforming growth factor Fβ-regulated and epithelial cell-enriched phosphatase *(TEP1)*, located on chr subband 10q23.3, was identified by three independent groups *(18–20)*. Mutational analysis of the nine coding exons of *PTEN/MMAC1/TEP1* (hereafter referred to as *PTEN*) demonstrated germ-line mutations in the related familial hamartoma syndromes, CD and Bannayan-Zonana syndrome (BZS), and LOH accompanied by somatic mutation of the remaining allele in multiple tumor types.

From: *Tumor Suppressor Genes in Human Cancer*
Edited by: D. E. Fisher © Humana Press Inc., Totowa, NJ

The protein product of the *PTEN* (PTEN) encodes a dual-specificity phosphatase. Recent evidence has demonstrated that PTEN functions, not only as a protein phosphatase, but as a lipid phosphatase as well. This latter activity is associated with the ability of PTEN to regulate signaling through the phosphoinositide 3-kinase (PI3K) pathway, and appears to be required for PTEN function as a tumor suppressor. This chapter reviews the biochemical, functional, and genetic data that have led to current understanding of PTEN tumor-suppressor function.

2. *PTEN*, THE GENE

2.1. Germ-line Mutations

CD is an autosomal dominantly inherited hamartoma syndrome that is characterized by hamartomas in multiple organ systems, as well as by an increased risk of breast and thyroid malignancies *(21)*. Hallmarks of CD include benign tumors of the hair follicle infundibulum, known as trichilemmomas and mucocutaneous papillomatosis. A subset of patients with CD are found to have Lherhmitte Duclos disease (LDD), characterized by an altered gait, epilepsy, and megencephaly secondary to dysplastic gangliocytoma of the cerebellum *(15)*. The clinical phenotype of BZS partially overlaps CD, and includes the presence of hamartomatous polyps of the intestine, macrocephaly, and lipomas. However, families afflicted with BZS are not found to have an increased risk of malignancy. Additional characteristics of BZS may include vascular malformations, Hashimoto's thyroiditis, speckled penis, mental retardation, and intracranial tumors *(21)*.

Germ-line mutations of *PTEN* have been found in 81% of families with CD, including those with LDD, and in 57% of BZS families *(15,21,22)*. Thus, mutations of *PTEN* appear to be associated with multiple phenotypes. This has led to speculation as to whether or not these syndromes represent a spectrum of a single entity that is modified by the developmental timing of the mutations or other epigenetic and/or genetic events.

Alternatively, certain PTEN mutations could give rise to specific clinical phenotypes; that is, there may be genotype–phenotype correlates. However, this latter possibility seems unlikely, because most *PTEN* mutations cluster within the minimal PTEN phosphatase domain (*see* Subheading 2.2.), and would therefore be predicted to disrupt its phosphatase activity. In addition, one mutation, R233X, has been found in the germ line of a family with BZS, and in two unrelated CD families *(21)*. These data would also argue against the possibility that differences in mechanisms of mutational inactivation of PTEN underlie the differences in the observed clinical spectra.

2.2. Somatic Mutations

Somatic mutations of *PTEN* have been identified in high-grade gliomas, advanced-stage prostate carcinomas, endometrial carcinomas, and malignant melanomas, and less commonly in small cell lung, thyroid, bladder, renal, and sporadic breast carcinomas and lymphoid and hematological malignancies *(23–49)*.

LOH on chr 10q has been observed in 69–82% of glioblastomas *(24,45,46,50)*. Intragenic mutations in the *PTEN* gene have been identified in 60% of primary glioblastoma cell lines *(18,20)* and in approx 20–44% of glioblastoma multiforme, the most aggressive subtype of astrocytic tumors *(18,20,24,39,45,51–53)*. In contrast, in low-grade gliomas (such as anaplastic astrocytomas) and meningiomas, although they also harbor chr 10q loss, *PTEN* alterations are comparatively rare *(24,39)*. *PTEN* mutations therefore appear to be correlated with the more advanced tumors.

Although two studies have found a low frequency of *PTEN* mutations in PC *(27,38)*, a number of additional studies have shown that *PTEN* alterations are more frequent in higher-grade tumors, and in metastatic tumors *(54–56)*. LOH on chr 10q is observed in about 30–60% of tumors, and is associated with tumor progression and metastasis *(57)*. In one series, 10/23 (43%) tumors with LOH of chr band 10q23 had a second inactivation event at *PTEN (55)*. Seven of the 10 tumors with *PTEN* alterations in this series had pelvic lymph node metastases. Similarly, homozygous deletions of *PTEN* were detected in 10–15% of stage B prostate carcinomas, which are reported to have between 0–40% chr 10q loss *(55)*. Study of PTEN expression, using immunohisto-chemistry, demonstrated that loss of PTEN expression correlated significantly with Gleason scores (a measure of the tumor grade) of 7 or greater *(56)*. Taken together, these data suggest that, as in glioblastoma, *PTEN* inactivation is a late event in prostate tumorigenesis, and thus might be associated with the progression from a primary to a metastatic tumor.

PTEN mutations can be found in 30–40% of malignant melanomas *(29,44)*, correlating with a LOH frequency of 30–50% on chr 10q *(1,58)*. In addition, in specific instances, PTEN mutations were found in metastatic foci, but not in the corresponding primary tumors, again suggesting that *PTEN* is involved in tumor progression *(44)*.

In contrast to findings in glioblastoma, prostate carcinoma, and melanoma, *PTEN* alterations are detected in 30–50% of endometrial carcinomas of the endometrioid type *(36,42,47,48.* Consistent with this data, chr 10q LOH can be found in approx 40% of endometrial cancers (Jones, 1994 #907; Nagase, 1996 #956; Peiffer, 1995 #967). In addition, *PTEN* mutations are found in approx 20% of endometrial hyper-plasias, which are thought to be premalignant precursor lesions of invasive endometrial carcinomas *(59)*, suggesting that here, *PTEN* mutations are not associated with advanced staging.

These findings suggest that *PTEN* may exhibit differential roles in specific tumor types. For example, in brain and prostate carcinomas, *PTEN* mutations probably occur as late events in tumorigenesis, and have been hypothesized to be associated with the metastatic phenotype. In contrast, in endometrial carcinomas, *PTEN* mutations may instead be initiating events. The latter role of *PTEN* as an initiator of transformation is also supported by the predisposition to breast and thyroid cancers conferred to certain individuals bearing a germ-line *PTEN* mutation. This role of *PTEN* as an initiator or as a progression factor is not a unique finding among TSGs. Loss of the retinoblastoma gene *(Rb)* predisposes to retinoblastoma and sarcoma, but somatic *Rb* mutations are found as late events in a number of tumors *(60)*. Likewise, *p53* mutations can be initiating events, as in Li-Fraumeni syndrome, or can occur late in the evolution of certain sporadic cancers: Therefore, it is not necessary to ascribe different biological roles of the protein product to the action of a tumor suppressor in initiation or progression.

2.3. Evidence for Second TSG Near PTEN?

In certain tumors, the frequency of chr 10q23 LOH exceeds the frequency of intra-genic *PTEN* mutations. The most striking example is found in primary breast tumors. Although the frequency of chr 10q LOH in primary breast tumors is as high as 50%, most series report only rare *PTEN* mutations *(37,43,61,62)*. This was particularly surprising, because CD patients, who have germ-line mutations in *PTEN*, have an increased risk of breast tumors.

Other noted examples of this discordance include lung cancer, in which LOH on chr 10q near *PTEN* reaches 91% in small cell and 41% in nonsmall cell lung carcinomas *(6)*. Again, *PTEN* mutations or homozygous deletions are significantly less, at approx 10% *(31)*. Similarly, in thyroid tumors, LOH on chr bands 10q22–23 occurs at about 25% frequency, but *PTEN* mutations are comparatively rare *(13,35,49)*.

A number of possible explanations for the discordance between 10q23 LOH and PTEN mutation can be envisioned. First, there may be a systematic underestimation of PTEN inactivation. For example, mutations in noncoding regions, which have not yet been characterized, may play a role in transcript or protein stability. Alternatively, other epigenetic mechanisms, such as promoter methylation, may inactivate *PTEN*. Second, monoallelic inactivation or haploinsufficiency of *PTEN* may be sufficient in some tumors to promote growth and/or progression. Third, other TSGs may reside within this locus.

Currently, only indirect evidence exists for the role of methylation in transcriptional repression. For example, in advanced prostate tumor xenografts, loss of PTEN protein and mRNA was found in the absence of PTEN mutation. Treatment of these xenografts with the demethylating agent, 4-azadeoxycytidine, restored PTEN protein expression, suggesting the presence of inactivating methylation events *(63)*. On the other hand, one study *(55)* found no evidence of methylation in six primary prostate tumor samples. Whether transcriptional or posttranscriptional mechanisms for *PTEN* inactivation exist remains an open question.

Evidence for the possibility of a significant role for *PTEN* haploinsufficiency in tumor formation is supported by studies in *PTEN+/–* mice. Here, as will be discussed later, loss of one *PTEN* allele leads to diffuse proliferative abnormalities in the prostate, colon, skin, and lymph nodes *(64)*. The diffuse nature of these lesions makes it unlikely that each and every cell has sustained a second inactivating mutation. Therefore, it is likely that this proliferative abnormality arises solely as a consequence of the loss of one *PTEN* allele. Furthermore, heterozygous embryonic stem cells show partial activation of Akt, compared to their wild-type (wt) and homozygous-null counterparts *(65)*. Finally, heterozygous lymphocytes and embryonic fibroblasts are protected from Fas-induced apoptosis, and have partially deregulated Akt activity *(66)*. These data suggest that loss of one allele of PTEN can lead to a significant alteration in cell behavior.

Two candidate TSGs have been identified distal to *PTEN: MXI1* max interactor 1 *(MXI1)* and deleted in malignant brain tumors *(DMBT1) (67,68)*. Although MXI1 is capable of suppressing the growth of glioblastoma cells in culture lines *(68)*, most studies of glioblastomas and prostate and lung carcinomas have found few to no mutations or homozygous deletions in *MXI1 (11,69–75)*. Homozygous deletions of *DMBT1* appear to occur frequently in gliomas, and less frequently in lung and gastrointestinal carcinomas *(51,67,76–78)*. An analysis of the prognostic significance of *PTEN* and *DMBT1* alterations in gliomas suggested that, although *PTEN* alterations were associated with progression of disease, *DMBT1* may be involved in the early stages of tumorigenesis *(77)*.

In addition, *FAS,* the cell-surface receptor involved in death signaling, is located on chr subband 10q24.1, just distal to *PTEN*. However, alterations of *FAS* are rare in non-lymphoid malignancies *(79–83)*. An interesting possibility with respect to Fas is that combined haploinsufficiency of *FAS* and *PTEN* could combine to promote cell survival (discussed below). Most recently, another inositol phosphatase, multiple inositol polyphosphate phosphatase *(MINPP1)*, was identified and mapped to chr band 10q23 *(84)*, probably in close proximity to *PTEN*. Given its function and location, mutational analyses of *MINPP1* in tumorigenesis will be of interest.

2.4. PTEN Is a Tumor Suppressor

Reconstitution of wt PTEN in PTEN-deficient glioblastoma cells, either through adenoviral, retroviral, or transfection-based methods, results in suppression of colony formation in culture, soft-agar growth, and tumor formation in nude mice *(85,86)*. Likewise, reconstitution of PTEN in PTEN–/– PC cells suppresses the outgrowth of colonies in vitro *(87,88)*. Chr transfers of wt chr 10 into either glioblastoma or malignant melanoma cell lines found that the microcell hybrids that escaped the chr-mediated suppression selectively lost portions of the chr band where *PTEN* is located *(50,89)*. Furthermore, as described below, mice rendered heterozygous for PTEN, through gene targeting, develop a variety of tumors that are associated with loss of the wt *PTEN* allele *(90,91)*. These data provide functional evidence that *PTEN* is in fact a TSG.

3. PTEN, THE PROTEIN

3.1. Introduction

PTEN contains 403 amino acid residues, and is a member of the VHR family of dual specificity phosphatases, containing the signature motif common to protein tyrosine (Tyr) phosphatases: HCXXGXXRS/T. Although PTEN bears homology to the cytoskeletal protein, tensin, and to the secretory vesicle protein, auxilin *(19,20)*, their region of homology encompasses the phosphatase homology region. Tensin and auxilin were observed to have homology to phosphatases prior to the cloning of *PTEN* *(92)*. However, in each case, certain critical residues, typically required for phosphatase activity, are absent. This observation led to speculations that, for these proteins, this domain may serve as a novel phosphate-binding motif *(92)*. It is therefore likely that the homology between PTEN and auxilin and tensin is more a reflection of the presence of phosphatase-like domains in the latter, rather than of a parallel function.

PTEN has phosphatase activity toward serine (Ser), threonine (Thr), and Tyr phosphorylated substrates *(93)*. PTEN, however, does not dephosphorylate a number of typical phosphatase substrates, including myelin basic protein and RCML *(93)*; rather, the phosphatase activity appears to be restricted to acidic substrates, such as polyGlu$_4$ Tyr$_1$ This preference for acidic substrates prompted Maehama and Dixon *(94)* to ask whether PTEN could act as a lipid phosphatase.

PTEN specifically dephosphorylates the D3 position on the inositol ring of phosphatidylinositol and inositol phosphates *(94)*. It exhibits the highest activity against phosphatidylinositol-3,4,5-trisphosphate (PtdIns-3,4,5-P$_3$) and phosphatidylinositol-3,4-bisphosphate (PtdIns-3,4-P$_2$); and, in decreasing order, against phosphatidylinositol-3-phosphate (PtdIns3P) and inositol-1,3,4,5-tetrakisphosphate (Ins-1,3,4,5-P$_4$) *(87,94)*. The specificity for the 3 position led those authors to predict that PTEN may function as an antagonist of PI3K signaling.

3.2. PTEN as a Lipid Phosphatase

3.2.1. Overview of PI3K/Akt Pathway

Phosphoinositide kinases are a family of lipid kinases involved in intracellular signaling. There are three classes of PI3Ks. Class I kinases are typified by their association with regulatory subunits, and by the ability to utilize PtdIns-3,4-P_2 and PtdIns-3,4,5-P_3 as substrates (95). For the purposes of this chapter, PI3K will refer to PI3Kα, the first-identified and cloned member of its class. PI3K is a class I kinase, and exists as a heterodimer with an 85-kDa regulatory subunit (p85) and an 110-kDa catalytic subunit (p110). Activation of PI3K by receptor and nonreceptor tyrosine kinases (TK), growth factors, and activated Ras initiates an intracellular signaling cascade that impacts cell proliferation and survival (96–99). In addition, this pathway has also been implicated in vesicle trafficking, exocytosis, and actin assembly (99). PI3K catalyzes the formation of PtdIns-3,4-P_2 and PtdIns-3,4,5-P_3, both of which can be dephosphorylated in the presence of PTEN (87,94). In quiescent cells, PtdIns-3,4,-P_2 and, particularly PtdIns-3,4,5-P_3 are normally absent. However, in response to certain growth stimuli, such as platelet-derived growth factor (PDGF), insulin-like growth factor 1, and nerve growth factor, this signaling cascade is initiated. Typically, receptor TKs undergo an auto- or transphosphorylation reaction, leading to the creation of docking sites for the p85 subunit. Docking of PI3K to the receptor serves two purposes: first, PI3K is brought into proximity with the membrane-localized lipid substrates, and, second, it is likely that this docking activates PI3K. These steps lead to the production of PtdIns-3,4-P_2 and PtdIns-3,4,5-P_3 (97). These phosphoinositides in turn recruit a variety of cytosolic proteins, including Grp-1, Akt, Bruton's TK, adapter protein-2, and synaptotagimen, to the membrane (99). This interaction between the membrane-bound phospholipid and specific proteins is mostly mediated by binding of the phospholipid to either pleckstrin homology (PH) domains or other lipid-binding domains, such as Src-homology-2 domains (99).

One cytosolic protein that has stimulated great interest is the Ser/Thr kinase, Akt. Binding of Akt to PtdIns-3,4-P_2 and/or PtdIns-3,4,5-P_3 is mediated by its amino-terminal PH domain (97,100). Binding of the lipid to the Akt PH domain may result in a conformational change in Akt. More importantly, the PH domain binding translocates Akt to the plasma membrane, where Akt can then be activated by phosphorylation (97,101). This activation requires phosphorylation at two sites, Thr308 and Ser473. These phosphorylations are mediated by the PtdIns-3,4,5-P_3-dependent protein kinase-1 (PDK-1) and a second unknown kinase, designated PDK2, respectively (100,102). PDK2 may in fact be a form of PDK1 that has altered substrate specificity (103).

3.2.2. Akt and its Downstream Targets

Akt/RAC/PKB was first identified as a retroviral oncogene isolated from rat T-cell lymphomas, as well as by homology to the related PKA and PKC (104–106). The retroviral fusion protein, Gag-Akt, creates a constitutively active form of Akt that is targeted to the membrane independent of PtdIns-3,4-P_2. Likewise, a constitutively active form of Akt, created by insertion of a myristylation signal that targets Akt to the membrane, is capable of transforming chicken embryo fibroblasts in vitro, and producing hemangiosarcomas in chickens in vivo (107). Intense interest has centered on the identification of downstream targets of Akt that may be essential for cell growth, survival, and/or transformation.

Table 1
Downstream Targets of Akt

Target	Action	Ref(s).
Bad	Prevents binding to Bcl-X$_L$	Datta et al., 1997
Caspase-9	Inhibits protease activity	Cardone et al., 1998
GSK3	Inhibits kinase activity	Cross et al., 1995
		Dudek et al., 1997
		Vanhesebroeck et al., 1997
PhasI/4E-BP1	Inhibits binding to eIF4E	Sonnenberg et al., 1997, 1999
FKHRL1	Prevents nuclear localization	Brunet et al., 1999
AFX		Kops et al., 1999
FKHR		Tang et al., 1999
		Nakase et al., 1999
		Guo et al., 1999
CREB	Stimulates transcriptional activity	Keyong et al., 1998
eNOS	Enhances NO production	Dimmeler et al., 1999
		Fulton et al., 1999
		Michell et al., 1999
IKKα	Prevents binding to NF-κB	Ozes et al., 1999
		Romashkova and Makarov, 1999
		Kane et al., 1999

A number of Akt substrates have been identified (Table 1; Fig. 1), including proapoptotic factors, such as Bad, caspase-9, IKKα, and members of the forkhead transcription factor family, as well as other substrates, such as glycogen synthase kinase-3 (GSK3), p70^{S6K}, eIF4E-binding protein-1 (4E-BP1), and endothelial nitrogen oxide synthase (eNOS) *(100,108–121)*. Of note, it is unclear whether the interaction between Akt and p70^{S6K} is a direct one *(100)*. A general, but not universal finding, is that Akt phosphorylations render the target substrate inactive. For example, Akt phosphorylation of Bad creates 14-3-3 binding sites. Once bound to 14-3-3, Bad is unable to interact with Bcl-X$_L$. Bcl-X$_L$ is an antiapoptotic Bcl-2 family member, and thus the net effect of Akt activity is promoting cell survival *(121)*. Similarly, Akt phosphorylation of the forkhead transcription factors FKHR and FKHRL1, results in either 14-3-3 binding or alterations in nucleocytoplasmic shuttling. By either mechanism, these proteins are localized to the cytoplasm after Akt phosphorylation, and are thus rendered transcriptionally silent *(109,110)*. Akt likewise phosphorylates and inactivates caspase-9, GSK3 and IKKα. Caspase-9 is involved in apoptosis *(120)*; GSK3 is involved in glycogen synthesis *(111)*; IKKα regulates nuclear factor-κB, which is a transcription factor *(114)*. 4EBP-1 is a translational repressor that is inactivated by Akt *(113)*. However, full inhibition of this protein requires phosphorylation by a second kinase, FRAP/mTOR, as well (122). In contrast to the other substrates discussed thus far, Akt phosphorylation activates eNOS, and leads to increased nitric oxide (NO) production *(117–119)*.

Downregulation of the PI3K–Akt pathway would therefore predictably promote proapoptotic activity and growth suppression. Thus, the ability of PTEN to dephospho-

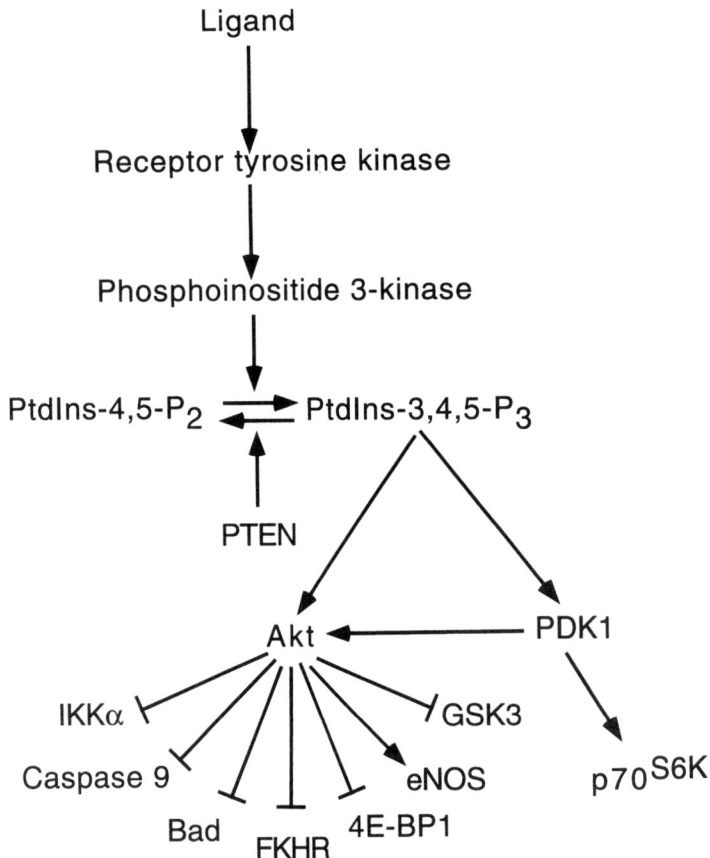

Fig. 1. PI3K–Akt pathway. Growth factor and survival factors bind and activate receptor TKs. Auto- or transphosphorylation allows receptor-mediated activation of PI3K. PI3K catalyzes the phosphorylation of PtdIns-4,5-P$_2$ to PtdIns-3,4,5-P$_3$. The latter activates both Akt family members and PDK1. PDK1 also participates in the activation of Akt. Akt phosphorylates and inactivates a number of substrates, as shown. In the case of eNOS, phosphorylation is activating. PDK1 phosphorylates and activates p70^{S6K}.

rylate PtdIns-3,4,5-P$_3$ and PtdIns-3,4-P$_2$ upstream of Akt would serve as a potent restraint to uncontrolled proliferation and prolonged cell survival.

3.2.3. PI3K–PTEN–AKT PATHWAY

Functional evidence for the role of PTEN in the regulation of the PI3K–Akt pathway was initially found by transfecting wt PTEN or a substrate-trapping mutant of PTEN (C124S) into a human embryonic kidney cell line (293) *(94)*. Introduction of wt PTEN led to decreased levels of PtdIns-3,4,5-P$_3$; introduction of the substrate-trapping mutant resulted in higher levels. Overexpression of PTEN also reduced the levels of PtdIns-3,4,5-P$_3$ in the presence of a constitutively active PI3K, suggesting that PTEN exerts its phosphatase activity on the products of PI3K, and not on PI3K itself *(87)*. In keeping with these data, loss of one or both copies of the murine *PTEN* gene was found to result in elevated PtdIns-3,4,5-P$_3$ levels *(123)*.

Multiple PTEN-deficient tumors cells have elevated basal levels of the phosphory-lated form of Akt (i.e., the activated form). Reintroduction of wt but not phosphatase-defective forms, of PTEN into such cells reduces both phosphorylated Akt and Akt kinase activity *(87,124–126)*. Override experiments have also demonstrated PTEN's placement in the PI3K–Akt pathway between PI3K and Akt. Specifically, phosphoryla-tion of 4EBP1, a downstream target of Akt, is blocked by PTEN in the presence of a constitutively active PI3K, but not a myristylated form of Akt *(88)*, and PTEN-medi-ated cell cycle arrest or apoptosis (discussed below) can be overridden by a constitu-tively active form of Akt *(124,127)*. Together, these data suggest that PTEN functions downstream of PI3K and upstream of Akt.

Finally, although most disease-derived mutations of *PTEN* ablate both its protein and lipid phosphatase activity, a particular missense mutation in the catalytic motif (G129E) retains protein phosphatase, but not lipid phosphatase activity. This mutant was isolated from two independent CD families, one BZS family, and one endometrial carcinoma *(21,48)*. This particular mutant fails to induce apoptosis in LNCaP cells, and fails to induce a G1 arrest in PTEN-null 786-O or U87-MG cells *(87,124,128)*. Addi-tional diseases- and tumor-derived mutations, which selectively retain activity for the phosphoinositides, have also been identified *(128)*. These data indicate that protein phosphatase activity is not sufficient for these PTEN functions, nor is it sufficient for the suppression of the CD phenotype.

PTEN can regulate apoptosis in some cell lines *(87,127,129)* and cell cycle arrest in others *(86,124,128)*, indicating that the PI3K–PTEN–Akt pathway may perform at least two distinct functions: one of cell survival and one of cell growth.

3.2.4. PTEN AS A CELL CYCLE REGULATOR

Several studies have demonstrated that PTEN is sufficient to arrest certain cells in G1 *(65,86,124,128)*. Reintroduction of PTEN into PTEN-null glioblastoma, or renal carcinoma cells deficient in PTEN, resulted in growth suppression with a substantial increase in cells in the G1 phase of the cell cycle *(86,124,128)*. This cell cycle block is overcome by expression of myristylated, and hence constitutively active, Akt, but not wt Akt, implying further that Akt exists downstream of PTEN. In cells arrested by PTEN, apoptosis was not observed *(124,128)*.

Introduction of the G129E PTEN mutant or the G129R PTEN mutant, which is characterized by loss of both protein and lipid phosphatase activity into 786-O renal carcinoma cells, failed to produce a G1 arrest *(124,128)*. Because the G129E PTEN mutant retained its protein phosphatase activity, this suggested that the presence of pro-tein phosphatase activity, as measured in vitro, was insufficient to arrest cells in G1.

The cell cycle is regulated at specific checkpoints, most of which are controlled by cyclins and cyclin-dependent kinase (CDK) complexes, along with a set of cyclin-dependent kinase inhibitors (CDKIS) *(130)*. Cyclin D–CDK4 acts during the G1 phase and is opposed by the CDKI, p16; cyclin E–CDK2 acts at the G1–S transition, and is probably primarily opposed by the CDKI, p27; cyclin A–CDK2 acts through S phase and the G2–M transition *(130)*. In cells susceptible to a PTEN-mediated G1 arrest, re-expression of PTEN leads to decreases in the activities of CDK2, cyclin A, and cyclin E *(86)*. In addition, p27, which can arrest cells in G1 when overexpressed *(130)*, levels are significantly increased *(86)*. Treatment of the cells with PI3K inhibitors also increases levels of p27, and decreases levels of phosphorylated Akt, suggesting that p27 may be another target of the PI3K–PTEN–Akt pathway *(86)*. Whether these are direct effects, or, alternatively, are related to changes in the cell cycle profile of the treated cells, is not yet known.

In keeping with the data above, *PTEN–/–* ES cells exhibit decreased levels of p27 and an associated increase in CDK2 and cyclin E and A activity *(65)*. Only levels of p27 were affected when asynchronously growing cells were examined. These data suggest that the loss of *PTEN* and consequent activation of the PI3K–PTEN–Akt pathway may selectively target p27, thereby decreasing time in G1, accelerating entry into S phase, and, ultimately, increasing cell proliferation.

3.2.5. PTEN AS REGULATOR OF APOPTOSIS

As mentioned previously, a number of Akt substrates are components of apoptotic signaling pathways. Another mechanism by which PTEN acts as a tumor suppressor is as a mediator of apoptosis. Certain tumor-derived cell lines are susceptible to PTEN-induced apoptosis. For instance, in some PTEN-deficient breast cancer cell lines (MCF7, MDA-MB-468, and ZR75-1), adenoviral transfer of PTEN results in apoptosis in 75–80% and >90% of the cell populations, respectively *(127,131)*. This PTEN-mediated apoptosis can be rescued by co-expression of myristylated Akt or Bcl-2, but not FAK or PI3K *(127)*. In one study, apoptosis was associated with reduced endogenous levels of the c-*myc* oncogene; PTEN was shown to inhibit c-*myc* promoter activity in vitro *(131)*.

PTEN-mediated cell death or inhibition of cell survival may not simply reflect the induction of apoptosis. For example, adenoviral introduction of PTEN into LNCaP prostate carcinoma cells results in a dose-dependent decrease in Akt phosphorylation and the induction of cell death. However, the number of cells undergoing apoptotic cell death is significantly less, compared to the expression of p53 in the same cell line. Both p53 and PTEN infection result in decreases in caspase precursors, and treatment with caspase inhibitors partially rescues cells from apoptosis *(129)*. However, in experiments that measure cell growth, using a metabolic assay, LNCaP cells infected with wt PTEN were found to be significantly more growth-inhibited than their wt p53-infected counterparts *(129)*. Furthermore, although overexpression of Bcl-2 decreases PTEN-mediated apoptosis, the growth inhibition that is observed remains unperturbed. These data raise the possibility that alternative pathways to cell death may be important, or that growth inhibition in these cells reflects both antiproliferative and proapoptotic functions of PTEN.

In keeping with the above data, *PTEN–/–* mouse embryonic fibroblasts (MEFs) have a reduced sensitivity to apoptotic stimuli such as UV irradiation and osmotic stress. This resistance can be restored either by re-expression of PTEN or by treatment with PI3K inhibitors *(123)*. Furthermore, certain *PTEN–/–* mice develop lymphoid aggregates. Lymphocytes from these mice have diminished Annexin V staining, suggesting that there is an in vivo defect in lymphoid apoptosis *(90)*. These data suggest that, in certain tissues, *PTEN* is required for the appropriate regulation of cell survival signals.

3.2.6. EVIDENCE FOR A PI3K-INDEPENDENT SURVIVAL PATHWAY?

In one study of PC cells (LNCaP) deficient in PTEN, serum, androgen, or epidermal growth factor stimulation was able to rescue cells that were inhibited by PI3K inhibitors *(131)*. In another study of glioblastoma cells, growth factor stimulation with

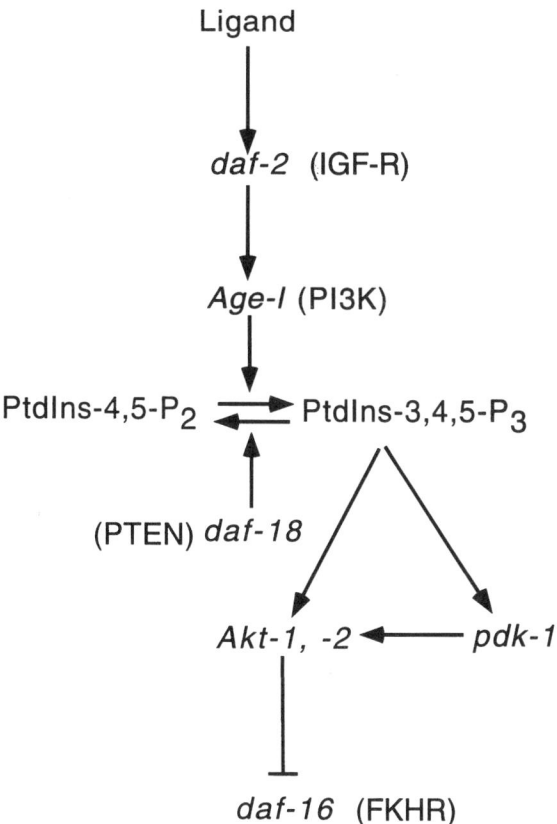

Fig. 2. PI3K–Akt pathway in *C. elegans.* The corresponding genetic alleles from *C. elegans* are shown positioned in the PI3K–Akt pathway shown in Fig. 1. The homologous mammalian proteins are shown in parentheses.

PDGF or insulin was able to bypass the PI3K–PTEN–Akt pathway, and generate activated Akt in the presence of PTEN *(87)*. In contrast, Wu et al. *(88)* and Davies et al. *(129)* demonstrated that, in Rat-1 fibroblast cells and LNCaP cells, respectively, transfection of wt PTEN was unable to produce phosphorylated Akt, despite serum stimulation. Growth factor concentrations and length of stimulation differed in those studies. These data may suggest that there exists either a threshold level of certain growth factors that may be required to bypass the PI3K–PTEN–Akt pathway, or that an alternate pathway for PI3K activation by growth factors exists, and that such a pathway could be cell-type-specific.

3.2.7. PI3K/AKT PATHWAY IN *CAENORHABDITIS ELEGANS*

Studies in *C. elegans* have provided genetic evidence that links the components of the PI3K–Akt pathway and PTEN (Fig. 2). In *C. elegans,* this pathway is involved in insulin receptor-like signaling, longevity, and dauer arrest (Fig. 2). Briefly, in *C. elegans,* the PI3K–Akt pathway involves the insulin receptor-like factor, DAF-2 *(132),* which interacts with an unidentified ligand and activates the PI3K homolog, AGE-1 *(133).* This signaling may be stimulated under reproductive growth conditions, and activates the Akt homologs, AKT-1 and AKT-2, which subsequently inactivate DAF-16 *(134)*, a homolog of the mammalian forkhead family of transcription factors. DAF-16 probably acts as a regulator of transcription *(135)*. Null mutations in *daf-2* and/or *age-1*, which increase longevity and produce the dauer state, a state of hibernation that is normally entered upon under circumstances of starvation or overcrowding, may be suppressed by null mutations of *daf-16 (135,136)*.

In addition, *daf-18* mutations have also been found to partially or completely suppress the *daf-2* and/or *age-1* mutants, respectively *(137–139)*. *daf-18* is the homolog of *PTEN,* and has been shown to antagonize the DAF-2–AGE-1 pathway *(140–143)*. Furthermore, loss of function mutations of *age-1* and *daf-2,* which produce a constitutive dauer phenotype, may be completely suppressed by certain *daf-18* mutations *(141–143)*. In *C.elegans,* levels of PtdIns-3,4,5-P_3 prevent dauer arrest. Certain *daf-18* mutations, which suppress either of the loss-of-function mutants, rescue reduced levels of PtdIns-3,4,5-P_3 and prevent dauer arrest *(140)*. In addition, although *daf-2* and *age-1* mutations increase life-span, mutations of *daf-18* decrease longevity *(141)*.

Overall, these studies corroborate the role of PTEN as a downregulator of the PI3K–Akt pathway, acting between PI3K and Akt. They also suggest that PTEN may, in addition, also modulate insulin signaling and longevity in mammalian cells, as well, although no studies have demonstrated this to date *(144)*.

3.2.8. SUMMARY

Current data support the notion that one of the primary mechanisms by which PTEN acts as a tumor suppressor is through its inhibition of PI3K signaling. Both regulation of cell survival and regulation of cell cycle progression can be ascribed to either loss of PTEN in primary tissues of re-expression of PTEN in tumor cells. This suggests that PTEN is both necessary and sufficient for these processes in specific contexts. Remaining questions include: What downstream targets of Akt or PI3K mediate these effects? And, what determines whether a particular cell is more inclined to depend on this pathway for survival, as opposed to proliferation? With respect to the latter question, it is clear that this is not simply a reflection of tumor-specific mutational events, because primary tissues in the mouse differ with respect to whether proliferation or cell survival is the net effect. Furthermore, a strict tissue-specific dependence does not exist because certain tumor cell lines, derived from different glioblastoma tumors, undergo either anoikis or cell cycle arrest, with reintroduction of PTEN *(86,128,145)*. One possibility is that, during the process of differentiation, cells undergo a transition from a reliance on PI3K signaling for proliferation to a reliance on PI3K for survival. Differences in cell lines derived from similar tissue types could be accounted for if such tumors arose from different populations of committed stem cells. For example, such a switch could revolve around the loss or gain of specific Akt substrates. This model would not necessarily preclude the possibility that additional tumor-derived mutations could further influence this balance.

3.3. PTEN as Protein Phosphatase

The protein phosphatase activity of PTEN, although not sufficient for PTEN function in regulating apoptosis or growth suppression, is linked to alterations in cell migration, spreading, and motility, and focal adhesion formation *(146–148)*. Overexpression of PTEN in fibroblasts or glioblastoma lines has been shown to reduce or inhibit integrin-mediated cell spreading and focal adhesion formation *(146)*. This is associated with a decrease in tyrosine phosphorylation of (FAK), as well as its downstream target, p130 Crk-associated substrate (p130^{Cas}); PTEN can dephosphorylate FAK in vitro. Furthermore, in contrast to the studies of apoptosis and cell cycle regulation, the G129E mutant of PTEN which lacks lipid phosphatase activity but retains protein phosphatase activity, is comparable to wt PTEN in these experiments *(146)*. Thus, this PTEN function is genetically distinguishable from the regulation of P13K signaling.

Subsequent studies have revealed that PTEN also decreases tyrosine phosphorylation of the adapter protein, Shc, and downregulates its downstream activation of mitogen-activated protein kinase (MAPK) *(148)*. This particular function appears to be involved in random cell motility *(147,148)*. In contrast, the FAK–p130^{Cas} pathway appears to be involved in a parallel pathway that directs persistent directional cell migration *(148)*. Tamura et al. *(149)* found that overexpression of FAK in PTEN-reconstituted cells inhibited P13K activity, but did not fully restore PtdIns-3,4,5-P$_3$ activity, compared to PTEN-deficient cells. This suggested that PTEN might have multiple effects on cellular physiology, and possibly on the P13K–Akt pathway: the regulation of PtdIns-3,4,5-p$_3$, through its lipid phosphatase activity, and the ability to dephosphorylate FAK and Shc, via its activity as a protein phosphatase.

In contrast to these data, *PTEN–/–* ES cells do not have excessive phosphorylation of FAK, nor obvious alterations in MAPK signaling *(65)*. Thus, if PTEN is involved in the regulation of these pathways, it is not necessary for appropriate regulation of the pathway. To date, mutants retaining protein phosphatase activity have been noted, but no mutant retaining only lipid phosphatase activity has been described. Therefore, although protein phosphatase activity is clearly not sufficient for inhibition of cell growth, it has not been possible to ask whether such activity is necessary.

3.4. Structure and Function

The minimal phosphatase domain of PTEN has been mapped to the region between amino acid residues 10 and 353 (Ramaswamy et al., unpublished data). This domain not only inhibits Akt kinase activity, but is also sufficient to induce a G1 arrest. The carboxyl-terminal (C-terminal) 50 residues (354–403) contains a PDZ-binding site *(94)*. However, this domain is not required for suppression of colony formation in soft agar, nor for the induction of apoptosis in LNCaP cells (Ramaswamy et al., unpublished data).

Although most mutations isolated occur in or near the catalytic motif *(87)*, there are a small number of mutations that are found within these last 50 amino acids. The C-terminal PDZ-binding site would be predicted to interact with proteins containing PDZ domains. PDZ-domain proteins, named for the first three members of the class (PSD-95, disks-large, and ZO-1), are protein–protein interaction domains that are thought to direct the assembly of multiprotein complexes at the cell membrane *(97)*. More recently, these domains have been implicated in the regulation of membranetrafficking of signaling receptors *(150)*. The presence of tumor-derived mutations raises the possibility that the C-terminus is required for tumor suppression in vivo. In keeping with this notion, deletion of the C-terminus leads to an approx fourfold decrease in the half-life of PTEN (Vasquez et al., unpublished data, and ref. *151*). Taken together, these data suggest that, although the C-terminus may be unnecessary for phosphatase function alone, it is important for regulating the stability and activity of the protein.

Whether the PDZ-binding domain and/or interaction with a PDZ domain containing protein is required for PTEN stability, and in turn for PTEN function as a tumor suppressor, remains to be determined.

4. MOUSE MODELS

The importance of *PTEN* in development and tumorigenesis has been elucidated by the study of PTEN–/– mice generated by multiple groups *(64,90,91,123)*. These mice do not survive beyond embryonic d 9.5, and are characterized by abnormal patterning and growth of cephalic and caudal regions *(123,152)*. Bromodeoxyuridine (BrdU) labeling of *PTEN*–/– embryos demonstrates marked increases in BrdU incorporation, suggesting that lethality and alterations in the patterning and growth of the cephalic and caudal regions may arise as a consequence of increased levels of cell proliferation *(123)*.

Studies of *PTEN*+/– chimeras and MEFs, as noted earlier, have demonstrated that they have an increased resistance to apoptotic stimuli *(123)*, and behave with the ability to proliferate in an anchorage-independent manner *(64)*. These cells also appear to have higher levels of activated Akt *(90,123)*.

Studies of *PTEN*+/– mice have reported different phenotypes. Suzuki et al. *(91)* and Stambolic et al. *(123)* have found a high incidence of T-cell lymphomas associated with LOH of the wt *PTEN* allele. Podyspanina et al. *(90)* have reported phenotypes similar to CD in humans with gastrointestinal polyps, follicular and papillary thyroid carcinomas, endometrial hyperplasia, prostate intraepithelial neoplasia, and lymphadenopathy associated with an apoptotic defect. Similarly, Di Cristofano et al. *(64)* have reported dysplastic changes of the colon, prostate, and skin, as well as marked lymphadenopathy. In that particular study, the observed changes appear to be associated with the retention of the wt allele.

These varying phenotypes may be explained by differences in breeding backgrounds or *PTEN* genotypes. However, all groups have reported a high incidence of lymphomas. Recent data from Di Cristofano et al. *(64)* have suggested that, in *PTEN*+/– mice, the development of lymphomas is secondary to a defect in Fas-mediated apoptosis, and possibly haploinsufficiency. This provides further evidence that PTEN is necessary to regulate the cell survival pathway, and, in mice, one of these downstream targets includes the death receptor, Fas.

5. CONCLUSION

The identification of *PTEN* as a TSG has led to the surprising discovery that it functions as both a protein and lipid phosphatase *(94,146)*. Through its specific activity against the phosphoinositides, PtdIns-3,4,5-P_3 and PtdIns-3,4-P_2, it regulates the P13K–Akt pathway, controlling both cell survival and cell proliferation *(97,144)*. Its protein phosphatase function appears to be important in another aspect of tumorigenesis, i.e., FAK-mediated cell spreading and motility *(146)*.

PTEN mutations occur in many tumor types, and are associated with different stages of tumorigenesis in certain tumor types (e.g., glioblastomas vs endometrial carcinomas). Currently, PTEN functions to regulate at least two distinct pathways: P13K–Akt and FAK, for at least three different purposes: cell cycle, cell survival, and cell motility regulation. PTEN antagonizes the P13K–Akt pathway, and, therefore, indirectly, through Akt, modulates its many downstream targets, many of which are proapoptotic factors, as well. Thus, several questions remain. What are the critical downstream targets of Akt that are required for tumorigenesis in the absence of PTEN? What are the mechanisms that underlie PTEN's role as a regulator of cell survival and cell cycle progression? Does PTEN differentially modulate the numerous targets of Akt in a tumor- and/or tissue-specific manner? Is PTEN itself regulated? And, what are the genetic alterations that, together with PTEN loss, cooperate to fully transform cells? Answers to these and other questions will probably be forthcoming in the near future.

REFERENCES

1. Isshiki K, Elder DE, Guerry D, Linnenbach AJ. Chromosome 10 allelic loss in malignant melanoma. *Genes Chromosomes Cancer* 1993; 8:178–184.
2. Thrash-Bingham CA, Greenberg RE, Howard S, Bruzel A, Bremer M, Goll A, et al. Comprehensive allelotyping of human renal cell carcinomas using microsatellite DNA probes. *Proc Natl Acad Sci USA* 1995; 92:2854–2858.
3. Simon M, von Deimling A, Larson JJ, Wellenreuther R, Kaskel P, Waha A, et al. Allelic losses on chromosomes 14, 10, and 1 in atypical and malignant meningiomas: a genetic model of meningioma progression. *Cancer Res* 1995; 55:4696–4701.
4. Zedenius J, Wallin G, Svensson A, Bovee J, Hoog A, Backdahl M, Larsson C. Deletions of the long arm of chromosome 10 in progression of follicular thyroid tumors. *Hum Genet* 1996; 97:299–303.
5. Petersen I, Langreck H, Wolf G, Schwendel A, Psille R, Vogt P, et al. Small-cell lung cancer is characterized by a high incidence of deletions on chromosomes 3p, 4q, 5q, 10q, 13q and 17p. *Br J Cancer* 1997; 75:79–86.
6. Virmani AK, Fong KM, Kodagoda D, McIntire D, Hung J, Tonk V, Minna JD, Gazdar AF. Allelotyping demonstrates common and distinct patterns of chromosomal loss in human lung cancer types. *Genes Chromosomes Cancer* 1998; 21:308–319.
7. Lamszus K, Kluwe L, Matschke J, Meissner H, Laas R, Westphal M. Allelic losses at 1p, 9q, 10q, 14q, and 22q in the progression of aggressive meningiomas and undifferentiated meningeal sarcomas. *Cancer Genet Cytogenet* 1999; 110:103–110.
8. Rebbeck TR, Godwin AK, Buetow KH. Variability in loss of constitutional heterozygosity across loci and among individuals: association with candidate genes in ductal breast carcinoma. *Mol Carcinog* 1996; 17:117–125.
9. Sourvinos G, Kiaris H, Tsikkinis A, Vassilaros S, Spandidos DA. Microsatellite instability and loss of heterozygosity in primary breast tumours. *Tumour Biol* 1997; 18:157–166.
10. Rasheed BK, McLendon RE, Friedman HS, Friedman AH, Fuchs HE, Bigner DD, Bigner SH. Chromosome 10 deletion mapping in human gliomas: a common deletion region in 10q25. *Oncogene* 1995; 10:2243–2246.
11. Gray IC, Phillips SM, Lee SJ, Neoptolemos JP, Weissenbach J, Spurr NK. Loss of the chromosomal region 10q23–25 in prostate cancer. *Cancer Res* 1995; 55:4800–4803.
12. Jones MH, Koi S, Fujimoto I, Hasumi K, Kato K, Nakamura Y. Allelotype of uterine cancer by analysis of RFLP and microsatellite polymorphisms: frequent loss of heterozygosity on chromosome arms 3p, 9q, 10q, and 17p. *Genes Chromosomes Cancer* 1994; 9:119–123.
13. Marsh DJ, Zheng Z, Zedenius J, Kremer H, Padberg GW, Larsson C, Longy M, Eng C. Differential loss of heterozygosity in the region of the Cowden locus within 10q22–23 in follicular thyroid adenomas and carcinomas. *Cancer Res* 1997; 57:500–503.
14. Bose S, Wang SI, Terry MB, Hibshoosh H, Parsons R. Allelic loss of chromosome 10q23 is associated with tumor progression in breast carcinomas. *Oncogene* 1998; 17:123–127.
15. Nelen MR, van Staveren WC, Peeters EA, Hassel MB, Gorlin RJ, Hamm H, et al. Germline mutations in the PTEN/MMAC1 gene in patients with Cowden disease. *Hum Mol Genet* 1997; 6:1383–1387.
16. Pershouse MA, Stubblefield E, Hadi A, Killary AM, Yung WK, Steck PA. Analysis of the functional role of chromosome 10 loss in human glioblastomas. *Cancer Res* 1993; 53:5043–5050.
17. Nihei N, Ichikawa T, Kawana Y, Kuramochi H, Kugo H, Oshimura M, et al. Localization of metastasis suppressor gene(s) for rat prostatic cancer to the long arm of human chromosome 10. *Genes Chromosomes Cancer* 1995; 14:112–119.

18. Steck PA, Pershouse MA, Jasser SA, Yung WK, Lin H, Ligon AH, et al. Identification of a candidate tumour suppressor gene, MMAC1, at chromosome 10q23.3 that is mutated in multiple advanced cancers. *Nat Genet* 1997; 15:356–362.

19. Li DM, Sun H. TEP1, encoded by a candidate tumor suppressor locus, is a novel protein tyrosine phosphatase regulated by transforming growth factor beta. *Cancer Res* 1997; 57:2124–2129.

20. Li J, Yen C, Liaw D, Podsypanina K, Bose S, Wang SI, et al. PTEN, a putative protein tyrosine phosphatase gene mutated in human brain, breast, and prostate cancer. *Science* 1997; 275:1943–1947.

21. Marsh DJ, Coulon V, Lunetta KL, Rocca-Serra P, Dahia PL, Zheng Z, et al. Mutation spectrum and genotype-phenotype analyses in Cowden disease and Bannayan-Zonana syndrome, two hamartoma syndromes with germline PTEN mutation. *Hum Mol Genet* 1998; 7:507–515.

22. Liaw D, Marsh DJ, Li J, Dahia PL, Wang SI, Zheng Z, et al. Germline mutations of the PTEN gene in Cowden disease, an inherited breast and thyroid cancer syndrome. *Nat Genet* 1997; 16:64–67.

23. Boni R, Vortmeyer AO, Burg G, Hofbauer G, Zhuang Z. The PTEN tumour suppressor gene and malignant melanoma. *Melanoma Res* 1998; 8:300–302.

24. Bostrom J, Cobbers JM, Wolter M, Tabatabai G, Weber RG, Lichter P, Collins VP, Reifenberger G. Mutation of the PTEN (MMAC1) tumor suppressor gene in a subset of glioblastomas but not in meningiomas with loss of chromosome arm 10q. *Cancer Res* 1998; 58:29–33.

25. Ueda K, Nishijima M, Inui H, Watatani M, Yayoi E, Okamura J, et al. Infrequent mutations in the PTEN/MMAC1 gene among primary breast cancers. *Jpn J Cancer Res* 1998; 89:17–21.

26. Peters N, Wellenreuther R, Rollbrocker B, Hayashi Y, Meyer-Puttlitz B, Duerr EM, et al. Analysis of the PTEN gene in human meningiomas. *Neuropathol Appl Neurobiol* 1998; 24:3–8.

27. Pesche S, Latil A, Muzeau F, Cussenot O, Fournier G, Longy M, Eng C, Lidereau R. PTEN/MMAC1/TEP1 involvement in primary prostate cancers. *Oncogene* 1998; 16:2879–2883.

28. Yokomizo A, Tindall DJ, Drabkin H, Gemmill R, Franklin W, Yang P, et al. PTEN/MMAC1 mutations identified in small cell, but not in non-small cell lung cancers. *Oncogene* 1998; 17:475–479.

29. Tsao H, Zhang X, Benoit E, Haluska FG. Identification of PTEN/MMAC1 alterations in uncultured melanomas and melanoma cell lines. *Oncogene* 1998; 16:3397–3402.

30. Sakai A, Thieblemont C, Wellmann A, Jaffe ES, Raffeld M. PTEN gene alterations in lymphoid neoplasms. *Blood* 1998; 92:3410–3415.

31. Forgacs E, Biesterveld EJ, Sekido Y, Fong K, Muneer S, Wistuba, II, et al. Mutation analysis of the PTEN/MMAC1 gene in lung cancer. *Oncogene* 1998; 17:1557–1565.

32. Davies MP, Gibbs FE, Halliwell N, Joyce KA, Roebuck MM, Rossi ML, et al. Mutation in the PTEN/MMAC1 gene in archival low grade and high grade gliomas. *Br J Cancer* 1999; 79:1542–1548.

33. Aveyard JS, Skilleter A, Habuchi T, Knowles MA. Somatic mutation of PTEN in bladder carcinoma. *Br J Cancer* 1999; 80:904–908.

34. Chen ST, Yu SY, Tsai M, Yeh KT, Wang JC, Kao MC, Shih MC, Chang JG. Mutation analysis of the putative tumor suppression gene PTEN/MMAC1 in sporadic breast cancer. *Breast Cancer Res Treat* 1999; 55:85–89.

35. Dahia PL, Marsh DJ, Zheng Z, Zedenius J, Komminoth P, Frisk T, et al. Somatic deletions and mutations in the Cowden disease gene, PTEN, in sporadic thyroid tumors. *Cancer Res* 1997; 57:4710–4713.

36. Kurose K, Bando K, Fukino K, Sugisaki Y, Araki T, Emi M. Somatic mutations of the PTEN/MMAC1 gene in fifteen Japanese endometrial cancers: evidence for inactivation of both alleles. *Jpn J Cancer Res* 1998; 89:842–848.

37. Freihoff D, Kempe A, Beste B, Wappenschmidt B, Kreyer E, Hayashi Y, et al. Exclusion of a major role for the PTEN tumour-suppressor gene in breast carcinomas. *Br J Cancer* 1999; 79:754–758.

38. Feilotter HE, Nagai MA, Boag AH, Eng C, Mulligan LM. Analysis of PTEN and the 10q23 region in primary prostate carcinomas. *Oncogene* 1998; 16:1743–1748.

39. Duerr EM, Rollbrocker B, Hayashi Y, Peters N, Meyer-Puttlitz B, Louis DN, et al. PTEN mutations in gliomas and glioneuronal tumors. *Oncogene* 1998; 16:2259–2264.

40. Butler MP, Wang SI, Chaganti RS, Parsons R, Dalla-Favera R. Analysis of PTEN mutations and deletions in B-cell non-Hodgkin's lymphomas. *Genes Chromosomes Cancer* 1999; 24:322–327.

41. Cairns P, Evron E, Okami K, Halachmi N, Esteller M, Herman JG, et al. Point mutation and homozygous deletion of PTEN/MMAC1 in primary bladder cancers. *Oncogene* 1998; 16:3215–3218.

42. Simpkins SB, Peiffer-Schneider S, Mutch DG, Gersell D, Goodfellow PJ. PTEN mutations in endometrial cancers with 10q LOH: additional evidence for the involvement of multiple tumor suppressors. *Gynecol Oncol* 1998; 71:391–395.

43. Rhei E, Kang L, Bogomolniy F, Federici MG, Borgen PI, Boyd J. Mutation analysis of the putative tumor suppressor gene PTEN/MMAC1 in primary breast carcinomas. *Cancer Res* 1997; 57:3657–3659.

44. Guldberg P, thor Straten P, Birck A, Ahrenkiel V, Kirkin AF, Zeuthen J. Disruption of the MMAC1/PTEN gene by deletion or mutation is a frequent event in malignant melanoma. *Cancer Res* 1997; 57:3660–3663.

45. Wang SI, Puc J, Li J, Bruce JN, Cairns P, Sidransky D, Parsons R. Somatic mutations of PTEN in glioblastoma multiforme. *Cancer Res* 1997; 57:4183–4186.

46. Rasheed BK, Stenzel TT, McLendon RE, Parsons R, Friedman AH, Friedman HS, Bigner DD, Bigner SH. PTEN gene mutations are seen in high-grade but not in low-grade gliomas. *Cancer Res* 1997; 57:4187–4190.

47. Tashiro H, Blazes MS, Wu R, Cho KR, Bose S, Wang SI, et al. Mutations in PTEN are frequent in endometrial carcinoma but rare in other common gynecological malignancies. *Cancer Res* 1997; 57:3935–3940.

48. Risinger JI, Hayes AK, Berchuck A, Barrett JC. PTEN/MMAC1 mutations in endometrial cancers. *Cancer Res* 1997; 57:4736–4738.

49. Halachmi N, Halachmi S, Evron E, Cairns P, Okami K, Saji M, et al. Somatic mutations of the PTEN tumor suppressor gene in sporadic follicular thyroid tumors. *Genes Chromosomes Cancer* 1998; 23:239–243.

50. Steck PA, Lin H, Langford LA, Jasser SA, Koul D, Yung WK, Pershouse MA. Functional and molecular analyses of 10q deletions in human gliomas. *Genes Chromosomes Cancer* 1999; 24:135–143.

51. Somerville RP, Shoshan Y, Eng C, Barnett G, Miller D, Cowell JK. Molecular analysis of two putative tumour suppressor genes, PTEN and DMBT, which have been implicated in glioblastoma multiforme disease progression.*Oncogene* 1998; 17:1755–1757.

52. Rasheed BK, Wiltshire RN, Bigner SH, Bigner DD. Molecular pathogenesis of malignant gliomas. *Curr Opin Oncol* 1999; 11:162–167.

53. Liu W, James CD, Frederick L, Alderete BE, Jenkins RB. PTEN/MMAC1 mutations and EGFR amplification in glioblastomas. *Cancer Res* 1997; 57:5254–5257.

54. Gray IC, Stewart LM, Phillips SM, Hamilton JA, Gray NE, Watson GJ, Spurr NK, Snary D. Mutation and expression analysis of the putative prostate tumour-suppressor gene PTEN. *Br J Cancer* 1998; 78:1296–1300.

55. Cairns P, Okami K, Halachmi S, Halachmi N, Esteller M, Herman JG, et al. Frequent inactivation of PTEN/MMAC1 in primary prostate cancer. *Cancer Res* 1997; 57:4997–5000.

56. McMenamin ME, Soung P, Perera S, Kaplan I, Loda M, Sellers WR. Loss of PTEN expression in paraffin-embedded primary prostate cancer correlates with high Gleason score and advanced stage. *Cancer Res* 1999; 59:4291–4296.

57. Komiya A, Suzuki H, Ueda T, Yatani R, Emi M, Ito H, Shimazaki J. Allelic losses at loci on chromosome 10 are associated with metastasis and progression of human prostate cancer. *Genes Chromosomes Cancer* 1996; 17:245–253.

58. Healy E, Belgaid C, Takata M, Harrison D, Zhu NW, Burd DA, et al. Prognostic significance of allelic losses in primary melanoma. *Oncogene* 1998; 16:2213–2218.

59. Maxwell GL, Risinger JI, Gumbs C, Shaw H, Bentley RC, Barrett JC, Berchuck A, Futreal PA. Mutation of the PTEN tumor suppressor gene in endometrial hyperplasias. *Cancer Res* 1998; 58:2500–2503.

60. Sellers WR, Kaelin WG, Jr. Role of the retinoblastoma protein in the pathogenesis of human cancer. *J Clin Oncol* 1997; 15:3301–3312.

61. Feilotter HE, Coulon V, McVeigh JL, Boag AH, Dorion-Bonnet F, Duboue B, et al. Analysis of the 10q23 chromosomal region and the PTEN gene in human sporadic breast carcinoma. *Br J Cancer* 1999; 79:718–723.

62. FitzGerald MG, Marsh DJ, Wahrer D, Bell D, Caron S, Shannon KE, et al. Germline mutations in PTEN are an infrequent cause of genetic predisposition to breast cancer. *Oncogene* 1998; 17:727–731.

63. Whang YE, Wu X, Suzuki H, Reiter RE, Tran C, Vessella RL, et al. Inactivation of the tumor suppressor PTEN/MMAC1 in advanced human prostate cancer through loss of expression. *Proc Natl Acad Sci USA* 1998; 95:5246–5250.

64. Di Cristofano A, Pesce B, Cordon-Cardo C, Pandolfi PP. Pten is essential for embryonic development and tumour suppression. *Nat Genet* 1998; 19:348–355.

65. Sun H, Lesche R, Li DM, Liliental J, Zhang H, Gao J, et al. PTEN modulates cell cycle progression and cell survival by regulating phosphatidylinositol 3,4,5,-trisphosphate and Akt/protein kinase B signaling pathway. *Proc Natl Acad Sci USA* 1999; 96:6199–6204.

66. Di Cristofano A, Kotsi P, Peng YF, Cordon-Cardo C, Elkon KB, Pandolfi PP. Impaired fas response and autoimmunity in Pten(+/–) mice. *Science* 1999; 285:2122–2125.

67. Mollenhauer J, Wiemann S, Scheurlen W, Korn B, Hayashi Y, Wilgenbus KK, von Deimling A, Poustka A. DMBT1, a new member of the SRCR superfamily, on chromosome 10q25.3–26.1 is deleted in malignant brain tumours. *Nat Genet* 1997; 17:32–39.

68. Wechsler DS, Hawkins AL, Li X, Jabs EW, Griffin CA, Dang CV. Localization of the human Mxil transcription factor gene (MXI1) to chromosome 10q24–q25. *Genomics* 1994; 21:669–672.

69. Prochownik EV, Eagle Grove L, Deubler D, Zhu XL, Stephenson RA, Rohr LR, Yin X, Brothman AR. Commonly occurring loss and mutation of the MXI1 gene in prostate cancer. *Genes Chromosomes Cancer* 1998; 22:295–304.

70. Kuczyk MA, Serth J, Bokemeyer C, Schwede J, Herrmann R, Machtens S, et al. The MXI1 tumor suppressor gene is not mutated in primary prostate cancer. *Oncol Rep* 1998; 5:213–216.

71. Edwards SM, Dearnaley DP, Ardern-Jones A, Hamoudi RA, Easton DF, Ford D, et al. No germline mutations in the dimerization domain of MXI1 in prostate cancer clusters. The CRC/BPG UK Familial Prostate Cancer Study Collaborators. Cancer Research Campaign/British Prostate Group. *Br J Cancer* 1997; 76:992–1000.

72. Eagle LR, Yin X, Brothman AR, Williams BJ, Atkin NB, Prochownik EV. Mutation of the MXI1 gene in prostate cancer. *Nat Genet* 1995; 9:249–255.

73. Kawamata N, Park D, Wilczynski S, Yokota J, Koeffler HP. Point mutations of the Mxil gene are rare in prostate cancers. *Prostate* 1996; 29:191–193.

74. Kim SK, Ro JY, Kemp BL, Lee JS, Kwon TJ, Hong WK, Mao L. Identification of two distinct tumor-suppressor loci on the long arm of chromosome 10 in small cell lung cancer. *Oncogene* 1998; 17:1749–1753.

75. Fults D, Pedone CA, Thompson GE, Uchiyama CM, Gumpper KL, Iliev D, et al. Microsatellite deletion mapping on chromosome 10q and mutation analysis of MMAC1, FAS, and MXI1 in human glioblastoma multiforme. *Int J Oncol* 1998; 12:905–910.

76. Wu W, Kemp BL, Proctor ML, Gazdar AF, Minna JD, Hong WK, Mao L. Expression of DMBT1, a candidate tumor suppressor gene, is frequently lost in lung cancer. *Cancer Res* 1999; 59:1846–1851.

77. Lin H, Bondy ML, Langford LA, Hess KR, Delclos GL, Wu X, et al. Allelic deletion analyses of MMAC/PTEN and DMBT1 loci in gliomas: relationship to prognostic significance. *Clin Cancer Res* 1998; 4:2447–2454.

78. Mori M, Shiraishi T, Tanaka S, Yamagata M, Mafune K, Tanaka Y, et al. Lack of DMBT1 expression in oesophageal, gastric and colon cancers. *Br J Cancer* 1999; 79:211–213.

79. Lee SH, Shin MS, Park WS, Kim SY, Dong SM, Pi JH, et al. Alterations of Fas (APO-1/CD95) gene in transitional cell carcinomas of urinary bladder. *Cancer Res* 1999; 59:3068–3072.

80. Lee SH, Shin MS, Park WS, Kim SY, Kim HS, Han JY, et al. Alterations of Fas (Apo-1/CD95) gene in non-small cell lung cancer. *Oncogene* 1999; 18:3754–3760.

81. Shin MS, Park WS, Kim SY, Kim HS, Kang SJ, Song KY, et al. Alterations of Fas (Apo-1/CD95) gene in cutaneous malignant melanoma. *Am J Pathol* 1999; 154:1785–1791.

82. Abdel-Rahman W, Arends M, Morris R, Ramadan M, Wyllie A. Death pathway genes Fas (Apo-1/CD95) and Bik (Nbk) show no mutations in colorectal carcinomas [letter]. *Cell Death Differ* 1999; 6:387–388.

83. Butler LM, Hewett PJ, Butler WJ, Cowled PA. Down-regulation of Fas gene expression in colon cancer is not a result of allelic loss or gene rearrangement. *Br J Cancer* 1998; 77:1454–1459.

84. Chi H, Tiller GE, Dasouki MJ, Romano PR, Wang J, O'Keefe RJ, et al. Multiple inositol polyphosphate phosphatase: evolution as a distinct group within the histidine phosphatase family and chromosomal localization of the human and mouse genes to chromosomes 10q23 and 19. *Genomics* 1999; 56:324–336.

85. Cheney IW, Johnson DE, Vaillancourt MT, Avanzini J, Morimoto A, Demers GW, et al. Suppression of tumorigenicity of glioblastoma cells by adenovirus-mediated MMAC1/PTEN gene transfer. *Cancer Res* 1998; 58:2331–2334.

86. Li DM, Sun H. PTEN/MMAC1/TEP1 suppresses the tumorigenicity and induces G1 cell cycle arrest in human glioblastoma cells. *Proc Natl Acad Sci USA* 1998; 95:15,406–15,411.

87. Myers MP, Pass I, Batty IH, Van der Kaay J, Stolarov JP, Hemmings BA, et al. The lipid phosphatase activity of PTEN is critical for its tumor suppressor function. *Proc Natl Acad Sci USA* 1998; 95:13,513–13,518.

88. Wu X, Senechal K, Neshat MS, Whang YE, Sawyers CL. The PTEN/MMAC1 tumor suppressor phosphatase functions as a negative regulator of the phosphoinositide 3-kinase/Akt pathway. *Proc Natl Acad Sci USA* 1998; 95:15,587–15,591.

89. Robertson GP, Furnari FB, Miele ME, Glendening MJ, Welch DR, Fountain JW, et al. In vitro loss of heterozygosity targets the PTEN/MMAC1 gene in melanoma. *Proc Natl Acad Sci USA* 1998; 95:9418–9423.

90. Podsypanina K, Ellenson LH, Nemes A, Gu J, Tamura M, Yamada KM, et al. Mutation of Pten/Mmac1 in mice causes neoplasia in multiple organ systems. *Proc Natl Acad Sci USA* 1999; 96:1563–1568.

91. Suzuki A, de la Pompa JL, Stambolic V, Elia AJ, Sasaki T, del Barco Barrantes I, et al. High cancer susceptibility and embryonic lethality associated with mutation of the PTEN tumor suppressor gene in mice. *Curr Biol* 1998; 8:1169–1178.

92. Haynie DT, Ponting CP. The N-terminal domains of tensin and auxilin are phosphatase homologues. *Protein Science* 1996; 5:2643–2646.

93. Myers MP, Stolarov JP, Eng C, Li J, Wang SI, Wigler MH, Parsons R, Tonks NK. P-TEN, the tumor suppressor from human chromosome 10q23, is a dual-specificity phosphatase. *Proc Natl Acad Sci USA* 1997; 94:9052–9057.

94. Maehama T, Dixon JE. The tumor suppressor, PTEN/MMAC1, dephosphorylates the lipid second messenger, phosphatidylinositol 3,4,5-trisphosphate. *J Biol Chem* 1998; 273:13,375–13,378.

95. Fruman DA, Rameh LE, Cantley LC. Phosphoinositide binding domains: embracing 3-phosphate. *Cell* 1999; 97:817–820.

96. Anderson RA, Boronenkov IV, Doughman SD, Kunz J, Loijens JC. Phosphatidylinositol phosphate kinases, a multifaceted family of signaling enzymes. *J Biol Chem* 1999; 274:9907–9910.

97. Cantley LC, Neel BG. New insights into tumor suppression: PTEN suppresses tumor formation by restraining the phosphoinositide 3-kinase/AKT pathway. *Proc Natl Acad Sci USA* 1999; 96:4240–4245.

98. Khwaja A, Rodriguez-Viciana P, Wennstrom S, Warne PH, Downward J. Matrix adhesion and Ras transformation both activate a phosphoinositide 3-OH kinase and protein kinase B/Akt cellular survival pathway. *EMBO J* 1997; 16:2783–2793.

99. Toker A, Cantley LC. Signalling through the lipid products of phosphoinositide-3-OH kinase. *Nature* 1997; 387:673–676.

100. Downward J. Mechanisms and consequences of activation of protein kinase B/Akt. *Curr Opin Cell Biol* 1998; 10:262–267.

101. Kohn AD, Takeuchi F, Roth RA. Akt, a pleckstrin homology domain containing kinase, is activated primarily by phosphorylation. *J Biol Chem* 1996; 271:21,920–21,926.

102. Alessi DR, Cohen P. Mechanism of activation and function of protein kinase B. *Curr Opin Genet Dev* 1998; 8:55–62.

103. Balendran A, Casamayor A, Deak M, Paterson A, Gaffney P, Currie R, Downes CP, Alessi DR. PDK1 acquires PDK2 activity in the presence of a synthetic peptide derived from the carboxyl terminus of PRK2. *Curr Biol* 1999; 9:393–404.

104. Bellacosa A, Testa JR, Staal SP, Tsichlis PN. A retroviral oncogene, akt, encoding a serine-threonine kinase containing an SH2-like region. *Science* 1991; 254:274–277.

105. Coffer PJ, Woodgett JR. Molecular cloning and characterisation of a novel putative protein-serine kinase related to the cAMP-dependent and protein kinase C families [published erratum appears in Eur J Biochem 1992; 205:1217]. *Eur J Biochem* 1991; 201:475–481.

106. Jones PF, Jakubowicz T, Pitossi FJ, Maurer F, Hemmings BA. Molecular cloning and identification of a serine/threonine protein kinase of the second-messenger subfamily. *Proc Natl Acad Sci USA* 1991; 88:4171–4175.

107. Aoki M, Batista O, Bellacosa A, Tsichlis P, Vogt PK. The akt kinase: molecular determinants of oncogenicity. *Proc Natl Acad Sci USA* 1998; 95:14,950–14,955.

108. Kops GJ, de Ruiter ND, De Vries-Smits AM, Powell DR, Bos JL, Burgering BM. Direct control of the Forkhead transcription factor AFX by protein kinase B. *Nature* 1999; 398:630–634.

109. Brunet A, Bonni A, Zigmond MJ, Lin MZ, Juo P, Hu LS, et al. Akt promotes cell survival by phosphorylating and inhibiting a Forkhead transcription factor. *Cell* 1999; 96:857–868.

110. Tang ED, Nunez G, Barr FG, Guan KL. Negative regulation of the forkhead transcription factor FKHR by Akt. *J Biol Chem* 1999; 274:16,741–16,746.

111. Cross DA, Alessi DR, Cohen P, Andjelkovich M, Hemmings BA. Inhibition of glycogen synthase kinase-3 by insulin mediated by protein kinase B. *Nature* 1995; 378:785–789.

112. Burgering BM, Coffer PJ. Protein kinase B (c-Akt) in phosphatidylinositol-3-OH kinase signal transduction. *Nature* 1995; 376:599–602.
113. Gingras AC, Kennedy SG, O'Leary MA, Sonenberg N, Hay N. 4E-BP1, a repressor of mRNA translation, is phosphorylated and inactivated by the Akt(PKB) signaling pathway. *Genes Dev* 1998; 12:502–513.
114. Ozes ON, Mayo LD, Gustin JA, Pfeffer SR, Pfeffer LM, Donner DB. NF-KappaB activation by tumor necrosis factor requires the Akt serine-threonine kinase. *Nature* 1999; 401:82–85.
115. Romashkova JA, Makarov SS. NF-KappaB is a target of anti-apoptotic PDGF signalling. *Nature* 1999; 401:86–90.
116. Kane LP, Shapiro VS, Stokoe D, Weiss A. Induction of NF-kappaB by the Akt/PKB kinase. *Curr Biol* 1999; 9:601–604.
117. Dimmeler S, Fleming I, Fisslthaler B, Hermann C, Busse R, Zeiher AM. Activation of nitric oxide synthase in endothelial cells by Akt-dependent phosphorylation. *Nature* 1999; 399:601–605.
118. Fulton D, Gratton JP, McCabe TJ, Fontana J, Fujio Y, Walsh K, Franke TF, Papapetropoulos A, Sessa WC. Regulation of endothelium-derived nitric oxide production by the protein kinase Akt. *Nature* 1999; 399:597–601.
119. Michell BJ, Griffiths JE, Mitchelhill KI, Rodriguez-Crespo I, Tiganis T, Bozinovski S, et al. The Akt kinase signals directly to endothelial nitric oxide synthase. *Curr Biol* 1999; 12:845–848.
120. Cardone MH, Roy N, Stennicke HR, Salvesen GS, Franke TF, Stanbridge E, Frisch S, Reed JC. Regulation of cell death protease caspase-9 by phosphorylation. *Science* 1998; 282:1318–1321.
121. Datta SR, Dudek H, Tao X, Masters S, Fu H, Gotoh Y, Greenberg ME. Akt phosphorylation of BAD couples survival signals to the cell-intrinsic death machinery. *Cell* 1997; 91:231–241.
122. Gingras AC, Gygi SP, Raught B, Polakiewicz RD, Abraham RT, Hoekstra MF, Aebersold R, Sonenberg N. Regulation of 4E-BP1 phosphorylation: a novel two-step mechanism. *Genes Dev* 1999; 13:1422–1437.
123. Stambolic V, Suzuki A, de la Pompa JL, Brothers GM, Mirtsos C, Sasaki T, et al. Negative regulation of PKB/Akt-dependent cell survival by the tumor suppressor PTEN. *Cell* 1998; 95:29–39.
124. Ramaswamy S, Nakamura N, Vazquez F, Batt DB, Perera S, Roberts TM, Sellers WR. Regulation of G1 progression by the PTEN tumor suppressor protein is linked to inhibition of the phosphatidylinositol 3-kinase/Akt pathway. *Proc Natl Acad Sci USA* 1999; 96:2110–2115.
125. Haas-Kogan D, Shalev N, Wong M, Mills G, Yount G, Stokoe D. Protein kinase B (PKB/Akt) activity is elevated in glioblastoma cells due to mutation of the tumor suppressor PTEN/MMAC. *Curr Biol* 1998; 8:1195–1198.
126. Dahia PLM, Aguiar RCT, Alberta J, Kum JB, Caron S, Sill H, et al. PTEN is inversely correlated with the cell survival factor Akt/PKB and is inactivated via multiple mechanisms in haematological malignancies. *Hum Mol Genet* 1999; 8:185–193.
127. Li J, Simpson L, Takahashi M, Miliaresis C, Myers MP, Tonks N, Parsons R. The PTEN/MMAC1 tumor suppressor induces cell death that is rescued by the AKT/protein kinase B oncogene. *Cancer Res* 1998; 58:5667–5672.
128. Furnari FB, Huang HJ, Cavenee WK. The phosphoinositol phosphatase activity of PTEN mediates a serum-sensitive G1 growth arrest in glioma cells. *Cancer Res* 1998; 58:5002–5008.
129. Davies MA, Koul D, Dhesi H, Berman R, McDonnell TJ, McConkey D, Yung WK, Steck PA. Regulation of Akt/PKB activity, cellular growth, and apoptosis in prostate carcinoma cells by MMAC/PTEN. *Cancer Res* 1999; 59:2551–2556.
130. Hunter T, Pines J. Cyclins and cancer. II: Cyclin D and CDK inhibitors come of age. *Cell* 1994; 79:573–582.
131. Carson JP, Kulik G, Weber MJ. Antiapoptotic signaling in LNCaP prostate cancer cells: a survival signaling pathway independent of phosphatidylinositol 3'-kinase and Akt/protein kinase B. *Cancer Res* 1999; 59:1449–1453.
132. Kimura KD, Tissenbaum HA, Liu Y, Ruvkun G. daf-2, an insulin receptor-like gene that regulates longevity and diapause in Caenorhabditis elegans. *Science* 1997; 277:942–946.
133. Morris JZ, Tissenbaum HA, Ruvkun G. A phosphatidylinositol-3-OH kinase family member regulating longevity and diapause in Caenorhabditis elegans. *Nature* 1996; 382:536–539.
134. Paradis S, Ruvkun G. Caenorhabditis elegans Akt/PKB transduces insulin receptor-like signals from AGE-1 P13 kinase to the DAF-16 transcription factor. *Genes Dev* 1998; 12:2488–2498.
135. Ogg S, Paradis S, Gottlieb S, Patterson GI, Lee L, Tissenbaum HA, Ruvkun G. The Fork head transcription factor DAF-16 transduces insulin-like metabolic and longevity signals in C. elegans. *Nature* 1997; 389:994–999.

136. Kenyon C, Chang J, Gensch E, Rudner A, Tabtiang R. A C. elegans mutant that lives twice as long as wild type. *Nature* 1993; 366:461–464.

137. Gottlieb S, Ruvkun G. daf-2, daf-16 and daf-23: genetically interacting genes controlling Dauer formation in Caenorhabditis elegans. *Genetics* 1994; 137:107–120.

138. Larsen PL, Albert PS, Riddle DL. Genes that regulate both development and longevity in Caenorhabditis elegans. *Genetics* 1995; 139:1567–1583.

139. Dorman JB, Albinder B, Shroyer T, Kenyon C. The age-1 and daf-2 genes function in a common pathway to control the lifespan of Caenorhabditis elegans. *Genetics* 1995; 141:1399–1406.

140. Gil EB, Malone Link E, Liu LX, Johnson CD, Lees JA. Regulation of the insulin-like developmental pathway of Caenorhabditis elegans by a homolog of the PTEN tumor suppressor gene. *Proc Natl Acad Sci USA* 1999; 96:2925–2930.

141. Mihaylova VT, Borland CZ, Manjarrez L, Stern MJ, Sun H. The PTEN tumor suppressor homolog in Caenorhabditis elegans regulates longevity and dauer formation in an insulin receptor-like signaling pathway. *Proc Natl Acad Sci USA* 1999; 96:7427–7432.

142. Ogg S, Ruvkun G. The C. elegans PTEN homolog, DAF-18, acts in the insulin receptor-like metabolic signaling pathway. *Mol Cell* 1998; 2:887–893.

143. Rouault JP, Kuwabara PE, Sinilnikova OM, Duret L, Thierry-Mieg D, Billaud M. Regulation of dauer larva development in Caenorhabditis elegans by daf-18, a homologue of the tumour suppressor PTEN. *Curr Biol* 1999; 9:329–332.

144. Maehama T, Dixon JE. PTEN: a tumour suppressor that functions as a phospholipid phosphatase. *Trends Cell Biol* 1999; 9:125–128.

145. Davies MA, Lu Y, Sano T, Fang X, Tang P, LaPushin R, et al. Adenoviral transgene expression of MMAC/PTEN in human glioma cells inhibits Akt activation and induces anoikis [published erratum appears in Cancer Res 1999; 59:1167]. *Cancer Res* 1998; 58:5285–5290.

146. Tamura M, Gu J, Matsumoto K, Aota S, Parsons R, Yamada KM. Inhibition of cell migration, spreading, and focal adhesions by tumor suppressor PTEN. *Science* 1998; 280:1614–1617.

147. Gu J, Tamura M, Yamada KM. Tumor suppressor PTEN inhibits integrin- and growth factor-mediated mitogen-activated protein (MAP) kinase signaling pathways. *J Cell Biol* 1998; 143:1375–1383.

148. Gu J, Tamura M, Pankov R, Danen EH, Takino T, Matsumoto K, Yamada KM. Shc and FAK differentially regulate cell motility and directionality modulated by PTEN. *J Cell Biol* 1999; 146:389–403.

149. Tamura M, Gu J, Takino T, Yamada KM. Tumor suppressor PTEN inhibition of cell invasion, migration, and growth: differential involvement of focal adhesion kinase and p130Cas. *Cancer Res* 1999; 59:442–449.

150. Cao TT, Deacon HW, Reczek D, Bretscher A, von Zastrow M. A Kinase-regulated PDZ-domain interaction controls endocytic sorting of the beta2-adrenergic receptor. *Nature* 1999; 401:286–290.

151. Georgescu MM, Kirsch KH, Akagi T, Shishido T, Hanafusa H. The tumor-suppressor activity of PTEN is regulated by its carboxyl-terminal region. *Proc Natl Acad Sci USA* 1999; 96:10,182–10,187.

152. Suzuki H, Freije D, Nusskern DR, Okami K, Cairns P, Sidransky D, Isaacs WB, Bova GS. Interfocal heterogeneity of PTEN/MMAC1 gene alterations in multiple metastatic prostate cancer tissues. *Cancer Res* 1998; 58:204–209.

11   Neurofibromatoses

André Bernards, PhD and
Andrea I. McClatchey, PhD

CONTENTS

1. INTRODUCTION

Neurofibromatosis type 1 (NF1, or von Recklinghausen's NF; formerly, also, peripheral NF), and neurofibromatosis type 2 (NF2, formerly, bilateral acoustic or central NF) are inherited neurocutaneous disorders or phakomatoses. Diagnostic criteria for the two syndromes were decided upon at an National Institutes of Health Consensus Development Conference in 1987, and are listed in Table 1. The recognition of NF1 and NF2 as distinct clinical entities has since been validated by the mapping of the *Nf1* and *Nf2* genes to different loci, and by the identification of different functions for the encoded proteins. Thus, the protein encoded by the chromosome (chr) 17q11.2 *Nf1* gene, termed "neurofibromin," functions as a guanosine triphosphatases GTPase, activating protein (GAP) for Ras, and may also play a role in signaling mediated by adenosine-3′,5′-monophosphate (cyclic AMP, or cAMP); the product of the chr 22q12.2 *Nf2* gene, named "merlin" or "schwannomin," is a member of the ezrin-radixin-moesin (ERM) family of membrane–cytoskeleton linker proteins.

This chapter summarizes what is currently known about the functions of neurofibromin and merlin-schwannomin, and discusses how this emerging understanding may lead to novel therapies. The work leading up to the identification of the *Nf1* and *Nf2* genes is discussed only briefly. This and other early work has been covered by several prior reviews (1–5).

2. NEUROFIBROMATOSIS TYPE 1 (NF1)

2.1. The Disease

Although descriptions and illustrations of individuals with NF1-like symptoms are known from throughout history, the first comprehensive description of NF1 was pro-

From: *Tumor Suppressor Genes in Human Cancer*
Edited by: D. E. Fisher © Humana Press Inc., Totowa, NJ

Table 1
Diagnostic Criteria for NF1 and NF2

Diagnostic criteria for NF1 are met if an individual has two or more of the following:

Six or more café-au-lait spots
  0.5 cm or larger before puberty
  1.5 cm or larger after puberty

Two or more cutaneous or subcutaneous neurofibromas, or one plexiform neurofibroma

Freckling in the axillary or inguinal regions

Optic glioma

Two or more Lisch nodules (benign iris hamartomas)

Distinctive bony lesion
  Sphenoid wing dysplasia
  Dysplasia, or thinning of long bone cortex, with or without pseudoarthrosis
First-degree relative with NF1 by the above criteria

Diagnostic criteria for NF2 are met if an individual has one of the following:

Bilateral vestibular schwannoma (VS)

Family history of NF2 plus a unilateral VS diagnosed before 30 yr of age, or any two of the following: meningioma, glioma, schwannoma, juvenile posterior subcapsular lenticular opacities, or juvenile cortical cataract

Individuals with the following should be evaluated for NF2:

Unilateral VS before 30 yr, plus at least one of the following: meningioma, glioma, schwannoma, juvenile posterior subcapsular lenticular opacities, or juvenile cortical cataract.

Two or more meningiomas, plus unilateral VS diagnosed before 30 yr, or one of the following: glioma, schwannoma, juvenile posterior subcapsular lenticular opacities, or juvenile cortical cataract

Rare related syndromes include familial café-au-lait spots, schwannomatosis, Watson syndrome (WS), and multiple meningiomas. NF1 and NF2 also occur in a nongeneralized form, called segmental NF1 or NF2, reflecting somatic mosaicism. Adapted with permission from ref. 6.

vided by von Recklinghausen toward the end of the nineteenth century *(7)*. Several subsequent studies contributed to current understanding of NF1 as a dominantly inherited disorder with complete penetrance, but highly variable expressivity *(8,9)*. The disease has an estimated prevalence of 1/3500 in all ethnic groups *(10)*, making it about 10-fold more common than NF2, and ranking NF1 among the most frequent genetic diseases of man. 30–50% of NF1 cases occur sporadically, suggesting that approx 1/10,000 gametes harbor a *de novo Nf1* mutation *(10)*. The large size of the *Nf1* gene (*see* Subheading 2.2.) provides at least a partial explanation for this high mutation rate. The hypothesis that gene conversion between the *Nf1* gene and several *Nf1* pseudogenes also contributes remains speculative *(11)*. Similar to what has been found for several other genetic diseases, the majority of *Nf1* mutations occurs in the male germ line *(12,13)*. However, large deletions are predominantly of maternal origin *(14,15)*.

Loss of *Nf1* gene function is probably the most common cause of inherited increased cancer risk. NF1 is more than just a tumor predisposition syndrome, however, because between 30 and 65% of patients exhibit learning disabilities, and an estimated 20%

have skeletal defects *(see____)*. Prenatal diagnosis in familial cases has been possible since the location of the *Nf1* gene was established *(16,17)*, but the utility of genetic counseling is limited by the highly variable expressivity of NF1 symptoms. This unpredictability is among the most troubling aspects of the disease, from a patient's perspective, and does not reflect the nature of the *Nf1* mutation, because most studies have failed to reveal obvious genotype–phenotype correlations *(9)*. Rather, the clustering of uncommon symptoms in monozygotic twins, and, to a lesser extent, in first-degree relatives with NF1, supports the notion that much variability reflects the action of symptom-specific modifier genes *(18)*.

The most common symptoms of NF1 are areas of abnormal skin pigmentation, called café-au-lait spots; abnormal freckling in skin-fold areas; pigmented iris hamartomas, known as Lisch nodules; and cutaneous or subcutaneous neurofibromas *(8,9)*. Each of these symptoms appears in an age-dependent manner. Thus, although café-au-lait spots are also found in 10% of normal individuals, and occur at increased frequency in NF2, McCune-Albright syndrome (MAS), and several other genetic disorders, the presence of six or more café-au-lait spots over 0.5 cm in size, in prepubertal individuals, is highly suggestive of NF1 (Table 1). Like café-au-lait spots and abnormal freckling, Lisch nodules have no associated morbidity, and are present in the eyes of most NF1 patients at puberty. Neurofibromas typically first appear in young adolescents, and increase in number later in life. Thus, adult NF1 patients can have from just a few to hundreds or even thousands of these benign tumors, which arise from peripheral nerve sheaths, and which consist of Schwann cells, (perineurial) fibroblasts, infiltrating mast cells, and other cellular elements *(8)*. NF1 has been called a neurocristopathy *(19)*, to stress that several of its symptoms involve cells originating from the embryonic neural crest. However, NF1 patients also suffer from skeletal abnormalities, and are at increased risk of developing juvenile malignant myeloid disorders. Moreover, it remains controversial whether defects in Schwann cells, nerves, or perineurial fibroblasts are responsible for neurofibroma formation *(20–22)*.

Although the most common symptoms of NF1 are in large part benign, patients also suffer from a variety of less frequent, but potentially more serious, problems *(8,9; see* ref. *23* for frequencies of the most common symptoms). Thus, an estimated 50% of patients exhibit macrocephaly (head circumference above the ninety-fifth percentile), and about 30% are below the fifth percentile for height. Patients are also at increased risk of exhibiting distinct skeletal defects, notably scoliosis, sphenoid wing dysplasia, or thinning of the long bones, with or without pseudoarthrosis. Learning and behavioral problems also occur with high frequency in NF1, with different studies finding 30–65% of children affected *(24)*. The cognitive phenotype includes both a specific learning defect (a significant discrepancy between ability and performance, sometimes likened to attention deficit disorder), and a somewhat lower full-scale IQ (typically around 90). No consensus exists whether the cognitive deficit is linked to the occurrence of T_2-weighted hyperintense lesions in brain magrative resonance images *(24)*. These transient lesions, which can be detected in the brains of up to two-thirds of children with NF1, appear to reflect the presence of intramyelinic edema *(25)*. Mental retardation (IQ <70) is relatively uncommon in NF1. Gliomas (pilocytic astrocytomas), most frequently of the optic pathway, are another feature of NF1. Optic gliomas occur in up to 20% of children with NF1, but do not usually lead to visual loss. One recent study *(26)* reported an association between the occurrence of optic gliomas and other central nervous system (CNS) tumors. Among the most serious complications, 20–30% of patients develop plexiform neurofibromas alongside deeper-lying nerves. These lesions are often congenital, and can cause severe overgrowth of surrounding soft tissue and bone. Malignant peripheral nerve sheath tumors (MPNSTs), previously called "neurofibrosarcomas" or "malignant schwannomas," are strongly associated with NF1, and have a tendency to arise from plexiform neurofibromas. Other tumors, notably juvenile malignant myeloid disorders, pheochromocytoma, and rhabdomyosarcoma, also occur at increased frequency, and the lifelong risk of malignant tumors in NF1 has been estimated to be around 5% (*see* ref. *23* and references therein).

The variability of NF1 symptoms remains a source of confusion, especially because other genetic disorders have symptoms that overlap with those of NF1. Thus, a proportion of NF1 patients have facial features that resemble those seen in mild cases of Noonan syndrome, but current evidence argues against the existence of a distinct NF1–Noonan syndrome *(27)*. Watson syndrome (WS) is a rare genetic disease characterized by café-au-lait spots, short stature, dull intelligence, and pulmonary valve stenosis. The genes responsible for WS and NF1 are tightly linked *(28)*, and a 42-bp duplication in the *Nf1* coding region has been found in a WS family *(29)*. Thus, although Lisch nodules and neurofibromas occur infrequently in WS *(28)*, and although heart defects are rare in NF1 *(23)*, WS is probably a subtype of NF1. It is interesting to speculate that WS mutations result in a mutant form of neurofibromin with partial dominant-negative characteristics, since murine *Nf1* knockouts also exhibit defective heart development (*see* below). A patient diagnosed with multiple lentigines syndrome, and a patient with encephalocraniocutaneous lipomatosis, also harbored *Nf1* mutations *(30,31)*. However, these patients may have been misdiagnosed, or may have had NF1 in addition to these other diseases. Finally, a recurring theme is the equation of NF1 with Elephant Man Disease, even though the severely disfigured "Elephant Man," Joseph Merrick, is more likely to have suffered from Proteus syndrome *(32)*.

2.2. Characterization of Nf1 Gene

The gene responsible for NF1 was mapped to chr 17q11.2 in 1987 *(16,17)*, and identified in 1990 *(33–35)*. The gene consists of 57 constitutive and 3 alternatively spliced exons, spans close to 350 kb of genomic DNA, and is transcribed into two 11–13-kb mRNAs with different 3′-untranslated segments *(36)*. *Nf1* mRNA, without any of the alternatively spliced exons, encodes a protein of 2818 amino acids (aas), with a calculated molecular mass of 317 kDa and an apparent molecular mass in sodium dodecyl sulfate (SDS) gels of 220–280 kDa. This protein, termed "neurofibromin," harbors a centrally located 360-aa segment related to the catalytic domains of Ras-specific GTPase activating proteins (RasGAPs) *(37,38)*, and shares more extensive, but relatively low-level, similarity, extending over nearly 1500 aas with two RasGAPs (Ira1p and Ira2p) from *Saccharomyces cerevisiae* (Fig. 1). The subsequent confirmation that neurofibromin has RasGAP activity suggested that defective Ras regulation may be responsible for at least some of the diverse symptoms of NF1 (Fig. 2).

The large size of the *Nf1* gene, and the presence of several related pseudogenes, have complicated the search for mutations, and, in most early studies, only small segments of the 8454 bp open reading frame were analyzed. More recently, protein-based assays have allowed the detection of mutations throughout the coding region. In two such studies *(39,40)*, around 70% of patients were found to harbor truncating mutations. A database

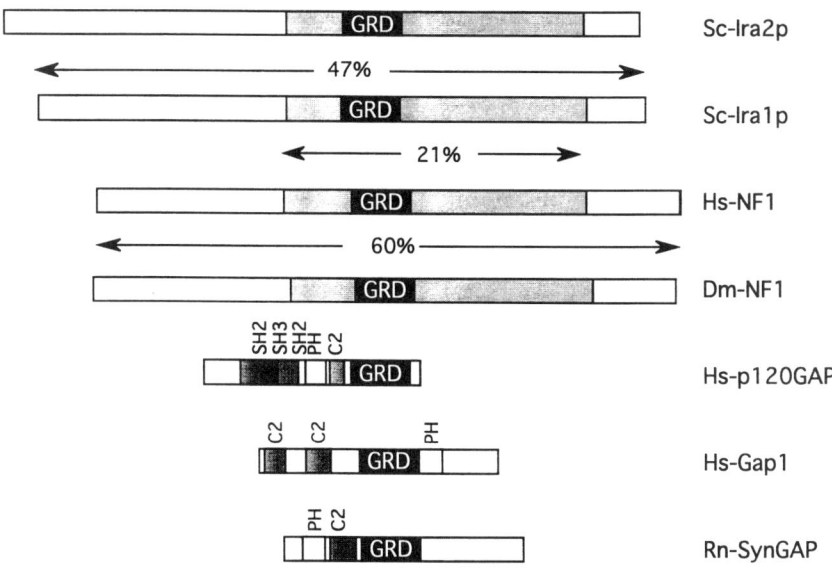

Fig. 1. Schematic structure and sequence similarity between RasGAPs. The proteins shown are (from top to bottom) *S. cerevisiae* Ira2p and Ira1p *(196); Homo sapiens* NF1; *D. melanogaster* NF1; human p120GAP *(197)*, a representative member of a small family of human proteins related to *Drosophila* Gap1 *(198)*; and *Rattus norvegicus* SynGAP *(199,200)*. Human and *Drosophila* neurofibromin are 60% identical over their entire length. The human protein also shares around 20% identity with Ira1p and Ira2p between residues 900 and 2350.

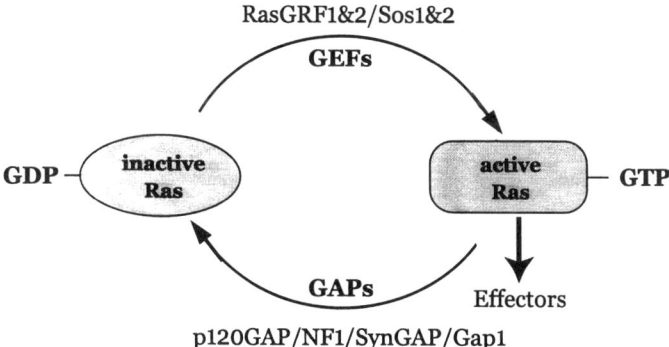

Fig. 2. Ras GTPases function as binary switches in signal transduction, by cycling between inactive guanosine diphosphate (GDP)- and active GTP-bound conformations. Ras proteins have a high affinity for both GDP and GTP. Cycling of mammalian Ras is promoted by at least four guanine nucleotide exchange factors, which activate Ras by promoting GTP for GDP exchange, and by as many as six GAPs, which stimulate the low intrinsic GTPase activity of Ras, thus promoting its inactivation. Active Ras interacts with several effector proteins, including Raf, phosphatidylinositol-3 kinase, and Ral-GDS, among others. Evidence exists that p120GAP and other RasGAPs may have effector functions in addition to their role as negative regulators of Ras-mediated signal transduction (*see* ref. *3* for further details).

maintained by the NF1 Genetic Analysis Consortium (http://www.nf.org/nf1gene) currently lists over 240 mutations, with no apparent hot spots among them. More than half of
the mutations are deletions and insertions, which probably represents a ascertainment
bias. Also listed are over 40 nonsense and around 30 missense mutations scattered
throughout the coding region. Missense mutations are interesting from a functional point
of view, although some may represent rare polymorphisms and others may affect protein
stability. No clear genotype–phenotype correlations can be distilled from the mutation
data, except that patients with large deletions appear more likely to exhibit a severe phenotype that includes facial dysmorphic features and large numbers of early-onset neurofibromas *(41,42)*. Genetic linkage to the *Nf1* locus has been established in some, but not all,
families with spinal NF *(43)*, or familial café-au-lait spots *(44)*, and a typical truncating
Nf1 mutation has been found in a family with spinal NF *(45)*. Whether the predominant
spinal location of neurofibromas in this family reflects a genetic background effect
remains to be established.

The *Nf1* gene harbors three other genes transcribed in the opposite direction within
one of its introns *(35)*. One embedded gene encodes oligodendrocyte-myelin glycoprotein (OMG), a 110-kDa glycosylphosphatidylinositol-linked protein expressed in the
CNS *(46)*. Expression of OMG inhibits the proliferation of fibroblasts, suggesting a
role in growth control *(47)*. The other intronic genes are *EVI2A* and *EVI2B,* whose
murine homologs were first identified in the midst of a cluster of viral integration sites
in retrovirus-induced murine myeloid leukemia (ML). This initially suggested a role
for *Evi2a* or *Evi2b* as myeloid leukemia oncogenes, but it is currently believed that
retrovirus-mediated inactivation of *Nf1* is responsible *(48)*. Whether loss of the embedded genes contributes to the severe phenotype in Nf1 patients, who harbor large deletions, remains unknown.

Tissue-specific alternatively spliced mRNAs predict neurofibromin variants that
include 10 extra aas upstream of the IRA-related segment *(49)*, 21 additional residues
within the GAP-related domain (GRD) *(50)*, or 18 extra aas close to the C-terminus
(51). The 21-aa GRD insert reduces, but does not abolish, GAP activity *(52)*, and the
functional significance of any neurofibromin variant remains to be established. Human
Nf1 mRNA is also subject to RNA editing *(53)*, and unequal expression of *Nf1* alleles
in patient-derived cell lines, possibly reflecting nonsense-mediated mRNA decay, has
been reported *(54)*. Whether these phenomena contribute to NF1 pathogenesis remains
unclear.

2.3. Structure and Function of Neurofibromin

Nf1 homologs from mice, the Japanese puffer fish, *Fugu rubripes,* and the fruit fly,
Drosophila melanogaster, encode proteins that share, respectively, 98.5, 91.5, and 60%
overall sequence identity with human neurofibromin *(55–57)*. Although participation in
protein complexes is among the most obvious reasons for such a high degree of evolutionary conservation, no obvious protein interaction motifs are evident in neurofibromin, and only Ras-related GTPases (*see* below), α- and β-tubulin *(58)*, and an
as-yet-unidentified 400–500 kDa protein *(59)* have been found to interact with human
neurofibromin.

Many cell lines express neurofibromin, and the protein is widely expressed during
mouse fetal development *(60,61)*. By contrast, neurofibromin is most abundant in the
peripheral nervous system and CNS of adult rats, with highest expression observed in

neurons, oligodendrocytes, and nonmyelinating Schwann cells *(62)*. No consensus exists as to neurofibromin's subcellular localization, with different groups reporting localization to the smooth endoplasmic reticulum in Purkinje cells *(63)*, to capped immunoglobulin in activated B-lymphocytes *(64)*, to microtubules in fibroblasts *(65)*, or to mitochondria *(66)*. Mitochondrial staining was seen in six cell lines, using four antibodies and two fixation methods. An estimated 20% of neurofibromin co-purified with mitochondria *(66)*, consistent with findings that neurofibromin is distributed between cytosolic and particulate fractions *(67)*. Neurofibromin not only acts as a GAP for H-, K-, and N-Ras, but also inactivates R-Ras *(68)*. This is especially intriguing, because both conventional p21 Ras and p23 R-Ras have been reported to interact with Bcl-2 *(69,70)*, an antiapoptotic protein that localizes to mitochondrial, microsomal, and nuclear membranes *(71)*. However, although a role in apoptotic signaling may help explain why *Nf1*-deficient neurons survive in the absence of neurotropic factors *(22)*, or why it is hard to express neurofibromin in transfected cells *(72)*, it cannot be ruled out that neurofibromin may yet have other GTPase targets. p120 RasGAP also stimulates GTP hydrolysis by Rab5 *(73)*.

Notwithstanding this speculation, there is ample evidence that downregulation of conventional Ras is among the important functions of neurofibromin. Thus, baculovirus-produced neurofibromin stimulated the GTPase activity of H- or N-Ras in vitro *(58,74)*, and expression of an NF1-GRD protein rescued the heat shock sensitivity of yeast *ira1* or *ira2* mutants *(75,76)*. Moreover, three NF1 patient-derived MPNST cell lines expressed either very little or no neurofibromin. These cells contained up to 50% nonmutant Ras in the active GTP-bound conformation, and expression of a RasGAP catalytic segment inhibited their proliferation concomitant with a reduction in Ras-GTP *(77,78)*. Less drastically elevated Ras-GTP levels were also found in neurofibromas *(79)*, and in neurofibromin-deficient ML bone marrow *(80)*. Mutations that activate Ras or inactivate neurofibromin define mutually exclusive subsets of ML, providing genetic evidence that loss of *Nf1* contributes to myeloid leukemogenesis by activating Ras *(81)*.

A more complex situation exists in melanoma (mel) and neuroblastoma (NB). Although neither of these neural-crest-derived tumors occurs at increased frequency in NF1, several NB or mel cell lines did not express neurofibromin, which in some cases reflected loss of both *Nf1* alleles *(82–84)*. Although loss of heterocygosty (LOH) at the *Nf1* locus was not found in primary mels *(85)*, and is uncommon in primary NBs (G. Cowley and Bemards, unpublished observation), a NB tumor from an NF1 patient harbored a homozygous *Nf1* gene deletion *(86)*, and heterozygous loss of *Nf1* strongly enhanced NB formation in MYCN overexpressing transgenic mice *(87)*. Unlike MPN-STs, neurofibromin-deficient NBs and mels showed no increased levels of Ras-GTP *(82,83)*, and re-expression of neurofibromin inhibited the growth of a mel line without altering its Ras-GTP content *(72)*. Taken together with findings that Ras mutations are infrequent in neural-crest-derived tumors *(88)*, which may reflect the fact that oncogenic Ras induces cell cycle arrest and differentiation in these cells *(89)*, these findings suggest that loss of neurofibromin may contribute to NB and mel transformation through a mechanism that does not involve Ras, possibly by disrupting differentiation *(72,82)*. However, it is important to note that only steady-state Ras-GTP levels were measured in these studies. Thus, it remains possible that *Nf1*-deficient NBs or mels exhibit transient defects in Ras regulation, similar to what has been observed in neurofibromin-deficient myeloid cells *(80,90,91)*.

2.4. Nf1 Gene as Tumor Suppressor

The focal nature of many NF1 lesions suggests that additional genetic or epigenetic events at the somatic level are required to trigger their development. There is accumulating evidence that, at least in neurofibromas and NF1-associated malignant tumors, one additional genetic hit involves the inactivation of the remaining *Nf1* allele. Thus, an NF1 patient-derived neurofibrosarcoma showed LOH of all chr 17 polymorphisms and a 200-kb deletion affecting the remaining *Nf1* allele *(92)*. This observation followed several reports of LOH of closely linked genetic markers in NF1-associated MPNSTs *(93,94)*. Pheochromocytomas occur in approx 1% of NF1 patients, and LOH at the *Nf1* locus was found in each of seven NF1-associated tumors. In all cases in which this could be determined, LOH involved the wild-type (wt) *Nf1* allele, providing strong support for a tumor suppressor role in these adrenal tumors *(95)*. Children with NF1 have a 200–500-fold enhanced risk of developing malignant myeloid disorders, particularly juvenile myelomonocytic leukemia (formerly known as juvenile chronic myelogenous leukemia) and monosomy 7 syndrome *(96,97)*. Providing support for a tumor suppressor role, bone marrow cells from several children with NF1 and ML showed homozygous inactivation of the *Nf1* gene *(98)*. Moreover, approx 10–20% of heterozygous *Nf1* knockout mice developed pheochromocytoma; a similar fraction developed ML (in all cases, with loss of the second *Nf1* allele; *99*), and irradiated mice whose hematopoietic system was reconstituted with *Nf1–/–* fetal liver cells developed granulocyte-macrophage colony-stimulating factor hypersensitive ML at high frequency *(90)*.

Whether *Nf1* serves as a frequent tumor suppressor in other cancers remains uncertain. Of special interest is a possible role in astrocytoma, the most common and malignant of human brain cancers. A role in astrocytoma was initially suggested by the frequent occurrence of optic pathway and other gliomas in NF1, and by the detection of aa substitutions at residue 1423 (a conserved lysine within the GRD), which strongly reduced the catalytic activity of neurofibromin, in one case each of anaplastic astrocytoma, colon adenocarcinoma, and myelodysplastic syndrome *(100)*. However, a subsequent study detected LOH at the *Nf1* locus in only 1/22 astrocytomas *(101)*; others detected elevated *Nf1* mRNA or protein levels, rather than loss of expression, in pilocytic astrocytomas and reactive astrocytes *(102–104)*. Thus, although other studies detected more frequent LOH *(105,106)*, a role for *Nf1* as a common astrocytoma tumor suppressor remains to be proven.

Lisch nodules and optic gliomas are not easily accessible, and the analysis of neurofibromas is complicated by the fact that these tumors consist of several cell types, not all of which may be transformed *(107,108)*. However, contrasting numerous earlier studies *(93,94,109–111)*, which had failed to detect loss of chr 17 polymorphisms in neurofibromas, three groups recently reported evidence favoring a double-hit mechanism in these benign tumors. Thus, Colman et al. *(112)* detected LOH of *Nf1* intragenic and other chr 17q polymorphisms in eight neurofibromas from two sporadic patients, out of a total of 22 tumors from five patients. That study did not identify the constitutive mutation in these patients, although the deleted allele was always of maternal origin, consistent with the idea that most sporadic *Nf1* mutations are of paternal origin *(12,13)*. Subsequently, Sawada et al. *(113)* identified a 4-bp deletion in *Nf1* exon 4b in a neurofibroma from a patient, which harbored a large constitutive deletion. DNA from microdissected tumor tissue showed both deleted and wt polymerase chain reaction products, consistent with the idea that neurofibromas consist of an admixture of normal and mutant cells. Finally, Serra et al. *(114)* detected LOH in 25% of 60 neurofibromas from 17 patients. Why patients, with large deletions centered on the *Nf1* gene, have a tendency to develop large numbers of early-onset neurofibromas remains to be explained.

The basis of NF1-associated pigmentation defects remains incompletely understood. Early studies noted an increased occurrence of macromelanophores or melanin macroglobules in café-au-lait patches of NF1 patients *(115,116)*, but similar structures also occur in normal skin *(117)*. Moreover, although melanin macroglobules can be observed in melanocytes, it remains possible that abnormal clearance of these structures by dentritic Langerhans cells or macrophages also contributes to the pigmentation defects *(118)*. Cultured melanocytes from NF1 patient café-au-lait spots expressed neurofibromin, and showed no evidence of LOH at the *Nf1* locus *(119)*.

2.5. Genetic Models

Of the many avenues of research aimed at ultimately designing rational therapy for human genetic disorders, few have been as fruitful as the generation and study of animal models. Thus, the first generation of *Nf1*-mutant knockout mice has proved invaluable, among several other things, in confirming a tumor suppressor role for neurofibromin; genetic analysis of *Drosophila Nf1* has provided the most compelling evidence so far for a Ras-independent role for neurofibromin in neuropeptide-mediated signal transduction.

Mice heterozygous for *Nf1*-null alleles do not exhibit pigmentation defects, iris abnormalities, or benign neurofibromas, and thus do not provide an accurate model for several aspects of the human disease *(99,120)*. However, 75% of heterozygotes succumbed to a variety of tumors by 27 mo of age, compared to 15% of a comparable set of wt animals, confirming a tumor suppressor function for murine *Nf1 (99)*. *Nf1* heterozygotes also showed learning and memory defects resembling those of NF1 patients. Thus, an impaired performance in the Morris water maze, which assesses spatial learning, was not fully penetrant, and could be overcome by increased training. The defect in spatial learning appeared specific, because associative learning was not impaired *(121)*.

Nf1 homozygous mutant mice die around d 13.5–14.5 of fetal development. A complex heart defect, resulting in generalized edema, is believed to be the proximal cause of embryonic death *(99,120)*. A developmental delay and hyperproliferation of sympathetic and parasympathetic ganglia was observed in only one study, perhaps reflecting a difference in genetic background *(120)*. Underlying the heart defect may be a failure of endocardial cushions to properly differentiate. Endocardial cushions are the precursors of heart valves, and form by a process of epithelial-to-mesenchymal conversion. Neurofibromin appears necessary to regulate this process, since, in its absence, mouse hearts develop overabundant endocardial cushions, obstructing the normal flow of blood *(122)*. Enhanced invasiveness of *Nf1*-deficient cardiac cushion cells was rescued by expression of a dominant-negative RasN17 mutant, and mimicked in wt cells by expressing activated Ras *(122)*.

Several explanations as to why murine *Nf1* heterozygotes do not develop neurofibromas or other common signs of human NF1 can be envisaged. Among other possibilities, the number of target cells that can give rise to neurofibromas or other lesions may be smaller in the mouse, or the animals may not live long enough to develop the lesions. Differences in the mutation rates of the human and murine *Nf1* homologs may also contribute, as may differences in synteny between human and murine genes. To circumvent these limitations, more sophisticated second-generation conditional *Nf1* mutants are being generated. Such mice may provide a more accurate model of human NF1, given that neurofibroma-like lesions have been found in chimeric mice made with homozygous *Nf1*-deficient embryonic stem cells (T. Jacks, personal communication).

The fruit fly, *D. melanogaster,* and the nematode, *Caenorhabditis elegans,* have been used widely to genetically dissect complex biological processes. The recently completed *C. elegans* genome sequence predicts several RasGAP-like proteins, but no *Nf1* homolog. In contrast, a *Drosophila Nf1* homolog consists of 17 common and two alternative C-terminal exons, and predicts proteins of 2764 and 2802 aas that share 60% overall sequence identity with human neurofibromin *(57).* However, unlike the widespread developmental defects observed in transgenic flies expressing activated Ras1 or Ras2 (a *Drosophila* R-Ras homolog) *(123,124),* homozygous *Drosophila Nf1*-null mutants are 15–20% reduced in size during postembryonic development, but are otherwise normal. Biochemical analysis confirmed that *Drosophila* NF1 is a potent RasGAP, but genetic studies revealed no defects in Ras-mediated signaling downstream of several receptors with tyrosine kinase activity, even in sensitized genetic backgrounds. Moreover, although expression of an inducible *Nf1* transgene restored the size of mutants to that of wt flies, manipulating *Ras1* or *Ras2* gene dosage or activity did not modify the size defect. Thus, the most obvious defect observed in *Drosophila Nf1* mutants appears to reflect a Ras-independent function of the protein *(57).*

Drosophila Nf1 mutants are also somewhat sluggish, and exhibit a diminished escape response, which did not reflect any obvious anatomical defect of the musculature or the nervous system *(57).* The behavioral defect may, however, result from abnormal neurotransmission, because a neuropeptide-elicited delayed rectifying potassium current at the larval body wall neuromuscular junction was absent in *Nf1*-null mutants. The authors analyzed signaling by the PACAP38 neuropeptide, because an increase in potassium conductance mediated by PACAP38 had been found to require both intact Ras-Raf and cAMP-mediated signaling pathways *(125).* However, the electrophysiological defect in *Nf1* mutants was not mimicked or modified by manipulating Ras, but was rescued by forskolin-mediated activation of adenylyl cyclase, or by providing cAMP *(126).* The reduced size of *Nf1* mutants was similarly rescued by expressing a constitutively active cAMP-dependent protein kinase A (PKA) transgene, or mimicked in flies carrying hypomorphic PKA alleles *(57).* Thus, *Drosophila Nf1* functions either upstream of adenylyl cyclase or in a parallel convergent pathway. In this context, it is worth noting that *Saccharomyces cerevisiae* cerevisiae Ras directly activates adenylyl cyclase *(127),* and that yeast adenylyl cyclase may be mislocalized in an *iral* mutant *(128).* However, yeast and metazoan adenylyl cyclases are very different in structure, and the significance of this observation remains unclear.

2.6. NF1: Current Picture and Future Directions

The fact that all *Nf1*-deficient phenotypes in *Drosophila* are rescued by increasing cAMP–PKA signals is a departure from what had been found in mammals, in which most studies had emphasized the Ras-related role of neurofibromin. However, the idea that neurofibromin may play dual roles in Ras and cAMP signaling is attractive for at least two reasons: First, the RasGAP catalytic domain accounts for only 10% of neurofibromin, and the rest of the protein is equally well conserved (55,57); second, although inappropriate Ras activity provides a ready explanation for NF1-associated tumors, other symptoms may be more easily understood in terms of defective cAMP signaling. This is especially true for NF1-associated learning disabilities, since mutations that affect cAMP signaling are known to cause learning and memory defects in *Drosophila,* mammals, and several other organisms (129,130). Defects in cAMP signaling may also contribute to other NF1 symptoms, however, and patients with MAS exhibit café-au-lait spots, skeletal defects (polystotic fibrous dysplasia, frequently leading to pseudoarthrosis), and endocrinologic abnormalities, including female precocious puberty, all of which are also seen in NF1. However, although *Drosophila Nf1*-deficient phenotypes are rescued by increasing cAMP–PKA signaling, MAS is caused by somatic mutations that constitutively activate the α-subunit of the adenylyl-cyclase-stimulating heterotrimeric Gs protein (131,132). Perhaps similar defects are caused by either too little or too much cAMP signaling: *D. rutabaga* (adenylyl cyclase) and *dunce* (cAMP phosphodiesterase) mutants have similar learning and memory defects.

The question whether defective cAMP signals explain some features of human NF1 is an important one, because different therapeutic strategies may be used to downregulate Ras or increase cAMP. Thus, farnesyl transferase inhibitors show great promise as anti-Ras agents (133), and have already been used to modify the in vitro properties of *Nf1*-deficient murine Schwann and human MPNST cells (134,135); cAMP phosphodiesterase inhibitors are among agents that may be employed to restore cAMP signals. So far, however, most research has focused on the Ras-related role of neurofibromin, with a recent study (136) suggesting that defective Ras regulation may provide an adequate explanation for all aspects of NF1. The authors of that study based their conclusion on the identification of a disease-associated missense mutation, which altered the catalytically important arginine finger residue in the NF1–GRD (137). The mutant protein had severely impaired GAP activity, but its interaction with Ras, and its stability, appeared normal. Several family members with this mutant allele had severe NF1, including learning disabilities, suggesting that loss of GAP activity may explain all features of NF1 (136). However, others found that missense mutations within the GRD may affect the subcellular localization of neurofibromin by disrupting its association with microtubules (138). Thus, although the cumulative evidence indicates that loss of RasGAP activity plays an important role in several NF1 symptoms, it cannot be concluded that other functions of neurofibromin are not also important. Most of the 30 or so disease-associated missense mutations occur outside of the RasGAP catalytic domain of neurofibromin.

Nine yr after the *Nf1* gene was identified, a global understanding about neurofibromin's role in signal transduction has been achieved. Among the most important remaining questions is the identity of the specific Ras- and/or cAMP-mediated signaling pathways that are responsible for NF1 symptoms. Another important question concerns the identity of modifier genes that determine the clinical severity of NF1. By employing a combination of biochemical, cell biological, and genetic approaches in man, mice, and flies, it should be possible to provide answers to these questions in the not too distant future.

3. NEUROFIBROMATOSIS TYPE 2

3.1. The Disease

NF2 (non-von Recklinghausen disease) is less common than NF1, affecting about 1/35,000 individuals, and is symptomatically distinct (*139*; Table 1). However, the disease features a similar dominantly inherited predisposition to developing nervous system tumors, indicating that the gene responsible is also a tumor suppressor. The symptoms of NF2 generally appear during the second and third decades of life, and, like those of NF1, are progressive. The hallmark of NF2 is the development of Schwann cell tumors of the eighth cranial (acoustic) nerves, usually causing loss of hearing. NF2 patients frequently develop multiple schwannomas of the cranial and spinal nerves, and are also predisposed to developing meningiomas, ependymomas, and posterior subcapsular cataracts. In contrast to neurofibromas, which are mixed tumors composed of several cell types including Schwann cells, schwannomas are homogeneous tumors derived from Schwann cells or their precursors (*140*). Moreover, schwannomas are benign tumors that rarely, if ever, progress to malignancy. However, they are associated with a high degree of morbidity in NF2, because of their multiplicity and intractable location at the boundary between the peripheral nervous system and CNS (*139*). Unfortunately, inconsistent histological diagnosis of schwannomas and neurofibromas has historically led to some diagnostic confusion; there is, in fact, probably no true overlap between NF1 and NF2.

3.2. The Gene

LOH studies localized the *Nf2* locus to human chr 22q12: The gene was subsequently identified by positional cloning in 1993 (*17,141–144*). Genetic studies have consistently revealed loss of the remaining wt allele in NF2-associated tumors, and somatic inactivation of both *Nf2* alleles in sporadic schwannomas and approx one-half of sporadically occurring meningiomas (*5*). Most mutations in the *Nf2* gene are deletions, or frameshift or nonsense mutations predicted to encode nonfunctional protein. Moreover, immunohistochemical absence of the NF2 antigen has been detected in several NF2-associated tumors, even when the mutation was a missense or truncating mutation predicted to encode detectable protein, suggesting that mutant *Nf2* protein is particularly unstable (*145*). Moreover, investigation of the nature of the mRNA transcript, produced from alleles carrying *Nf2* splice-site mutations, revealed the use of cryptic splice sites that resulted in frameshifts and protein truncation, instead of the expected excision of a single exon (*146*). Thus, no clear genotype–phenotype correlations have been made. Taken together, this genetic behavior is consistent with the notion that the *Nf2* locus encodes a tumor suppressor whose normal growth-suppressing function must be lost, for tumor formation to occur. The normal growth-suppressing function of NF2 appears to be specific for, or specifically relied upon in, the Schwann and meningeal cell lineages, because *Nf2* mutations have not consistently been identified in other types of human cancers, with one exception; Cell lines derived from malignant mesotheliomas of the lung lining, which are associated with asbestos exposure, frequently harbor *Nf2* mutations (*147,148*). The fact the NF2 patients are not predisposed to mesothelioma development may reflect the rare co-occurrence of asbestos exposure and germ-line *Nf2* mutation or the requirement for other genetic alterations, perhaps preceding Nf2 loss in mesothelioma formation.

In general, there is much less variability in the expressivity of NF2, compared to NF1, rendering disease course more predictable, and suggesting less genetic modification. However, two related disorders have been described: Schwannomatosis is likely to involve loss of *Nf2* gene function, either through somatic mosaicism (segmental NF2), or, when familial, through an apparently inherited predisposition to develop mutations at the *Nf2* locus *(149)*. On the other hand, it has been reported that meningiomatosis, or multiple meningiomas, as inherited in an autosomal dominant fashion, is not linked to the *Nf2* locus on chr 22, and may result from mutation of an as-yet-unidentified tumor suppressor gene *(150)*. This is supported by the fact that a large subset of sporadic meningiomas do not exhibit obvious mutation at the *Nf2* locus *(151)*.

Transcription from the *Nf2* gene produces transcripts of 2.6, 4.4, and 7 kb, which probably differ with respect to the lengths of 5′ and 3′ untranslated regions *(143,144)*. These transcripts encode a 595- or 590-aa protein *(see* below) with an apparent mol wt of ~69 kDa on SDS-polyacrylamide gel electrophorotic. The *Nf2* mRNA and protein is widely expressed during development and in the adult, with particularly high levels found in the developing nervous system *(61,152,153)*. Although several alternatively spliced forms have been reported, the expression of only one has been studied extensively. Alternative use of the extreme carboxy, (C)-terminal exon is controlled in a tissue-specific and temporal fashion; isoform I lacks exon 16 and uses exon 17 as its final exon; isoform II contains exon 16, which carries its own stop codon (Fig. 3). Isoform II is the predominant isoform expressed in most adult tissues, including the adult brain; isoform I is expressed in the fetal brain and some adult renewal tissues, such as the spleen, kidney, and ovary *(152;* McClatchey, unpublished observations). Recent experimentation suggests that the biochemical properties of the proteins encoded by these two mRNAs may be functionally distinct, supporting the significance of their different expression patterns *(154; see* below).

3.3. Structure and Function of Merlin/Schwannomin

The cloning of the *Nf2* gene led to the surprising discovery that the encoded protein is a member of the band 4.1 superfamily of cytoskeleton–membrane linker proteins *(155)*. This places the NF2 protein in an intriguing physical niche within the cell for a tumor suppressor. Many tumor suppressors function as transcription factors or components of the cell cycle machinery controlling cellular proliferation from within the nucleus, or, like neurofibromin, function in the cytoplasm to regulate signal transduction pathways that directly control growth-promoting genes in the nucleus. In contrast, band 4.1 family members were previously thought to function primarily in providing cytoskeletal integrity, and perhaps reorganization, secondary to nuclear decision-making. In particular, the NF2 protein is most closely related to the ERM proteins, and was thus dubbed merlin (moesin, ezrin, radixin-like protein) by one group, and schwannomin by another *(156)*. The ERM proteins associate with membrane partners via their amino (N)-terminus, and directly with actin via their C-terminus; they are localized to areas of active cortical cytoskeletal remodeling. Merlin also localizes similarly to cortical actin structures, particularly membrane ruffles, microvilli, and the cleavage furrow, and can bind to some ERM membrane partners via its N-terminus, but lacks the C-terminal actin-binding domain of the ERMs *(157–160)*.

What little is known about the molecular function of merlin is modeled on an emerging, but still relatively poor, understanding of ERM function. For example, a number of

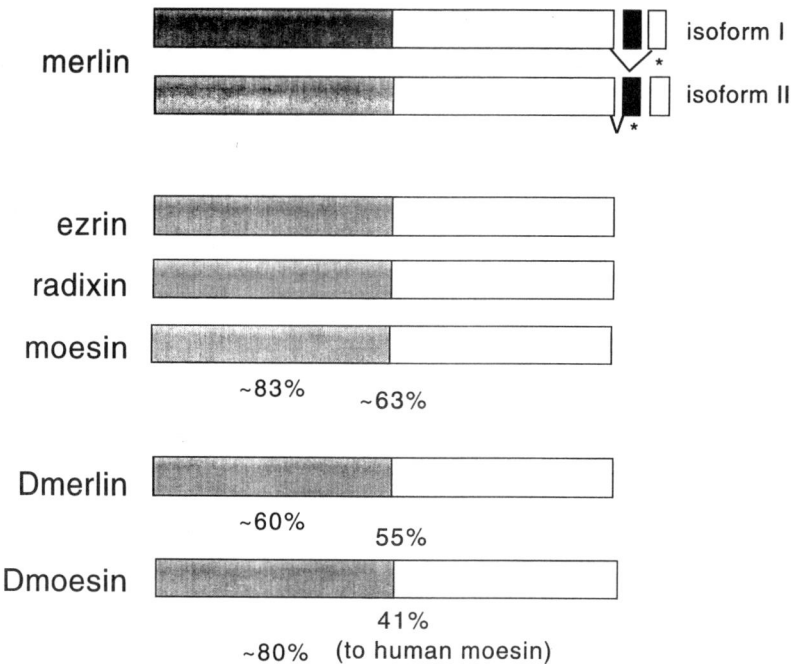

Fig. 3. The NF2 protein, merlin, belongs to the ERM subfamily of the band 4.1 superfamily of cytoskeleton–membrane linkers. These proteins share similarity primarily within their N-terminal halves (gray boxes). Thus, the ERMs share 83% aa identity to one another within their N-terminal halves, and ~63% identity throughout their entire lengths. Similarity to the *Drosophila* proteins, Dmerlin and Dmoesin, are also shown *(194)*. Dmerlin and Dmoesin share 55 and 41% aa identity to human merlin across their entire lengths; Dmerlin shares 60% identity to the human merlin N-terminus, and Dmoesin shares 80% identity to the human moesin N-terminus.

ERM membrane-interacting proteins have recently been described, including the hyaluronate receptor, CD44; CD43; the integrins, ICAM-2 and ICAM-3; the myosin-binding subunit of myosin phosphatase; PKA; and a co-factor for the Na^+/H^+ exchanger, hNHE-RF/EBP50 *(161–166)*. Thus far, merlin has similarly been reported to bind to CD44 and hNHE-RF/EBP50 in vitro *(167,168)*. In addition, the C-terminus of merlin isoform II specifically has been reported *(169)* to bind to the actin-binding protein βII-spectrin/fodrin, suggesting an indirect mechanism whereby merlin associates with the actin. Although the consequences of these interactions are not yet known, an emerging theme posits that the ERM proteins function to locally organize signaling components for rapid and local response. For example, several lines of experimentation suggest that the ERM proteins control the formation of specialized membrane protrusions, such as microvilli and uropods, and the relocalization and concentration of cell surface molecules to them *(163,170,171)*. Similarly, delocalization of the ERMs has been suggested to causally precede microvillar breakdown at the initial stages of apoptosis *(172)*.

The ERM proteins are capable of intramolecular, monomeric or intermolecular, oligomeric head-to-tail association *(173–175)*. In addition, phosphorylation of the ERMs on serine/threonine and tyrosine (Tyr) residues occurs in response to various stimuli, and phospholipid (4,5-PIP$_2$) binds to the ERM N-terminus *(156)*. Phosphorylation, and/or 4,5-PIP$_2$ binding, is thought to activate the ERMs, relieving this conformation, and causing relocalization to the membrane and interaction with membrane and cytoskeletal partners. In support of this idea, only the N-terminal halves, and not full-length ERM proteins, can bind to CD44 in vitro; however, interaction between full-length ERM and CD44 can occur in vitro in the presence of 4,5-PIP$_2$ *(161)*. Similarly, threonine phosphorylation of the C-terminus of the ERMs disrupts the intramolecular association of their N- and C-terminal halves *(176)*. Tyr phosphorylation of the ERMs occurs in response to epidermal growth factor, platelet-derived growth factor, and hepatocyte growth factor (HGF) treatment *(177–179)*. In fact, ezrin has been reported to be a direct substrate of the HGF receptor, Met *(179)*. Despite the strong aa identity between the ERMs, they were differentially phosphorylated in response to each growth factor *(178)*. Regulated localization of the ERMs may lead to their differential use as substrates. The relative contributions of these posttranslational modifications to ERM activation remains an important area of investigation.

Merlin is phosphorylated on serine/threonine residues, but, to date, Tyr phosphorylation of merlin has not been detected *(180)*. Indeed, the known ERM Tyr phosphorylation sites are not conserved in merlin. In contrast to the ERMs, unphosphorylated merlin is associated with various forms of growth arrest, and thus with the active growth-suppressive form of merlin *(180)*. For example, total merlin levels, and particularly levels of unphosphorylated merlin, increase with increasing cell density, serum deprivation, or loss of adhesion. This unphosphorylated form is predominantly found in the detergent insoluble (cytoskeletal) compartment; stimuli that lead to phosphorylation of merlin lead to the appearance of phosphorylated merlin in the cytosol. Merlin isoforms I and II have C-termini with very different predicted secondary structures. In fact, emerging evidence suggests that merlin isoform I, and not II, is capable of head-to-tail interaction *(154)*. Moreover, some evidence suggests that isoform I specifically can suppress the growth of rat Schwann cells in vitro. Together, these observations lead to the similar, yet contrasting models for ERM/merlin function depicted in Fig. 4. Much more information is necessary to complete these simplistic models.

Several lines of recent evidence link ERM function to signaling pathways controled by the Rho GTPases, a subfamily of the Ras GTPase superfamily *(181)*. The prototypes of this subfamily, RhoA, Rac 1, and Cdc42Hs, effect specific forms of cytoskeletal remodeling, in response to various stimuli. Thus, activation of RhoA, Rac 1, and Cdc42 lead to the formation of stress fibers, membrane ruffles, and filopodia, respectively *(181)*. In addition, the function of these proteins has been linked to such cellular activities as adhesion, migration, proliferation, and transformation. A requirement for the ERM proteins in Rho- and Rac-dependent actin remodeling in permeabilized cells has been demonstrated *(182)*. Activation of RhoA, but not Rac 1 or Cdc42Hs, effects phosphorylation and relocalization of the ERM proteins *(183)*. In fact, the ERM proteins have recently been identified as direct substrates of the RhoA effector, Rho kinase (RhoK) *(176)*. Coordinated regulation of the phosphorylation state of the ERM C-terminus by RhoK and myosin light chain phosphatase, whose regulatory subunit can bind to the ERMs, has been proposed *(164)*. Finally, it has been reported that, in vitro, the ERM proteins can bind directly to both negative (RhoGDI) and positive (Db1) regulators of RhoA, suggesting that the ERMs may function to control the localization of regulators of Rho GTPase signaling pathways *(184,185)*. The consequences of RhoA-

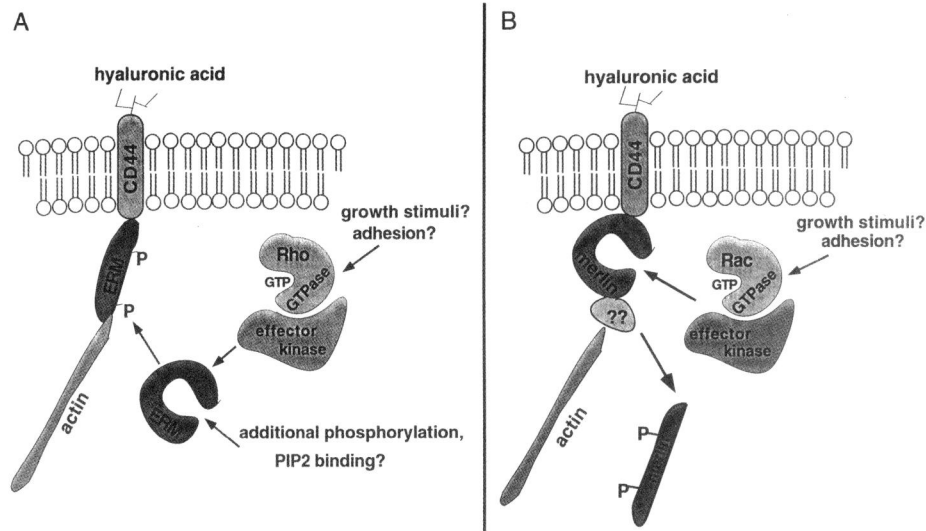

Fig. 4. Contrasting hypothetical models of ERM (**A**) and merlin (**B**) regulation. The ERM proteins and merlin isoform I can exist in an intramolecular/monomeric or intermolecular oligomeric head-to-tail conformation. Phosphorylation, and possibly 4,5-PIP2 binding, is thought to relieve this conformation. For the ERM proteins, this activates them, and strengthens their association with the cytoskeleton, and membrane partners, such as CD44. In contrast, phosphorylation appears to inactivate merlin, decreasing the strength of its cytoskeletal association. Rho kinase is one known effector of ERM phosphorylation; in contrast, merlin appears to be phosphorylated in response to Rac activation.

mediated ERM phosphorylation to the cytoskeleton, cellular movement, and/or proliferative signaling remain to be determined.

Preliminary evidence indicates that merlin is phosphorylated in response to activation of Rac, and not Rho (R.J. Shaw, A. Yaktine, T. Jacks, McClatchey, unpublished); Rac induces the formation of membrane ruffles, a structure to which merlin preferentially localizes *(158)*. In this case, phosphorylation would be predicted to inactivate merlin, removing it from tight cytoskeletal association; indeed, preliminary experimental evidence supports this (R.J. Shaw, A. Yaktine, T. Jacks, McClatchey, unpublished). Much attention is currently focused on determining the role of merlin and merlin phosphorylation in Rho GTPase signaling, and on identifying the kinase(s) responsible for merlin phosphorylation. It will be particularly interesting to determine whether merlin functions to regulate any of the known downstream components of Rac (or Rho or Cdc42) signaling, such as jun-N-terminal kinase or nuclear factor F-κB activation, cell cycle entry, or targets of the serum response element.

At the cellular level, roles for the ERMs have been identified in cell adhesion, migration, and morphology *(156)*. Antisense oligonucleotide impairment of ERM expression suggests redundant functions for the three proteins *(186)*. Inactivation of single family members produces little effect; however, loss of expression of all three ERM proteins (as determined by Western blot analysis) leads to the disappearance of microvilli from the surface of cells, and impaired cell–cell and cell–substrate adhesion. A positive role for the ERMs in cell migration is suggested by evidence that ezrin function is both necessary and sufficient for HGF-mediated cell migration in vitro *(179)*. This is also consistent with a relationship between the ERMs and the Rho GTPases, which are known to promote cell migration *(181)*.

Both antisense studies and the study of the properties of Nf2-deficient cells suggest similar cellular functions for merlin. Antisense disruption of merlin expression leads to loss of adhesion and altered morphology of Schwann-like and glioma cells in vitro *(187)*. Primary Nf2-deficient mouse embryo fibroblasts (MEFs) exhibit defects in adhesion, manifested by reduced spreading upon replating and by withdrawal of serum (R.J. Shaw, I. Saotome, T. Jacks, McClatchey, unpublished). Furthermore, in contrast to the ERMs, the study of *Nf2–/–* MEFs reveals a negative role for merlin in cell migration. In culture, *Nf2–/–* MEFs migrate to close monolayer wounds significantly faster than their wt counterparts (R.J. Shaw, I. Saotome, T. Jacks, McClatchey, unpublished).

3.4. Genetic Models

As for NF1, the generation and study of animal models has proven a powerful approach to the study of NF2. In the case of NF2, mouse modeling has confirmed the tumor suppressor function of merlin, revealed several developmental contexts requiring merlin function, and implied a greater role for NF2, and perhaps its ERM family members, in cancer development and progression in humans, than previously thought *(188,189)*.

The mouse NF2 protein is 97% identical to human NF2: There are only 7 aa differences between the two proteins *(190)*. *Nf2* heterozygous mice are cancer-prone, but do not develop schwannomas; instead, they develop malignant osteosarcomas, fibrosarcomas, and hepatocellular carcinomas that exhibit *Nf2* LOH *(189)*. A large percentage of these tumors metastasize to distant sites, a phenomenon that is unusual in the mouse. This observation raises the possibility that loss of Nf2 contributes to metastatic potential in this setting. Indeed, initial experiments support a role for merlin loss in metastatic progression. This is particularly interesting, given the interaction of the ERM proteins and merlin with CD44, a molecule that has long been associated with metastatic progression in humans *(191)*. A role for merlin loss in metastatic progression is also consistent with the increased motility exhibited by *Nf2–/–* MEFs, and a relationship between merlin and Rac, which is known to promote cellular invasion and metastasis *(181)*. It will obviously be important to investigate whether merlin or the ERMs are altered during metastatic progression in humans.

Nf2 heterozygous-mutant mice do not develop schwannomas, and *Nf2* mutations have not been found in human osteosarcomas, but explanations for differing human–mouse phenotypes have emerged from other studies *(192)*. In some cases, such as those of *Rb+/–*, humans and mice, it is clear that closely related family members are able to compensate for retinal loss of Rb in the mouse, and not in humans. In other cases, such as *Nf1+/–* humans and mice, it is clear that the rate of loss of the wt allele in mice is limiting. Thus, although *Nf1+/–* mice do not develop neurofibromas, chimeric mice, composed in part of *Nf1–/–* cells, develop many neurofibroma-like lesions (K. Cichowski, S. Shih, T. Jacks, in preparation). An extension of this idea is supported by the observation that mice carrying *Nf2* and *p53* or *Nf1* mutations in *cis* on the same chr (these three loci are all linked in mice, and not in humans), leads to dramatic acceleration of tumor formation *(189;* McClatchey, I. Saotome, T. Jacks, unpublished). These studies not only reveal cooperation between a *Nf2* mutation and mutations in both the *p53* and *Nf1* genes in cancer development, but they also reveal the importance of linkage of cancer-predisposing mutations. Differences in mouse and human genome configuration may dramatically influence combinatorial mutation rates, and thereby account for differences in the spectra of tumors that develop. The identification of the molecular pathway, in which merlin functions, may reveal other targets of mutation during the development of schwannomas in humans or osteosarcomas in mice.

Cooperation between *Nf1* and *Nf2* mutations in mouse tumor development was somewhat unexpected, given the growing symptomatic distinction between NF1 and NF2 in humans. Mice carrying *Nf2* and *p53* mutations in *cis* exhibit a dramatic acceleration of osteosarcoma formation, developing multiple, early-onset osteosarcomas, but *Nf2–Nf1* compound *(cis)* heterozygotes reproducibly develop a tumor type resembling a MPNST (McClatchey, I. Saotome, T. Jacks, unpublished). The fact that Rac functions downstream of Ras signaling provides one possible explanation for this cooperativity *(181)*. Rac function has been shown to be required for Ras-mediated transformation, and Ras and Rac activation cooperate in cell transformation. Loss of merlin and neurofibromin may similarly cooperate in activating the two connected pathways. In fact, merlin has been reported to suppress Ras-mediated transformation *(193)*.

Gene targeting in the mouse has also revealed that the normal function of merlin is manifest during several stages of mouse embryonic development *(188)*. A homozygous *Nf2* mutation in the mouse leads to developmental failure early, during embryonic development at gastrulation. The defect resides in the extraembryonic lineage, and may result from abnormal cell adhesion/invasion. In addition, mosaic embryos composed of *Nf2–/–* and wt cells reveal additional developmental requirements for Nf2, including during cardiac development (I. Saotome, T. Jacks, McClatchey, unpublished). In the developing myocardium, loss of Nf2 function is sufficient for hyperproliferation of myocardial precursors; chimeric hearts contain tumor-like lesions that are composed entirely of *Nf2–/–* cells, the nature of which is reminiscent of NF2-associated schwannomas. These developmental contexts reflect the growth-controlling situations that merlin evolved to perform, and thus important ones in which to study merlin function.

In *D. melanogaster,* homologs for merlin (Dmerlin) and, apparently, a single ERM protein (Dmoesin), have been identified *(194)*. In addition, a merlin homolog can be identified in the recently sequenced *C. elegans* genome. Dmerlin shares 55% aa identity to that of human merlin, with particularly concentrated relatedness throughout the N-terminal halves (Fig. 3). Dmoesin shares ~58% aa identity with the ERM proteins. Conditional expression of exogenous Dmerlin in S2 cells reveals that the protein is initially localized at the membrane, but is rapidly internalized and localized to endocytic vesicles. Similarly, detection of endogenous protein in *Drosophila* embryos reveals both membrane and punctate, intracellular staining. This raises the intriguing possibility that merlin functions in the cytoskeletal reorganization that accompanies endocytic recycling. Impaired recycling of active signaling molecules, such as growth factor receptors, could lead to hyperactivation of the signaling pathways that they govern, and to uncontrolled cellular growth. In addition, somatic mosaic analysis in *Drosophila* does, in fact, reveal a cell-autonomous growth-suppressive function for Dmerlin, supporting its identity as a tumor suppressor *(195)*. Additional studies have examined the ability of a series of mutant Dmerlin isoforms to localize subcellularly in vitro, and to transgenically rescue a null phenotype in vivo *(195)*. These studies identified both potentially dominant-active and dominant-negative merlin isoforms. Examination of the mammalian counterparts of these mutants may be particularly enlightening.

3.5. NF2: Current View and Future Directions

Most current effort is placed on the identification of signaling pathway(s) in which merlin functions. Particular attention is focused on an assessment of the function of merlin in the signaling pathways governed by the Rho family of GTPases. This is being pursued chiefly through the attempted identification of additional interacting proteins, and through the comparison of Nf2-expressing and Nf2-deficient cells. Many published studies have compared primary human Schwann cells and cells derived from Nf2-deficient schwannomas, which suffer from the probability of additional mutations present in the tumor cells. The generation of cell lines that stably overexpress merlin, and the reintroduction of merlin expression to Nf2-deficient tumor cells, has proven difficult, probably because excess merlin is not tolerated in most cell types. The use of inducible systems, and the generation of primary Nf2-deficient cells and their wt counterparts, should allow the controlled evaluation of the cellular and molecular consequences of merlin function.

A determination of the relationship between merlin and the ERM proteins remains an important area of investigation. Although modeling of merlin function upon ERM function may be fruitful, several lines of evidence suggest that ERM and merlin functions are different, and may, in fact, be antagonistic. It will be interesting to compare the status of the ERM proteins in Nf2-expressing and Nf2-deficient cells. For example, are the ERM proteins phosphorylated, localized, or oligomerized differently under various conditions in Nf2-expressing vs Nf2-deficient cells?

Future investigations will also focus on the development of better models of Nf2-associated schwannoma and meningioma formation in mice. For example, the use of conditional targeting and chimeric analysis will allow the generation of adult animals that lack Nf2 in the Schwann and meningial cell lineages, and a determination of whether loss of the wt *Nf2* allele is rate-limiting for schwannoma formation. Alternatively, intercrossing of *Nf2*+/– mice with other mutant mouse strains may reveal genetic interaction in schwannoma or meningioma formation. The use of both mouse and invertebrate models will be invaluable toward identifying genetic modifier loci and other members of the merlin pathway.

The placement of merlin, within known signal transduction pathways, will suggest rational treatment possibilities. For example, given the anti-Ras function of merlin and the placement of Rho GTPases downstream of Ras, farnesyl transferase inhibitors may be expected to be effective in treating NF2-deficient tumors. Alternatively, geranyl transferase inhibitors, which would be predicted to disable some Rho family members, including Rac, may be effective. The identification of genes that are upregulated or activated in Nf2-deficient cells may also be targets. For example, if merlin normally antagonizes ERM function, then targeting the inactivation of the ERMs may have therapeutic value. Mouse models of Nf2-deficient tumor formation provide important vehicles in which to test these therapeutic strategies. Finally, the investigation of a broader role for NF2 and the ERMs in human cancer is an important avenue of inquiry that may help to decipher the role of the poorly understood membrane–cytoskeleton interface in cancer development and progression.

REFERENCES

1. Gutmann DH, Collins FS. The neurofibromatosis type 1 gene and its protein product, neurofibromin. *Neuron* 1993; 10:335–343.

2. Viskochil D, White R, Cawthon R. The neurofibromatosis type 1 gene. *Annu Rev Neurosci* 1993; 16:183–205.

3. Bernards A. Neurofibromatosis type 1 and Ras-mediated signaling: filling in the GAPs. *Biochim Biophys Acta* 1995; 1242:43–59.

4. Shen MH, Harper PS, Upadhyaya M. Molecular genetics of neurofibromatosis type 1 (NF1). *J Med Genet* 1996; 33:2–17.

5. Gusella JF, Ramesh V, MacCollin M, Jacoby LB. Neurofibromatosis 2: loss of merlin's protective spell. *Curr Opin Genet Dev* 1996; 6:87–92.

6. Gutmann D, Aylsworth A, Carey J, Korf B, Marks J, Pyeritz R, Rubenstein A, Viskochil D. The diagnostic evaluation and multidisciplinary management of neurofibromatosis 1 and neurofibromatosis 2. *JAMA* 1997; 278:51–57.

7. von Recklinghausen F. *Ueber die multiplen Fibrome der Haut und ihre Beziehung zu den multiplen Neuromen.* Berlin: August Hirschwald. 1882.

8. Riccardi VM. *Neurofibromatosis: Phenotype, Natural History and Pathogenesis,* 2nd ed. Baltimore, MD: Johns Hopkins University Press. 1992.

9. Upadhyaya M, Cooper DN. *Neurofibromatosis Type 1 from Genotype to Phenotype.* Oxford: Bios Scientific. 1998:1–230.

10. Huson SM, Compston DAS, Clark P, Harper PS. A genetic study of von Recklinghausen neurofibromatosis in South East Wales. I. Prevalence, fitness, mutation rate, and effect of parental transmission on severity. *J Med Genet* 1989; 26:704–711.

11. Marchuk DA, Tavakkol R, Wallace MR, Brownstein BH, Taillon-Miller P, Fong CT, et al. A yeast artificial chromosome contig encompassing the type 1 neurofibromatosis gene. *Genomics* 1992; 13:672–680.

12. Jayadel D, Fain P, Upadhyaya M, Ponder MA, Huson SM, Carey J, et al. Paternal origin of new mutations in Von Recklinghausen neurofibromatosis. *Nature* 1990; 343:558–559.

13. Stephens K, Kayes L, Riccardi VM, Rising M, Sybert VP, Pagon RA. Preferential mutation of the NF1 gene in paternally derived chromosomes. *Hum Genet* 1992; 88:279–282.

14. Lazaro C, Gaona A, Ainsworth P, Tenconi R, Vidaud D, Kruyer H, et al. Sex differences in mutational rate and mutational mechanism in the NF1 gene in neurofibromatosis type 1 patients. *Hum Genet* 1996; 98:696–699.

15. Upadhyaya M, Ruggieri M, Maynard J, Osborn M, Hartog C, Mudd S, et al. Gross deletions of the neurofibromatosis type 1 (NF1) gene are predominantly of maternal origin and commonly associated with a learning disability, dysmorphic features and developmental delay. *Hum Genet* 1998; 102:591–597.

16. Barker D, Wright E, Nguyen K, Cannon L, Fain P, Goldgar D, et al. Gene for von Recklinghausen neurofibromatosis is in the pericentric region of chromosome 17. *Science* 1987; 236:1100–1102.

17. Seizinger BR, Rouleau GA, Ozelius LJ, Lane AH, Faryniarz AG, Chao MV, et al. Genetic linkage of von Recklinghausen neurofibromatosis to the nerve growth factor receptor gene. *Cell* 1987; 49:589–594.

18. Easton DF, Ponder MA, Huson SM, Ponder BAJ. An analysis of variation in expression of neurofibromatosis (NF) type I (NFI): evidence for modifying genes. *Am J Hum Genet* 1993; 53:305–313.

19. Bolande RP. Neurofibromatosis: the quintessential neurocristopathy: pathogenetic concepts and relationships. *Adv Neurol* 1981; 29:67–75.

20. Sheela S, Riccardi VM, Ratner N. Schwann cells in neurofibromas show premalignant properties. *J Cell Biol* 1990; 111:645–653.

21. Rosenbaum T, Boissy YL, Kombrinck K, Brannan CI, Jenkins NA, Copeland NG, Ratner N. Neurofibromin-deficient fibroblasts fail to form perineurium in vitro. *Development* 1995; 121:3583–3592.

22. Vogel KS, Brannan CI, Jenkins NA, Copeland NG, Parada LF. Loss of neurofibromin results in neurotropin-independent survival of embryonic sensory and sympathetic neurons. *Cell* 1995; 82:733–742.

23. Friedman JM, Birch PH. Type 1 neurofibromatosis: a descriptive analysis of the disorder in 1,728 patients. *Am J Med Genet* 1997; 70:138–143.

24. North KN, Riccardi V, Samango-Sprouse C, Ferner R, Moore B, Legius E, Ratner N, Denckla MB. Cognitive function and academic performance in neurofibromatosis. 1: consensus statement from the NF1 Cognitive Disorders Task Force. *Neurology* 1997; 48:1121–1127.

25. DiPaolo D, Zimmerman R, Rorke L, Zackai E, Bilaniuk L, Yachnis A. Neurofibromatosis type 1: pathologic substrate of high-signal-intensity foci in the brain. *Radiology* 1995; 195:721–724.

26. Friedman JM, Birch P. An association between optic glioma and other tumours of the central nervous system in neurofibromatosis type 1. *Neuropediatrics* 1997; 28:131–132.

27. Bahuau M, Flintoff W, Assouline B, Lyonnet S, Le Merrer M, Prieur M, et al. Exclusion of allelism of Noonan syndrome and neurofibromatosis-type 1 in a large family with Noonan syndrome-neurofibromatosis association. *Am J Med Genet* 1996; 66:347–355.

28. Allanson JE, Upadhyaya M, Watson GH, Partington M, MacKenzie A, Lahey D, et al. Watson syndrome: is it a subtype of type 1 neurofibromatosis? *J Med Genet* 1991; 28:752–756.

29. Tassabehji M, Strachan T, Sharland M, Colley A, Donnai D, Harris R, Thakker N. Tandem duplication within a neurofibromatosis type 1 (NF1) gene exon in a family with features of Watson syndrome and Noonan syndrome. *Am J Hum Genet* 1993; 53:90–95.

30. Wu R, Legius E, Robberecht W, Dumoulin M, Cassiman JJ, Fryns JP. Neurofibromatosis type I gene mutation in a patient with features of LEOPARD syndrome. *Hum Mutat* 1996; 8:51–56.

31. Legius E, Wu R, Eyssen M, Marynen P, Fryns JP, Cassiman JJ. Encephalocraniocutaneous lipomatosis with a mutation in the NF1 gene. *J Med Genet* 1995; 32:316–319.

32. Tibbles JA, Cohen MM, Jr. The proteus syndrome: the Elephant Man diagnosed. *Br Med J (Clin Res Ed)* 1986; 293:683–685.

33. Viskochil D, Buchberg AM, Xu G, Cawthon RM, Stevens J, Wolff RK, et al. Deletions and a translocation interrupt a cloned gene at the neurofibromatosis type I locus. *Cell* 1990; 62:187–192.

34. Cawthon RM, Weiss R, Xu G, Viskochil D, Culver M, Stevens J, et al. A major segment of the neurofibromatosis type 1 gene: cDNA sequence, genomic structure, and point mutations. *Cell* 1990; 62:193–201.

35. Wallace MR, Marchuk DA, Andersen LB, Letcher R, Odeh HM, Saulino AM, et al. Type 1 neurofibromatosis gene: identification of a large transcript disrupted in three NF1 patients. *Science* 1990; 240:181–186.

36. Li Y, O'Connell P, Huntsman-Breidenbach H, Cawthon R, Stevens J, Xu G, et al. Genomic organization of the neurofibromatosis 1 gene *(NF1)*. *Genomics* 1995; 25:9–18.

37. Buchberg AM, Cleveland LS, Jenkins NA, Copeland NG. Sequence homology shared by neurofibromatosis type-1 gene and IRA-1 and IRA-2 negative regulators of the RAS cyclic AMP pathway. *Nature* 1990; 347:291–294.

38. Xu G, O'Connell P, Viskochil D, Cawthon R, Robertson M, Culver M, et al. The neurofibromatosis type 1 gene encodes a protein related to GAP. *Cell* 1990; 62:599–608.

39. Heim RA, Kam-Morgan LN, Binnie CG, Corns DD, Cayouette MC, Farber RA, et al. Distribution of 13 truncating mutations in the neurofibromatosis 1 gene. *Hum Mol Genet* 1995; 4:975–981.

40. Park VM, Pivnick EK. Neurofibromatosis type 1 (NF1): a protein truncation assay yielding identification of mutations in 73% of patients. *J Med Genet* 1998; 35:813–820.

41. Wu BL, Schneider GH, Korf BR. Deletion of the entire NF1 gene causing distinct manifestations in a family. *Am J Med Genet* 1997; 69:98–101.

42. Tonsgard JH, Yelavarthi KK, Cushner S, Short MP, Lindgren V. Do NF1 gene deletions result in a characteristic phenotype? *Am J Med Genet* 1997; 73:80–86.

43. Poyhonen M, Leisti EL, Kytola S, Leisti J. Hereditary spinal neurofibromatosis: a rare form of NF1? *J Med Genet* 1997; 34:184–187.

44. Abeliovich D, Gelman-Kohan Z, Silverstein S, Lerer I, Chemke J, Merin S, Zlotogora J. Familial cafe au lait spots: a variant of neurofibromatosis type 1. *J Med Genet* 1995; 32:985–986.

45. Ars E, Kruyer H, Gaona A, Casquero P, Rosell J, Volpini V, Serra E, Lazaro C, Estivill X. A clinical variant of neurofibromatosis type 1: familial spinal neurofibromatosis with a frameshift mutation in the NF1 gene. *Am J Hum Genet* 1998; 62:834–841.

46. Viskochil D, Cawthon R, P. OC, Xu GF, Stevens J, Culver M, Carey J, White R. The gene encoding the oligodendrocyte-myelin glycoprotein is embedded within the neurofibromatosis type 1 gene. *Mol Cell Biol* 1991; 11:906–912.

47. Habib AA, Gulcher JR, Hognason T, Zheng L, Stefansson K. The OMgp gene, a second growth suppressor within the NF1 gene. *Oncogene* 1998; 16:1525–1531.

48. Cho BC, Shaughnessy JD, Jr, Largaespada DA, Bedigian HG, Buchberg AM, Jenkins NA, Copeland NG. Frequent disruption of the Nf1 gene by a novel murine AIDS virus-related provirus in BXH-2 murine myeloid lymphomas. *J Virol* 1995; 69:7138–7146.

49. Danglot G, Régnier V, Fauvet D, Vassal G, Kujas M, Bernheim A. Neurofibromatosis 1 *(NF1)* mRNAs expressed in the central nervous system are differentially spliced in the 5′ part of the gene. *Hum Mol Gen* 1995; 4:915–920.

50. Nishi T, Lee PS, Oka K, Levin VA, Tanase S, Morino Y, Saya H. Differential expression of two types of the neurofibromatosis type 1 (NF1) gene transcripts related to neuronal differentiation. *Oncogene* 1991; 6:1555–1559.

51. Gutmann DH, Andersen LB, Cole JL, Swaroop M, Collins FS. An alternatively-spliced mRNA in the carboxy terminus of the neurofibromatosis type 1 *(NF1)* gene is expressed in muscle. *Hum Mol Genet* 1993; 2:989–992.

52. Andersen LB, Ballester R, Marchuk DA, Chang E, Gutmann DH, Saulino AM, et al. A conserved alternative splice in the von Recklinghausen neurofibromatosis *(NF1)* gene produces two neurofibromin isoforms, both of which have GTPase-activating protein activity. *Mol Cell Biol* 1993; 13:487–495.

53. Cappione AJ, French BL, Skuse GR. A potential role for NF1 mRNA editing in the pathogenesis of NF1 tumors. *Am J Hum Genet* 1997; 60:305–312.

54. Hoffmeyer S, Assum G, Griesser J, Kaufmann D, Nurnberg P, Krone W. On unequal allelic expression of the neurofibromin gene in neurofibromatosis type 1. *Hum Mol Genet* 1995; 4:1267–1272.

55. Bernards A, Snijders AJ, Hannigan GE, Murthy AE, Gusella JF. Mouse neurofibromatosis type 1 cDNA sequence reveals high degree of conservation of both coding and non-coding mRNA segments. *Hum Mol Genet* 1993; 2:645–650.

56. Kehrer-Sawatzki H, Maier C, Moschgath E, Elgar G, Krone W. Genomic characterization of the Neurofibromatosis Type 1 gene of Fugu rubripes. *Gene* 1998; 222:145–153.

57. The I, Hannigan GE, Cowley GS, Reginald S, Zhong Y, Gusella JF, Hariharan IK, Bernards A. Rescue of a Drosophila NF1 mutant phenotype by protein kinase A. *Science* 1997; 276:791–794.

58. Bollag G, McCormick F, Clark R. Characterization of full-length neurofibromin: tubulin inhibits Ras GAP activity. *EMBO J* 1993; 12:1923–1927.

59. DeClue JE, Cohen BD, Lowy DR. Identification and characterization of the neurofibromatosis type 1 protein product. *Proc Natl Acad Sci USA* 1991; 88:9914–9918.

60. Daston MM, Ratner N. Neurofibromin, a predominantly neuronal GTPase activating protein in the adult, is ubiquitously expressed during development. *Dev Dynamics* 1992; 195:216–226.

61. Huynh DP, Nechiporuk T, Pulst SM. Differential expression and tissue distribution of type I and type II neurofibromins during mouse fetal development. *Dev Biol* 1994; 161:538–551.

62. Daston MM, Scrable H, Nordlund M, Sturbaum AK, Nissen LM, Ratner N. The protein product of the neurofibromatosis type 1 gene is expressed at highest abundance in neurons, Schwann cells, and oligodendrocytes. *Neuron* 1992; 8:415–428.

63. Nordlund M, Gu X, Shipley MT, Ratner N. Neurofibromin is enriched in the endoplasmic reticulum of CNS neurons. *J Neurosci* 1993; 13:1588–1600.

64. Boyer MJ, Gutmann DH, Collins FS, Bar-Sagi D. Crosslinking of the surface immunoglobulin receptor in B lymphocytes induces a redistribution of neurofibromin but not p120-GAP. *Oncogene* 1994; 9:349–357.

65. Gregory PE, Gutmann DH, Mitchell A, Park S, Boguski M, Jacks T, Wood DL, Jove R, Collins FS. Neurofibromatosis type 1 gene product (neurofibromin) associates with microtubules. *Somat Cell Mol Genet* 1993; 19:265–274.

66. Roudebush M, Slabe T, Sundaram V, Hoppel CL, Golubic M, Stacey DW. Neurofibromin colocalizes with mitochondria in cultured cells. *Exp Cell Res* 1997; 236:161–172.

67. Golubic M, Roudebush M, Dobrowolski S, Wolfman A, Stacey DW. Catalytic properties, tissue and intracellular distribution of neurofibromin. *Oncogene* 1992; 7:2151–2159.

68. Rey I, Taylor-Harris P, van Erp H, Hall A. R-ras interacts with rasGAP, neurofibromin and c-raf but does not regulate cell growth or differentiation. *Oncogene* 1994; 9:685–692.

69. Fernandez-Sarabia MJ, Bischoff JR. Bcl-2 associates with the *ras*-related protein R-ras p23. *Nature* 1993; 366:274–275.

70. Chen CY, Faller DV. Phosphorylation of Bcl-2 protein and association with p21Ras in Ras-induced apoptosis. *J Biol Chem* 1996; 271:2376–2379.

71. Hockenbery DM, Oltvai ZN, Yin XM, Milliman CL, Korsmeyer SJ. Bcl-2 functions in an antioxidant pathway to prevent apoptosis. *Cell* 1993; 75:241–251.

72. Johnson MR, DeClue JE, Felzmann S, Vass WC, Xu G, White R, Lowy DR. Neurofibromin can inhibit ras-dependent growth by a mechanism independent of its GTPase-accelerating function. *Mol Cell Biol* 1994; 14:641–645.

73. Liu K, Li G. Catalytic domain of the p120 Ras GAP binds to Rab5 and stimulates its GTPase activity. *J Biol Chem* 1998; 273:10,087–10,090.

74. Martin GA, Viskochil D, Bollag G, McCabe PC, Crosier WJ, Haubruck H, et al. The GAP-related domain of the neurofibromatosis type 1 gene product interacts with *ras* p21. *Cell* 1990; 63:843–849.

75. Ballester R, Marchuk D, Boguski M, Saulino A, Letcher R, Wigler M, Collins F. The *NF1* locus encodes a protein functionally related to mammalian GAP and yeast *IRA* proteins. *Cell* 1990; 63:851–859.

76. Xu GF, Lin B, Tanaka K, Dunn D, Wood D, Gesteland R, White R, Weiss R, Tamanoi F. The catalytic domain of the neurofibromatosis type 1 gene product stimulates ras GTPase and complements ira mutants of S. cerevisiae. *Cell* 1990; 63:835–841.

77. Basu TN, Gutmann DH, Fletcher JA, Glover TW, Collins FS, Downward J. Aberrant regulation of ras proteins in malignant tumour cells from type 1 neurofibromatosis patients. *Nature* 1992; 356:713–715.

78. DeClue JE, Papageorge AG, Fletcher JA, Diehl SR, Ratner N, Vass WC, Lowy DR. Abnormal regulation of mammalian p21ras contributes to malignant tumor growth in von Recklinghausen (type 1) neurofibromatosis. *Cell* 1992; 69:265–273.

79. Guha A, Lau N, Huvar I, Gutmann D, Provias J, Pawson T, Boss G. Ras-GTP levels are elevated in human NF1 peripheral nerve tumors. *Oncogene* 1996; 12:507–513.

80. Bollag G, Clapp DW, Shih S, Adler F, Zhang YY, Thompson P, et al. Loss of NF1 results in activation of the Ras signaling pathway and leads to aberrant growth in haematopoietic cells. *Nat Genet* 1996; 12:144–148.

81. Kalra R, Paderanga DC, Olson K, Shannon KM. Genetic analysis is consistent with the hypothesis that NF1 limits myeloid cell growth through p21ras. *Blood* 1994; 84:3435–3439.

82. The I, Murthy AE, Hannigan GE, Jacoby LB, Menon AG, Gusella JF, Bernards A. Neurofibromatosis type 1 gene mutations in neuroblastoma. *Nat Genet* 1993; 3:62–66.

83. Johnson MR, Look AT, DeClue JE, Valentine MB, Lowy DR. Inactivation of the NF1 gene in human melanoma and neuroblastoma cell lines without impaired regulation of GTP.Ras. *Proc Natl Acad Sci USA* 1993; 90:5539–5543.

84. Andersen LB, Fountain JW, Gutmann DH, Tarle SA, Glover TW, Dracopoli NC, Housman DE, Collins FS. Mutations in the neurofibromatosis 1 gene in sporadic malignant melanoma cell lines. *Nat Genet* 1993; 3:118–121.

85. Gomez L, Rubio MP, Martin MT, Vazquez JJ, Idoate M, Pastorfide G, et al. Chromosome 17 allelic loss and NF1-GRD mutations do not play a significant role as molecular mechanisms leading to melanoma tumorigenesis. *J Invest Dermatol* 1996; 106:432–436.

86. Martinsson T, Sjoberg RM, Hedborg F, Kogner P. Homozygous deletion of the neurofibromatosis-1 gene in the tumor of a patient with neuroblastoma. *Cancer Genet Cytogenet* 1997; 95:183–189.

87. Weiss WA, Aldape K, Mohapatra G, Feuerstein BG, Bishop JM. Targeted expression of MYCN causes neuroblastoma in transgenic mice. *EMBO J* 1997; 16:2985–2995.

88. Bos JL. ras oncogenes in human cancer: a review. *Cancer Res* 1989; 49:4682–4689.

89. Ridley AJ, Paterson HF, Noble M, Land H. Ras-mediated cell cycle arrest is altered by nuclear oncogenes to induce Schwann cell transformation. *EMBO J* 1988; 7:1635–1645.

90. Largaespada DA, Brannan CI, Jenkins NA, Copeland NG. Nf1 deficiency causes Ras-mediated granulocyte/macrophage colony stimulating factor hypersensitivity and chronic myeloid leukaemia. *Nat Genet* 1996; 12:137–143.

91. Zhang YY, Vik TA, Ryder JW, Srour EF, Jacks T, Shannon K, Clapp DW. Nf1 regulates hematopoietic progenitor cell growth and ras signaling in response to multiple cytokines. *J Exp Med* 1998; 187:1893–1902.

92. Legius E, Marchuk DA, Collins FS, Glover TW. Somatic deletion of the neurofibromatosis type 1 gene in a neurofibrosarcoma supports a tumour suppressor gene hypothesis. *Nat Genet* 1993; 3:122–126.

93. Skuse GR, Kosciolek BA, Rowley PT. Molecular genetic analysis of tumors in von Recklinghausen neurofibromatosis: loss of heterozygosity for chromosome 17. *Genes Chromosome Cancer* 1989; 1:36–41.

94. Menon AG, Anderson KM, Riccardi VM, Chung RY, Whaley JM, Yandell DW, et al. Chromosome 17p deletions and p53 gene mutations associated with the formation of malignant neurofibrosarcomas in von Recklinghausen neurofibromatosis. *Proc Natl Acad Sci USA* 1990; 87:5435–5439.

95. Xu W, Mulligan LM, Ponder MA, Liu L, Smith BA, Mathew CG, Ponder BA. Loss of NF1 alleles in phaeochromocytomas from patients with type I neurofibromatosis. *Genes Chromosome Cancer* 1992; 4:337–342.

96. Bader JL, Miller RW. Neurofibromatosis and childhood leukemia. *J Pediatr* 1978; 92:925–929.

97. Shannon KM, Watterson J, Johnson P, P. OC, Lange B, Shah N, Steinherz P, Kan YW, Priest JR. Monosomy 7 myeloproliferative disease in children with neurofibromatosis, type 1: epidemiology and molecular analysis. *Blood* 1992; 79:1311–1318.

98. Side L, Taylor B, Cayouette M, Conner E, Thompson P, Luce M, Shannon K. Homozygous inactivation of the NF1 gene in bone marrow cells from children with neurofibromatosis type 1 and malignant myeloid disorders. *N Engl J Med* 1997; 336:1713–1720.

99. Jacks T, Shih TS, Schmitt EM, Bronson RT, Bernards A, Weinberg RA. Tumour predisposition in mice heterozygous for a targeted mutation in *Nf1*. *Nat Genet* 1994; 7:353–361.

100. Li Y, Bollag G, Clark R, Stevens J, Conroy L, Fults D, et al. Somatic mutations in the neurofibromatosis 1 gene in human tumors. *Cell* 1992; 69:275–281.

101. Jensen S, Paderanga DC, Chen P, Olson K, Edwards M, Iavorone A, Israel MA, Shannon K. Molecular analysis at the NF1 locus in astrocytic brain tumors. *Cancer* 1995; 76:674–677.

102. Hewett SJ, Choi DW, Gutmann DH. Expression of the neurofibromatosis 1 (NF1) gene in reactive astrocytes in vitro. *Neuroreport* 1995; 6:1565–1568.

103. Gutmann DH, Giordano MJ, Mahadeo DK, Lau N, Silbergeld D, Guha A. Increased neurofibromatosis 1 gene expression in astrocytic tumors: positive regulation by p21-ras. *Oncogene* 1996; 12:2121–2127.

104. Platten M, Giordano MJ, Dirven CM, Gutmann DH, Louis DN. Up-regulation of specific NF 1 gene transcripts in sporadic pilocytic astrocytomas. *Am J Pathol* 1996; 149:621–627.

105. Thiel G, Marczinek K, Neumann R, Witkowski R, Marchuk DA, Nurnberg P. Somatic mutations in the neurofibromatosis 1 gene in gliomas and primitive neuroectodermal tumours. *Anticancer Res* 1995; 15:2495–2499.

106. Muhammad AK, Yoshimine T, Maruno M, Tokiyoshi K, Takemoto O, Ninomiya H, Hayakawa T. Chromosome 17 allelic loss in astrocytic tumors and its clinico-pathologic implications. *Clin Neuropathol* 1997; 16:220–226.

107. Fialkow PJ, Sagebiel RW, Gartler SM, Rimoin DL. Multiple cell origin of hereditary neurofibromas. *N Eng J Med* 1971; 297:696–698.

108. Skuse GR, Kosciolek BA, Rowley PT. The neurofibroma in von Recklinghausen neurofibromatosis has a unicellular origin. *Am J Hum Genet* 1991; 49:600–607.

109. Glover TW, Stein CK, Legius E, Andersen LB, Brereton A, Johnson S. Molecular and cytogenetic analysis of tumors in von Recklinghausen neurofibromatosis. *Genes Chromosomes Cancer* 1991; 3:62–70.

110. Shimizu E, Shinohara T, Mori N, Yokota J, Tani K, Izumi K, Obashi A, Ogura T. Loss of heterozygosity on chromosome arm 17p in small cell lung carcinomas, but not in neurofibromas, in a patient with von Recklinghausen neurofibromatosis. *Cancer* 1993; 71:725–728.

111. Lothe RA, Slettan A, Saeter G, Brogger A, Borresen AL, Nesland JM. Alterations at chromosome 17 loci in peripheral nerve sheath tumors. *J Neuropathol Exp Neurol* 1995; 54:65–73.

112. Colman SD, Williams CA, Wallace MR. Benign neurofibromas in type 1 neurofibromatosis (NF1) show somatic deletions of the NF1 gene. *Nat Genet* 1995; 11:90–92.

113. Sawada S, Florell S, Purandare SM, Ota M, Stephens K, Viskochil D. Identification of NF1 mutations in both alleles of a dermal neurofibroma. *Nat Genet* 1996; 14:110–112.

114. Serra E, Puig S, Otero D, Gaona A, Kruyer H, Ars E, Estivill X, Lazaro C. Confirmation of a double-hit model for the NF1 gene in benign neurofibromas. *Am J Hum Genet* 1997; 61:512–519.

115. Benedict P, Szabó G, Fitzpatrick T, Sinesi S. Melanotic macules in Albright's syndrome and in neurofibromatosis. *JAMA* 1968; 205:618–626.

116. Martuza RL, Philippe I, Fitzpatrick TB, Zwaan J, Seki Y, Lederman J. Melanin macrogobules as a cellular marker of neurofibromatosis: a quantitative study. *J Invest Dermatol* 1985; 85:347–350.

117. Bartosik J. Melanosome complexes and melanin macroglobules in normal human skin. *Acta Derm Venereol* 1991; 71:283–286.

118. Nakagawa H, Hori Y, Sato S, Fitzpatrick TB, Martuza RL. The nature and origin of the melanin macroglobule. *J Invest Dermatol* 1984; 83:134–139.

119. Eisenbarth I, Assum G, Kaufmann D, Krone W. Evidence for the presence of the second allele of the neurofibromatosis type 1 gene in melanocytes derived from café-au-lait macules of NF1 patients. *Biochem Biophys Res Commun* 1997; 237:138–141.

120. Brannan CI, Perkins AS, Vogel KS, Ratner N, Nordlund ML, Reid SW, et al. Targeted disruption of the neurofibromatosis type-1 gene leads to developmental abnormalities in heart and various neural crest-derived tissues. *Genes Dev* 1994; 8:1019–1029.

121. Silva AJ, Frankland PW, Marowitz Z, Friedman E, Lazlo G, Cioffi D, Jacks T, Bourtchuladze R. A mouse model for the learning and memory deficits associated with neurofibromatosis type I. *Nat Genet* 1997; 15:281–284.

122. Lakkis MM, Epstein JA. Neurofibromin modulation of ras activity is required for normal endocardial-mesenchymal transformation in the developing heart. *Development* 1998; 125:4359–4367.

123. Bishop III JG, Corces VG. Expression of an activated *ras* gene causes developmental abnormalities in transgenic *Drosophila melanogaster. Genes Dev* 1988; 2:567–577.

124. Fortini ME, Simon MA, Rubin GM. Signalling by the sevenless protein tyrosine kinase is mimicked by Ras1 activation. *Nature* 1992; 355:559–561.

125. Zhong Y. Mediation of PACAP-like neuropeptide transmission by coactivation of Ras/Raf and cAMP signal transduction pathways in Drosophila. *Nature* 1995; 375:588–592.

126. Guo HF, The I, Hannan F, Bernards A, Zhong Y. Requirement of Drosophila NF1 for activation of adenylyl cyclase by PACAP38-like neuropeptides. *Science* 1997; 276:795–798.

127. Toda T, Uno I, Ishikawa T, Powers S, Kataoka T, Broek D, et al. In yeast, *RAS* proteins are controlling elements of the adenylate cyclase. *Cell* 1985; 40:27–36.

128. Mitts MR, Bradshaw-Rouse J, Heideman W. Interactions between adenylate cyclase and the yeast GTPase-activating protein IRA1. *Mol Cell Biol* 1991; 11:4591–4598.

129. Silva AJ, Smith AM, Giese KP. Gene targeting and the biology of learning and memory. *Annu Rev Genet* 1997; 31:527–546.

130. Dubnau J, Tully T. Gene discovery in Drosophila: new insights for learning and memory. *Annu Rev Neurosci* 1998; 21:407–444.

131. Weinstein LS, Shenker A, Gejman PV, Merino MJ, Friedman E, Spiegel AM. Activating mutations of the stimulatory G protein in the McCune-Albright syndrome. *N Engl J Med* 1991; 325:1688–1695.

132. Schwindinger WF, Francomano CA, Levine MA. Identification of a mutation in the gene encoding the alpha subunit of the stimulatory G protein of adenylyl cyclase in McCune-Albright syndrome. *Proc Natl Acad Sci USA* 1992;89:5152–5156.

133. Gibbs JB, Oliff A, Kohl NE. Farnesyltransferase inhibitors: Ras research yields a potential cancer therapeutic. *Cell* 1994; 77:175–178.

134. Kim HA, Ling B, Ratner N. Nf1-deficient mouse Schwann cells are angiogenic and invasive and can be induced to hyperproliferate: reversion of some phenotypes by an inhibitor of farnesyl protein transferase. *Mol Cell Biol* 1997; 17:862–872.

135. Yan N, Ricca C, Fletcher J, Glover T, Seizinger BR, Manne V. Farnesyltransferase inhibitors block the neurofibromatosis type I (NF1) malignant phenotype. *Cancer Res* 1995; 55:3569–3575.

136. Klose A, Ahmadian MR, Schuelke M, Scheffzek K, Hoffmeyer S, Gewies A, et al. Selective disactivation of neurofibromin GAP activity in neurofibromatosis type 1. *Hum Mol Genet* 1998; 7:1261–1268.

137. Scheffzek K, Ahmadian MR, Wiesmuller L, Kabsch W, Stege P, Schmitz F, Wittinghofer A. Structural analysis of the GAP-related domain from neurofibromin and its implications. *EMBO J* 1998; 17:4313–4327.

138. Xu H, Gutmann DH. Mutations in the GAP-related domain impair the ability of neurofibromin to associate with microtubules. *Brain Res* 1997; 759:149–152.

139. Huson SM, Neurofibromatosis 2: clinical features, genetic counseling and management issues. In Huson SM, Hughes RAC, eds. *The Neurofibromatoses: A Practical and Clinical Overview.* London: Chapman and Hall. 1994:211–233.

140. Woodruff JM. Pathology of the major peripheral nerve sheath neoplasms. *Monogr Pathol* 1996; 38:129–161.

141. Seizinger BR, Martuza RL, Gusella JF. Loss of genes on chromosome 22 in tumorigenesis of human acoustic neuroma. *Nature* 1986; 322:644–647.

142. Seizinger BR, Rouleau GA, Lane AH, Farmer G, Ozelius LJ, Haines JL, et al. Linkage analysis in von Recklinghausen neurofibromatosis (NF1) with DNA markers for chromosome 17. *Genomics* 1987; 1:346–348.

143. Rouleau GA, Merel P, Lutchman M, Sanson M, Zucman J, Marineau C, et al. Alteration in a new gene encoding a putative membrane-organizing protein causes neuro-fibromatosis type 2. *Nature* 1993; 363:515–521.

144. Trofatter JA, MacCollin MM, Rutter JL, Murrell JR, Duyao MP, Parry DM, et al. A novel moesin-, ezrin-, radixin-like gene is a candidate for the neurofibromatosis 2 tumor suppressor [published erratum appears in *Cell* 1993; 75:826]. *Cell* 1993; 72:791–800.

145. Stemmer-Rachamimov AO, Xu L, Gonzalez-Agosti C, Burwick JA, Pinney D, Beauchamp R, et al. Universal absence of merlin, but not other ERM family members, in schwannomas. *Am J Pathol* 1997; 151:1649–1654.

146. Jacoby LB, MacCollin M, Barone R, Ramesh V, Gusella JF. Frequency and distribution of NF2 mutations in schwannomas. *Genes Chromosomes Cancer* 1996; 17:45–55.

147. Bianchi AB, Mitsunaga SI, Cheng JQ, Klein WM, Jhanwar SC, Seizinger B, et al. High frequency of inactivating mutations in the neurofibromatosis type 2 gene (NF2) in primary malignant mesotheliomas. *Proc Natl Acad Sci USA* 1995; 92:10,854–10,858.

148. Sekido Y, Pass HI, Bader S, Mew DJ, Christman MF, Gazdar AF, Minna JD. Neurofibromatosis type 2 (NF2) gene is somatically mutated in mesothelioma but not in lung cancer. *Cancer Res* 1995; 55:1227–1231.

149. Jacoby LB, Jones D, Davis K, Kronn D, Short MP, Gusella J, MacCollin M. Molecular analysis of the NF2 tumor-suppressor gene in schwannomatosis. *Am J Hum Genet* 1997; 61:1293–1302.

150. Pulst SM, Rouleau GA, Marineau C, Fain P, Sieb JP. Familial meningioma is not allelic to neurofibromatosis 2. *Neurology* 1993; 43:2096–2098.

151. Wellenreuther R, Kraus JA, Lenartz D, Menon AG, Schramm J, Louis DN, et al. Analysis of the neurofibromatosis 2 gene reveals molecular variants of meningioma. *Am J Pathol* 1995; 146:827–832.

152. Gutmann DH, Wright DE, Geist RT, Snider WD. Expression of the neurofibromatosis 2 (NF2) gene isoforms during rat embryonic development. *Hum Mol Genet* 1995; 4:471–478.

153. Huynh DP, Tran TM, Nechiporuk T, Pulst SM. Expression of neurofibromatosis 2 transcript and gene product during mouse fetal development. *Cell Growth Differ* 1996; 7:1551–1561.

154. Sherman L, Xu HM, Geist RT, Saporito-Irwin S, Howells N, Ponta H, Herrlich P, Gutmann DH. Interdomain binding mediates tumor growth suppression by the NF2 gene product. *Oncogene* 1997; 15:2505–2509.

155. Takeuchi K, Kawashima A, Nagafuchi A, Tsukita S. Structural diversity of band 4.1 superfamily members. *J Cell Sci* 1994; 107:1921–1928.

156. Tsukita S, Yonemura S. ERM (ezrin/radixin/moesin) family: from cytoskeleton to signal transduction. *Curr Opin Cell Biol* 1997; 9:70–75.

157. den Bakker MA, Riegman PH, Hekman RA, Boersma W, Janssen PJ, van der Kwast TH, Zwarthoff EC. The product of the NF2 tumour suppressor gene localizes near the plasma membrane and is highly expressed in muscle cells. *Oncogene* 1995; 10:757–763.

158. Gonzalez-Agosti C, Xu L, Pinney D, Beauchamp R, Hobbs W, Gusella J, Ramesh V. The merlin tumor suppressor localizes preferentially in membrane ruffles. *Oncogene* 1996; 13:1239–1247.

159. Sainio M, Zhao F, Heiska L, Turunen O, Bakker M, Zwarthoff E, et al. Neurofibromatosis 2 tumor suppressor protein colocalizes with ezrin and CD44 and associates with actin-containing cytoskeleton. *J Cell Sci* 1997; 110:2249–2260.

160. Shaw RJ, McClatchey AI, Jacks T. Localization and functional domains of the neurofibromatosis type II tumor suppressor, merlin. *Cell Growth Differ* 1998; 9:287–296.

161. Tsukita S, Oishi K, Sato N, Sagara J, Kawai A, Tsukita S. ERM family members as molecular linkers between the cell surface glycoprotein CD44 and actin-based cytoskeletons. *J Cell Biol* 1994; 126:391–401.

162. Yonemura S, Hirao M, Doi Y, Takahashi N, Kondo T, Tsukita S. Ezrin/radixin/moesin (ERM) proteins bind to a positively charged amino acid cluster in the juxta-membrane cytoplasmic domain of CD44, CD43, and ICAM-2. *J Cell Biol* 1998; 140:885–895.

163. Serrador JM, Alonso-Lebrero JL, del Pozo MA, Furthmayr H, Schwartz-Albiez R, Calvo J, Lozano F, Sanchez-Madrid F. Moesin interacts with the cytoplasmic region of intercellular adhesion molecule-3 and is redistributed to the uropod of T lymphocytes during cell polarization. *J Cell Biol* 1997; 138:1409–1423.

164. Fukata Y, Kimura K, Oshiro N, Saya H, Matsuura Y, Kaibuchi K. Association of the myosin-binding subunit of myosin phosphatase and moesin: dual regulation of moesin phosphorylation by Rho-associated kinase and myosin phosphatase. *J Cell Biol* 1998; 141:409–418.

165. Dransfield DT, Bradford AJ, Smith J, Martin M, Roy C, Mangeat PH, Goldenring JR. Ezrin is a cyclic AMP-dependent protein kinase anchoring protein. *EMBO J* 1997; 16:35–43.

166. Reczek D, Berryman M, Bretscher A. Identification of EBP50: a PDZ-containing phosphoprotein that associates with members of the ezrin-radixin-moesin family. *J Cell Biol* 1997; 139:169–179.
167. Sainio M, Zhao F, Heiska L, Turunen O, den Bakker M, Zwarthoff E, et al. Neurofibromatosis 2 tumor suppressor protein colocalizes with ezrin and CD44 and associates with actin-containing cytoskeleton. *J Cell Sci* 1997; 110:2249–2260.
168. Murthy A, Gonzalez-Agosti C, Cordero E, Pinney D, Candia C, Solomon F, Gusella J, Ramesh V. NHE-RF, a regulatory cofactor for Na(+)-H+ exchange, is a common interactor for merlin and ERM (MERM) proteins. *J Biol Chem* 1998; 273:1273–1276.
169. Scoles DR, Huynh DP, Morcos PA, Coulsell ER, Robinson NG, Tamanoi F, Pulst SM. Neurofibromatosis 2 tumour suppressor schwannomin interacts with betaII- spectrin. *Nat Genet* 1998; 18:354–359.
170. Helander TS, Carpen O, Turunen O, Kovanen PE, Vaheri A, Timonen T. ICAM-2 redistributed by ezrin as a target for killer cells. *Nature* 1996; 382:265–268.
171. Serrador JM, Nieto M, Alonso-Lebrero JL, del Pozo MA, Calvo J, Furthmayr H, et al. CD43 interacts with moesin and ezrin and regulates its redistribution to the uropods of T lymphocytes at the cell-cell contacts. *Blood* 1998; 91:4632–4644.
172. Kondo T, Takeuchi K, Doi Y, Yonemura S, Nagata S, Tsukita S. ERM (ezrin/radixin/moesin)-based based molecular mechanism of microvillar breakdown at an early stage of apoptosis. *J Cell Biol* 1997; 139:749–758.
173. Berryman M, Gary R, Bretscher A. Ezrin oligomers are major cytoskeletal components of placental microvilli: a proposal for their involvement in cortical morphogenesis. *J Cell Biol* 1995; 131:1231–1242.
174. Gary R, Bretscher A. Heterotypic and homotypic associations between ezrin and moesin, two putative membrane-cytoskeletal linking proteins. *Proc Natl Acad Sci USA* 1993; 90:10,846–10,850.
175. Gary R, Bretscher A. Ezrin self-association involves binding of an N-terminal domain to a normally masked C-terminal domain that includes the F-actin binding site. *Mol Biol Cell* 1995; 6:1061–1075.
176. Matsui T, Maeda M, Doi Y, Yonemura S, Amano M, Kaibuchi K, Tsukita S, Tsukita S. Rho-kinase phosphorylates COOH-terminal threonines of ezrin/radixin/moesin (ERM) proteins and regulates their head-to-tail association. *J Cell Biol* 1998; 140:647–657.
177. Krieg J, Hunter T. Identification of the two major epidermal growth factor-induced tyrosine phosphorylation sites in the microvillar core protein ezrin. *J Biol Chem* 1992; 267:19,258–19,265.
178. Fazioli F, Wong WT, Ullrich SJ, Sakaguchi K, Appella E, Di Fiore PP. The ezrin-like family of tyrosine kinase substrates: receptor-specific pattern of tyrosine phosphorylation and relationship to malignant transformation. *Oncogene* 1993; 8:1335–1345.
179. Crepaldi T, Gautreau A, Comoglio PM, Louvard D, Arpin M. Ezrin is an effector of hepatocyte growth factor-mediated migration and morphogenesis in epithelial cells. *J Cell Biol* 1997; 138:423–434.
180. Shaw RJ, McClatchey AI, Jacks T. Regulation of the neurofibromatosis type 2 tumor suppressor protein, Merlin, by adhesion and growth arrest stimuli. *J Biol Chem* 1998; 273:7757–7764.
181. Van Aelst L, D'Souza-Schorey C. Rho GTPases and signaling networks. *Genes Dev* 1997; 11:2295–2322.
182. Mackay DJ, Esch F, Furthmayr H, Hall A. Rho- and rac-dependent assembly of focal adhesion complexes and actin filaments in permeabilized fibroblasts: an essential role for ezrin/radixin/moesin proteins. *J Cell Biol* 1997; 138:927–938.
183. Shaw RJ, Henry M, Solomon F, Jacks T. RhoA-dependent phosphorylation and relocalization of ERM proteins into apical membrane/actin protrusions in fibroblasts. *Mol Biol Cell* 1998; 9:403–419.
184. Takahashi K, Sasaki T, Mammoto A, Takaishi K, Kameyama T, Tsukita S, Takai Y. Direct interaction of the Rho GDP dissociation inhibitor with ezrin/radixin/moesin initiates the activation of the Rho small G protein. *J Biol Chem* 1997; 272:23,371–23,375.
185. Takahashi K, Sasaki T, Mammoto A, Hotta I, Takaishi K, Imamura H, et al. Interaction of radixin with Rho small G protein GDP/GTP exchange protein Dbl. *Oncogene* 1998; 16:3279–3284.
186. Takeuchi K, Sato N, Kasahara H, Funayama N, Nagafuchi A, Yonemura S, Tsukita S, Tsukita S. Perturbation of cell adhesion and microvilli formation by antisense oligonucleotides to ERM family members. *J Cell Biol* 1994; 125:1371–1384.
187. Huynh DP, Pulst SM. Neurofibromatosis 2 antisense oligodeoxynucleotides induce reversible inhibition of schwannomin synthesis and cell adhesion in STS26T and T98G cells. *Oncogene* 1996; 13:73–84.

188. McClatchey AI, Saotome I, Ramesh V, Gusella JF, Jacks T. The Nf2 tumor suppressor gene product is essential for extraembryonic development immediately prior to gastrulation. *Genes Dev* 1997; 11:1253–1265.

189. McClatchey AI, Saotome I, Mercer K, Crowley D, Gusella JF, Bronson RT, Jacks T. Mice heterozygous for a mutation at the Nf2 tumor suppressor locus develop a range of highly metastatic tumors. *Genes Dev* 1998; in press.

190. Haase VH, Trofatter JA, MacCollin M, Tarttelin E, Gusella JF, Ramesh V. The murine NF2 homologue encodes a highly conserved merlin protein with alternative forms. *Hum Mol Genet* 1994; 3:407–411.

191. Kincade PW, Zheng Z, Katoh S, Hanson L. The importance of cellular environment to function of the CD44 martix receptor. *Curr Opin Cell Biol* 1997; 9:635–642.

192. McClatchey AI, Jacks T. Tumor suppressor mutations in mice: the next generation. *Curr Opin Genet Dev* 1998; 8:304–310.

193. Tikoo A, Varga M, Ramesh V, Gusella J, Maruta H. An anti-Ras function of neurofibromatosis type 2 gene product (NF2/Merlin). *J Biol Chem* 1994; 269:23,387–23,390.

194. McCartney BM, Fehon RG. Distinct cellular and subcellular patterns of expression imply distinct functions for the Drosophila homologues of moesin and the neurofibromatosis 2 tumor suppressor, merlin. *J Cell Biol* 1996; 133:843–852.

195. LaJeunesse DR, McCartney BM, Fehon RG. Structural analysis of Drosophila merlin reveals functional domains important for growth control and subcellular localization. *J Cell Biol* 1998; 141:1589–1599.

196. Tanaka K, Nakafuku M, Satoh T, Marshall MS, Gibbs JB, Matsumoto K, Kaziro Y, Toh-e. A. S. cerevisiae genes *IRA1* and *IRA2* encode proteins that may be functionally equivalent to mammalian *ras* GTPase activating protein. *Cell* 1990; 60:803–807.

197. Trahey M, McCormick F. A cytoplasmic protein stimulates normal N-*ras* p21 GTPase, but does not affect oncogenic mutants. *Science* 1987; 238:542–545.

198. Cullen JP, Hsuan JJ, Truong O, Letcher AJ, Jackson TR, Dawson AP, Irvine RF. Identification of a specific Ins(1,3,4,5)P_4-binding protein as a member of the GAP1 family. *Nature* 1995; 376:527–530.

199. Chen HJ, Rojas-Soto M, Oguni A, Kennedy MB. A synaptic Ras-GTPase activating protein (p135 SynGAP) inhibited by CaM kinase II. *Neuron* 1998; 20:895–904.

200. Kim JH, Liao D, Lau LF, Huganir RL. SynGAP: a synaptic RasGAP that associates with the PSD-95/SAP90 protein family. *Neuron* 1998; 20:683–691.

12 Von Hippel-Lindau Disease
Clinical and Molecular Aspects

Othon Iliopoulos, MD and
William G. Kaelin, Jr., MD

1. INTRODUCTION

von Hippel-Lindau (VHL) disease (OMIM 193300) is a hereditary cancer syndrome characterized by the development of central nervous system (CNS) and retinal hemangioblastomas, renal cell carcinomas (RCCs), pheochromocytomas, and other lesions *(1)*. The first description of retinal angiomata affecting two siblings can be traced to Treacher Collins (1894) *(2)*. The German opthalmologist, Eugene von Hippel, described two unrelated kindreds with retinal angiomata in 1904 *(3)*. The familial nature of this disease was fully appreciated by neurophysiologist John Fulton in 1920 *(4)*. The Swedish opthalmologist, Arvin Lindau, observed the connection between retinal lesions and larger angiomatous lesions of the CNS in 1927 *(5)*. The term "Lindau disease" was coined by Melmon and Rosen in 1964, in a comprehensive review of the literature *(6)*.

Significant advancements in understanding VHL disease have occurred over the past decade. Cloning of the tumor suppressor gene (TSG), which, when altered, gives rise to VHL disease, was reported in 1993. The availability of the VHL gene has allowed for genetic, rather than purely clinical, diagnosis of VHL disease. As a result, the spectrum of clinical abnormalities associated with VHL disease is now better appreciated, and certain genotype–phenotype correlations are emerging. In addition, the VHL gene

From: *Tumor Suppressor Genes in Human Cancer*
Edited by: D. E. Fisher © Humana Press Inc., Totowa, NJ

product (pVHL) has been identified in mammalian cells, and studies of its biochemical and biological functions have already yielded clues as to why VHL patients are cancer-prone. This chapter highlights classic and recently appreciated clinical features of the disease, describes new molecular insights into the function of pVHL, and suggests potential strategies for therapeutic intervention based on this information.

2. CLINICAL FEATURES OF VHL DISEASE

2.1. Hemangioblastoma and Retinal Angiomatosis

CNS hemangioblastomas and retinal angiomas are histologically identical lesions. They are the commonest, earliest, and most characteristic lesions of VHL disease, occurring in 60–80% of VHL patients *(9–10)*.

VHL-associated hemangioblastomas are typically multifocal, and may occur synchronously or metachronously. They involve the cerebellar hemispheres (75%), spinal cord (20%), and brain stem (5%). Spinal hemangioblastomas occur most often in the cervical and lumbar area, are intradural, and often produce syringomyelia *(11,12)*. Rare cases of supratentorial, nerve root, peripheral nerve, and hepatic hemangioblastomas, as well as spinal leptomeningeal hemangioblastosis, have been reported *(13)*. Despite the fact that hemangioblastomas are nonmetastatic lesions, they cause considerable morbidity and mortality because of their space-occupying nature. Symptoms depend on the affected area, and develop once the tumor reaches a size critical for its location: headache, vomiting and cerebellar ataxia (infratentorial lesions, with increased intracranial pressure), orthostatic hypotension (brain stem lesions), spinal motor and sensory disturbances (spinal cord lesions), and pain (peripheral nerve lesions). Hemangioblastomas may be associated with polycythemia as a result of tumor-derived erythropoietin *(14)*. Sporadic, non-VHL-associated hemangioblastomas, in contrast to VHL-associated lesions, are typically unifocal, and rarely involve the spinal cord *(15,16)*.

Macroscopically, a typical cerebellar hemangioblastoma, presents as a cystic lesion harboring a solid component attached to its wall. The solid: cystic ratio varies. Purely cystic lesions do exist, but they are rare. Spinal hemangioblastomas are more often solid tumors or mixed solid tumors bearing small cysts *(11)*.

Histologically, hemangioblastomas consist of a mixture of stromal cells embedded within a rich and well-defined capillary network. Electron microscopy and immunohistochemical studies have addressed the origin of the stromal cell *(17)*. It appears that these cells are probably of mesenchymal origin. They may share a common ancestor with endothelial cells, but it is still unclear at which stage stromal and epithelial cells become committed to distinct differentiation programs. Stromal cells stain positively for neuron-specific enolase and vimentin, but negatively for epithelial membrane antigen and Ki-67 *(17)*. This immunocytochemical profile supports the mesenchymal origin of the stromal cell, and may help differentiate hemangioblastomas from RCCs metastatic to the CNS.

Retinal hemangioblastomas in VHL disease are also typically multifocal and bilateral. They occur, by decreasing order of frequency, in the temporal periphery, the nasal periphery of the retina, and in the posterior pole next to the optic nerve and the optic disk *(18)*. Left untreated, they cause retinal exudates, hemorrhage, and, ultimately, glaucoma and retinal detachment, leading to blindness. Decreased visual acuity in a VHL patient may also result from optic nerve atrophy caused by chronic obstructive hydrocephalus related to a cerebellar hemangioblastoma.

2.2. Renal Cysts and Clear Cell Renal Carcinomas

Between 60 and 85% of VHL patients will develop hundreds of bilateral renal cysts during their lifetime *(9,19,20)*. The majority of these cysts will grow slowly over a period of years. Nevertheless, they usually remain clinically silent, rarely causing impairment of renal function *(21)*. Macroscopically, they appear purely cystic, mixed cystic with a solid component, or multiloculated *(19)*. It is currently believed that mixed and multiloculated cysts arise from the proximal renal tubule, and that purely cystic ones arise from the distal renal tubule *(22)*.

RCC and CNS hemangioblastomas are the major causes of mortality in VHL patients *(8,23)*. VHL-associated RCCs are clear cell or predominantly clear cell carcinomas, and arise from the epithelial lining of the proximal renal tubule *(24)*. They are often multicentric and bilateral, and may form within a pre-existing mixed or multiloculated cyst, or *de novo* in the renal parenchyma *(25)*.

2.3. Pheochromocytoma

VHL is phenotypically a heterogeneous disease. The development of pheochromocytomas tends to cluster within certain families (*see* Subheading 3.2.). VHL disease appears to be the most common cause of hereditary pheochromocytoma. Neumann et al. *(26)* examined the frequency and clinical characteristics of familial-type pheochromocytomas (MEN-2 and VHL) in an unselected group of 82 consecutive patients with documented symptomatic pheochromocytoma: 19% of patients had VHL, and 4% had MEN-2 disease. Compared to those with sporadic pheochromocytomas, VHL patients often have multifocal and bilateral lesions, were diagnosed at an earlier age, and were less likely to have histologically malignant disease. Prospective screening of a population at risk (VHL patients and their first-degree relatives), presented in the same study, revealed the presence of pheochromocytoma in 46% of the cases.

Many VHL-associated pheochromocytomas are asymptomatic *(26)*. When symptoms occur, they include sweating, palpitations, anxiety, and headaches caused by catecholamine hypersecretion, and, if left untreated, may result in sequelae of uncontrolled hypertension, such as stroke or myocardial infarction.

2.4. Pancreatic Cysts and Neoplasms

The commonest pancreatic lesions associated with VHL disease include pancreatic cysts, microcystic serous adenomas (cystadenomas), and islet cell tumors. These lesions tend to cause less morbidity and mortality, compared to hemangioblastomas and RCCs. Many of these lesions may be clinically silent. Their estimated frequency varies from 10 to 60% among various series, depending primarily on the mode of detection (clinical, radiological, pathological) *(8,9,27–29)*.

In VHL disease, simple cysts occur throughout the pancreas, without predilection for specific site. Their size ranges from a few millimeters to more than 10 cm. The serous adenomas are benign clusters of grape-like cystic lesions *(13,28,30)*. Cysts and cystadenomas are asymptomatic in most cases. Excessive local growth may cause biliary duct obstruction, pancreatitis, and, in rare cases, exocrine and endocrine pancreatic insufficiency caused by cystic replacement of the parenchyma *(28–30)*. VHL patients do not appear to be at risk for classic adenocarcinoma of the pancreas.

Islet cell tumors are unrelated to pancreatic cystic disease. These tumors are of neural-crest origin *(31)* and occur more frequently in patients with pheochromocytoma

(27,33). In most cases, these tumors grow slowly and remain asymptomatic *(27,29,31)*. More rarely, they may exhibit more aggressive biology, manifested by rapid growth and metastasis, coupled with secretion of biologically active peptides, such as vasoactive intestinal peptide, calcitonin, insulin, glucagon, gastrin, or somatostatin *(13)*. In the latter case, the type of symptoms depends on the endocrine function of the hypersecreted peptide.

2.5. Endolymphatic Sac Tumors

Endolymphatic sac tumors (ELSTs) are locally invasive, but nonmetastatic, papillary adenocarcinomas arising from the endolymphatic sac, an ectodermal extension of the membranous labyrinth of the internal ear to the posterior surface of the petrous bone. Early symptoms consist of gradual or acute onset of decreased auditory acuity, tinnitus, and vertigo *(33)*. Progressive tumor growth and invasion of local structures results in facial paresis and anesthesia, vocal cord paralysis manifested as hoarseness, and sternocleidomastoid muscle atrophy *(33,34)*. Prospective audiologic evaluation at the National Cancer Institute revealed that 65% of VHL patients had pure tone threshold abnormalities, which, in one-half of cases, was bilateral *(33)*. Many VHL patients with hearing loss do not have a radiographically detectable ELST at the time of examination. Nonetheless, it is possible that early auditory abnormalities herald the development of a microscopic ELST.

2.6. Papillary Cystadenomas of Mesonephric Origin

The mesonephric duct forms during embryologic development. In males, it gives rise to ductus deferens and epididymis. In females, it involutes to the nonfunctional structures of the duct of epoöphoron and the duct of Gartner, extending along the adnexal area from the ovaries to the lateral wall of the vagina. These structures are the targets of tumor development in VHL patients.

Up to 60% of male VHL patients develop papillary cystadenomas of the epididymis *(8,35)*. These are nonmetastatic tumors, ranging in size from 1 to 5 cm. They are mostly asymptomatic, unless they become inflamed or rupture. Infertility, although a rare complication, can arise if sperm delivery is impaired by bilateral tumors. Simple epididymal cysts are common in the general population, and do not raise suspicion for VHL disease. Papillary cystadenoma of the epididymis, however, is rare in the general population, and, if bilateral, is almost diagnostic for VHL disease.

Development of adnexal papillary cystadenomas of mesonephric origin has only recently been appreciated in female VHL patients *(36,37)*. Consequently, it is still too early to estimate its real incidence in this patient population. Tumors may develop at any site along the mesonephric duct.

3. *VHL* GENE

3.1. Cloning, Structure, and Expression of VHL Gene

The *VHL* gene was mapped to human chromosome (chr) 3p25 region, using linkage analysis *(38,39)*. The identification of three unrelated patients, with small constitutional deletions in this area, was helpful in this regard *(38)*. The gene consists of three exons (Fig. 1; *40*). Northern blot and reverse transcriptase-polymerase chain reaction analysis of mRNA, obtained from various cell lines with wild-type (wt) VHL, reveals

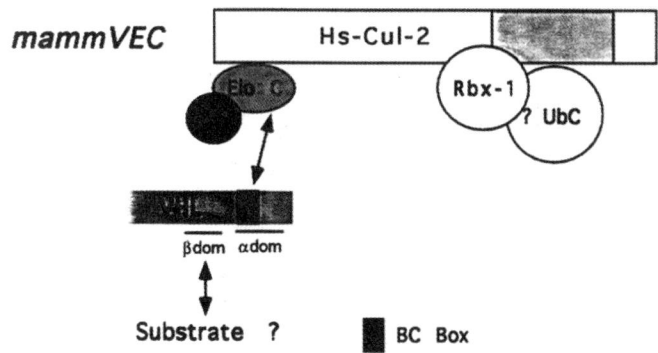

Fig. 1. **(A)** Structure of human *VHL* gene open reading frame (ORF), with representative mutations. Exon 1 encodes amino acids 1–113, exon 2 (shaded area) amino acids 114–153, and exon 3 aa 154–213. Arrowheads indicate missense mutations; vertical lines, nucleotide insertions; filled circles, stop codons; and letter D, deletions. Arrows point to the codon corresponding to the indicated aa. **(B)** Human pVHL30 and pVHL19. The crystal structure of pVHL 54–213 (third row) reveals two domains, α and β. BC box extends between aa 157 and 172, and corresponds to the H1 helix of the α-domain. Transcription-dependent nuclear localization domain (TDNL) encompasses exon 2. **(C)** Mouse and rat proteins drawn to scale with the human pVHL. The degree of homology between the two rodent species and the corresponding segments of the human protein are shown.

the presence of two mRNA species. The larger of these, which is approx 4.6 kb in size, corresponds to exons 1–3, and the smaller to an alternative spliced form consisting of exons 1 and 3 only *(41)*.

The VHL promoter contains putative binding sites for several transcription factors, including Sp1, AP-2, PAX, nuclear respiratory factor-1 and retinoic acid receptor, but neither a TATA nor a CCAAT box. Transcription is initiated around a putative Sp1 site located 60 bp upstream of the first AUG codon in the VHL mRNA *(42)*. The 3'UTR of *VHL* gene is ~3.6 kB in length, and contains multiple *Alu* repeat elements *(43)*.

VHL gene is expressed in every adult human tissue examined, and in all three germ layers during human and mouse embryonic development *(44,45)*. In adults, the *VHL* gene is strongly expressed in tumor-developing tissues (such as the renal proximal tubular epithelium and the CNS), as well as in tissues not known to be targets of tumor development, such as testis, lung, and liver *(46)*. Mouse and rat VHL have high homology to the human VHL gene *(47,48;* Fig. 1). No clear VHL homologs have been reported so far in yeast.

3.2. Germ-line Mutations and Genotype–Phenotype Correlations

Germ-line mutations of the VHL gene can be detected in almost 100% of patients who carry a diagnosis of VHL disease, based on clinical criteria *(49)* (Table 1). Examples of missense or nonsense mutations, microdeletions, or microinsertions resulting in frameshifts, as well as large deletions encompassing the entire VHL gene, have been described *(50–58)*. A continuously updated list of VHL mutations is accessible through the internet at: http://www.umd.necker.fr *(59)*. All intragenic mutations reported thus far map downstream of the codon corresponding to amino acid (aa) Met 54. These mutations occur at several sites across the three exons. A cluster of nonsense and mis-

Table 1
Major Clinical Manifestations of VHL Disease

Horton (170)	Lamiell (8) (n = 50)	Maher (9) n = 554	Maddock (171) n = 152	n = 83
	%Frequency (mean age of diagnosis)			
Hemangioblastomas				
CNS	44 (31)	61 (28)	72 (29)	72 (30)
Retinal	58 (28)	57 (25)	59 (25)	41 (21)
Renal cancer	28 (41)	24 (39)	28 (44)	25 (38)
Pheochromocytomas	10 (34)	19 (27)	7 (nr)	12 (nr)

Mean age at the time of diagnosis as reported in several series is shown. Diagnosis was made by clinical examination and/or imaging studies in both symptomatic patients and asymptomatic relatives at risk (8,9,170,171). nr, not reported.

sense mutations is noticeable at the 3′ end of the exon 1 and the 5′ end of the exon 3 (Fig. 1).

Hemangioblastomas (56), RCCs (60), and islet cell tumors of the pancreas (31), which develop in VHL patients, exhibit loss of heterozygosity (LOH) at the VHL locus, because of loss of the remaining wt allele. In the case of hemangioblastomas, it is the stromal cell component of the tumor that exhibits biallelic inactivation of the VHL gene (61). Lubensky et al. (60,62), using laser-capture microdissection, were able to detect LOH at the VHL locus in cells lining atypical renal cysts of VHL patients. This observation places loss of VHL function at the very initial steps of renal tumor development.

Clinical heterogeneity is an established feature of VHL disease (52,55,63,64). Patients with type I disease are predisposed to develop clear cell carcinoma, or CNS and retinal hemangioblastoma, but not pheochromocytoma. Families with type II disease are also predisposed to develop pheochromocytoma. Type II families can be subdivided into those with low (type IIA) or high (type IIB) risk for development of renal carcinoma. Several reports have established a correlation between specific VHL germline mutations and the clinical phenotype of the disease (Table 2). More than 90% of type II patients have missense mutations; large deletions, microdeletions/microinsertions, and nonsense mutations are typical in type I disease (52,58). Recently, missense germ-line VHL mutations that result in familial predisposition to pheochromocytoma, without the other stigmata of VHL disease, were described (54,65–69). These patients may constitute an additional VHL disease subtype (type IIC).

3.3. VHL Gene in Sporadic Neoplasms

The Knudson two-hit genetic model for TSG predicts that familial cancers result from the inheritance of a germ-line mutation in a TSG, followed by somatic inactivation of the remaining wt allele (70). This is the case for several familial cancer syndromes, such as those arising from mutations in Rb1, p53, WT1, and, as described above, for VHL. A corollary to this model predicts that these genes would also be altered in the sporadic counterparts of the tumors encountered in their respective hereditary cancer syndromes. The VHL gene obeys this rule.

LOH at the VHL locus, with mutational inactivation of the remaining allele, was detected in 50–60% of sporadic clear-cell-type renal carcinomas (41,71–73). Inactiva-

Table 2
Genotype–Phenotype Correlations

	Classification	Phenotype	Germ-line mutations
Type I	H + RCC		Deletions 46%
			Nonsense 10%
			Missense 44%
			N78H/S/T S80R/I S111R/N
			R161P C162P/Y/T L184P/R
Type II			Missense 96%
IIA	H + Pheo + RCC (low)		Y98H, Y112H
IIB	H + Pheo + RCC (high)		V74G, R161G, R167W/Q/G
IIC	Pheo only		L188V, G114S, F119S,
			V84L, V166F

Left column: VHL subtype. Middle column: Clinical phenotype. H, CNS and retinal hemangioblastoma; Pheo, pheochromocytoma; RCC, renal cell carcinoma; (low), low risk of RCC; (high), high risk of RCC. Right column: most frequent mutations encountered for each phenotype.

tion of *VHL* gene bears histologic specificity: VHL mutations do not occur in papillary RCCs, a histologic subtype not encountered in VHL families *(41)*. In an additional 25% of sporadic clear cell renal cancers, the *VHL* gene is silenced by hypermethylation in CpG islands spanning the area around the promoter and the 5′ end of the first exon. In total, both copies of *VHL* gene are inactivated in approx 75% of sporadic clear cell renal carcinomas.

The *VHL* gene was also found to be mutationally inactivated in 50–60% of sporadic hemangioblastomas *(74–77)*. No evidence of hypermethylation has been reported so far in sporadic hemangioblastomas. The *VHL* gene in sporadic pheochromocytomas, unlike the situation for RCCs and hemangioblastomas, does not appear to conform to Knudson two-hit model: so far, only two cases of VHL inactivation in sporadic pheochromocytomas have been reported *(66,78)*. Whether the *VHL* gene is hypermethylated in sporadic pheochromocytoma is not known. Lastly, LOH at the VHL locus was reported in 7/10 studied sporadic pancreatic microcystic adenomas *(79)*.

The status of *VHL* gene was also examined in several cell lines and tissue obtained from non-VHL-associated tumors (including breast, colon, prostate, ovarian, nonsmall cell lung cancer) *(41,80)*, and in tumors suspected of harboring inactivation of a putative TSG on chr 3p (small cell, squamous cancer of the head and neck, esophageal) *(80–82)*. With rare exceptions, corresponding usually to advance tumor stage and consequent loss of genetic material because of genetic instability *(83)*, neither VHL mutation nor LOH at the VHL locus have been reported in these settings.

4. VHL TUMOR SUPPRESSOR PROTEIN

4.1. Identification of pVHL Isoforms

The human *VHL* gene encodes a 213-aa protein that migrates with an apparent mol wt of 28–30 kDa (pVHL30), following sodium dodecyl sulfate-polyacrylamide gel electrophoresis *(38,47,84)*. A second pVHL isoform, migrating with an apparent mol wt of 19 kDa (pVHL19), has also been identified *(25,85,86)*. Mutational analysis and pulse-chase experiments indicate that pVHL19 is not a proteolytic product of pVHL30, but is probably generated by translational initiation at methionine 54. pVHL19 has been detected in every cell line producing pVHL30. pVHL19, as described below, is a biologically active molecule, and almost all tumor-derived VHL mutations map C-terminal to codon 54. This underscores the notion that both pVHL isoforms need to be inactivated in order for tumor formation to ensue.

It is possible that additional pVHL isoforms exist. Immunoprecipitation of cellular extracts, followed by Western blot analysis with monoclonal anti-VHL antibodies, reveals the presence of specific bands in addition to the ones corresponding to the isoforms mentioned above *(84,87)*. As described above, cells produce an alternatively spliced VHL mRNA consisting of exons 1 and 3 *(41)*. Whether this mRNA encodes a stable protein product is not known. Some intronic VHL mutations result in the exclusive production of this mRNA, suggesting that its protein product, if made, is defective as a tumor suppressor.

Several groups investigated the cellular localization of pVHL. Biochemical fractionation of expotentially growing cells suggested that endogenous pVHL30 resides primarily in the cytoplasm and, to a lesser extent, in the nucleus *(84,88)*. In the cytoplasm, pVHL30 was detected both in the cytosol and in fractions enriched for membranes, including the Golgi apparatus and the endoplasmic reticulum (ER). Immunocytochemical studies of paraffin-embedded human tissue and mammalian cell lines corroborated these observations *(89,90)*. pVHL19, on the other hand, appears equally distributed between the cytosol and the nucleus, and is not measurably present in the membranous fraction *(85)*. Recently, Lee et al. *(91)*. reported that pVHL30 shuttles between the cytoplasm and the nucleus. Exposure of cells to actinomycin D or 5,6-dichlorobenzimidazole, both general inhibitors of polymerase II-mediated transcription, inhibited pVHL30 nuclear export, and resulted in relocalization of pVHL30 to the nucleus. Exon 2-encoded aa are necessary for this function, and exon 2-deleted mutants appear biologically impaired, at least regarding their ability to regulate hypoxia-inducible proteins (*see* Subheading 6). The nature of physiologic stimuli that regulate this nucleocytoplasmic shuttling is currently under investigation.

In asynchronously growing cells, pVHL30, but not pVHL19, is phosphorylated on serine residues *(92)*. The sites of phosphorylation, the relevant kinase(s), and the functional consequences of pVHL phosphorylation are currently unknown.

4.2. pVHL Is a Gatekeeper Tumor Suppressor Protein

Molecular studies in colon cancer support the notion that the development of cancer in humans is a multistep process *(93)*. It appears that the development of a fully malignant clone is shaped by the accumulation of mutations that are characteristic for a particular tumor type. Thus, it is anticipated that there will be a molecular signature for different tumors derived from different tissues. For a given tissue, there may be a critical gene, or genes, which regulate cell growth, and which must be inactivated early during the course of carcinogenesis, if a fully malignant clone is to emerge. For example, it appears that inactivation of the *APC* gene plays such a role in colorectal cancer. Reconstitution of the function of these genes, referred to as "gatekeepers," suppresses the malignant phenotype *(94)*. In contrast, inactivation of another family of TSGs (such as *MSH1* and *MLH2*) leads to genomic instability and tumor development through secondary mutations affecting other genes (including gatekeepers). Reconstitution of the function of these caretaker genes cannot override these secondary effects, and fails to suppress the malignant phenotype. The distinction between these classes of TSGs has obvious implications with respect to therapeutic strategies based on reconstitution of tumor suppressor function.

Several groups studied the effects of reintroducing wt pVHL into renal carcinoma cell lines that lacked a wt VHL allele. With the exception of one report *(95)*, restoration of pVHL function to such cells did not alter their ability to form colonies in soft agar, or to grow in monolayer culture under serum-rich conditions *(84,96,97)*. In contrast to these cell culture experiments, restoration of pVHL function dramatically suppressed the ability of VHL-defective renal carcinoma cells to form tumor xenografts in nude mice *(84,96,97)*. This disparity between the effects of pVHL in vitro and in vivo led to the hypothesis that pVHL mediates tumor suppression, at least in part, by mechanisms involving interactions between tumor cells and their surrounding microenvironment. The subsequent demonstration that pVHL plays a role in angiogenesis and cell matrix formation (*see* Subheading 5.3 and 6.1.) is consistent with this view.

Some cell–microenvironment interactions can be, in part, recapitulated by growing cells as multicellular spheroids. Under such conditions, Lieubau-Teillet at al. *(97)* showed that VHL–/– RCC formed compact spheroids, and exhibited a higher proliferative index (by H3 thymidine and bromodeoxyuridine incorporation) than their pVHL-expressing counterparts. In addition, RCC clones producing pVHL formed loosely arranged spheroids that exhibited morphological characteristics of renal tubular differentiation. In addition, these spheroids were associated with increased deposition of a fibronectin matrix, compared with spheroids formed by pVHL-defective cells (*see* Subheading 5.3).

The in vitro effect of pVHL on cellular proliferation was revisited by Pause et al. *(98)*. RCC clones, expressing wt pVHL, exited the cell cycle upon serum withdrawal, when grown under certain cell-density conditions. In contrast, VHL–/– RCC clones continued to cycle under these conditions. The reason that cell density is critical for the ability of pVHL to mediate cell cycle arrest is unclear, but again underscores the potential importance of cell-microenvironment communication regarding tumor suppression by pVHL.

These series of experiments, along with the observation that pVHL inactivation is an early event in the generation of RCC, place pVHL in the gatekeeper group of tumor suppressor proteins. From the therapeutic standpoint, they also support the concept that pharmacological restoration of pVHL function(s) in an established clear cell renal carcinoma may lead to tumor regression.

4.3. Mice Models for VHL Disease

Inheritance of an inactivated copy of *VHL* gene predisposes patients to develop VHL disease. To examine the role of pVHL in development and tumor suppression, Gnarra et al. *(96)*, using gene knockout techniques, generated VHL +/– mice, thus simulating the germ-line genotype of the *VHL* gene in VHL patients. These heterozygous VHL+/– mice develop normally, and have been followed, up to the age of 2 yr, without any evidence of tumors or other stigmata of VHL disease. The response of heterozygous mice to carcinogenic stimuli is currently under investigation.

Mice with homozygous deletion of both VHL loci die *in utero* at 10.5–12.5 d of gestation. These VHL–/– embryos exhibit no developmental abnormalities at the time of death. Instead, the cause of death appears to be lack of endothelium and blood vessel formation in the embryonic component of the placenta, leading to whole embryo resorption. Failure of embryonic placenta vasculogenesis correlates with great reduction of vascular endothelial growth factor (VEGF) levels in the trophoblast. Because of the role of pVHL in negatively regulating VEGF expression in renal carcinoma cells (*see* Subheading 6.1.), these results may indicate that either VEGF regulation by pVHL differs in trophoblasts, compared to renal carcinoma cells, or that the temporal expression of a host of vasculogenic factors regulated by pVHL is critical for the mature formation of vessels.

The next generation of transgenic mice experiments is expected to focus on rescuing the VHL–/– embryos by applying embryo transfer techniques. This may allow the generation of viable VHL–/– offspring, and the study of disease development and progression in a mouse model. In parallel, efforts to develop mice, in which VHL can be conditionally inactivated in selected target tissues (e.g., kidney, cerebellum), are currently underway.

5. MECHANISMS OF TUMOR SUPPRESSION AND pVHL-ASSOCIATED PROTEINS

5.1. Elongins C and B

In order to understand the biochemical mechanism(s) leading to tumor suppression by pVHL, several groups looked for cellular proteins that stably bind to pVHL. Initially, it was found that pVHL binds to elongins C and B *(87,99,100),* which form a tripartite enzymatic complex when bound to elongin A, called elongin or SIII. These three proteins were initially purified from rat liver nuclei as components of a multiprotein complex that enhances the overall rate of RNA polymerase II transcriptional elongation in vitro *(101).* This complex is evolutionary conserved, and members of it have been identified in yeast, *C. elegans,* and *Drosophila.*

The biochemistry of transcriptional elongation has been extensively studied, using in vitro assays of transcript elongation rates from artificial templates. In such assays, elongin A constitutes the catalytic subunit of the complex *(102,103).* The minimal domain required for the catalytic activity of elongin A maps between aa 521 and 690, although sequences extending between aa 400 and 773 are required for maximal catalytic activity *(102).* Elongin C binds directly to elongin A, and enhances its catalytic activity *(104).* Elongin B binds, in turn, through its N-terminal ubiquitin-like domain, to elongin C, and appears to stabilize the complex *(105).* Elongin/SIII is thought to suppress transient pausing of RNA polymerase II as it encounters impediments along DNA templates. Whether elongin/SIII is a general factor, or regulates specific genes in vivo, is not known.

Elongin C binds directly to pVHL aa 157–172 *(87).* This region constitutes a hotshot of naturally occurring, tumor-associated, mutations, supporting the concept that elongin binding is connected to the ability of pVHL to act as a tumor suppressor. pVHL (157–172) contains an elongin C/B-binding motif (termed "BC box") shared between pVHL and elongin A: (T,S)Lxxx(C,S)xxV(L,I) *(106).* Alanine scanning showed that the conserved aa of this motif are critical for binding of pVHL to elongins C/B *(107)*. These findings led to early speculation that pVHL tumor suppressor function was related to its ability to inhibit elongin/SIII. This model was initially supported by the observations that a) binding of elongins C/B to elongin A or pVHL was mutually exclusive and b) pVHL inhibited elongin/SIII transcriptional elongation activity in vitro.

Although it is still possible that some pVHL function(s) may relate to transcriptional elongation inhibition, possibly through nuclear–cytoplasmic shuttling, several recent observations, outlined below, challenge this view. First, elongins C/B are in vast excess of pVHL and elongin A, rendering sequestration of elongin B/C by pVHL improbable *(107a)*. Second, elongins C/B reside primarily in the cytoplasm, as pVHL does, suggesting that they may perform additional function(s) unrelated to transcriptional elongation *(107b)*. This hypothesis is further supported by the recent observation that pVHL/elongin B/C, but not elongin/SIII, binds to human cullin-2, as described below *(107b)*.

5.2. Cullin-2

The first member of the cullin family of proteins, *Caenorhabditis Elegans* cullin-1, was cloned by Kipreos et al. *(108)* as a gene that, when inactivated, gives rise to small, but differentiated, larvae. Database searches revealed five *C. elegans* and six human genes that are highly homologous to *C. Elegans* cullin-1. *(108)*.

pVHL forms complexes with human cullin-2 (Cul-2) *(109,110)*. The N-terminus of Cul-2 binds to elongin C *(92)*, which, in turn, binds to pVHL and elongin B, as described in subheading 5.1. *(94)*. Elongin A, in contrast to pVHL, does not associate with Cul-2 *(107b)*.

Human Cul-1 and Cul-2 are highly homologous to yeast Cdc53, a protein that plays a role in ubiquitination of specific substrates (E3) *(108,109)*. Based on this homology, it is reasonable to hypothesize that Cul-2 also plays a role in ubiquitination. Such a function, shown for human Cul-1 *(111–113)*, remains to be proven regarding Cul-2. Protein degradation through ubiquitin-dependent proteolysis is a rapid and irreversible process regulating diverse cellular functions, such as progression through the cell cycle, kinetochore function, orderly sister chromatid separation, DNA repair, and response to nutritional starvation *(114,115)*. Protein ubiquitination may lead to outcomes other than degradation, including receptor internalization or alteration in subcellar compartmentalization. Ubiquitin is activated by the ubiquitin activating enzyme (E1), and subsequently transferred to one of many ubiquitin-conjugating enzymes (E2s). In a third step, ubiquitin is covalently linked to the target protein by a ubiquitin ligase (E3).

Multiprotein complexes mediating the ubiquitination of various substrates have been identified in eukaryotes. Substrate specificity depends on members of the complex acting as receptors. These substrate receptors associate with specific E2 and E3 partner proteins *(116)*.

In *Saccharomyces cerevisiae*, Cdc53 serves as a scaffold protein for the assembly of such complexes *(117)*. These multiprotein complexes, referred to as SCF complexes, contain Skp1, Cdc53, and a receptor protein containing a co-linear motif known as an F-box (so named because the prototypical sequence is present in cyclin F). The substrate specificity and nomenclature for a given SCF complex is determined by the F-box protein (e.g., SCF^{Cdc4}, SCF^{Grr1}, or SCF^{Met30}) *(118,119)*. SCF complexes containing Cdc4 bind and degrade yeast G1 cyclins (cln 1 and 2); complexes containing Grr1 bind and degrade the Clb5/Cdc28 inhibitor, p40/Sic1 *(119–121)*. Met 30 con-

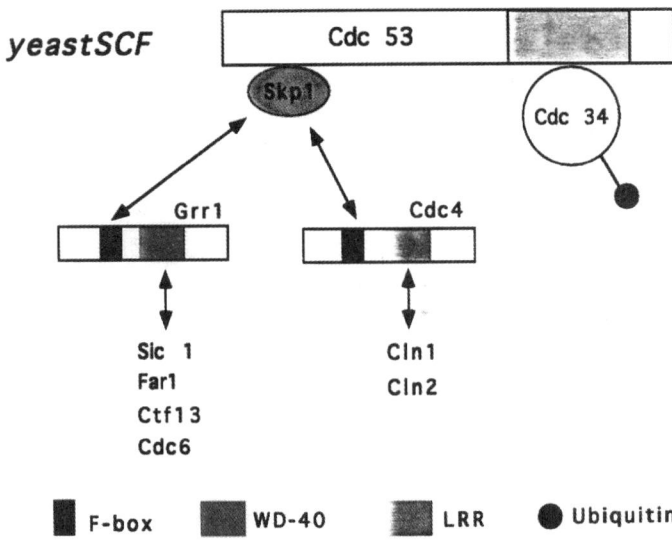

Fig. 2. Comparison of yeast SCF (ySCF) to mammalian SCF and VHL/elongins/Cul-2 (VEC) complexes. **(A)** Yeast SCF^Grr1 and SCF^Cdc4 complexes. Skp1 binds to the F-box domain of the receptors. Substrates bind to a second protein–protein interaction domain of the receptor. The shaded area corresponds to the cullin box. **(B)** Mammalian SCF^SKP2 and SCF^Trcp. Cyclin A/CDK2 bind to SKP2 through the indicated motif. **(C)** Mammalian VEC complex. Notice the modular similarities between VEC and the complexes depicted in (A) and (B).

taining SCF complexes target Swe 1 for degradation *(122)*. A structurally similar multiprotein complex (anaphase promoting complex, [APC]) guides progression through mitosis. One of its subunits, APC2, bears C-terminal homology to cdc53, and it is regarded as the equivalent E3 scaffold ligase *(116)*.

In human cells, complexes containing the F-box protein SKP2 and the human homologs of Skp1, Cdc34, and Cul-1, have been identified *(123,124)*. An intriguing homology exists between such SCF complexes and the pVHL-containing complexes described above (Fig. 2). As mentioned above, the C-terminus of Cul-2 is similar to the C-termini of Cul-1 and Cdc53. This region contains a highly conserved subdomain referred to as a "cullin-box." The N-terminus of elongin C is homologous to Skp1 *(109,118)*. Elongin B, bears an N-terminal ubiquitin-like domain. pVHL does not contain an F-box as originally defined by Bai et al. *(118)*. Nonetheless, pVHL, by virtue of its association with elongin C, may play the role of an F-box protein, namely, to serve as the substrate recognition determinant in the complex.

From this point of view, the crystal structure of the ternary complex formed between pVHL and elongins C/B provides interesting insights. VHL has two domains: an amino-terminal β-domain (residues 63–154), rich in β sheet, and a smaller α-domain (residues 155–192), consisting of three helices, H1–H3. These three helices form a groove that binds to a helix formed by the C-terminus of elongin C. The previously mentioned BC box corresponds to the first H1 helix of the α-domain, making extensive contacts with elongin C. Residues from the other two helices contribute to the stability of the complex. Half of the reported tumor-derived mutations map to the residues mediating contact between the α-domain and elongin C. Structural comparison reveals some loose similarity between the F box and the α-domain of pVHL. In addition, it appears that several residues located in the N-terminal portion of β-domain form a macromolecular patch that is likely to participate in protein–protein interactions. This patch may mediate interactions between pVHL and the substrate(s) of the pVHL–Elongins C/B–cul-2 (VEC) complex.

These observations raise the possibility that a primary role of VEC complexes is to regulate ubiquitination of specific protein substrates. Two recent reports support this hypothesis. Gorospe et al. *(125)* showed that, after prolonged exposure to low-glucose, VHL–/– cells undergo apoptosis, but their VHL+/+ counterparts do not. This differential response was attributed to the formation of hypoglycosylated and misfolded proteins. It was reasoned that VHL+/+ cells effectively degrade these misfolded proteins through the ubiquitination–proteasome pathway, but VHL–/– do not, resulting in apoptosis. Second, pVHL co-purifies with a protein called Rbx1 or ROC1 *(126,127)*. This protein interacts with both pVHL and Cul2, and has been implicated in the recruitment of ubiquitin-conjugating enzymes (E2) to SCF complexes *(128,129)*. If the rate of VEC complex in ubiquitination is validated, it will be important to identify their protein substrates, and to understand the stimuli that regulate this activity.

5.3. Fibronectin

Wild-type pVHL30, but not tumor-derived pVHL mutants, bind (at least indirectly) to fibronectin in intact cells *(130)*. Immunofluoresence using confocal microscopy, as well as cellular fractionation experiments, suggest that the interaction of pVHL and fibronectin takes place in association with the ER. pVHL19, in contrast to pVHL30, is not found in the membranous compartment of cells (which includes the ER), and does not bind to pVHL *(85)*.

The association of fibronectin with pVHL has functional consequences. Renal carcinoma cells lacking wt pVHL fail to assemble an extracellular fibronectin matrix, as determined by immunofluoresence and ELISA assays. This defect can be corrected by reintroduction of wt, but not mutant, pVHL. In addition, extracellular fibronectin matrix assembly in VHL–/– embryos and VHL–/– mouse embryo fibroblasts is likewise defective, compared with their VHL+/+ counterparts *(130)*. This pVHL effect is specific for fibronectin, since it does not extend to other macromolecules of the extracellular matrix, such as laminin or collagen.

Several lines of evidence suggest that alterations in fibronectin play a role in cellular transformation. The assembly of fibronectin in multimeric complexes has been shown to exert antiproliferative and antimetastatic effects in various model systems *(131–133)*. Conversely, many transformed cells produce diminished amounts of fibronectin matrix, compared to their nontransformed counterparts. A potential role for fibronectin in renal carcinogenesis is also supported by the finding that ACHN cells, a clear cell, renal carcinoma cell line retaining wt VHL, harbors a frameshift mutation of the fibronectin gene. This mutation results in the production of a truncated fibronectin molecule able to stimulate RCC growth in vitro *(134)*.

In a series of experiments described in Section 6, pVHL has been linked to tumor neoangiogenesis, through its ability to downregulate the production of angiogenic peptides. Moreover, rearrangement of the extracellular matrix is a critical biologic process implicated in tumor neoangiogenesis and metastasis. The effect of pVHL on fibronectin matrix formation may be an additional mechanism by which pVHL regulates neoangiogenesis.

How does pVHL affect extracellular fibronectin deposition? Western blot analysis and pulse-chase experiments suggest that VHL-defective cells are able to synthesize and secrete fibronectin, but the fibronectin fails to oligomerize properly, once secreted. pVHL itself does not appear to be secreted, but, as described above, can be found in association with the ER. Several reports showed that appropriate protein-folding and maturation is surveyed within the ER *(135,136)*. Misfolded/misprocessed proteins are actively transported to the cytoplasmic surface of the ER, and degraded by proteasomes following ubiquitination *(137,138)*. If, indeed, VEC complexes are involved in protein degradation, as described above, missfolded fibronectin may provide an example of one of the substrates specifically recognized and targeted for destruction by the complex. This remains to be tested.

5.4. Sp1 and PKC

VHL–/– cells have inappropriately high levels of VEGF mRNA, under conditions of normal ambient oxygen tension (*see* Subheading 6.). A number of stimuli upregulate VEGF expression (e.g., hypoxia, certain cytokines, and activated oncogenes such as *ras* and *src*) through either transcriptional or posttranscriptional mechanisms (or both). Mukhopadhyay et al. *(139)* showed that pVHL moderately repressed the activity of a transfected reporter plasmid containing the VEGF promoter. The minimal area of VEGF promoter necessary for this suppression contained multiple Sp1-binding sites. Furthermore, those investigators showed that pVHL binds to Sp1 in vitro, raising the possibility that the abovementioned suppression of the VEGF promoter was linked to complex formation between pVHL and Sp1.

Following this observation, Pal et al. *(140)* showed that pVHL specifically co-immunoprecipitates, in vivo, with the δ and ζ isoforms of protein kinase (PKC). Furthermore, they showed that pVHL inhibits the phosphorylation of Sp1 by PKC ζ, in vitro *(141)*. Taken together, these results indicate that pVHL may inhibit certain signal transduction pathways that involve activation of PKC isoforms and Sp1. Experiments addressing the biological relevance of these observations are needed.

6. DOWNSTREAM TARGETS OF PVHL

6.1. Negative Regulation of Hypoxia-inducible Genes and Implications for VHL Disease

The initial observation that pVHL binds to elongins C/B, two factors thought to be involved in transcriptional elongation, led to the hypothesis that pVHL regulated the transcription of certain target genes. Two observations helped to identify such genes: subset of patients with renal carcinoma, hemangioblastoma, or pheochromocytoma (sporadic or VHL-associated) manifest paraneoplastic erythrocytosis caused by inappropriate secretion of erythropoetin from the tumor cells *(147);* renal carcinoma and hemangioblastoma are highly vascular tumors, shown by immunohistochemical studies to overexpress the angiogenic peptide, VEGF *(143–145)*. Both erythropoetin and VEGF are normally induced by hypoxia. These observations led to the testable hypothesis that pVHL plays a role in a biochemical pathway that senses and/or responds to changes in ambient oxygen tension.

With this in mind, several groups examined the effect of pVHL on hypoxia-inducible genes, which are normally repressed when cells are exposed to well-oxygenated (normoxic) conditions (21% O_2 for cells grown in vitro). In contrast, cells lacking pVHL produce high levels of hypoxia-inducible mRNAs, such as the mRNAs encoding VEGF, platelet-derived growth factor-β, transforming growth factor (TGF-α), and the glucose transporter, Glut-1, under both oxygen-poor and oxygen-rich conditions *(96,145,148)*. Reintroduction of wt, but not mutant, pVHL restores the normal response of VHL-defective cells to hypoxia. These differences in mRNA levels translate into overproduction of the corresponding proteins *(96,146)*.

Given the putative link between pVHL and transcriptional elongation, the effect of pVHL on hypoxia-inducible mRNAs was found to be largely posttranscriptional *(96,146,147)*. Specifically, cells lacking pVHL fail to degrade hypoxia-inducible mRNAs following shift to well-oxygenated conditions. It was shown previously that changes in mRNA stability play an important role in the changes in the abundance of these mRNAs that accompany changes in ambient oxygen conditions.

The ability to rapidly alter mRNA stability, and hence the production of specific proteins, in response to various stimuli is important to cellular homeostatesis. Considerable progress has been made in understanding how mRNA decay is regulated in yeast and mammalian cells *(149,150)*. mRNA bound to ribonucleoproteins is actively exported from the nucleus through an as-yet-unidentified receptor in mammalian cells *(151,152)*. Ribosomal-bound mRNA targeted for decay undergoes polyA tail-shortening, followed by uncapping and exo- and endoribonuclease digestion. mRNA sequences implicated in regulating mRNA stability have been mapped in 3'UTR, coding and 5'UTR regions. Cellular proteins bound to these regions, as well as to the polyA tail, are suspected of being involved in regulating the rate of mRNA decay *(150)*.

Regulation of VEGF mRNA involves both transcriptional and posttranscriptional mechanisms *(153–155)*. The former involves promoter-binding sites for hypoxia-inducible factors *(156)*. The latter depends on a 3'UTR element rich in AUUUA repeats (ARE) *(157,158)*. This ARE forms in vivo complexes, with multiple cellular proteins believed to confer mRNA stability *(159,160)*. One of them, HuR, was recently cloned and shown to upregulate VEGF expression when transfected into cells *(161,162)*. Levy et al. *(163)* showed that ARE–protein complexes are attenuated in VHL+/+ cells growing in normoxic conditions, compared to VHL–/– cells. It is therefore possible to hypothesize, in light of the recent notion that VEC complexes are involved in protein destruction, that RNA-binding proteins involved in mRNA stabilization are degraded by pVHL. Support for this hypothesis is provided by the observation that binding to elongins and Cul-2 is necessary for pVHL to downregulate hypoxia-inducible genes *(109)*.

6.2. TGF-β

Renal carcinoma cells overproduce TGF-β, a growth factor with proliferative, antiproliferative, and proangiogenic functions, depending upon the cells examined *(164)*. Ananth et al. examined the regulation of TGF-β by pVHL, and showed that VHL–/– cells express high levels of TGF-β mRNA and protein *(165)*. As is the case with the hypoxia-inducible genes, pVHL appears to regulate TGF-β mRNA at the posttranscriptional level *(165)*. Reintroduction of wt pVHL into these cells suppressed TGF-β production *(165)*. The cells examined by those investigators lack TGF-β receptor II, and their growth was not altered by TGF-β in vitro. In contrast, administration of anti-TGF-—neutralizing antibodies inhibited the growth of VHL–/– cells in vivo, presumably by antagonizing some paracrine function of TGF-β.

6.3. Carbonic Anhydrase

Large-scale differential display analysis revealed that wt pVHL downregulates the expression of the transmembrane proteins, carbonic anhydrase 2 (CA12) and 9 (CA9) *(166)*. In contrast to CA12, which is expressed in several normal and tumor tissues, including normal adult kidney, CA9 is normally expressed only in adult gastric mucosa *(167)*. CA9 may, therefore, be viewed as a tumorassociated antigen, when present in kidney cancer. Its specific detection in clear cell renal carcinoma may prove useful for diagnostic purposes.

CA9 and CA12 belong to the larger family of CAs *(168)*. Both CA9 and CA12 isoforms are catalytically active, promoting the extracellular hydration of CO_2, and the unstable formation of carbonic acid, which subsequently decomposes into HCO_3^- and H_3O^+. This activity is expected to affect the function of several ionic channels that shift protons and HCO_3^- across the cellular membrane *(161)*. The net result of CA9 and CA12 should therefore be to acidify the tumor microenvironment. Acidification of the extracellular environment was reported to enhance invasive behavior of tumor cells in vitro, and may also affect cell growth *(169)*.

The mechanism by which pVHL inhibits the CA12 and CA9 mRNAs is currently unknown. The elongin-binding domain of pVHL appears necessary for regulation of both isoenzymes. An additional domain, spanning the boundary of exon 1 and 2, appears to contribute to the regulation of CA12 by pVHL *(159)*.

7. CLINICAL IMPLICATIONS

7.1. Diagnostic Criteria and Surveillance Guidelines

The cloning of *VHL* gene has significantly advanced understanding of VHL disease at both the clinical and molecular level. Prior to the cloning of the *VHL* gene, the diagnosis of VHL disease was based purely on clinical criteria. It is now possible, using semiquantitative Southern blot analysis, to detect VHL mutations in ~100% of patients who meet the clinical criteria for VHL disease. Relatives of affected individuals and patients without family history with VHL-associated tumors (such as hemangioblastoma), who do not yet fulfill clinical criteria for VHL disease (see Subheading 2.) are considered individuals at risk. Such individuals can now be accurately genotyped and entered into appropriate surveillance programs.

There is mounting evidence that specific VHL mutations are associated with specific VHL disease phenotypes (see Subheading 3.2.). The establishment of an international database for VHL mutations, along with progress in understanding of the biochemical functions of pVHL, may ultimately allow the accurate prediction of the expected disease phenotype for a given mutation. This may lead to the development of risk-oriented screening and preventive strategies. For now, a uniform program of surveillance is proposed, based on current understanding of the disease, and the availability of modern, noninterventional, imaging technology. Summary of this program is provided in Table 3.

Table 3

Screening Guidelines for VHL Patients

From conception:	Inform obstetrician of VHL family history.
From birth:	Physical examination and neurological assessment
	Fundoscopic examination
Ages 2–10	
Annual:	Physical examination and neurological assessment
	Fundoscopic examination
	Blood and 24-h urine catecholamine levels
Ages 11–19	
Every 6 mo:	Fundoscopic examination
Annual:	Physical examination and neurological assessment
	Blood and 24-h urine catecholamine levels
	Abdominal US (If abnormal, MRI or CT of abdomen, except in pregnancy)
Every 2 yr:	Brain and spine MRI with gadolinium.
	Annually at onset of puberty, or before and after pregnancy (not during pregnancy, except in medical emergencies.)
Age 20 yr and beyond	
Annual:	Physical examination and neurological assessment
	Blood and 24-h urine catecholamine levels
	Abdominal US.
Every 2 yr:	Abdominal CT with contrast (or US for females in reproductive age)
	Brain and spine MRI with gadolinium.
	Audiometric examination (MRI of the internal auditory canal, in case of signs or symptoms of auditory abnormalities)

These guidelines are based on consensus conferences organized by the VHL Family Alliance Foundation, and are published in the information booklet of the Alliance (http://www.vhl.org). These general guidelines should be individualized according to the clinical judgment of the caring physician, and they are expected to change with time, as new medical knowledge is generated.

7.2. Molecular Targets for Therapeutic Interventions

The ultimate goal of VHL research is the development of effective therapies for VHL disease, based on the functions of pVHL. Rational approaches to therapy can now be envisioned, based on admittedly limited understanding of how pVHL suppresses tumor growth. Examples of such strategies are provided below:

1. pVHL negatively regulates the expression of hypoxia-inducible polypeptides. This family of proteins includes growth and angiogenic factors likely to be critical in tumor establishment and progression. It is reasonable to hypothesize that VHL patients will benefit from treatment with antiangiogenic drugs. Drugs that inhibit VEGF signaling will be of particular interest.
2. The VEC complex may target specific substrates for degradation through the ubiquitin–proteosme pathway. There is strong evidence that such substrates regulate hypoxia-inducible mRNAs and their corresponding polypeptides. Elucidation of the molecules involved in this specific recognition and degradation process may lead to the identification of novel drug targets. In short, drugs that affected the stability of pVHL substrates may substitute for pVHL function.

3. Upregulation of growth and angiogenic factors in VHL–/– tumors may correlate with concommitant upregulation of their corresponding receptors, thus creating a positive feedback loop promoting tumor development. Identification of specific receptor antagonists may provide an additional strategy for specific treatment of VHL–/– tumors.

4. Further understanding of pVHL biochemical functions may, in time, facilitate the design of effective and specific gene transfer strategies.

REFERENCES

1. McKusick VA. *Mendelian Inheritance in Man.* Baltimore; The Johns Hopkins University Press. 1992:
2. Collins ET. Intra-ocular growths (two cases, brother and sister, with peculiar vascular new growth, probably retinal, affecting both eyes). *Trans Ophthal Soc UK* 1894; 14:141–149.
3. von Hippel E. Ueber eine sehr seltene Erkrankung der Nethaut. *Graefe Arch Ophthal* 1904; 59:83–106.
4. Fulton JF. *Harvey Cushing: A Biography* Springfield, Il: Charles C Thomas.
5. Lindau A. Zur Frage der Angiomatosis Retinae und Ihrer Hirncomplikation. *Acta Opthal* 1927; 4:193–226.
6. Melmon K, Rosen S. Lindau's disease. *Am J Med* 1964; 36:595–617.
7. Filling-Katz M, Choyke P, Oldfield E, Charnas L, Patronas N, Glenn G, et al. Central nervous system involvement in Von Hippel-Lindau disease. *Neurology* 1991; 41:41–46.
8. Lamiell J, Salazar F, Hsia Y. von Hippel-Lindau disease affecting 43 members of a single kindred. *Medicine* 1989; 68:1–29.
9. Maher ER, Yates JRW, Harries R, Benjamin C, Harris R, Moore AT, Ferguson-Smith MA. Clinical features and natural history of von Hippel-Lindau disease. *Q J Med* 1990; 77:1151–1163.
10. Neumann HPH, Lips CJM, Hsia YE, Zbar B. von Hippel-Lindau syndrome. *Brain Pathol* 1995; 5:181–193.
11. Resche F, Moisan JP, Mantoura J, de Kersaint-Gilly A, Andre MJ, Perrin-Resche I. Hemangioblastomas, hemangioblastomatosis and von Hippel-Lindau disease. *Adv Tech Stand Neurosurg* 1993; 20:197–303.
12. Richard S, Campello C, Taillandier L, Parker F, Resche F. Hemangioblastoma of the central nervous system in von Hippel-Lindau disease. *J Intern Med* 1998; 243:547–553.
13. Choyke PL, Glenn GM, Walther MM, Patronas NJ, Linehan WM, Zbar B. von Hippel-Lindau disease: genetic, clinical and imaging features. *Radiology* 1995; 194:629–642.
14. Krieg M, Marti HH, Plate KH. Coexpression of erythropoietin and vascular endothelial growth factor in nervous system tumors associated with von Hippel-Lindau tumor suppressor gene loss of function. *Blood* 1998; 92:3388–3393.
15. Richard S, Beigelman C, Gerber S, Van Effenterre R, Gaudric A, Sahel M, et al. L hemangioblastome existe-t-il en dehors de la maladie de von Hippel-Lindau? *Neurochirurgie* 1994; 40:145–154.
16. Sharma RR, Cast IP, O'Brien C. Supratentorial haemangioblastoma not associated with Von Hippel Lindau complex or polycythaemia: case report and literature review. *Br J Neurosurg* 1995; 9:81–84.
17. Wizigmann-Voos S, Plate KH. Pathology, genetics and cell biology of hemangioblastomas. *Histol Histopathol* 1996; 11:1049–1061.
18. Wittebol-Post D, Hes FJ, Lips CJM. The Eye in von Hippel-Lindau disease. Long term follow up of screening and treatment recommendations. *J Intern Med* 1998; 243:555–561.
19. Choyke P, Glenn G, Walther M, Zbar B, Weiss G, RB A, et al. The natural history of renal lesions in von Hippel-Lindau disease: a serial CT study in 28 patients. *Am J Radiol* 1992; 159:1229–1234.
20. Walther MM, Lubensky IA, Vezdon D, Zbar B, Linehan WM. Prevalence of microscopic lesions in grossly normal renal parenchyma from patients with VHL disease, sporadic renal cell carcinoma and no renal disease:clinical implications. *J Urol* 1995; 154:2010–2014.
21. Richard S, Chauveau D, Chretien Y, Beigelman C, Denys A, Fendler J, et al. Renal lesions and pheochromocytoma in von Hippel-Lindau disease. *Adv Nephrol* 1994; 23:1–27.
22. Kragel PJ, MM W, JP P, MR F-K. Simple renal cysts, atypical renal cysts and renal carcinoma in situ in VHL disease: alectin and immunohistochemical study in six patients. *Modern Pathol* 1991; 4:210–214.
23. Malek RS, Omess PJ, Benson RC, Zincke H. Renal carcinoma in VHL syndrome. *Am J Med* 1987; 82:236–238.

24. Nelson JB, Oyasu R, Dalton DP. The clinical and pathologic manifestations of renal tumors in VHL disease. *J Urol* 1994; 156:2221.

25. Blankenship C, Naglich JG, Whaley JM, et al. Alternate choice of initaial codon produces a biologically active product of the von Hippe-Lindau gene with tumor suppressor activity. Oncogene 1999; 18(8):1529–1535.

26. Neumann HPH, Berger DP, Sigmund G, Blum U, Schmidt D, Parmer RJ, Volk B, Kirste G. Pheochromocytomas, multiple endocrine neoplasia type 2, and von Hippel-Landau Disease. *N Engl J Med* 1993; 329:1531–1538.

27. Binkovitz L, Johnson C, Stephens D. Islet cell tumors in von Hippel-Lindau disease: increased prevalence and relationship to the multiple endocrine neoplasias. *Am J Roentgen* 1990; 155:501–505.

28. Hough DM, Stephens DH, Johnson CD, Binkovitz LA. Pancreatic lesions in VHL disease: prevalence, clinical significance and CT findings. *AJR* 1994; 162:1091–1094.

29. Neumann HP, Dinkel E, Brambs H, Wimmer B, Friedburg H, Volk B. et al. Pancreatic lesions in the von Hippel-Lindau syndrome. *Gastrenterology* 1991; 101:465–471.

30. Fishman RS, Bartholomew LG. Severe pancreatic involvement in three generations in von Hippel-Lindau disease. *Mayo Clin Proc* 1979; 54:329–331.

31. Lubensky IA, Pack S, Ault D, Vortmeyer AO, Libutti SK, Choyke PL. Multiple neuroendocrine tumors of the pancreas in von Hippel-Lindau disease patients: histopathological and molecular genetic analysis. *Am J Pathol* 1998; 153:223–231.

32. Mulshire JL, Tubbs R, Sheeler LR, Gifford RW. Case report: clinical significance of the association of the von Hippel-Lindau disease with pheochromocytoma and pancreatic apudoma. *Am J Med Sci* 1984; 288:212–214.

33. Manski TJ, Heffner DK, Glenn GM, et al. Endolymphatic sac tumors: a source of morbid hearing loss in VHL disease. *JAMA* 1997; 277:1461–1466.

34. Megerian CA, McKenna MJ, Nuss RC, Maniglia AJ, Ojemann RG, Pilch BZ, Nadol JB. Endolymphatic sac tumors: histopathologic confirmation, clinical characterization and implication in VHL disease. *Laryngoscope* 1995; 105:801–808.

35. Choyke PL, Glenn GM, Wagner JP, Lubensky IA, Thakore K, Zbar B, Linehan WM, Walther MM. Epididymal cystadenomas in von Hippel-Lindau disease. *Urology* 1997; 49:926–931.

36. Karsdorp N, Elderson A, Wittwbol-Post D, et al. von Hippel-Lindau disease: new strategies in early detection and treatment. *Am J Med* 1994; 97:158–168.

37. Korn WT, Schatzki SC, DiSciullo AJ, Scully RE. Papillary cystadenoma of the broad ligament in von Hippel-Lindau disease. *Am J Obstet Gynecol* 1990; 163:596–598.

38. Latif F, Tory K, Gnarra J, Yao M, Duh F-M, Orcutt ML, et al. Identification of the von Hippel-Lindau disease tumor suppressor gene. *Science* 1993; 260:1317–1320.

39. Seizinger BR, Rouleau GA, Ozelius LJ, Lane AH, Farmer GE, Lamiell JM, et al. Von-Hippel Lindau disease maps to the region of chromosome 3 associated with renal cell carcinoma. *Nature* 1988; 332:268–269.

40. Gnarra JR, Duan DR, Weng Y, Humphrey JS, Chen DYT, Lee S, et al. Molecular cloning of the von Hippel-Lindau tumor suppressor gene and its role in renal cell carcinoma. *Biochim Biophys Acta* 1996; 1242:201–210.

41. Gnarra JR, Tory K, Weng Y, Schmidt L, Wei MH, Li H, et al. Mutations of the VHL tumour suppressor gene in renal carcinoma. *Nature Genet* 1994; 7:85–90.

42. Kuzmin I, Duh F-M, Latif F, Geil L, Zbar B, Lerman MI. Identification of the promoter of the human von Hippel-Lindau disease tumor suppressor gene. *Oncogene* 1995; 10:2185–2194.

43. Renbaum P, Duh F-M, Latif F, Zbar B, Lerman M, Kuzmin I. Isolation and characterization of the full-length 3′ untranslated region of the human von Hippel-Lindau tumor suppressor gene. *Hum Genet* 1996; 98:666–671.

44. Kessler P, Vasavada S, Rackley R, Stackhouse T, Duh F, Latif F, et al. Expression of the von Hippel-Lindau tumor-suppressor gene, VHL, in human fetal kidney and during mouse embryogenesis. *Mol Med* 1995; 1:457–466.

45. Richards F, Schofield P, Fleming S, Maher E. Expression of the von Hippel-Lindau disease tumour suppressor gene during human embryogenesis. *Hum Mol Genet* 1996; 5:639–644.

46. Nagashima Y, Miyagi Y, Udagawa K, Taki A, Misugi K, Sakai N. von Hippel-Lindau tumor suppressor gene. Localization of expression by in situ hybridization. *J Pathol* 1996; 180:271–274.

47. Duan DR, Humphrey JS, Chen DYT, Weng Y, Sukegawa J, Lee S, et al. Characterization of the VHL tumor suppressor gene product: localization, complex formation, and the effect of natural inactivating mutations. *Proc Natl Acad Sci USA* 1995; 92:6495–6499.

48. Gao J, Naglich JG, Laidlaw J, Whaley JM, Seizinger BR, Kley N. Cloning and characterization of a mouse gene with homology to the human von Hippel-Lindau disease tumor suppressor gene: implications for the potential organization of the human von Hippel-Lindau disease gene. *Cancer Res* 1995; 55:743–747.

49. Stolle C, Glenn G, Zbar B, Humphrey J, Choyke P, Walther M, et al. Improved detection of germline mutations in the von Hippel-Lindau disease tumor suppressor gene. *Hum Mutat* 1998; 12:417–423.

50. Bailly M, Bain C, Favrot M-C, Ozturk M. Somatic mutations of VHL tumor suppressor gene in European kidney cancers. *Int J Cancer* 1995; 63:660–664.

51. Brauch H, Kishida T, Glavac D, Chen F, Pausch F, Hofler H, et al. von Hippel-Lindau disease with pheochromocytoma in the Black Forest region in Germany: evidence for a founder effect. *Hum Genet* 1995; 95:551–556.

52. Chen F, Kishida T, Yao M, Hustad T, Glavac D, Dean M, et al. Germline mutations in the von Hippel-Lindau disease tumor suppressor gene:correlations with phenotype. *Hum Mutat* 1995; 5:66–75.

53. Crossey PA, Richards FM, Foster K, Green JS, Prowse A, Latif F, et al. Identification of intragenic mutations in the von Hippel-Lindau disease tumor suppressor gene and correlation with disease phenotype. *Hum Mol Gen* 1994; 3:1303–1308.

54. Glavac D, Neumann HPH, Wittke C, Jaening H, Masek O, Streicher T, et al. Mutations in the VHL tunor suppressor gene and associated lesions in families with von Hippel-Lindau disease from central Europe. *Hum Genet* 1996; 98:271–280.

55. Maher E, Webster A, Richards F, Green J, Crossey P, Payne S, Moore A. Phenotypic expression in von Hippel-Lindau disease: correlations with germline VHL gene mutations. *J Med Genet* 1996; 33:328–332.

56. Prowse A, Webster A, Richards F, Richard S, Olschwang S, Resche F, Affara N, Maher E. Somatic inactivation of the VHL gene in Von Hippel-Lindau disease tumors. *Am J Hum Gene* 1997; 60:765–761.

57. Richards F, Crossey P, Phipps M, Foster K, Latif F, Evans G, et al. Detailed mapping of germline deletions of the von Hippel-Lindau disease tumour suppressor gene. *Hum Mol Genet* 1994; 3:595–598.

58. Zbar B, Kishida T, Chen F, Schmidt L, Maher ER, Richards FM, et al. Germline mutations in the von Hippel-Lindau (VHL) gene in families from North America, Europe, and Japan. *Hum Mutat* 1996; 8:348–357.

59. Beroud C, Joly D, Gallou C, Staroz F, Orfanelli M, Junien C. Software and database for the analysis of mutations in the VHL gene. *Nucleic Acids Res* 1998; 26:256–258.

60. Lubensky IA, Gnarra JR, Bertheau P, Walther MM, Linehan WM, Zhuang Z. Allelic deletions of the VHL gene detected in multiple microscopic clear cell renal lesions in von Hippel-Lindau disease patients. *Am J Pathol* 1996; 149:2089–2094.

61. Vortmeyer A, Gnarra J, Emmert-Buck M, Katz D, Linehan W, Oldfield E, Zhuang Z. von Hippel-Lindau gene deletion detected in the stromal cell component of a cerebellar hemangioblastoma associated with von Hippel-Lindau disease. *Hum Pathol* 1997; 28:540–543.

62. Zhuang Z, Bertheau P, Emmert-Buck M, Liotta L, Gnarra J, Linehan W, Lubensky I. A microscopic dissection technique for archival DNA analysis of specific cell populations in lesions < 1mm in size. *Am J Pathol* 1995; 146:620–625.

63. Chen F, Slife L, Kishida T, Mulvihill S, Tisherman E, Zbar B. Genotype-phenotype correlation in von Hippel-Lindau disease: identification of a mutation associated with VHL type 2A. *J Med Genet* 1996; 33:716–717.

64. Neumann H, Bender B. Genotype-phenotype correlations in von Hippel-Lindau disease. *J Intern Med* 1998; 243:541–545.

65. Crossey P, Eng C, Ginalska-Malinowska M, Lennard T, Wheeler D, Ponder B, Maher E. Molecular genetic diagnosis of von Hippel-Lindau disease in familial phaeochromocytoma. *J Med Genet* 1995; 32:885–886.

66. Eng C, Crossey P, Mulligan L, Healey C, Houghton C, Prowse A, et al. Mutations in the RET proto-oncogene and the von Hippel-Lindau disease tumour suppressor gene in sporadic and syndromic phaeochromocytomas. *J Med Genet* 1995; 32:934–937.

67. Gross D, Avishai N, Meiner V, Filon D, Zbar B, Abeliovich D. Familial phenochromoccytoma associated with a novel mutation in the von Hippel-Lindau gene. *J Clin Endocrinol Met* 1996; 81:147–149.

68. Neumann H, Eng C, Mulligan L, Glavac D, Ponder B, Crossey P, Maher E, Brauch H. Consequences of direct genetic testing for germline mutations in the clinical management of families with multiple endocrine neoplasia type 2. *JAMA* 1995; 274:1149–1151.

69. Ritter M, Frilling A, Crossey P, Hoppner W, Maher E, Mulligan L, Ponder B, Engelhardt D. Isolated familial pheochromocytoma as a variant of von Hippel-Lindau disease. *J Clin Endocrinol Met* 1996; 81:1035–1037.

70. Knudson AG, Jr. Mutation and cancer: statistical study of retinoblastoma. *Proc Natl Acad Sci USA* 1971; 68:820–823.

71. Foster K, Prowse A, van den Berg A, Fleming S, Hulsbeek MMF, Crossey PA, et al. Somatic mutations of the von Hippel-Lindau disease tumor suppressor gene in non-familial clear cell renal carcinoma. *Hum Mol Gen* 1994; 3:2169–2173.

72. Shuin T, Kondo K, Torigoe S, Kishida T, Kubota Y, Hosaka M, et al. Frequent somatic mutations and loss of heterozygosity of the von Hippel-Lindau tumor suppressor gene in primary human renal cell carcinomas. *Cancer Res* 1994; 54:2852–2855.

73. Whaley JM, Naglich J, Gelbert L, Hsia YE, Lamiell JM, Green JS, et al. Germ-line mutations in the von Hippel-Lindau tumor suppressor gene are similar to somatic von Hippel-Lindau abberations in sporadic renal cell carcinoma. *Am J Hum Genet* 1994; 55:1092–1102.

74. Kanno H, Kondo K, Ito S, Yamamoto I, Fujii S, Torigoe S, et al. Somatic mutations of the von Hippel-Lindau Tumor supressor gene in sporadic central nervous systems hemangioblastomas. *Cancer Res* 1994; 54:4845–4847.

75. Lee J-Y, Dong S-M, Park W-S, Yoo N-J, Kim C-S, Jang J-J, et al. Loss of heterozygosity and somatic mutations of the VHL tumor suppressor gene in sporadic cerebellar hemangioblastomas. *Cancer Res* 1998; 58:504–508.

76. Oberstarb J, Reifenberger G, Reifenberger J, Weschler W, Collins VP. Mutations of the von Hippel-Lindau tumor suppressor gene in capillary hemangioblastomas of the central nervous system. *J Pathol* 1996; 179:151–156.

77. Tse J, Wong J, Lo K-W, Poon W-S, Huang D, Ng H-K. Molecular genetic analysis of the von Hippel-Lindau disease tumor suppressor gene in familial and sporadic cerebellar hemangioblastomas. *Am J Clin Pathol* 1997; 107:459–466.

78. Hofstra RMW, Stelwagen T, Stulp RP, de Jong D, Hulsbeek M, Kamsteeg EJ, et al. Extensive mutation scanning of RET in sporadic medullary thyroid carcinoma and of RET and VHL in sporadic pheochromocytoma reveals involvement of these genes in only a minority of cases. *J Clin Endocrinol Metab* 1996; 81:2881–2884.

79. Vortmeyer AO, Lubensky IA, Fogt F, Linehan WM, Khettry U, Zhuang Z. Allelic deletion and mutation of the von Hippel-Lindau tumor suppressor gene in pancreatic microcystic adenomas. *Am J Pathol* 1997; 151:951–956.

80. Sekido Y, Bader S, Latif F, Gnarra J, Gazdar A, Linehan W, et al. Molecular analysis of the von Hippel-Lindau disease tumor suppressor gene in human lung cancer cell lines. *Oncogene* 1994; 9:1599–1604.

81. Dolan K, Garde J, Gosney J, Sissons M, Wright T, Kingsnorth AN, et al. Allelotype analysis of oesophageal adenocarcinoma: loss of heterozygosity occurs at multiple sites. *Br J Cancer* 1998; 78:950–957.

82. Waber PG, Lee NK, Nisen PD. Frequent allelic loss at chromosome arm 3p is distinct from genetic alterations of the VHL tumor suppressor gene in head and neck cancer. *Oncogene* 1996; 12:365–369.

83. Zhuang Z, Emmert-Buck MR, Roth MJ, Gnarra JR, Linehan WM, Liotta LA, Lubensky IA. VHL gene deletion detected in microdissected sporadic human colon carcinoma specimen. *Hum Pathol* 1996; 27:152–156.

84. Iliopoulos O, Kibel A, Gray S, Kaelin WG. Tumor suppression by the human von Hippel-Lindau gene product. *Nat Med* 1995; 1:822–826.

85. Iliopoulos O, Ohh M, Kaelin W. pVHL19 is a biologically active product of the von Hippel-Lindau gene arising from internal translation initiation. *Proc Natl Acad Sci USA* 1998; 95:11,661–11,666.

86. Schoenfeld A, Davidowitz E, Burk R. A second major native von Hippel-Lindau gene product, initiated from an internal translation start site, functions as a tumor suppressor. *Proc Natl Acad Sci USA* 1998; 195:8817–8822.

87. Kibel A, Iliopoulos O, DeCaprio JD, Kaelin WG. Binding of the von Hippel-Lindau tumor suppressor protein to elongin B and C. *Science* 1995; 269:1444–1446.

88. Iliopoulos O, Kaelin WG. The Molecular basis of von Hippel-Lindau disease. *Mol Med* 1997; 3:289–293.

89. Corless CL, Kibel A, Iliopoulos O, Kaelin WGJ. Immunostaining of the von Hippel-Lindau Gene Product (pVHL) in normal and neoplastic human tissues. *Hum Pathol* 1997; 28:459–464.

90. Los M, Jansen GH, Kaelin WG, Lips CJM, Blijham GH, Voest EE. Expression pattern of the von Hippel-Lindau protein in human tissues. *Lab Invest* 1996; 75:231–238.

91. Lee S, Neumann M, Stearman R, Stauber R, Pause A, Pavlakis G, Klausner R. Transcription-dependent nuclear-cytoplasmic trafficking is required for the function of the von Hippel-Lindau tumor suppressor protein. *Mol Cell Biol* 1999; 19:1486–1497.

92. Iliopoulos O, Kaelin WG. Unpublished data.

93. Fearon ER, Vogelstein B. A Genetic model of colorectal tumorigenesis. *Cell* 1990; 61:759–767.

94. Kinzler K, Vogelstein B. Lessons from hereditary colorectal cancer. *Cell* 1996; 87:159–170.

95. Kishida T, Stackhouse TM, Chen F, Lerman MI, Zbar B. Cellular proteins that bind the von Hippel-Lindau Disease gene product: mapping of binding domains and the effect of missense mutations. *Cancer Res* 1995; 55:4544–4548.

96. Gnarra JR, Zhou S, Merrill MJ, Wagner J, Krumm A, Papavassiliou E, et al. Post-transcriptional regulation of vascular endothelial growth factor mRNA by the VHL tumor suppressor gene product. *Proc Natl Acad Sci* 1996; 93:10,589–10,594.

97. Lieubeau-Teillet B, Rak J, Jothy S, Iliopoulos O, Kaelin W, Kerbel RS. VHL gene mediated growth suppression and induction of differentiation in renal cell carcinoma cell grown as multicellular tumor spheroids. *Cancer Res* 1998; 58:4957–4962.

98. Pause A, Lee S, Lonergan KM, Klausner RD. The von Hippel-Lindau tumor suppressor gene is required for cell cycle exit upon serum withdrawal. *Proc Natl Acad Sci* 1998; 95:993–998.

99. Aso T, Lane WS, Conaway JW, Conaway RC Elongin (SIII): a multisubunit regulator of elongation by RNA polymerase II. *Science* 1995; 269:1439–1443.

100. Duan DR, Pause A, Burgress W, Aso T, Chen DYT, Garrett KP, et al. Inhibition of transcriptional elongation by the VHL tumor suppressor protein. *Science* 1995; 269:1402–1406.

101. Bradsher JN, Jackson KW, Conaway RC, Conaway JW. RNA polymerase II transcription factor SIII. *J Biol Chem* 1993; 268:25,587–25,593.

102. Aso T, Haque D, Barstead R, Conaway R, Conaway J. The inducible elongin A elongation activation domain: structure, function and interaction with elongin BC complex. *EMBO J* 1996; 15:101–110.

103. Reines D, Conaway JW, Conaway RC. The RNA polymerase II general elongation factors. *Trends Biochem Sci* 1996; 21:351–355.

104. Takagi Y, Conaway R, Conaway J. Characterization of elongin C functional domain required for interaction with elongin B and activation of elongin A. *J Biol Chem* 1996; 271:1–7.

105. Brower CS, Shilatifard A, Mather T, Kamura T, Takagi Y, Haque D, et al. Elongin B ubiquitin homology domain. *J Biol Chem* 1999; 274:13,629–13,636.

106. Conaway JW, Kamura T, Conaway RC. Elongin BC complex and the von Hippel-Lindau tumor suppressor protein. *Biochim Biophys Acta* 1998; 1377:M49–54.

107a. Kamura T, Sato S, Haque D, Liu L, Kaelin WJ, Conaway R, Conaway J. The Elongin BC complex interacts with the conserved SOCS-box motif present in members of the SOCS, ras, WD-40 repeat, and ankyrin repeat families. *Genes Dev* 1998; 12:3872–3881.

107b. Iliopoulos O, Kaelin WG, Conaway JW, Conaway RC. Unpublished observations.

108. Kipreos ET, Lander LE, Wing JP, He WW, Hedgecock EM. cul-1 is required for cell cycle exit in C. elegans and identifies a novel gene family. *Cell* 1996; 85:829–839.

109. Lonergan KM, Iliopoulos O, Ohh M, Kamura T, Conaway RC, Conaway JW, Kaelin WG. Regulation of hypoxia-inducible mRNAs by the von Hippel-Lindau protein requires binding to complexes containing elongins B/C and Cul2. *Mol Cell Biol* 1998; 18:732–741.

110. Pause A, Lee S, Worrell RA, Chen DYT, Burgess WH, Linehan WM, Klausner RD. The von Hippel-Lindau tumor-suppressor gene product forms a stable complex with human CUL-2, a member of the Cdc53 family of proteins. *Proc Natl Acad Sci USA* 1997; 94:2156–2161.

111. Latres E, Chiaur DS, Pagano M. The human F box protein beta-Trcp associates with the Cull/Skp1 complex and regulates the stability of beta-catenin. *Oncogene* 1999; 18:849–854.

112. Spencer E, Jiang J, Chen ZJ. Signal-induced ubiquitination of IkBa by the F-box protein Slimb/β-TrCP. *Genes Dev* 1999; 13:284–294.

113. Winston JT, Strack P, Beer-Romero P, Chu CY, Elledge SJ, Harper JW. The SCFβTrcp–ubiquitin ligase complex associates specifically with phosphorylated destruction motifs in IkBa and bcatenin and stimulates IkBa ubiquitination in vitro. *Genes Dev* 1999; 13:270–283.

114. Hochstrasser M. Ubiquitin-dependent protein degradation. *Ann Rev Genet* 1996; 30:405–439.

115. Jentsch S. The ubiquitin-conjugation system. *Ann Rev Genet* 1992; 26:179–207.

116. Peters J-M. SCF and APC: the Yin and the Yang of cell cycle regulated proteolysis. *Curr Opin Cell Biol* 1998; 10:759–768.
117. Patton EE, Willems AR, Sa D, Kuras L, Thomas D, Craig KL, Tyers M. Cdc53 is a scaffold for multiple Cdc34/Skp1/F-box protein complexes that regulate cell division and methionine biosynthesis in yeast. *Genes Dev* 1998; 12:692–705.
118. Bai C, Sen P, Hofmann K, Ma L, Goebl M, Harper JW, Elledge SJ. SKP1 connects cell cycle regulators to the ubiquitin proteolysis machinery through a novel motif, the F-box. *Cell* 1996; 86:263–274.
119. Skowyra D, Craig KL, Tyers M, Elledge SJ, Harper JW. F-Box proteins are receptors that recruit phosphorylated substrates to the SCF ubiquitin ligase complex. *Cell* 1998; 91:209–219.
120. Feldman RMR, Correll CC, Kaplan KB, Deshaies RJ. A complex of cdc4p, skp1p and Cdc53/cullin catalyzes ubiquitination of the phosphorylated CDK inhibitor Sic1p. *Cell* 1997; 91:2210–2230.
121. Patton E, Willems AR, Tyers M. Combinatorial control in ubiquitin dependent proteolysis: don't Skp the F-box hypothesis. *Trends Genet* 1998; 14:236–243.
122. Kaiser P, Sia RAL, Bardes EGS, Lew DJ, Reed SI. Cdc34 and the F-box protein Met30 are required for degradation of the Cdk-inhibitory kinase Swe1. *Genes Dev* 1998; 12:2587–2597.
123. Lisztwan J, Marti A, Sutterluty H, Gstaiger M, Wirbelauer C, Krek W. Association of human CUL-1 and ubiquitin-conjugating enzyme CDC34 with the F-box protein p45(SKP2): evidence for evolutionary conservation in the subunit composition of the CDC34-SCF pathway. *EMBO J* 1998; 17:368–383.
124. Michel JJ, Xiong Y. Human CUL-1, but not other cullin family members, selectively interact with SKP1 to form a complex with SKP2 and Cyclin A. *Cell Growth Differ* 1998; 9:445–449.
125. Gorospe M, Egan JM, Zbar B, Lerman M, Geil L, Kuzmin I, Holbrook NJ. Protective function of von Hippel-Lindau protein against impaired protein processing in renal carcinoma cells. *Mol Cell Biol* 1999; 19:1289–1300.
126. Kamura T, Koepp DM, Conrad MN, Skowyra D, Moreland RJ, Iliopoulos O, et al. Rbx-1, a component of the VHL tumor suppressor complex and SCF ubiquitin ligase. *Science* 1999; 284:657–661.
127. Skowyra D, Koepp DM, Kamura T, Conrad MN, Moreland RJ, Conaway RC, et al. Reconstitution of G1 Cyclin ubiquitination with complexes containing SCF^{Grr1} and Rbx-1. *Science* 1999; 284:662–665.
128. Ohta T, Michel JJ, Schottelius AJ, Xiong Y. ROC1, a homolog of APC11, represents a family of cullin partners with an associated ubiquitin ligase activity. *Cell* 1999; 3:535–541.
129. Tan P, Fuchs SY, Chen A, Wu K, Gomez C, Ronai Z, Pan Z-Q. Recruitment of a ROC1-CUL1 ubiquitin ligase by Skp1 and HOS to catalyze the ubiquitination of IκBα. *Cell* 1999; 3:527–533.
130. Ohh M, Yauch RL, Lonergan KM, Whaley JM, Stemmer-Rachamimov AO, Louis DN, et al. The von Hippel-Lindau tumor suppressor protein is required for proper assembly of an extracellular fibronectin matrix. *Mol Cell* 1998; 1:959–968.
131. Giancotti F, Mainiero F. Integrin-mediated adhesion and signaling in tumorigenesis. *Biochim Biophys Acta* 1994; 1198:47–64.
132. Giancotti FG, Ruoslahti E. Elevated levels of the a5b1 fibronectin receptor suppress the transformed phenotype of chinese hamster ovary cells. *Cell* 1990; 60:849–859.
133. Pasqualini R, Bourdoulous S, Koivunen E, Woods Jr V, Ruoslahti E. A polymeric form of fibronectin has antimetastatic effects against multiple tumor types. *Nat Med* 1996; 2:1197–1203.
134. Kochevar G, Stanek J, Rucker E. Truncated fibronectin. *Cancer* 1992; 69:2311–2315.
135. Gottesman S, Wickner S, Maurizi MR. Protein quality control: triage by chaperones and proteases. *Genes Dev* 1997; 11:815–823.
136. Kopito RR. ER quality control: the cytoplasmic connection. *Cell* 1997; 88:427–430.
137. Biederer T, Volkwein C, Sommer T. Role of Cue1p in ubiquitination and dedgradation at the ER surface. *Science* 1997; 278:1806–1809.
138. Hiller MM, Finger A, Schweiger M, Wolf DH. ER degradation of a misfolded luminal protein by the cytosolic ubiquitin-proteasome pathway. *Science* 1996; 273:1725–1728.
139. Mukhopadhyay D, Knebelmann B, Cohen H, Ananth S, Sukhatme V. von Hippel-Lindau tumor suppressor gene product interacts with Sp1 to repress vascular endothelial growth factor promoter activity. *Mol Cell Biol* 1997; 17:5629–5639.
140. Pal S, Claffey K, Cohen H, Mukhopadhyay D. Activation of Sp1-mediated vascular permeability factor/vascular endothelial growth factor transcription requires specific interaction with protein kinase C ζ. *J Biol Chem* 1998; 273:26,277–26,280.

141. Pal S, Claffey K, Cohen H, Mukhopadhyay D. The von Hippel-Lindau gene product inhibits vascular permeability factor/vascular endothelial growth factor expression in renal cell carcinoma by blocking protein kinase C pathways. *J Biol Chem* 1997; 272:27,509–27,512.

142. Golde DW, Hocking WG. Polycythemia: mechanisms and management. *Ann Intern Med* 1981; 95:71–87.

143. Sato K, Terada K, Sugiyama T, Takahashi S, Saito M, Moriyama M, et al. Frequent overexpression of vascular endothelial growth factor gene in human renal cell carcinoma. *Tohoku J. Exp. Med.* 1994; 173:355–360.

144. Takahashi A, Sasaki H, Kim S, Tobisu K, Kakizoe T, Tsukamoto T, et al. Markedly increased amounts of messenger RNAs for vascular endothelial growth factor and placenta growth factor in renal cell carcinoma associated with angiogenesis. *Cancer Res* 1994; 54:4233–4237.

145. Wizigmann-Voos S, Breier G, Risau W, Plate K. Up-regulation of vascular endothelial growth factor and its receptors in von Hippel-Lindau disease-associated and sporadic hemangioblastomas. *Cancer Res* 1995; 55:1358–1364.

146. Iliopoulos O, Jiang C, Levy AP, Kaelin WG, Goldberg MA. Negative regulation of hypoxia-inducible genes by the von Hippel-Lindau protein. *Proc Natl Acad Sci* 1996; 93:10,595–10,599.

147. Knebelmann B, Ananth S, Cohen HT, Sukhatme VP. Transforming growth factor a is a target for the von Hippel-Lindau tumor suppressor protein. *Cancer Res* 1998; 58:226–231.

148. Siemeister G, Weindel K, Mohrs K, Barleon B, Martiny-Baron G, Marme D. Reversion of deregulated expression of vascular endothelial growth factor in human renal carcinoma cells by von Hippel-Lindau tumor suppressor protein. *Cancer Res* 1996; 56:2299–2301.

149. Beelman CA, Parker R. Degradation of mRNA in eukaryotes. *Cell* 1995; 81:179–183.

150. Jacobson A. Interrelationships of the pathways of mRNA decay and translation in eukaryotic cells. *Annu Rev Biochem* 1996; 65:693–739.

151. Izzauralde E, Mattaj I. RNA export. *Cell* 1995; 81:153–159.

152. Stutz F, Rosbash M. Nuclear RNA export. *Genes Dev* 1998; 12:3303–3319.

153. Kolch W, Martiny-Baron G, Kieser A, Marme D. Regulation of the expression of the VEGF/VPS and its receptors: role in tumor angiogenesis. *Breast Cancer Res Treat* 1995; 36:139–155.

154. Levy AP, Levy NS, Wegner S, Goldberg MA. Transcriptional regulation of the rat vascular endothelial growth factor gene by hypoxia. *J Biol Chem* 1995; 270:13,333–13,340.

155. Stein I, Neeman M, Shweiki D, Itin A, Keshet E. Stabilization of vascular endothelial growth factor mRNA by hypoxia and hypoglycemia and coregulation with other ischemia-induced genes. *Mol Cell Biol* 1995; 15:5363–5368.

156. Ikeda E, Achen MG, Breier G, Risau W. Hypoxia-induced transcriptional activation and increased mRNA stability of vascular endothelial growth factor in C6 glioma cells. *J Biol Chem* 1995; 270:19,761–19,766.

157. Claffey KP, Shih S-C, Mullen A, Dziennis S, Cusick JL, Abrams KR, Lee SW, Detmar M. Identification of a human VPF/VEGF 3′ untranslated region mediating hypoxia-induced mRNA stability. *Mol Biol Cell* 1998; 9:469–481.

158. Levy AP, Levy NS, Goldberg MA. Post-transcriptional regulation of vascular endothelial growth factor by hypoxia. *J Biol Chem* 1996; 271:2746–2753.

159. Levy AP, Levy NS, Iliopoulos O, Jiang C, Kaelin WG, Goldberg MA. Regulation of vascular endothelial growth factor by hypoxia and its modulation by the von Hippel-Lindau tumor suppressor gene. *Kidney Int* 1997; 51:575–578.

160. Shih S-C, Claffey KP. Regulation of human vascular endothelial growth factor mRNA stability in hypoxia by heterogeneous nuclear ribonucleoprotein L. *J Biol Chem* 1999; 274:1359–1365.

161. Levy NS, Chung S, Furneaux H, Levy AP. Hypoxic stabilization of vascular endothelial growth factor mRNA by the RNA binding protein HuR. *J Biol Chem* 1998; 273:6417–6423.

162. WJ Ma, Campell CS, Wright A, Furneaux H. Cloning and characterization of HuR, a ubiquitously expressed Elav-like protein. *J Biol Chem* 1996; 271:8144–8155.

163. Levy AP, Levy NS, Goldberg MA. Hypoxia-inducible protein binding to to vascular endothelial growth factor mRNA and its modulation by the von Hippel-Lindau protein. *J Biol Chem* 1996; 271:25,492–25,497.

164. Massague J. TGF-beta signal transduction. *Annu Rev Biochem* 1998; 67:753–791.

165. Ananth S, Kenbelmann B, Gruning W, Dhonabal M, et cal. Transformating growth factor beta 1 is a target for the von Hippel-Lindau tumor suppressor and a critical growth factor for clear cell renal carcinoma. *Cancer Res* 1999; 59(9):2210–2216.

166. Ivanov S, Kuzmin I, Wei M-H, Pack S, Geil L, Johnson B, Stanbridge E, Lerman M. Down-regulation of transmembrance carbonic anhydrases in renal cell carcinoma cell lines by wild-type von Hippel-Lindau transgenes. *Proc Natl Acad Sci* 1998; 95:12,596–12,601.
167. Liao S, Brewer C, Zavada J, Pastorek S, Pastorekova S, Manetta A, et al. Identification of the MN antigen as a diagnostic biomarker of cervical intraepithelial squamous and glandular neoplasia and cervical carcinomas. *Am J Pathol* 1994; 145:598–609.
168. Sly WS, Hu PY. Human carbonic anhydrases and carbonic anhydrase deficiencies. *Annu Rev Biochem* 1995; 64:375–401.
169. Martinez-Zaguilan R, Seftor EA, Seftor RE, Chu YW, Gillies RJ, Hendrix MJ. Acidic pH enhances the invasive behavior of human melanoma cells. *Clin Exp Metastasis* 1996; 14:176–186.
170. Horton WA, Wong V, Eldridge R. Von Hippel-Lindau disease. *Arch Intern Med* 1976; 136:769–777.
171. Maddock I, Moran A, Maher E, Teare M, Norman A, Payne S. A genetic register for von Hippel-Lindau disease. *J Med Genet* 1996; 33:120–127.

13 Wilms' Tumor
A Developmental Anomaly

Aswin L. Menke, PhD and Nick D. Hastie, PhD

CONTENTS

INTRODUCTION
WILMS' TUMOR: A CASE OF DISRUPTED DEVELOPMENT
GENETICS OF WT
WT1 GENE AND ITS PRODUCTS
BIOCHEMICAL FUNCTIONS OF WT1 PROTEINS
BIOLOGICAL ACTIVITIES OF WT1 PROTEINS
DISRUPTION OF WT1 FUNCTION AND TUMORIGENESIS
CONCLUSIONS

INTRODUCTION

Wilms' tumor (WT), or nephroblastoma, is a pediatric kidney malignancy that was first described by the surgeon Max Wilms, in 1899. WT accounts for 8% of all childhood cancers, and is the fifth most common pediatric malignancy after central nervous system tumors, lymphomas, neuroblastomas, and soft tissue carcinomas. The random risk of developing WT is estimated to be 1/10,000 live births, and 90% of all WT patients are younger than 7 yr of age *(1)*. 93% of WT patients develop a tumor in just one of the kidneys; 7% develop tumors in both kidneys *(2)*. In the case of unilateral disease, the median age at diagnosis is 41.5 mo for males and 46.9 mo for females. Bilateral disease is noted to occur earlier, with a median age of 29.5 mo for males and 32.6 mo for females *(3)*.

As a result of the low incidence of the disease, etiologic epidemiological studies of WT have been difficult to carry out *(4)*. Geographical variations in incidence rates, however, have been reported. The highest incidence rate of WT has been found in the Delaware Valley, in the United States, with 13.7 cases per million children (0–14 yrs) per year. The lowest incidence rate in Shanghai, China, with 0.5 cases per million children (0–14 yrs) per year. Within the Delaware Valley, the incidence rate was appreciably higher among blacks (11.9) than among whites (6.2). Overall, the incidence rate of WT is the highest among blacks in the United States and in Nigeria, suggesting that, apart from environmental factors, genetic predisposition may be an important factor in

From: *Tumor Suppressor Genes in Human Cancer*
Edited by: D. E. Fisher © Humana Press Inc., Totowa, NJ

the etiology of WT. Several studies have been undertaken to identify risk factors before conception, during pregnancy, and after birth. A number of putative risk factors have been identified, such as exposure to X-radiation, pesticides, hydrocarbons, lead, and aromatic azo compounds. Other risk factors include the intake of oral contraceptives during pregnancy, antibiotics, alcohol, and coffee. The strongest risk factor identified has been exposure to pesticides. Children can be exposed to pesticides in many ways: insect extermination at home, contact with contaminated clothing of the parents, or via their food. Exposure may already have taken place *in utero*. The fetus may be exposed to substances to which the mother was exposed even before conception. A study of parental age in relation to WT development showed a relative risk of 1.4 for children born of mothers over 40 yr of age, compared to mothers younger than 20 yr of age. In addition, mothers of children with bilateral WT are older than the mothers of children with unilateral WT *(5)*. Whether the increased risk with age is the result of accumulation of harmful substances in the mother remains to be elucidated.

WT is treated with surgery, radiation, and chemotherapy (CT), depending on the stage of the disease at diagnosis. The first nephrectomy was performed by Jessop in 1877, and, although the patient died several months after the operation, his treatment laid the foundation for surgical management of WTs. In 1898, a first attempt to treat renal tumors pharmacologically was described by William Coley, and, in 1916, Alfred Friedlander reported a partial response to radiation therapy. Despite these developments, still no more than 10% of WT patients survived the disease in the 1920s. In 1956, the first specific anti-WT agent, dactinomycin, was described, followed by vincristine in 1963. In the late 1960s, patients were treated with a combination of surgery, radiation, and CT. Randomized clinical trials conducted by multidisciplinary groups, such as the National Wilms' Tumor Study (NWTS) and the International Society of Pediatric Oncology, further improved the treatment regimes, establishing the current survival rate for WT at over 80% *(6–8)*.

The treatment of WT is a success story. However, with the improved survival rate, the side effects of treatment have also become more apparent. Depending on the intensity of the CT and radiation treatment, the patient may develop a whole range of abnormalities (Table 1;9). Current studies are therefore focused on simplifying, shortening, and refining the therapy given to WT patients. To fine-tune the treatment regimes, it will be important to refine the staging of the disease. The identification of prognostic factors will be one step in the right direction. The development of other treatments more specific than CT and radiation will be another step. For both steps, it will be important to unravel the underlying mechanism of the disease, which will hopefully lead to a 100% cure rate.

2. WT: A CASE OF DISRUPTED DEVELOPMENT

WT is of embryonic origin, and is believed to be derived from metanephric blastemal tissue of the developing kidney that failed to undergo the normal maturation process *(10)*. Therefore, to understand the development of the tumor, it is important to understand the complex process of normal kidney formation *(11–13)*. The kidney is formed through the reciprocal interaction of two tissues: the metanephric mesenchyme and the ureteric bud epithelium. Similar mesenchyme–epithelium interactions have been described during the development of many other organs *(14)*.

Table 1
Late Effects of WT Treatment

Treatment	Affected tissue	Effect	Years[a]
Radiation/chemotherapy	Heart	Cardiotoxicity	
Radiation/chemotherapy	Gonads	Delayed pubertal development	
Radiation	Thyroid gland	Thyroid carcinoma	6
Radiation/chemotherapy	Liver	Hepatomegaly	
		Ascites	
		Jaundice	
		Hepatoma	0–13
Radiation	Intestine	Bowel obstruction	
		Colon carcinoma	11–26
Radiation	Lung	Reduced lung volume	
		Mesothelioma	16
Radiation	Skeletal system	Asymmetry vertebral bodies	
		Vertebral end plate irregularities	
		Scoliosis	
		Kyphosis	
		Platyspondyly	
		Hypoplasia ileum	
		Osteochondroma	5–16
Chemotherapy	Nervous system	Loss of reflexes	
		Muscle weakness	
		Gait disturbances	
		Sensory loss	
		Jaw pain	
		Neuropsychological abnormalities?	
Radiation		Neurofibrosarcoma	16
Radiation	Kidney	Nephritis	
		Renal cell carcinoma	21
Radiation	Hematopoietic system	Acute myeloid leukemia	3–17
Radiation		Acute lymphoid leukemia	4–5
		Chronic granulocytic leukemia	9
Radiation		Breast carcinoma	16

[a] Indicates the number of years between WT detection and the development of the second malignancy.

The mammalian kidney is derived from the intermediate mesoderm, which is first distinguishable around E18 in humans and E7.5 in mice. The first event in the differentiation of the intermediate mesoderm is the formation of the wolffian duct (or nephric duct) in a rostral–caudal direction (Fig. 1). The duct runs parallel to a longitudinal band of mesoderm, called the nephrogenic cord, which itself arises from the intermediate mesoderm. As the duct elongates, it induces a series of tubules in the nephrogenic cord. At first the pronephric tubules are formed around E22 in humans and E8 in mice. The pronephroi are transitory, nonfunctional, and analogous to the kidneys of primitive fish. Caudal of the pronephros, the mesonephroi are induced by

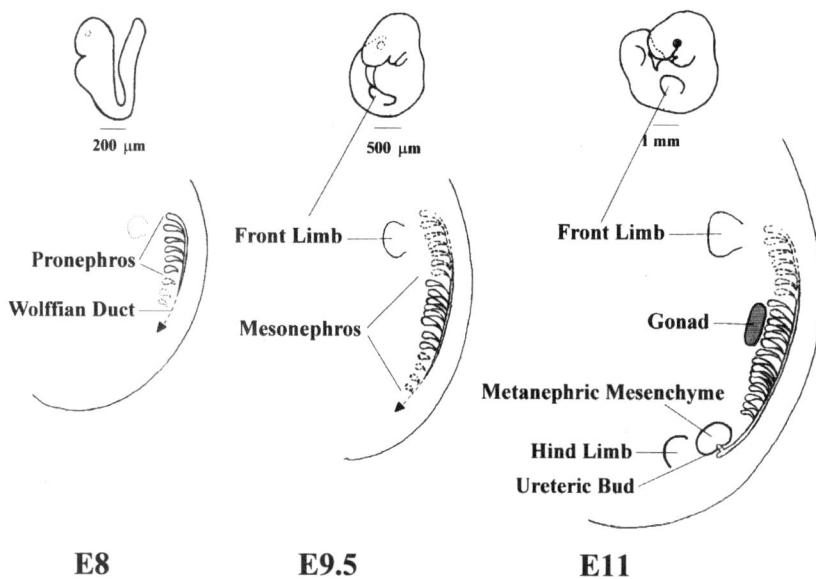

Fig. 1 An overall scheme of the development of the mouse kidney *(11,12)*.

the wolffian duct around E24 in humans and E9.5 in mice. The mesonephroi range from simple epithelium, more rostrally, to longer, convoluted tubules, complete with glomeruli, more caudally. The mesonephros is the permanent kidney in amphibians, and may function transiently in mammals. The human mesonephros disappears in females in the third month of development. In males, some caudal tubules and the wolffian duct persist, and contribute to the male genitals (efferent ductules and vasa deferentia).

Around E35 in humans and E11 in mice, the metanephric mesenchyme induces the wolffian duct, and the ureteric bud evaginates from the caudal end of the wolffian duct, and invades the metanephric mesenchyme and branches (Fig. 2A). In turn, the mesenchymal cells aggregate, undergo a burst of proliferation, form the blastema around the ureteric bud, and undergo an epithelial conversion (Fig. 2B). These early epithelial cells form a spherical cyst, called the renal vesicle. The renal vesicle matures further, via the comma- and s-shaped bodies, into a functional nephron, which consists of the glomerulus, the proximal convoluted tubule, the loop of Henle, and the distal convoluted tubule. Mesenchymal cells, which form a ring of cells around the induced metanephric mesenchyme, fail to generate epithelium, and are thought to give rise to stromal cells, or to undergo programmed cell death *(15)*.

2.1. Nephrogenic Rests and WT

In humans, normal kidney development is usually complete by 36 wk of gestation. Groups of renal blastema cells that persist after this period are known as nephrogenic rests, which are considered to be potential precursor lesions of WT. Nephrogenic rests are found to be associated with up to 40% of unilateral WTs and nearly 100% of bilateral WT *(16)*. In comparison, nephrogenic rests are found in only 0.2–0.95% of routine

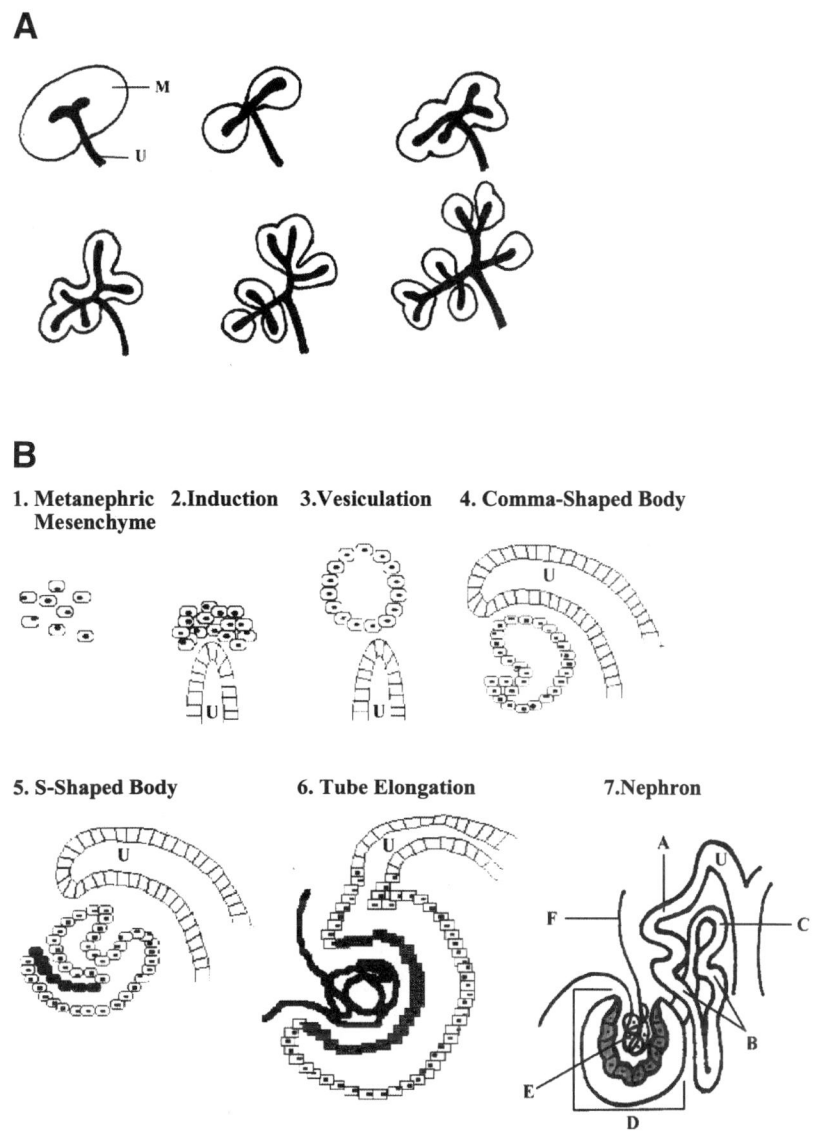

Fig. 2 Schematic diagrams representing stages of nephrogenesis. **(A)** Branching of the ureteric bud and mesenchymal condensation with its subsequent splitting into secondary and teriary condensates *(11)*. U = Ureter; M = Metanephric mesenchyme. **(B)** Upon induction by the ureteric bud *(1)*, the metanephric mesenchyme cells aggregate, ndergo a burst of proliferation and dorm the blastema around the uterteric bud *(2)*. These early epithelial cells develop into the renal vesicle *(3)* which further matures via the comma-*(4)* and s-shaped bodies *(5)* into a functional nephron *(6,7)*. The nephron consist of the distal convoluted tube (A), the loop of Henle (B), the proximal convoluted tube (C), and the glomerulus (D). Capillaries and megangial cells (E). Blood vessels, vas efferens, and vas efferens (F). The gray cells develop into podocytes.

pediatric autopsies *(17)*. Nephrogenic rests can be divided in two major subgroups: perilobar nehrogenic rests (or multifocal superficial nephroblastomatosis), which are the most common reported, and intralobar rests (or multifocal deep cortical nephroblastomatosis) *(10,16)*.

Perilobar rests are discontinuous collections of metanephric blastemal cells and differentiated derivatives in the form of renal tubules and stroma. They are located at the periphery of the lobules, that is, in the subcapsular region and in the columns of Bertin. There are several histological subtypes of perilobar rests that often co-exist in the same kidney:

1. Nodular renal blastema, which consists of nonmitotic nodules (diameter up to 300 μm) that can only be detected by microscopic examination. These are encapsulated groups of blastemal cells that sometimes show a mild epithelial differentiation.
2. Metanephric hamartomas, which are usually larger lesions. These are always differentiated, most commonly by the formation of epithelial tubules, but sometimes there are differentiated stromal elements. They show little or no mitotic activity.
3. Wilms' tumorlets, which are neoplastic nodules, as described under[2], that have reached a diameter of 1 cm or more. These nodules are monomorphous, consisting of primitive epithelium, in a rosette pattern or with some evidence of tubular differentiation. They do not have a malignant stromal pattern. The tumorlets are well circumscribed, and do not appear to infiltrate surrounding tissues.

Intralobar nephrogenic rests are lesions distributed at or near the corticomedullary junction, and may more readily be mistaken for WT proper. The nodules are of various sizes, and may be confluent. The lesions are ill-defined, often appearing to infiltrate into adjacent renal tissue. The lesions show a high mitotic rate. In addition, they have a significant stromal component, and may mimic triphasic WT, which contains primitive blastemal cells, more differentiated epithelial cells, and stromal cells.

WTs can be subdivided according to their association with nephrogenic rests. WTs, associated with residual intralobar nephrogenic rests, can be composed of tissues mimicking the whole range of nephrogenesis. They have a prominent stromal component, and sometimes contain heterologous stromal elements, such as striated muscle, cartilage, and bone. WTs associated with perilobar nephrogenic rests are mostly composed of blastemal and epithelial elements, mimicking later stages of nephrogenesis.

WTs may also be subdivided according to their histology: favorable or unfavorable. The latter subset is noted to have a poorer prognosis, and can be further divided into two categories: sarcomatous and anaplastic. Sarcomatous tumors include rhabdoid tumors and clear cell carcinomas, but, today, these tumors are classified separately from WT *(6)*. Anaplastic tumors comprise about 4.8% of all WTs *(18)*, and are characterized by the presence of hyperchromatic nuclei that are 3× larger than those of non-neoplastic cells, and have multiple mitotic figures.

3. GENETICS OF WT

WT has been the subject of intense clinical and basic scientific research, because it represents a model for cancer treatment, a model for the relationship between development and cancer, and it illustrates the impact of genetic alterations on development and tumorigenesis. As a genetic model, WT exhibits many of the same complexities as adult-onset tumors, such as genetic heterogeneity, and incomplete penetrance of predisposing mutations. However, unlike most adult tumors, WTs are in general euploid, and display only a very low (< 5%) frequency of loss of heterozygosity (LOH) throughout the genome *(19–21)*. This low frequency of somatic alterations suggests that the few alterations that are observed are likely to be significant in tumorigenesis.

WT develops in otherwise healthy children, but, in about 10% of the cases, the tumor occurs in individuals with recognized syndromes *(22,23)*. The presence of WTs in these syndromes provided a starting point for investigating the genetics of WT. They can be classified into overgrowth syndromes and nonovergrowth syndromes *(6)*. Overgrowth syndromes are the result of excessive prenatal and postnatal somatic growth, and result in macroglossia, nephromegaly, and hemihypertrophy. The two most common overgrowth disorders associated with WT are the Beckwith-Wiedemann syndrome (BWS) and isolated hemihypertrophy *(22)*. Others include the Simpson-Golabi-Behemel syndrome, a fetal overgrowth disorder that results from mutations in the glypican 3 gene at Xq26 *(24)*; the Perlman syndrome, a rare disorder of renal dysplasia, fetal gigantism, and multiple congenital abnormalities *(25)*; and the Sotos syndrome *(22)*.

Nonovergrowth syndromes associated with WT development include isolated aniridia (malformation or absence iris); aniridia in combination with urogenital abnormalities and mental retardation; urogenital malformations; the Denys Drash syndrome (DDS) *(22,26,27)*; trisomy 18 *(28)*; neurofibromatosis type 1, which is caused by mutations in the *Nf1* gene at 17q11 *(29)*, the Li-Fraumeni syndrome, which is attributed to mutations in the *p53* gene at 17p13 in some families *(30)*; the Bloom syndrome, a disorder of DNA repair caused by defects in the *BLM* gene at 15q26 *(31,32)*; the hereditary hyperparathyroid–jaw tumor syndrome, which maps to 1q21–q31 *(33)*; and the breast–ovarian cancer syndrome, which is caused by mutations in *BRCA1* in 17q21 in many families *(34)*. These observations imply that the genetic alterations observed in these syndromes may also play a role in the development of WT.

3.1. Wilms' Tumor 1 (WT1) Gene

In 1964, Miller et al. reported for the first time the association between aniridia and WT. Aniridia occurs in 1/70,000 children; WT arises in 1/10,000 children *(26)*. Despite the rarity of these two conditions, aniridia is detected in 1/70 children with WT and WT is detected in 1/3 children with aniridia.

Aniridia is often accompanied by varying degrees of mental retardation and genitourinary abnormalities. The developmental abnormalities of the genitourinary tract range from common conditions, such as hypospadias and cryptorchidism, to more extreme abnormalities, such as renal hypoplasia or agenesis, horseshoe kidney, ureteral atresia, or bifid ureters *(2,35)*. In cases in which the patient also develops WT, the syndrome is described as the WAGR-syndrome (WT, aniridia, genitourinary abnormalities, mental retardation). Cytogenetic analyses of these patients showed a common congenital deletion of chromosome (chr) band 11p13 *(36,37)*. Several studies indicated that the locus involved in aniridia was different from the locus involved in WT. A translocation, disrupting 11p13, was found in two families with hereditary aniridia *(38,39)*. The translocated chr segregated with aniridia, but the affected individuals did not develop WT. In addition, Turleau et al. *(40)* described a boy with a proximal 11p13 deletion, WT, and genitourinary abnormalities, but without aniridia or mental retardation. Molecular analysis of two other patients (one with WT and one with aniridia) showed that the loci for aniridia and WT were indeed distinct *(41)*.

The discovery of a sporadic WT, with two nested deletions within 11p13 *(42)*, made it possible to identify a genomic fragment of 350 kb that was homozygously deleted in this tumor *(43)*. Using a library of human genomic DNA derived from a somatic cell hybrid, clones were identified that were homozygously deleted in this tumor, and which showed cross-species hybridization. These genomic clones were then used to isolate a series of cDNAs, encoding a gene now called the Wilms' tumor 1 (WT1) gene (44). Simultaneously, using a chr-jumping cloning technique, the same gene was isolated by another group (45).

Several lines of evidence suggest that the WT1 gene plays an important role in the etiology of WT. First, WT1 is expressed in the developing human, rat, and mouse fetal kidney. The spatial and temporal expression of WT1 is in agreement with the hypothesis that disruption of its function leads to tumorigenesis (46–48). Second, WT1 is found to be mutated in 94% of all DDS patients (49). The DDS is a rare disease that encompasses three abnormalities: WT, nephropathy involving mesangial sclerosis, and genital/intersex anomalies. DDS patients were shown to be constitutionally heterozygous for WT1 mutations (50), with the mutation reaching homozygosity/hemizygosity in the WTs of almost all cases (51). Further evidence for the involvement of WT1 in WT development came from whole-animal experiments. Rats, injected with the alkylating agent, N-nitroso-N-methyl-urea (NMU), developed renal tumors that resemble WT, and WT1 was found to be mutated in almost 50% of these tumors (52). In another study, mice were generated in which one WT1 allele was targeted to make a specific DDS mutation (53). As a result, both heterozygous and chimeric adult mice showed mesangial sclerosis, male genital defects, and one mouse developed WT. In this tumor, the transcript of the nontargeted allele showed an exon 9-skipping event, implying a causal link between WT1 dysfunction and Wilms' tumorigenesis.

In humans, however, approx 90% of all WTs are sporadic, and only 5% of these show intragenic WT1 mutations. Cytogenetic, comparative genomic hybridization, and LOH studies suggest that, apart from WT1, several other genes may be involved in the development of WT.

3.2. WT 2,3,4,5. … Gene?

Patients with the BWS have an increased predisposition to develop cancer. About 4% of these patients develop WT; about 3.5% develop different types of childhood tumors, including adrenocortical carcinoma, hepatoblastoma, rhabdomyosarcoma, and neuroblastoma. Genetic evidence shows that defects in chr region 11p15.5 are associated with the development of both BWS and WT (54,55). Structural change, involving the short arm of chr 11, is one of the most common abnormalities found in WT. 11p variations have been reported for approx 30% of all WTs studied (20,56,57), and LOH is limited to 11p15 in 50% of the informative cases (56). The preferential loss of maternal 11p15 alleles in sporadic WT suggests that genomic imprinting may be involved in the development of the disease. The 11p15.5 region contains several genes that are imprinted, including IGF-II, H19 (58–62), IPL (63), KVLQT1 (64), and ORCTL (65). IGF-II is normally expressed from the paternal-derived allele, and acts as a mitogenic factor in a wide range of cell types (66). H19 is expressed from the maternally expressed gene, and is believed to play a role in growth suppression, directly or indirectly, by altering the expression of IGF-II (67–70). The cyclin-dependent kinase inhibitor (CDKI), p57^{KIP2}, is preferentially expressed from the maternally derived allele, and can induce cell cycle arrest in G1 (62,71). Loss of imprinting (LOI) and overexpression of IGF-II have been found in many WTs (72–74). In contrast, expression of p57^{KIP2} and H19 is repressed in a majority of WTs (60,75). Of all the genetic subclasses of the BWS, patients with paternal uniparental disomy, affecting 11p15, have the greatest incidence of tumors *(76)*. These patients express two copies of all paternally expressed imprinted genes (e.g., *IGF-II*); the maternally expressed genes (e.g., *H19* and *p57^KIP2*) are lost. This may result in aberrant proliferation. Mice have been generated that lack either *H19* or *p57^KIP2*, or which overexpress *IGF-II (68, 77–79)*. These mice exhibit most phenotypes of the BWS, but do not develop WTs. It is possible that the combination of *H19* and p57^KIP2 underexpression and *IGF-II* overexpression is necessary in order to see WT development.

Alternatively, other genes, like *IPL, KVLQT1, ORCTL,* or *NAP2 (80),* play a more important role in WT development than *H19, p57^KIP2,* and *IGF-II*. Recently, using constitutional and tumor DNA from 38 WT patients, the location of a putative WT gene within 11p15 has been refined to a region of about 800 kb *(81)*. Both *NAP2* and *ORCTL* map to this region. In addition, a novel locus at 11p15 was identified, which spans a distance of approx 336 kb. Integrin-linked kinase *p59^ILK* maps to this region. This gene has previously been shown to induce anchorage-independent growth and a tumorigenic phenotype in rodents *(82)*.

Genetic alterations of genes in 11p15 may not result in development of WT *per se.* The alterations may, however, predispose to the development of WT, e.g., overexpression of growth-promoting *IGF-II* and underexpression of the growth inhibitors *H19* and *p57^KIP2* may lead to overgrowth, which increases the chances of disruption of normal kidney development, leading in some cases to the development of WTs. The observation that *H19* expression is lost in nephrogenic rests supports this hypothesis *(83)*. The fact that WT can develop in overgrowth syndromes other than BWS, which are not linked to 11p15, also supports this hypothesis. In addition, other tumor types also display 11p15 LOH and LOI, such as rhabdomyosarcoma *(84,85)*, breast cancer *(81,86)*, ovarian carcinoma *(87)*, stomach carcinoma *(88)*, hepatoblastoma *(89)*, and glioma *(90)*. This suggests that altered expression of 11p15 genes plays a more widespread role in tumorigenesis.

Analyses of more than 500 WTs indicate that, apart from 11p13 and 11p15, several other chr regions may be involved in the development of WT *(20,56,57,91–93)*. In WT, chr gains are more common than chr loss. The most common gains involve chr 1q (~18%), 6 (~15%), 8 (~21%), 12 (~39%), 13 (~14%), 17 (~13%), 18 (~12%), and 20 (~14%). Losses are mostly detected in 1p (~10%), 4q (~5%), 11q (~10%), 17p (~4%), 16q (~14%), and 22q (2–14%). Because these changes are found in many other tumors, they could well be secondary events associated with tumor progression.

The LOH of 16q may be such a secondary event. Patients with tumor-specific LOH for chr 16q had a statistically poorer survival than those without LOH in this region *(57,93)*. Loss of 1p was also associated with a worse outcome, although this difference did not reach statistical significance *(57)*. The prognostic value of alterations in both 1p and 16q is currently being analyzed in the NWTS 5 *(94)*. A potential candidate gene at 16q is the uvomorulin gene, which maps to 16q22.2, and lies within the region of loss in WT *(57)*. Adhesion molecules, such as uvomorulin, may function as tumor suppressors. Concerning LOH in 1p, different translocation breakpoints in 1p35 have been detected in WT cells (Mannens, M, personal communications). The WT breakpoint has currently been located within a region of 1 CM.

Cytogenetic abnormalities involving chr 7p have been reported in about 13% of WTs *(93,96,97)*. Although the numbers are very small, there may be an association between LOH of markers on 7p with earlier onset and more advanced staging of WT *(93)*.

Loss of 17p has been detected in approx 4% of the WTs analyzed. The *p53* gene maps to region 17p13, and is frequently found to be mutated in anaplastic WTs *(98,99)*. Anaplasia is a feature of WT that is associated with resistance to CT *(17)*. Anaplastic WT represents 4.8% of WT cases *(18)*, a percentage similar to the LOH in 17p. Anaplastic changes are confined to specific regions of the primary tumor. Screening for LOH of 17p or expression of *p53* may help the pathologist to access anaplasia, and in this way determine the intensity of the treatment. Patients with WTs with diffuse anaplasia have been shown to benefit from more aggressive CT *(100)*.

3.3. Familial WT Genes

Although WT is usually sporadic, approx 1% occurs in families in which susceptibility appears to be inherited as an autosomal dominant trait with incomplete penetrance *(101)*. Utilizing comparative genomic hybridization, eight samples of familial WT showed several chr abnormalities *(102)*. As previously reported for sporadic WTs, gains were most frequently found in chr 6, 8, and 12 *(56,72)*. Most frequent sites of loss were found in chr 3, 4, 9, 16, and 20.

Linkage analyses have been performed in several WT families (Table 2). The analysis of some families excluded the possibility of a predisposing mutation in the previously described regions: 11p13, 11p15, and 16q *(103–105)*. However, three other families have been described in which the father, carrying a germ-line *WT1* mutation, transmitted the mutation to an offspring who developed WT *(107–109)*. The tumor showed loss of the wild-type (WT) allele, in agreement with the classical two-hit model by Knudson and Strong *(110)*. Two of the fathers were unaffected; the third had been successfully treated for WT. In one family, WT was linked to chr region 17q12–q21 *(105)*. A putative candidate gene in this region is the insulin-like growth factor binding protein 4 gene *(IGFBP4)*. IGFBP4 is a major binding protein for *IGF-II (111)*. Since deregulation of *IGF-II* expression is implicated in WT, IGFBP4 may play an important role in the development of the disease.

In five other families, WT was linked to chr region 19q13.3–q13.4 *(112)*, which harbors several potentially relevant genes, such as the apoptosis regulator *bax (113,114)*, the DNA repair genes, *Ercc1* and *Ercc2 (115)*, and the G0/G1 switch regulator, *Fos-B (116)*.

3.4. Working Model of WT Development

From the above paragraphs, it is clear that the genetics of WT is complicated and involves many different genes. Although many of these genes have not yet been cloned, they can be divided into three classes:

1. Genes whose disruption increases the chance of developing any type of tumor, including WT. This group encompasses genes that are involved in WT associated nonovergrowth syndromes and overgrowth syndromes. For example, disruption of the *BLM* gene, in the nonovergrowth Bloom syndrome, results in impaired DNA repair *(31,32)*, and loss of 11p15 in the overgrowth BWS may result in a general growth advantage. In agreement with the latter, analyses of both tumors and associated nephrogenic rests suggest that loss of 11p15 may be an early event in the pathogenesis of WT *(91,117)*.

Table 2
Linkage Analyses of WT Families

Family	16q	19q	11p13	11p15	17q	Refs.
WTX502	NL	–	NL	NL	–	104,306
WTX524	NL	L	NL	–	–	103,112
WTX637	NL	L	–	–	–	306,112
WTX149	?	NL	–	–	–	306,112
WTX593	NL	L	–	–	–	306,112
WTX917	–	L	–	–	–	112
WTX614	–	L	–	–	–	112
WTX668	–	NL	–	–	–	112
WTX480	NL	–	NL	NL	L	105
K1104	–	–	–	NL	NL	106

L, linkage; NL, No linkage.

2. Genes whose disruption plays a role in tumor progression, e.g., loss of genes on 16q and 1p appear to have an adverse effect on the relapse-free period and overall survival of the WT, patient (57,93). In agreement, analysis of WTs and associated nephrogenic rests suggest that LOH of 16q and 1p are late events in the pathogenesis of WT (91,117).

3. Genes that play a direct role in the disruption of normal nephrogenesis leading to WT. It appears that different chr abnormalities may lead to a similar phenotype, e.g., trisomy 12 has been reported as the only cytogenetic abnormality in one WT, and so have trisomy 6 and trisomy 18 (56). In order to explain this phenomenon, one must assume that all chr abnormalities disrupt the same molecular pathway. Concerning the molecular pathway, only one gene has been cloned that plays a direct role in the etiology of WT: the *WT1* gene. The authors therefore hypothesize that many of the observed chr abnormalities will affect genes that are part of the *WT1* pathway. Understanding the regulation and the function of the *WT1* gene will help to unravel the *WT1* pathway, and to identify the genes involved.

4. WT1 GENE AND ITS PRODUCTS

The *WT1* gene is localized at human chr 11p13. It contains 10 exons and spans about 50 kb (44,45; Fig. 3). Comparison of partial nucleotide sequences of the *WT1* gene, from chicken, alligator, *Xenopus laevis,* zebrafish, and pufferfish, reveals extensive conservation, suggesting an important role of the gene throughout evolution (118,119). The gene mainly encodes a transcript of about 3 kb, but transcripts of 2.5 and 1.8 kb have been reported (47,48,120). Translation of the *WT1* transcript reveals a protein that contains four zinc fingers of the Kruppel C2-H2 class in the carboxy (C)-terminus, and which has a proline/glutamine-rich amino (N)-terminus, features that are commonly found in bona fide transcription factors (TFs), such as Egr-1, Sp1, and CTF/NF1 (121,122).

As a result of alternative RNA splicing, each transcript can encode for four different proteins (123) with mol wt of 52–54 kDa (124). One alternative splicing event results in either inclusion or exclusion of exon 5, which encodes a stretch of 17 amino acids (±17 aa) just N-terminal of the four Zn fingers. The other event involves a selection of

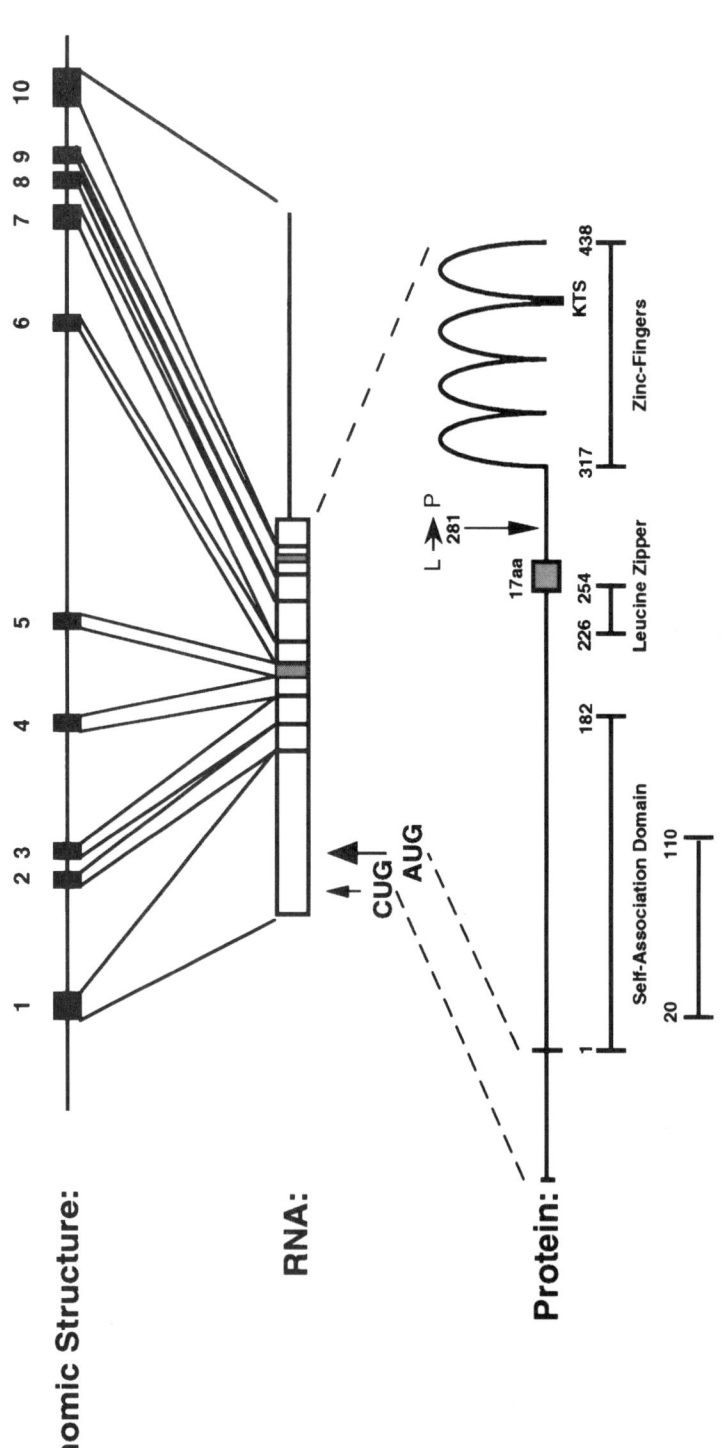

Fig. 3 Schemetic structure of the *WT1* gene, mRNA, and protein products.

alternative splice-acceptor sites in exon 9, resulting in the presence or absence of a 3-aa insert (lysine, threonine, serine [±KTS]) between Zn finger three and four. The presence of the KTS insert in chicken, alligator, and marsupial, and the conservation of the genomic sequence in the genomic DNA of the zebrafish, underlines the importance of this alternative splice form in vertebrate development *(118)*. The 17-aa insertion, however, is only present in mammals. Recently, an alternative non-AUG translational initiation site (CUG) was identified upstream of the major AUG translational initiation site, resulting in WT1 proteins with a higher mol wt *(125)*. Downstream of the major AUG translational initiation site, there may be yet another translation initiation site, resulting in WT1 proteins with a lower mol wt (Scharnhorst V, personal communication). Together with the observed RNA editing, leading to the replacement of a leucine by a proline at position 281 *(126)*, there could be as many as 24 different WT1 isoforms.

Little or no research has yet been done on the function of the larger or smaller WT1 proteins derived from the alternative translation initiation sites. The WT1 proteins discussed in this review are therefore solely derived from the major AUG translation initiation site. The authors refer to the WT1 splice variant, which lacks both inserts, as the WT1–/– isoform; the splice variant that only contains the 17-aa insert, as the WT1+/– isoform; the splice variant that only contains the KTS insert, as the WT1–/+ isoform, and the splice variant that contains both inserts, as the WT1+/+ isoform. Occasionally, the authors refer to the splice variants lacking or containing the KTS insert as the WT1(–KTS) or the WT1(+KTS) isoforms, respectively. Similarly, the authors refer to the splice variants lacking or containing the 17-aa insert isoforms as the WT1(–17aa) and the WT1(+17aa) isoforms, respectively.

5. BIOCHEMICAL FUNCTIONS OF WT1 PROTEINS

5.1. WT1 Proteins May Function as TFs

Because the predicted protein sequence suggested the WT1 isoforms to be transcription factors, several studies focused on identifying their target genes. Binding site selection with oligonucleotides, whole genomic polymerase chain reaction, and DNase I footprint analysis (127–131) revealed several WT1 DNA-binding sites. Extrapolation of these data, toward known promoter sequences, led to the identification of a number of putative WT1-target genes (Table 3). In support of the possible transcriptional regulation of these genes by WT1, it was found that WT1 could affect their promoter activity in transient transfection assays. However, whether these data reflect the normal physiological situation remains to be elucidated. As has been pointed out by Hastie *(132)*, crystallographic and biochemical studies predict that WT1 should interact with a 12-nucleotide G-rich stretch of the form GNGNGGGNGNGN *(133)*. These G-rich binding sequences are found in promoter regions of the many genes that have CpG islands. These include housekeeping genes, and it seems unlikely that WT1 regulates all these genes. Furthermore, it has been shown that the choice of the cell system *(134)*, the type of expression vector *(135)*, or the exact topology of the reporter construct *(136)* can influence the transcription-regulating activity of the WT1 proteins.

What role the different splice variants play in the regulation of gene expression is currently unknown. Initially, it was found that only the splice variants lacking the KTS insertion could bind to Egr-1-like consensus sequences *(127)*, and repress the transcription of reporter constructs containing these sequences in their promoter region *(137)*.

This suggested that the WT1 (–KTS) and the WT1(+KTS) isoforms may regulate different target genes. Later on, DNA sequences were identified to which both the WT1(+KTS) and the WT1(–KTS) forms could bind, albeit with different affinities *(128)*. It was concluded that both splice variants may differentially regulate the same target genes. The latter hypothesis was supported by studies showing that both the WT1(+KTS) and the WT1(–KTS) proteins could bind overlapping DNA sequences in the promoters of the *IGF-II* gene *(138)*, the gene for the Platelet-derived growth factor, a chain (PDGFA) *(136)*, the *WT1* gene *(139)*, and the *PAX-2* gene *(140)*. The presence or absence of the 17-aa stretch does not seem to affect the DNA-binding activity. Instead, this insert seems to have a transcriptional repressor function in addition to more N-terminally located sequences *(136,137,141)*. The WT1+/+ protein suppresses the activity of the WT1 promoter about 25-fold better than the WT1–/+ protein, in transient transfection assays *(139)*. It has even been observed that the WT1+/+ protein represses transcription of a certain modified PDGF-A promoter construct, but the WT1–/+ protein activates transcription *(136)*. Altogether, these data suggest that the WT1 splice variants may function as TFs with different transactivating activities on (partly) different target genes.

5.2. WT1 Products May Function as Co-factors

The WT1 isoforms may not function as TFs *per se,* but may exert their effect on transcription by binding to other proteins. Using the two-hybrid system and in vitro association, WT1–/– was found to bind steroid factor 1 (SF1) *(142)*. Transient transfection assays, using an artificial mullerian inhibiting substance (MIS)-promoter/Luc reporter construct, showed that WT1(–KTS) isoforms function as co-factors when cotransfected with SF1. The transcriptional activation was increased by more than 20-fold, compared to SF1 alone; WT1(–KTS) itself had no effect, and did not bind to the MIS promoter.

In another study *(143)*, WT1 was found to bind to the hsp70 protein, and expression of WT1 in U2OS cells led to the induction of hsp70. The (HSE) site in the hsp70 promoter was identified as being required for this induction. Transcriptional activation of hsp70 was observed following transfection of wild type WT1 as, well as the C-terminal deletion mutant WT1 delZ, demonstrating that this effect was independent of WT1's DNA-binding capacity. In addition, gel retardation assays showed no binding of WT1 to the HSE consensus sequence. These observations were consistent with current models of HSE-dependent hsp70 induction, suggesting that binding of WT1 to hsp70 induces the release of HSF from its complex with hsp70, resulting in HSF-mediated induction of hsp70 expression.

5.3. WT1 Proteins May Function as Posttranscriptional Regulators

In addition to their role as TFs, the WT1 proteins may regulate gene expression at the posttranscriptional level. It appears that Zn finger motifs, as are present in the WT1 protein, may not simply be DNA-binding modules, but may behave in a more complex multifunctional manner *(144)*, e.g., the *X. laevis* TFIIA protein contains nine Zn fingers, which are involved in transcriptional activation of the 5S RNA promoter, but also bind specifically to 5 SRNA *(145)*, implying a role in both gene transcription and RNA metabolism. The human and mouse MOK2 proteins have 10 and 7 Zn fingers, respectively, and show a similar ability to bind to both DNA and RNA *(146)*. It has been shown that the expression of reporter gene constructs, containing promoter 2 and exon 2 sequences of *IGF-II,* is downregulated when WT1-binding sites are present within 5′ untranslated leader sequences *(147).* Nuclear run-on assays indicate that this downregulation is mediated by posttranscriptional events. In agreement with such a function, both the WT1(+KTS) and the WT1(–KTS) proteins have been shown to bind to *IGF-II exon 2 RNA (148).* Whether this binding indeed downregulates the expression remains to be elucidated.

WT1 may also play a role in RNA splicing. Using M15 cells (WT1-expressing mouse mesonephros cells), Larsson et al. *(149)* observed that the subnuclear localization of WT1 is very similar to the clusters of interchromatin granules that contain components of spliceosomes *(150–153).* Subsequent immunofluorescence showed that WT1 and snRNPs indeed co-localize; disruption of the spliceosomes resulted in the increased concentration of both snRNPs and WT1 in similar enlarged foci. Nuclear staining of COS7 cells, transfected with four different WT1 splice variants, revealed that the WT1(+KTS) proteins localize in the nucleus in a more speckled pattern, compared to the diffuse pattern of the WT1(–KTS) isoforms. The WT1(+KTS) proteins co-localized with proteins of the splicing machinery, suggesting a role for these WT1 isoforms in splicing, whereas the WT1(–KTS) isoforms co-localized with TFs such as Pax-6 and Sp1. These differences in subnuclear localization may not always be observed. In adenovirus-transformed baby rat kidney cells *(154)* or hepatoma cells *(155),* the WT1(+KTS) isoforms were not present in clusters, and showed the same localization as the WT1(–KTS) isoforms. In osteosarcoma cells, the WT1(+KTS) isoforms did localize in clusters, but the pattern looked different from that in COS7 cells *(156).* In the latter study, the WT1(+KTS) isoforms did not co-localize with SC35, a spliceosome assembly factor that is required for the initial step of pre-mRNA splicing. WT1 mutants with a disrupted DNA-binding domain localized to the same clusters as the WT1(+KTS) isoform, and the authors therefore concluded that the observed subnuclear clusters may represent storage sites for WT1 isoforms, with reduced DNA-binding activity.

However, in support of a role for WT1 in splicing, an evolutionary conserved N-terminal RNA recognition motif has been identified in all known WT1 isoforms *(157).* In addition, Davies et al. *(158)* recently showed, with two-hybrid analysis, in vitro binding assays, and in vivo immunoprecipitation, that WT1 interacts with an essential splicing factor, U2AF65, and gel-filtration experiments indicate that the WT1 protein is present in spliceosomes. Furthermore, Ladomery et al. (in preparation) found that WT1 is enriched by oligo(dT) chromatography, as is U2AF65, suggesting again a role for WT1 in RNA processing. In order to prove that WT1 is indeed involved in RNA splicing, a functional assay must be developed. This may prove to be rather difficult, since WT1 may only affect the splicing of specific target RNAs, and no target RNAs have been identified thus far. Recently, Bardeesy and Pelletier *(159)* have tried to identify RNA-binding sites by systematic evolution of ligands by exponential enrichment. Using bacterially produced, truncated WT1 (–KTS) protein, several sequences were identified. The sequences were found in a number of genes, but their physiological significance remains to be established.

6. BIOLOGICAL ACTIVITIES OF WT1 PROTEINS

WT1 appears to function at three different stages of kidney development. The onset of nephrogenesis, the progression of nephrogenesis, and the maintenance of normal podocyte function. As described in subheading 2, the kidney is formed through the reciprocal interaction of two tissues: the metanephric mesenchyme and the epithelium of the ureteric bud (Fig. 2). Upon induction by the ureteric bud, the mesenchymal cells undergo a burst of proliferation, aggregate, and form the blastema around the ureteric bud. The blastema develops into the renal vesicle, and matures further, via the comma-and S-shaped bodies, into epithelial cells that form the proximal tubules, the distal tubules, and the glomerulus of the nephron. *WT1* expression can be detected in metanephric mesenchymal cells *(160)* at the onset of nephrogenesis. *WT1* expression is low in the developing blastema, and increases in the comma- and S-shaped bodies, as nephrogenesis progresses. Upon further differentiation, *WT1* expression is downregulated, except in the podocyte cells of the mature glomerulus, where it is maintained into adulthood.

In WT1-null mice, the ureteric bud is absent, indicating that WT1 is essential for the development of this structure *(161)*. Because *WT1* is not expressed in the ureteric ECs *(162)*, the lack of growth may be the result of the absence of growth factor expression in the metanephric mesenchymal cells. Organ culture experiments have demonstrated that mesenchymal factors are the major driving force for the development of epithelia *(14)*. Some of these mesenchymal factors have been identified, such as scatter factor/hepatic growth factor, and neuregulin/NDF. These factors can induce motility, growth, and morphogenesis of epithelial cells in culture, but whether these factors play a role in the outgrowth of the ureteric bud remains to be elucidated.

A factor that is involved in the development of the ureteric bud is giant cell line-derived neutrophilic factor (GDNF), a c-ret-binding ligand. GDNF is a member of the transforming growth factor-β (TGF-β) superfamily *(163)*, and is expressed in the metanephric mesenchyme *(164)*. The c-Ret protein, a member of the receptor tyrosine kinase family, is expressed in cells on the tip of the ureteric bud, and mice that lack the c-*ret* gene show ureter defects and severe renal agenesis *(165)*. In mice that lack the *GDNF* gene, the metanephric mesenchyme cells undergo apoptosis, with complete disappearance by E13.5 *(164)*. In the WT1-null mice, the metanephric mesenchyme cells also undergo apoptosis *(161)*. Since *TGF-β* is a putative target gene of WT1, it is tempting to speculate about the regulation of *GDNF* by WT1.

Induction of the mutant blastema by embryonic spinal cord tissue, the strongest inducer of metanephric mesenchyme to form kidney tubules *(11)*, did not result in the differentiation of the WT1-null mesenchyme *(161)*. This experiment indicated that the failure of the WT1-null blastema cells to differentiate is not only the result of the absence of the ureteric duct, but is also a cell-autonomous defect. During normal development, WT1 may exert its effect by rendering the metanephric mesenchymal cells sensitive to induction by the ureteric duct. In this connection, Wnt-1-expressing NIH3T3 cells can induce kidney mesenchyme to differentiate into epithelial tubules *(166)*. Expression of *WT1* may, therefore, upregulate the expression of receptors that recognize members of the Wnt family. Alternatively, *WT1*-expression may be essential for the early steps of differentiation, immediately after the induction by the ureteric bud. The *Wnt-4* gene and the *Pax-2* gene may be involved in this process. During normal nephrogenesis, both genes are expressed in condensing kidney mesenchymal cells, shortly after induction by the ureteric bud, and expression persists in the comma- and s-shaped bodies before being downregulated *(167,168)*. In Wnt-4-null mice, there appears to be a developmental arrest of the metanephric mesenchyme, so that no tubular epithelium is ever formed *(167)*. In homozygous Pax-2-mutant mice, there is a complete lack of kidneys, ureters, and genital tracts *(169)*. It has been shown that WT1 can regulate the expression of *PAX-2 (140)*. In turn, PAX-2 is also capable of regulating the expression of the *WT1* gene *(170)*, suggesting the presence of a feedback loop. Whether the presence of WT1 is essential for the proper regulation of the *Wnt-4* and the *PAX-2* gene during nephrogenesis remains to be elucidated.

The expression pattern of *WT1* suggests that the gene is also involved in the further progression of nephrogenesis. In support of this, Moore et al. *(171)* showed that the kidney phenotype of the WT1-null mouse could be partially rescued by introducing a Yeastartificial chromosome (YAC) containing the genomic human *WT1* sequence. Analysis of 43 embryos showed that, in 15 embryos, the ureter had repeatedly branched, and condensation of the metanephric mesenchyme was taking place. In 12 embryos, nephrogenesis had taken place, until the stage of the comma-shaped bodies. Formation of s-shaped bodies was rare, and glomeruli were never seen.

In the adult kidney, *WT1* continues to be expressed in the podocytes *(46,48,162)*. The podocytes, which line the blood vessels in the glomerulus, are involved in a variety of glomerular functions *(172)*. The podocytes can synthesize the glomerular basement membrane, and may also play a role in its degradation, since they show abundant endocytic activity. In DDS patients, in which *WT1* is heterozygously mutated, the podocytes are often underdeveloped *(173)*. Glomerular nephropathy is the most consistent finding in these patients, who suffer from hypertension, following the collapse of the arteries in the glomerulus. This is caused by the production of fibrotic material by the so-called mesangial cells. Regarding the role that podocytes may play in the degradation of the glomerular basement membrane, the authors hypothesize that proper expression of *WT1* may be essential for the normal development and maintenance of this membrane. The podocytes also provide structural support to the glomerular tuft, and may influence the filtration rate in the glomerulus. So WT1 appears to play an important role throughout nephrogenesis and the maintenance of normal podocyte function. But what does WT1 do at the cellular level?

6.1. Proliferation

Kudoh et al. *(174)* showed that microinjection of the WT1(+17aa) isoforms into synchronized NIH3T3 cells significantly blocked serum-induced cell cycle progression. Similar results were obtained when CV-1, Cos-7, F9, or P19 were injected with the WT1+/+ isoform. The inhibitory activity of the WT1-isoforms was abrogated by the overexpression of either cyclin E/CDK2 or cyclin D1/CDK4.

Hewitt and Saunders *(175)* found that expression of both the WT(+17aa) and the WT1(−17aa) isoforms in a rat renal carcinoma cell line (IFC), suppressed proliferation. In contrast to the observations by Kudoh et al. *(174)*, the WT1(−17aa) isoforms displayed a stronger growth-suppressive effect than the WT1(+17aa) isoforms. Accordingly, FACS analysis showed that a higher percentage of WT1(−17aa)-expressing cells was found in the G0/G1 phase. In agreement with these findings, Englert et al. *(176)* found that expression of a WT1(−KTS) isoform in Saos-2 cells resulted in a 25% increase of these cells in the G1 fraction. In addition, they found that the CDKI p21, was induced upon the expression of WT1(−KTS) in Saos-2, U2OS, and baby rat kidney cells. Because CDKIs are potent inducers of cell cycle arrest, the increase in p21 levels may well account for the observed prolongation in the G1 phase. Northern analysis and transient transfection assays suggest that p21 may be a direct target gene of WT1. In order to induce p21 in Saos-2 cells, binding of WT1 to hsp70 appears to be necessary *(143)*. Another gene may account for the observed suppression of proliferation upon WT1 expression. Stable expression of the WT1–/– isoform in 293 cells reduces the proliferation rate of these cells by a factor of 2 *(177)*. Using suppression-subtractive hybridization polymerase chain reaction, a WT1 target gene was isolated that was upregulated about 15-fold in cells expressing WT1. The gene was identified as the retinoblastoma suppressor-associated protein 46 (RbAp46), and expression of this gene in 293 cells reduced the proliferation rate again by about twofold. This observation suggested that RbAp46 may be a downstream mediator of WT1.

The above results suggest that WT1 only plays a role in growth suppression. However, Yamagami et al. *(178)*, showed recently that suppression of endogenous WT1 expression induced G2–M arrest in K562 cells. WT1 antisense oligomers exhibited significant inhibitory effects on the cell growth of K562 cells, in association with significant reductions in WT1 protein levels *(179)*. FACS analysis revealed that these cells were arrested in G2–M. WT1 may therefore play a role in the control of two cell cycle checkpoints. WT1 overexpression induces G1/S arrest; WT1 suppression induces G2–M arrest.

6.2. Differentiation

During nephrogenesis, the induced metanephric mesenchyme cells differentiate from mesenchymal to ECs *(180)*. Similar transitions have been observed in other tissues expressing WT1, suggesting that WT1 may play a role in this process *(46,50,161,181,182)*. Consistent with a prominent role for WT1 in the mesenchymal–epithelial transition, Hosono et al. *(183)* recently showed that expression of the WT1–/– isoform in NIH3T3 cells initiates features of epithelial differentiation.

Most WTs comprise a mixture of undifferentiated blastemal cells, differentiated epithelial cells, and mesenchymal stromal cells. However, ectopic components, not normally found in nephrogenesis (particularly skeletal muscle), are observed in 5–10% of WTs *(184)*. Metanephric-mesenchymal stem cells may have the capacity to differentiate into skeletal muscle cells, as well as ECs, and complete loss of WT1 function may lead to the activation of the myogenic program. High levels of myogenic gene expression were observed in 5/7 WTs in which a homozygous WT1 mutation was documented. In addition, expression of the wild type WT1 isoforms in C2 myoblasts, suppressed myogenesis upon dexamethasone treatment, but a mutant form did not.

The individual WT1-isoforms may have different effects on differentiation. Rat renal cells expressing the WT1(–17aa) isoforms grow widely scattered, lacking close cell–cell contact. In addition, they are smaller and have fewer processes, compared to nonexpressing cells. WT1(+17aa)-expressing renal rat cells have a greater cytoplasmic volume than nonexpressing cells, and display close cell–cell contacts *(175)*.

Other cell systems also imply a role for WT1 in differentiation. Myoblastic leukemic M1 cells treated with leukemia inhibitory factor (LIF) can be induced to undergo terminal macrophage differentiation, coupled to growth arrest and apoptosis of mature cells *(185)*. WT1 cannot be detected in M1 cells. However, 24–168 h after adding LIF, WT1 expression is high. To study the role of WT1 during differentiation of M1 cells, stable transfectants were made. No transfectant could be obtained expressing the WT1(–KTS) isoforms. Upon expression of the WT1(+KTS) isoforms, the M1 cells differentiated to various stages along the monocytic differentiation pathway. The cells, however, did not undergo terminal differentiation, suggesting that expression of the WT1(+KTS) isoforms are important for only part of the differentiation program. In contrast, Svedberg et al. *(186)* showed that expression of either the WT1–/– or the WT1–/+ isoform partially blocked the differentiation program of monoblastic U937 cells upon incubation with either retinoic acid (RA) or vitamin D_3.

Treatment of embryonic carcinoma cells (P19) and embryonic stem cells with RA resulted in differentiation of these cells, and the activation of WT1 expression. Immunohistochemical analysis showed that *WT1* is expressed in endodermal, glial, and epithelial cells, suggesting that WT1 plays a role in the differentiation of these cell types *(187)*.

In contrast, several other cell systems show that differentiation correlates with the downregulation of *WT1* expression. HL60 cells differentiate into granulocytes upon treatment with dimethyl sulfoxide or RA *(188)*. At the same time, expression levels of WT1 drop. When treated with TPA or vitamin D_3, HL60 cells differentiate into mature macrophages and monocytes, respectively, and again WT1 levels decrease. A similar downregulation of *WT1* was observed when chronic myelogenous K562 leukemia cells were induced by sodium butyrate or TPA to differentiate into erythroid cells or megakaryocytic cells, respectively *(189)*. Experiments with antisense WT1 indicated that downregulation of *WT1* is not sufficient for the observed differentiation of K562 cells *(190)*. The K562 cells, treated with antisense WT1, stop growing and subsequently undergo apoptosis, suggesting that WT1 plays an important role in proliferation and survival.

6.3. Apoptosis

During development of the rat kidney, 3% of the cells within nephrogenic areas are apoptotic at any given time, implying that large-scale apoptosis takes place during renal development *(191)*. In WT1-null mice, mesenchymal cells fail to differentiate, and degenerate via apoptosis, indicating that WT1 expression is required for their survival. WT1 may therefore function as a survival factor. In agreement with this hypothesis is the observation that WT1 is able to suppress p53-induced apoptosis *(192)*. In addition, when the WT1-expressing leukemia cell line, K562, is treated with antisense WT1, the cells stop growing, and subsequently undergo apoptosis *(190)*. However, WT1 is also able to induce apoptosis. Expression of each of the four splice variants in both Saos-2 and U2OS cells results in apoptosis *(193)*; expression of only the WT1(–KTS) isoforms induces apoptosis in HepG2 and Hep3B cells *(155)*.

Whether a cell will go into apoptosis is determined by a variety of signals of both extracellular and internal origin *(194–197)*, and WT1 may play an important role in the regulation of these signals (Fig. 4). As to the internal signals, it has been shown that WT1 can regulate the expression of *Bcl-2, c-myc,* and *c-myb* (Table 2). Bcl-2 can promote cell survival, and inhibits apoptosis in certain cell types *(198–202)*; overexpression of c-myc or c-myb can induce apoptosis *(197,201,203–206)*. WT1 may regulate the expression of *TGF-β*, which has been shown to induce apoptosis in Hep3B cells *(207,208)* and rat hepatocytes *(209)*. WT1 may also regulate the expression of *IGF-II*, which may function as a survival factor by binding to the insulin-like growth factor-1 receptor *(210)*. Cell-type-dependent transcriptional activity of the WT1 isoforms may explain the opposite effects of WT1 on apoptosis. In one cell type, e.g., *WT1* expres-

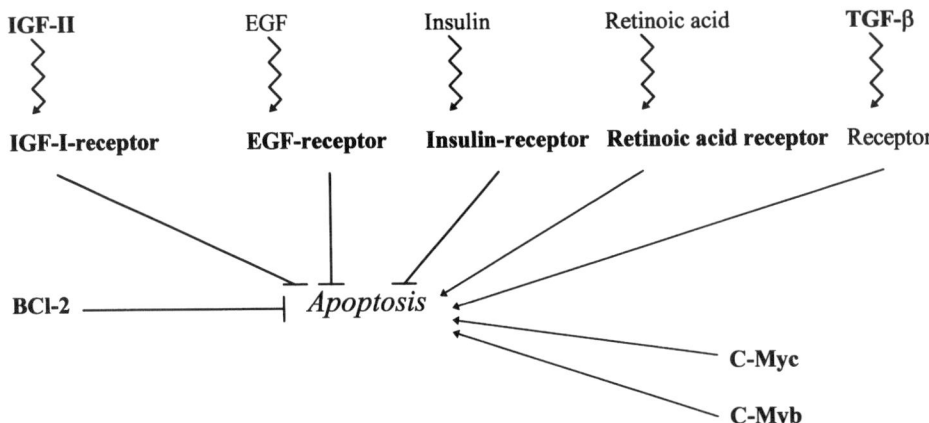

Fig. 4 Proposed roles of WT1 in apoptosis. Genes for IGF-II, IGF-I-receptor, EGF-receptor, Insulin-receptor, Bcl-2, Retinoic acid receptor, TGF-β, C-Myb are all putative target genes of WT1. Differential regulation of these genes may determine whether a cell will go into apoptosis.

Table 3
Potential Target Genes of *WT1* Gene Products

Gene	Ref
WT1	*(139,247,307)*
Egr-1	*(127,137,234)*
IGF-II	*(308,265)*
IGF-I receptor	*(309,310,287,288)*
PDGF-A	*(136,281,311,312)*
CSF-1	*(313)*
TGF-β	*(314)*
PAX-2	*(140)*
Nov-H	*(315)*
RAR-a	*(316)*
Inhibin-a	*(317)*
C-myb	*(286)*
ODC	*(284)*
G-protein ai-2	*(283)*
bcl-2	*(318)*
c-Myc	*(131,318)*
EGF-receptor	*(131,155,194)*
Ki-ras	*(131)*
Insulin-receptor	*(131,155,319)*
Syndecan-1	*(320)*
Midkine	*(321)*
p75 neurotrophin-R	*(322)*
MIS	*(323)*
SRY	*(323)*
Androgen	*(323)*
RbAp46	*(177)*
p21	*(176)*
Dax-1	*(324)*

sion may result in the downregulation of survival factors; in another cell type, WT1 may boost their expression. To what extent this will affect cell survival is dependent on the crosstalk between the cell and its cellular environment. WT1 may also play an important role in this crosstalk. Extracellular signals can be provided by growth factors, the extracellular matrix *(211)*, and direct cell–cell contact *(212)*. Concerning the first group, WT1 may regulate the expression of several growth factor receptors that play a role in the onset of apoptosis, such as the RA receptor-α, the IGF-I receptor *(IGF-IR)*, the EGF receptor, and the insulin receptor (Table 3; Fig. 4). Overexpression of WT1 in F9 embryonic carcinoma cells induces only apoptotic cell death in the presence of RA *(213)*. Downregulation of the IGF-IR correlates with increased apoptosis in C6 glioblastoma cells *(210);* overexpression of the EGF receptor rescues both U2OS *(193)* and Hep3B cells *(155)* from WT1-induced apoptosis. Overexpression of the insulin receptor also rescues Hep3B cells from WT1-induced apoptosis *(155)*. The mechanisms by which WT1 is involved in apoptosis may be complex, as is clear from the onset of atresia in granulosa cells *(134)*, in which the absence of WT1 may eventually lead to the abrogation of trophic support.

The above data indicate that WT1 plays a role in proliferation, differentiation, and apoptosis, three processes whose balance is important for normal development. Inactivation of several putative WT1-target genes, such as *IGF-II (214)*, *IGF-IR (214,215)*, Gαi-2 *(216)*, *Pax-2 (217)*, or the *Bcl-2 (218)* results in reduced and aberrant nephrogenesis. The balance between the expression levels of these target genes may determine whether a cell will proliferate, differentiate, or undergo apoptosis. Disruption of any of these three processes may result in tumor formation.

7. DISRUPTION OF WT1 FUNCTION AND TUMORIGENESIS

WT1 has been proven to be essential throughout the whole process of nephrogenesis. The involvement of WT1 in proliferation, differentiation, and apoptosis, and its temporal and spatial expression pattern, makes it a prime suspect in tumorigenesis. Disruption of the function of WT1 may explain the development of many WTs.

7.1. Homozygous Mutation/Deletion of WT1

A WT1-null mutation in mice results in failure of kidney development *(161)*. The metanephric mesenchyme cells fail to differentiate and undergo apoptosis. But what will happen if the expression of WT1 is switched later in development? In this case, the metanephric cells may already have been induced by the ureteric bud. The absence of WT1 prohibits the proliferating metanephric mesenchyme cells from differentiating into epithelial nephrons. Instead, the metanephric mesenchyme cells may develop into stromal cells or into heterologous elements, such as muscle tissue. Several studies support this hypothesis by showing that stromal cells or heterologous tissues in WTs do not express WT1 *(162,219–221)*. In a study by Schumacher et al. *(222)*, 67% of stromal predominant tumors with germ-line WT1-mutations, showed LOH, while in one WT a different somatic mutation was found in addition to the germ-line mutation. These data indicate that WTs, associated with intralobar nephrogenic rest (prominent stromal component), may represent a subset of WTs in which the classical two-hit inactivation model, with loss of functional WT1 protein, is the underlying cause of tumor development.

7.2. Heterozygous Mutation of WT1

WT1 is found to be homozygously mutated/deleted in only a few WTs *(223–229)*, and, since the *WT1* gene is not subjected to imprinting in the kidney *(230)*, the mutation of one allele will not be enough to silence expression. However, tumors may develop as a result of reduced WT1 expression, because suppression of tumorigenicity may be dose-dependent *(231)*. Contradicting this possibility is the observation that mice heterozygous for the WT1-null mutant, appear normal, and develop no tumors *(161)*.

However, heterozygous mutations of WT1 may in some cases be enough to disrupt normal development. Heterozygous point mutations in the *WT1* gene, as have been found in DDS patients, lead to a much more severe phenotype than the heterozygous deletions inherited by children with WAGR syndrome. Heterozygous deletions of the *WT1* gene can lead to WT and mild developmental defects of gonads and kidneys *(41,50)*. Children with DDS develop WTs and usually die of severe glomerular nephropathy *(232,233)*. The mutant WT1 proteins in DDS may therefore act in a dominant-negative way, abrogating the function of the wild type proteins. Transient co-transfection experiments demonstrate that mutant WT1 can affect the transcriptional activity and localization of the wild type protein, possibly by dimerization *(156,234)*.

The mutated protein may have a completely different spectrum of target genes, compared to the wild type protein. Alternatively, the mutated protein may compete with the wt protein for the same DNA-binding sites, but have a different transactivating activity. Considering the latter, a point mutation in exon 5 (codon 266) decreased the transcription-repressing activity of WT1 by 2–3-fold in transient transfection assays *(235)*. A point mutation in exon 3 (codon 201) or exon 6 (codon 273) resulted in a much more dramatic effect, converting the wild type protein from a repressor into an activator of transcription *(181,236)*.

Based on the tumors analyzed so far, it has been estimated that 5–25% of the WTs contain homozygous or heterozygous *WT1* mutations *(132,237–241)*. In fact, *WT1* expression can still be detected in most WTs, and has even been found to be highly expressed in a high percentage of these tumors *(46,219,221,242–244)*. Therefore, most WTs cannot be explained by inactivating intragenic mutations or deletions. However, there may be other mechanisms by which the function of the gene can be disrupted.

7.3. Disruption of Spatial and Temporal Expression

The *WT1* gene is expressed during normal development in a strict temporal and spatial pattern *(46,118,160,245)*. Extrapolation of the cell culture data to the in vivo situation predicts that spatial or temporal expression failures may have devastating effects, e.g., a cell may proliferate when it should go into apoptosis. Thus, proper regulation of *WT1* expression is essential. Several potential TF recognition sites have been identified in the *WT1* promoter (Table 4). Transient transfection assays show that WT1 *(246)*, SP1 *(247,248)*, nuclear-factor-κB *(249)*, Pax-2 and Pax-8 *(170,250)* may indeed regulate the promoter activity of the *WT1* gene. Disruption of their function may lead to loss or aberrant *WT1* expression. In addition to the TF binding sites, a silencer region has been identified in intron 3 *(251)*, and an enhancer region was localized 50 kb downstream of the *WT1* transcription start site *(252)*. The silencer region does not seem to function in cells of renal origin *(251)*, and the enhancer only appears to work in cells

Table 4
Potential TF Binding Sites in *WT1* Promoter and
in the 3' Enhancer Region

WT1 promoter	3' Enhancer region
CTCF	CTCF
E2A	E2A
NF-IL6	NF-IL6
PEA3	PEA3
TCF-1	TCF-1
WT1	AP-1
Egr-1	CF-1
Sp1	HINF-A
Sp2	TBP
Sp3	GATA
Pax-2	
Pax-8	
AP-2	
AP-4	
Ets-1	
F-ACT1	
GCF	
H4TF1	
PuF	
TCF-2a	
TEF-2	
NF-κB	

that express the hematopoietic transcription factor, GATA-1 *(253)*. Disruption of these regulatory sequences may lead to abnormal *WT1* expression, and subsequent tumorigenesis. No mutations have been detected in the *WT1*-promoter in WTs *(254)*, but only a very limited region has been analyzed.

The synthesis of the WT1 products may not only be regulated at the transcriptional level. Transcripts from the antisense strand of the *WT1* gene have been detected in 293 cells *(255)*, and recently an antisense *WT1* promoter has been identified in intron 1 of the *WT1* gene, which was activated by the expression of *WT1 (256)*. Expression of the anti-sense RNA in 293 cells resulted in a 50% decrease of WT1 at the protein level. Although Grubb et al. *(162)* showed that *WT1* transcripts and proteins are co-ordinately expressed, these results suggest that the antisense RNAs may affect the translation of WT1 mRNA into protein.

7.4. Disruption of Ratio Between Splice Variants

Besides disruption of spatial or temporal expression of *WT1*, changes in the ratio between the splice variants could also lead to tumor formation. Transient transfections indicate that each splice variant may have a different transcriptional activity on (partly) different target genes *(127,128,136–139)*. Tumorigenicity assays also show that the isoforms can have different effects *(154,257,258)*. Expression of the WT1–/– isoform, in adenovirus-transformed baby rat kidney cells, increased the tumorigenic potential of the cells; expression of the WT1–/+ isoform suppressed the tumorigenicity. These data indicate that disruption of the ratio between the splice variants may result in tumor formation. In support of this, Simms et al. *(259)* found that 7/10 WTs show an increase in the ratio between the WT1(–17aa) and the WT1(+17aa) transcripts, relative to the ratio in normal kidney tissue. Recently, another group showed a similar increase in the ratio between the WT1(–17aa) and the WT1(+17aa) transcripts in 4/7 sporadic unilateral WTs *(260)*. If this increased ratio reflects higher expression levels of the WT1–/– isoform, it could explain the abundant expression of the putative target gene, *IGF-II*, in WTs *(261–264)*. The WT1–/– isoform has been shown to induce the expression of the endogenous *IGF-II* gene in both RM1 cells *(265)* and NIH3T3 cells *(183)*. In turn, these findings could explain the observation that expression of the WT1–/– isoform increases the tumor growth rat-of adenovirus-transformed baby rat kidney cells *(154)*.

In some tumors, the splice donor/acceptor sites may be mutated; in other tumors, the splicing machinery may be disrupted. Which splicing factors are involved in the regulation of the ratio between the WT1 isoforms remains to be elucidated. It has been suggested that alternate splicing of certain genes may be regulated by the antagonistic activities of the serine arginine family of splicing factors and the heterogeneous nuclear ribonucleoprotein particle A1 (hnRNPA). Tissue-specific differences in the activities of these proteins may determine the splicing pattern *(266)*.

7.5. Posttranslational Modification

Phosphorylation can regulate the activity of several TFs by affecting the activity of the transactivation domain, the DNA-binding activity, or the nuclear translocation *(267–270)*. The activity of the WT1 protein may also be modulated by phosphorylation. The C-terminal domain of WT1 contains a number of potential phosphorylation sites for protein kinase A (PKA) and PKC *(271)*. Recently, it has been shown that the in vitro phosphorylation of WT1 by PKA or PKC inhibited DNA binding *(271,272;* Semba, K, personal communication). In support of this, transient transfection assays in NIH3T3 cells indicated that activation of PKA inhibits WT1's transcriptional repressor activity. Based on the crystal structure of the Egr-1 protein *(133)*, the Ser365 and the Ser393 residues in Zn fingers 2 and 3, respectively, are involved in making contacts with the guanine residues of the DNA-binding sites. One could envisage that phosphorylation of these sites could perturb DNA binding. In NMU-induced rat embryonic tumors, which resemble WTs, Ser365 was converted into phenylalanine (Phe) *(52)*, and electrophoretic mobility assays showed that this mutation severely impaired the DNA-binding activity of the WT1 protein *(128)*. Phosphorylation of both residue Ser365 and Ser393 has been observed in transient transfection assays *(272)*. Phosphorylation by PKA of full-length WT1, in WT1-transfected COS cells, abolished its DNA-binding activity. In vitro phosphorylation of WT1 also abolished its DNA-binding activity, which was restored upon alkaline phosphatase treatment. In CV-1 cells, the Ser365-Phe mutant still retained DNA-binding activity, even when PKA was expressed, indicating that this site is important in the regulation of DNA binding by phosphorylation. In agreement with the findings by Ye et al. *(271)*, PKA treatment of WT1-expressing CV-1 cells inhibited the transcriptional-repressing activity of the WT1 proteins. The transcriptional activity of the WT1 mutants, in which the Ser365 and/or the Ser393 had been converted, was not affected by PKA treatment.

In in vitro binding assays with the WT1 Zn-finger domain, however, phosphorylation still abolished DNA binding in a double mutant in which both Ser residues had been mutated. The latter result suggests that other phosphorylation sites may be of additional importance for the regulation of the DNA-binding activity of WT1. Concerning this, TPA-induced WT1 phosphorylation occurs at an as-yet-unknown site (Semba, K, et al., personal communication), e.g., whether the 17-aa insert may be a site of phosphorylation, important in the regulation of DNA binding activity remains to be elucidated. Deletion of four Ser residues in this aa stretch abolishes the repressor activity of this insert in transient transfection assays *(136)*. Phosphorylation of WT1 may also inhibit its transcriptional repressor activity by affecting the subcellular localization. Treatment of glomerular visceral ECs with forskolin (a strong PKA activator) resulted in a shift of WT1 from a nuclear localization to both nuclear and cytosolic localization *(271)*.

7.6. Proteins that Modify WT1 Function

Recent data suggest the existence of an interactive nuclear component in NIH3T3 cells that is required for the transcriptional repressor function of the WT1 protein *(273)*. Gel filtration assays show that the WT1 protein may be present in complexes ranging from 100 to 669 kDa *(274)*. Apart from the association of the WT1(+KTS) isoforms with factors of the splicing machinery *(149,158)*, several other proteins have been found to interact with WT1. Immunoprecipitation and gel filtration assays indicate that WT1 physically interacts with p53, and transient transfection assays show that p53 can affect the transcriptional activity of WT1 *(274)*. WT1(–KTS) acts as a transcriptional repressor in NIH3T3 cells, but functions as a transcriptional activator in Saos-2 cells, which lack the p53 protein *(274;* Table 3). Moreover, co-transfection of wild type p53 suppresses WT1-mediated transcriptional activation in this cell line, but mutant p53 fails to do so. However, transient transfection assays show that WT1–/– stimulates the Egr-1 promoter activity in both the p53-negative Saos-2 cells *(274)* and the p53-positive U2OS cells *(193)*, indicating that the effects of WT1 may be determined by other factors, in addition to p53.

Three additional WT1-binding proteins have been identified; human par-4 *(275)*, UBC9 *(276)*, and Ciao 1 *(277)*. Transient transfection assays show that par-4 reduces WT1-induced promoter activity in 293 cells. At the same time, it enhances the ability of WT1 to repress transcription. Because par-4 does not affect the DNA-binding activity of WT1, it has been suggested that par-4 may function as a repressor by binding to WT1, and in this way bringing an additional repression domain to the promoter. Ciao 1 reduces the transcriptional activity mediated by WT1. The function of UBC9 binding to WT1 is less clear. In yeast, UBC9 is involved in cell cycle progression, and is required for viability and the degradation of S- and M-phase cyclins *(278)*. In this context, it is important to note that WT1 can block cell cycle progression, and this block is relieved by expression of either cyclinE/CDK2 or cyclinD1/CDK4 *(174)*.

The function of the WT1 isoforms may also be affected indirectly by several other proteins. The Egr-1 protein belongs to the group of immediate-early transcription factors *(279)* and, like WT1, Egr-1 is a DNA-binding protein with highly conserved Zn finger structures of the cysteine 2–histidine 2 subclass toward the C-terminal end. The Zn fingers 2, 3, and 4 of the WT1 protein exhibit 61% aa sequence homology with Zn fingers 1, 2, and 3 of the Egr-1 protein *(280)*. Both proteins can bind to similar DNA sequences, suggesting that they may regulate the same target genes. In transient transfection assays, it has been shown that Egr-1 activates several promoters; WT1 suppresses the activity of these promoters *(137,140,281–284)*. It is conceivable that the expression of Egr-1 modulates the effect of WT1 by upregulating the expression of the same genes that are downregulated by WT1. Reciprocal expression of Egr-1 and WT1 has been observed during differentiation of LLC-PK1 cells *(282,283)*. Maximum Egr-1 expression coincides with maximum activation of the *Gαi-2* promoter, which contains an Egr-1 consensus sequence; maximum WT1 expression coincides with suppression of this promoter activity.

These data argue that the Egr-1 and the WT1 protein may compete for the same target genes, but have different transcription modulating activities. The same may be true for the Sp1 protein, another transcription factor of the Zn finger family. Recently, a 9-bp CTC repeat was identified in the 5′-flanking sequence of *WT1*, which accounted for approx 80% of WT1 transcription in stable transfected cells *(248)*. Enhancer activity of the element correlated completely with its ability to form a DNA–protein complex in gel shifts. Subsequent studies indicated that the CTC-binding factor is the transcriptional activator, Sp1. WT1 binds to a similar CTC repeat sequence, which has been found in the promoter region of the *Ki-ras*, the *EGF-R*, the *insulin-receptor*, the *c-myc*, and the *TGF-β* gene *(131)*. Whether the CTC repeat is a potential site for SP1–WT1 competition, as has been reported elsewhere for SP1 and Egr-1 *(285)*, remains to be elucidated. Sp1 recognizes the same DNA sequences as WT1 in the *ODC*, the *c-Myb*, and the *IGF-IR* promoter *(284,286,287)*, and functions as a strong activator of *IGF-IR* expression, whereas WT1 functions as a suppressor *(288)*. p53 may also compete for the same target genes as WT1. For example, p53 has been shown to activate the *EGF-R* promoter *(289)*; WT1 suppresses the activity of this promoter *(193)*. The TFs mentioned above may also act in concert with WT1, and different expression levels of both proteins may affect the phenotype of the cell. For example, WT1 and Egr-1 may act in concert during the differentiation of HL60 cells into macrophages or granulocytes. During this process, the expression of *WT1* is downregulated *(188)*, but only the cells that express *Egr-1* differentiate into macrophages. The other cells differentiate into granulocytes *(290)*. In further support of the importance of Egr-1, it has been shown that the addition of antisense *Egr-1* prevents the HL60 cells from differentiating into macrophages; the HL60 cells can no longer differentiate into granulocytes when *Egr-1* is overexpressed.

8. CONCLUSIONS

WT is a pediatric kidney malignancy that affects about 1/10,000 children. More than 80% of all cases can be successfully treated, but the treatment is traumatic for the patient, and, with improved survival rates, various side effects have become more apparent. To reduce these side effects, current studies are focused on shortening, simplifying, and refining the treatment strategies. Unraveling the underlying mechanism of the disease will be helpful in this process. Cytogenetic and molecular studies indicate that several chr regions are involved in the development of WT, but so far only one gene has been cloned and proven to play a role in the etiology of this type of tumor: the *WT1* gene. However, if this gene is so important, why are *WT1* mutations only found in about 5% of sporadic WT? And why are there only a few familial WT cases in which *WT1* mutations appear to play a role? This discrepancy could be explained by the fact that *WT1* does not only play an important role in kidney development, but is also essential for proper development of the genitalia *(134)*. *WT1*-null mice do not develop gonads *(161)* and WAGR patients may have genital abnormalities like cryptorchidism (undescended testis) and hypospadias (defect in the wall of male urethra or female vagina, resulting in an opening on the underside of the penis or in the vagina). In addition, DDS patients show genital/intersex anomalies that can be extremely variable *(132)*. At the most severe end of the spectrum, there may only be rudimentary streak gonads. At the mild end of the spectrum, XY individuals may have mild pseudohermaphroditism, such as that seen in boys with the WAGR syndrome.

WT1 is not only essential for proper gonad development, but expression studies indicate that it may also play a role in conception. *WT1* is expressed in the Sertoli and granulosa cells, which play an important role in spermatogenesis and maturation of the follicles, respectively *(134)*. *WT1* expression can also be detected in the antimesometrial region of the uterus, which undergoes decidualization upon implantation of the oocyte *(182)*. So disruption of the function of *WT1* could preclude reproduction, and hence the transmission of most germinal WT1 mutations, but there may be other contributory mechanisms. For example, several lines of evidence suggest that *WT1* plays a role in blood cell development *(134)*. One could envisage that disruption of its function in the hematopoietic system may lead to embryonic death, as well. Altogether, the above data can explain the low incidence of WT1 mutations found in WTs and the few familial WT cases with WT1 mutations. Only less disruptive alterations of the WT1 function will be transmitted, which may include mild mutations of WT1, but also other disruptions, occurring either upstream or downstream of WT1.

The previous subheading discussed several mechanisms by which the function of WT1 could be impaired, leading to the development of WT. However, nothing or relatively little is known about the molecular pathways involved. It will therefore be important to identify the gene products that regulate the transcription of WT1, the factors that determine the splicing of the WT1 transcripts, the proteins that modulate the activity of the WT1 isoforms, and the targets of the WT1 isoforms. Further analysis of the chromosomal regions implicated in the development of WT (subheading 2), may well reveal the identity of these factors. This could become a very heterogeneous group, encompassing TFs, splicing factors, phosphatases, kinases, interacting proteins, and, in addition perhaps, numerous target genes.

Experiments with cell culture systems have been the major source of information about the function of the *WT1* gene. *WT1* has been shown to play a role in differentiation, proliferation, and apoptosis. Disruption of any of these processes may result in the formation of WT. Whether WT1 is involved in only one or all of these processes in vivo remains to be elucidated. Notwithstanding the usefulness of cell culture systems, they have their limitations. Many of the functions attributed to WT1 appear to be cell-type dependent, e.g., expression of the WT1–/– isoform induces apoptosis in hepatoma cells *(155)*, differentiation of NIH3T3 cells *(183)*, and stimulates in vivo growth rate of adenovirus-transformed baby rat kidney cells *(154)*. In addition, in all ectopic expression studies, only one WT1 isoform is expressed at a time; studies of endogenous expression suggest that the expression of all isoforms is essential for normal functioning of the *WT1* gene *(123)*. Furthermore, the effect of *WT1* on a cell is determined by the environment of the cell, e.g., ectopic expression of *WT1* in F9 embryonic carcinoma cells only induces apoptotic cell death in the presence of RA *(213)*.

The above illustrates that, in order to get additional insight into the physiological function of WT1, analysis will have to be carried out in animal models. Only few animal models have been described so far. *WT1* knockout mice have been generated, but, since these mice die early in development, it is not possible to investigate the function of the *WT1* gene at later stages of development. The creation of conditional knockouts could circumvent this problem. Use may be made of the doxycycline-mediated, quantitative, and tissue-specific control of gene expression in transgenic mice *(291)*. With this system, it would be possible to switch *WT1*-expression on and off at different stages of development by adding doxycycline to the drinking water. In this way, the function of *WT1* during different stages of kidney development could be examined. At the same time, it will provide a system to test the validity of all the putative target genes identified so far. The system can also be used to boost the expression of just one isoform at different time-points, in order to examine how the ratio between the isoforms affects normal development. In the same way, one could induce the expression of mutated forms of *WT1*.

Further information about the function of *WT1* may come from experiments in which *WT1* is ectopically expressed. CMV-driven expression of *WT1* has been shown to be embryonically lethal *(292)*, but the use of tissue-specific promoters may circumvent this problem. Ectopic expression of *WT1* may prove or disprove certain hypotheses concerning the function of *WT1*. For example, *WT1* is believed to play a role in the suppression of muscle differentation *(184)*. Will mice develop muscles if *WT1* is ectopically expressed in these tissues?

8.1. Diagnostics and Therapy

Knowledge about the biological function of WT1 in Wilms' tumorigenesis is limited, but there may be a few applications in terms of diagnosis and therapy. In acute leukemia, a clear correlation has been observed between the relative expression levels of WT1 and prognosis *(293)*. Leukemia patients with relatively low expression levels had significantly higher rates of complete remission, disease-free survival, and overall survival, compared to patients with high expression levels. No such correlation has been investigated in WT, to the authors' knowledge. Since WT1 has been shown to inhibit p53-mediated apoptosis *(192)*, the expression levels of WT1 may well have implications for CT and radiation therapy of WTs. WT1 stabilizes p53, and the relative expression levels of p53 in WTs appear to correlate very well with *WT1* expression levels *(192)*. In addition, immunostaining showed that p53 expression follows that of *WT1*. Strongest immunoreactivity is seen in the epithelial tubular components of the tumor; the stromal elements were usually negative *(294)*.

There is a trend that *WT1* mutations in acute myeloid leukemias make these tumors more resistant to treatment. No such correlation has been investigated for WT. Because mutant *WT1* does not rescue cells from p53-induced apoptosis *(192)*, certain *WT1* mutations may be associated with increased sensitivity to treatment.

Male patients with pseudohermaphroditism should be screened for mutations in *WT1*. In cases in which the *WT1* sequence harbors one of the DDS mutations (Fig. 5), the patients are likely to develop WT. If a mutation in *WT1* disrupts alternative splicing at the exon 9 splice donor site, no WT1(+KTS) isoforms will be synthesized, and the

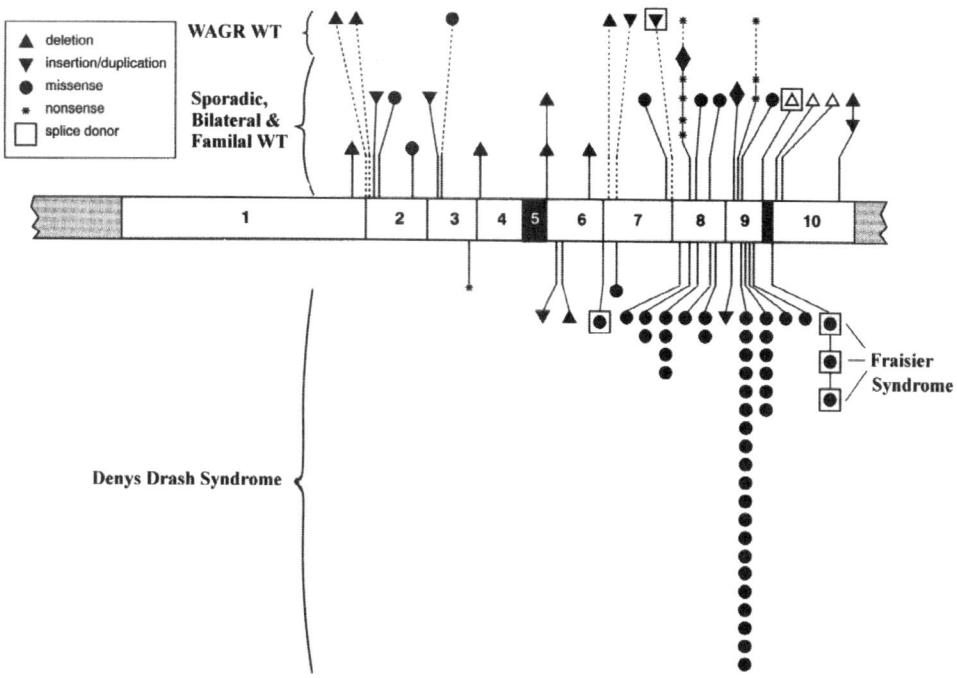

Fig. 5   Intragenic WT1 mutations (Williamson and Van Heyningen) *(134)*.

patient is likely to develop gonadoblastoma. These patients suffer from Frasier syndrome, characterized by focal glomerular sclerosis, delayed kidney failure, and complete gonadal dysgenesis *(295)*.

This review has focused on the role of *WT1* in Wilms' tumorigenesis, but *WT1* may also be involved in the development of several other tumors like Leukemia *(293,296–298)*, granulosa and Leydig cell tumors *(299)*, ovarian tumors *(300)*, leukemia's *(293,296,301)*, melanomas *(302)*, mesotheliomas *(303)*, breast cancer *(304)*, and desmoplastic small round cell tumors *(305)*. The authors hypothesize that the *WT1* gene allows a cell to respond appropriately to signals from its environment, and the many *WT1* isoforms allow the cell to fine-tune this response. Disruption of the function of *WT1* may result in a failure of the cell to differentiate or undergo apoptosis, and may lead to aberrant proliferation. Concerning the communication of a cell and its environment, it must be kept in mind that the kidney is formed through the reciprocal interaction of the metanephric mesenchyme and the ureteric bud epithelium. It is conceivable that WT is not the result of a cell-autonomous defect alone, but may also be the result of a communication breakdown between these tissues. Therefore, defects in cells that normally do not express *WT1,* may, in some cases, also cause WT formation. More than 200 genes/proteins are listed in the Kidney Development Database *(13)*. Expression studies, in combination with the analyses of knockout mice, have confirmed the importance of several of these genes in kidney development as a whole, and the signaling cascade between the metanephric mesenchyme and the ureteric bud

Table 5
Crucial Genes in Kidney Development

Gene	Kidney phenotype in null mice
WT1	No outgrowth ureteric bud *(161)*.
Pax-2	No outgrowth ureteric bud *(169)*.
Emx-2	No branching of ureteric bud, followed by degeneration *(325)*.
GDNF	No ureteric bud or reduced branching *(164,326,327)*.
Lim-1	No kidneys *(328)*.
c-ret	No ureteric bud or reduced branching *(165)*.
α8β1 integrin (329).	Reduced growth/branching of ureter
Pod-1	Reduced branching *(330)*.
Wnt-4	Undifferentiated mesenchyme, interspersed with branches of the ureter *(167)*.
BF-2	Inhibition of mesenchyme differentiation into comma and s-shaped bodies *(15)*.
BMP-7	Inhibition of differentiation of comma- and s-shaped bodies into glomeruli *(331,332)*.
PDGF-B	No mesangial cells or glomerular capillary tufts *(333)*.
PDGF-B-R	No mesangial cells or glomerular capillary tufts *(334)*.
α3β1 integrin	Abnormal podocyes and reduced number of collecting ducts *(335)*.

The genes are listed in the order in which their absence has an effect on kidney development. The expression pattern of each gene is visualized on the right. M, metanephric mesenchyme; U, ureter; W, wolffian duct.

epithelium (Table 5). Further analysis of this signaling cascade will help understanding of Wilms' tumorigenesis.

ACKNOWLEDGMENTS

The authors thank Dr. Davies for critically reading this manuscript.

REFERENCES

1. Greenwood MF, Holland P. Clinical and biochemical manifestations of Wilms' tumor. In: Pochedly C, Baum ES, eds., *Wilms' Tumor: Clinical and Biological Manifestations*. New York: Elsevier. 1984:215–250.
2. Breslow NE, Beckwith JB. Epidemiological features of Wilms' tumor: results of the National-Wilms-Tumor-Study. *J Nat Cancer Inst* 1982; 68:429–436.
3. Breslow NE, Langholz B. Childhood-cancer incidence: geographical and temporal. *Int J Cancer* 1983; 32:703–716.

4. Sharpe CR, Franco EL. Etiology of Wilms' tumor. *Epidemiol Rev* 1995; 17:415–432.
5. Bond JV. Bilateral Wilms' tumor; age at diagnosis, associated congentital anomalies and possible pattern of inheritance. *Lancet* 1975; 2:482–484.
6. Wiener JS, Coppes MJ, Ritchey ML. Current concepts in the biology and management of Wilms tumor. *J Urol* 1998; 159:1316–1325.
7. Metha MP, Bastin KT, Wiersma SR. Treatment of Wilms' tumour: current recommendations. *Drugs* 1991; 42:766–780.
8. Finklestein JZ. Wilms' tumor: a model of succes in cancer therapy. In: Pochedly C, Baum ES, eds., *Wilms' Tumor: clinical and Biological Manifestations*. New York: Elsevier. 1984:215–250.
9. Byrd RL, Levine AS. Late effects of treatment of Wilms' tumor. In: Pochedly C, Baum ES, eds., *Wilms' Tumor: Clinical and Biological Manifestations*. New York: Elsevier. 1984:215–250.
10. Machin GA. Persistent renal blastema as a precursor of Wilms' tumor In: Pochedly C, Baum ES, eds., *Wilms' Tumor: Clinical and Biological Manifestations*. New York: Elsevier. 1984:215–250.
11. Saxen L. *Organogenesis of the Kidney*. Cambridge University Press, 1987.
12. Lechner MS, Dressler GR. The molecular basis of embryonic development. *Mech Dev* 1997; 62:105–120.
13. Lipschutz JH. Molecular development of the kidney: a review of the results of gene disruption studies. *Am J Kidney Dis* 1998; 31:383–397.
14. Birchmeier C, Meyer D, Riethmacher D. Factors controlling growth, motility, and morphogenesis of normal and maligant epithelial cells. *Int Rev Cytol* 1995; 160:221–266.
15. Hatini V, Huh SO, Herzlinger D, Soares VC, Lai E. Essential role of stromal mesenchyme in kidney morphogenesis revealed by targeted disruption of Winged Helix transcription factor BF-2. *Genes Dev* 1996; 10:1467–1478.
16. Beckwith JB, Kiviat NB, Bonadio JF. Nephrogenic rests, and nephroblastomatosis. The pathogenesis of Wilms' tumor. *Pediatr Pathol* 1990; 10:1–36.
17. Beckwith JB, Zuppan CE, Browning NG, Moksness J, Breslo NE. Histological analysis of aggressiveness and responsiveness in Wilms' tumor. *Med Pediatr Oncol* 1996; 27:422–428.
18. Bonadio JF, Storer B, Norkool P, Farewell VT, Beckwith JB, D'Angio GJ. Anaplastic Wilms' tumor: clinical pathological studie. *J Clin Oncol* 1985; 3:513–520.
19. Huff V. Wilms tumor genetics. *Am J Med Genet* 1998; 79:260–267.
20. Maw MA, Grundy PE, Millow LJ, Eccles MR, Dunn RS, Smith PJ, et al. A third Wilms' tumor locus on chromosome 16q. *Cancer Res* 1992; 52:3094–3098.
21. Wang-Wuu S, Soukup S, Bove K, Gotwals B, Lampkin B. Chromosome analysis of 31 Wilms' tumors. *Cancer Res* 1990; 50:2786–2793.
22. Clericuzio CL. Clinical phenotypes of Wilms' tumor. *Med Pediatr Oncol* 1993; 21:182–187.
23. Clericuzio CL, Johnson C. Screening for Wilms' tumor in high risk individuals. *Hematol Oncol Clin North Am* 1995; 9:1253.
24. Pilia G, Hughes-Benzie RM, MacKenzie A, Baybayan P, Chen EY, Huber R, et al. Mutations in GPC3, a glypican gene, cause the Simpson-Golabi-Behmel overgrowth syndrome. *Nat Genet* 1996; 12:241–247.
25. Neri G, Martinineri ME, Katz BE, Opitz JM. The Perlman syndrome; familial renal dysplasia with Wilms' tumor, fetal gigantism and multiple congenital anomalies. *Am J Med Genet* 1984; 19:195–207.
26. Miller RW, Fraumeni JF Jr, Manning MD. Association of Wilms' tumor with aniridia, hemihyperthrophy and other congenital abnormalities. *N Engl J Med* 1964; 270:922.
27. Olson JM, Hamilton A, Breslow NE. Non-11p constitutional chromosome abnormalities in Wilms' tumor patients. *Med Pediatr Oncol* 1995; 24:305–309.
28. Geiser C, Schindler A. A long term survival in male with 18-trisomy syndrome and Wilms' tumour. *Paediatrics* 1969; 44:111–116.
29. Stay EJ, Vawter G. The relationship between nephroblastoma and neurofibromatosis. *Cancer* 1977; 39:2550–2555.
30. Hartley AL, Birch JM, Tricker K, Wallace SA, Kelsey AM, Harris M, Jones PHM. Wilms-tumor in the Li-Fraumeni cancer family syndrome. *Cancer Genet Cytogenet* 1993; 67:133–135.
31. Cairney A, Andrews M, Greenberg M, Smith D, Weksberg R. Wilms' tumour in three patients with Bloom syndrome. *J Pediatr* 1990; 111:414–416.
32. Ellis NA, Groden J, Ye TZ, Straughen J, Lennon DJ, Ciocci S, Proytcheva M, German J. Bloom syndrome gene product is homologous to RecQ helicases. *Cell* 1995; 83:655–666.

33. Szabo J, Heath B, Hill VM, Jackson CE, Zarbo RJ, Mallette LE, et al. Hereditary hyperparathy-roidism jaw tumor syndrome: the endocrine tumor gene HRPT2 maps to chromosome 1q21-q31. *Am J Hum Genet* 1995; 56:944–950.

34. Narod S. Genetics of breast and ovarian cancer. *B Med Bull* 1994; 50:656–676.

35. Pendergrass TW. *Cancer* 1976; 37:403–409.

36. Riccardi VM, Sujansky E., Smith AC, Francke U. Chromosomal imbalance in the Aniridia-Wilms' tumor association: 11p interstitial deletion. *Pediatrics* 1978; 61:604–610.

37. Francke U, Holmes LB, Atkins L, Riccardi VM. Aniridia-wilms' tumor association: evidence for spe-cific deletion of 11p13. *Cytogenet Cell Genet* 1979; 24:185–192.

38. Simola KOJ, Knuutila S, Kaitila I, Pirkola A, Pohja P. Familial aniridia and translocation t(4-11)(q22-p13) without Wilms' tumor. *Hum Genet* 1983; 63:158–161.

39. Moore JW, Hyman S, Antonarakis SE, Mules EH, Thomas GH. Familial isolated aniridia associated with a translocation involving chromosomes 11 and 22 [t(11,22)(p13, Q122)]. *Hum Genet* 1986; 72:297–302.

40. Turleau C, Dechrouchy J, Nihoulfekete C, Dufier JL, Chavincolin F, Junien C. Del 11p13/nephrob-lastoma without aniridia. *Hum Genet* 1984; 67:455–456.

41. Van Heyningen V, Bickmore WA, Seawright A, Fletcher JM, Maule J, Fekete G, et al. Role for the Wilms tumor gene in genital development? *Proc Natl Acad Sci USA* 1990; 87:5383–5386.

42. Lewis WH, Yeger H, Bonetta L, Chan HSL, Kang J, Junien C, et al. Homozygous deletion of a DNA marker from chromosome 11p13 in sporadic Wilms' tumor. *Genomics* 1988; 3:25–31.

43. Rose EA, Glaser T, Jones C, Smith CL, Lewis WH, Call KM, et al. Complete physical map of the WAGR region of 11p13 localizes a candidate Wilms' tumor gene. *Cell* 1990; 60:495–508.

44. Call KM, Glaser T, Ito CY, Buckler AJ, Pelletier J, Haber DA, et al. Isolation and characterization of a zinc finger polypeptide gene at the human chromosome 11 Wilms' tumor locus. *Cell* 1990; 60:509–520.

45. Gessler M, Poustka A, Cavenee W, Neve RL, Orkin SH, Bruns GAP. Homozygous deletion in Wilms' tumors of a zinc-finger gene identified by chromosome jumping. *Nature* 1990; 343:774–778.

46. Pritchard-Jones K, Flemming S, Davidson D, Bickmore W, Porteous D, Gosden C, et al. The candi-date Wilms' tumour gene is involved in genitoury development. *Nature* 1990; 346:194–197.

47. Buckler AJ, Pelletier J, Haber DA, Glaser T, Housman DE. Isolation, characterization, and expression of the murine Wilms' tumor gene (WT1) during kidney development. *Mol Cell Biol* 1991; 11:1707–1712.

48. Sharma PM, Yang X, Bowman M, Roberts V, Sukumar S. Molecular cloning of rat Wilms' tumor complementary DNA and study of of messenger RNA expression in the urogenital system and the brain. *Cancer Res* 1992; 52:6407–6412.

49. Little M, Wells C. A clinical overview of WT1 gene mutations. *Hum Mutat* 1997; 9:209–225.

50. Pelletier J, Bruening W, Kashtan CE, Mauer SM, Manivel JC, Striegel JE, et al. Germline mutations in the Wilms' tumor suppressor gene are associated with abnormal urogenital development in Denys-Drash syndrome. *Cell* 1991a; 67:437–447.

51. Little MH, Williamson KA, Mannens M, Kelsey A, Gosden C, Hastie ND, Van Heyningen V. Evi-dence that WT1 mutations in Denys-Drash syndrome patients may act in a dominant-negative fash-ion. *Hum Mol Genet* 1993; 2:259–264.

52. Sharma PM, Bowman M, Yu B-F, Sukumar S. A rodent model for Wilms tumors: embryonal kidney neoplasms induced by N-nitroso-N'-methylurea. *Proc Natl Acad Sci USA* 1994a, 91:9931–9935.

53. Patek CE, Little M, Flemming S, Miles C, Charlieu J-P, Clarke AR, et al. A zinc finger truncation of murine WT1 results in the characteristic urogenital abnormalities of Denys-Drash syndrome. *Proc Natl Acad Sci USA* 1999; 96:2931–2936.

54. Ping AJ, Reeve AE, Law DJ, Young MR, Boehnke M, Feinberg AP. Genetic-linkage of Beckwith-Wiedemann syndrome to 11p15. *Am J Hum Genet* 1989; 44:720–723.

55. Koufos A, Grundy P, Morgan K, Aleck KA, Hadro T, Lampkin BC, Kalbakji A, Cavenee WK. Famil-ial Wiedemann-Beckwith syndrome and a 2nd Wilms' tumor locus both map to 11p15.5. *Am J Hum Genet* 1989; 44:711–719.

56. Slater RM, Mannens MMAM. Cytogenetics and molecular genetics of Wilms' tumor of childhood. *Cancer Genet Cytogenet* 1992; 61:111–121.

57. Grundy PE, Telzerow PE, Breslow N, Mokness J, Huff V, Paterson MC. Loss of heterozygosity for chromosomes 16q and 1p in Wilms' tumors predicts an adverse outcome. *Cancer Res* 1994; 54:2331–2333.

58. Neumann B, Barlow DP. Multiple roles for DNA methylation in gametic imprinting. *Curr Opin Genet Dev* 1996; 6:159–163.

59. John RM, Surani MA. Imprinted genes regulation of gene expression by epigenetic inheritance. *Curr Opin Cell Biol* 1996; 8:348–353.

60. Hatada I, Inazawa J, Abe T, Nakayama M, Kaneko Y, Jinno Y, et al. Genomic imprinting of human p57(KIP2) and its reduced expression in Wilms' tumors. *Hum Mol Genet* 1996; 5:783–788.

61. Hatada I, Ohashi H, Fukushima Y, Kaneko Y, Inoue M, Komoto Y, et al. An imprinted gene p57(KIP2) is mutated in Beckwith-Wiedemann syndrome. *Nat Genet* 1996; 14:171–173.

62. Matsuoka S, Thompson JS, Edwards MC, Barletta JM, Grundy P, Kalikin LM, et al. Imprinting of the gene encoding a human cyclin-dependent kinase inhibitor, p57(KIP2), on chromosome 11p15. *Proc Natl Acad Sci USA* 1996; 93:3026–3030.

63. Qian N, Frank D, O'Keefe D, Dao D, Zhao L, Yuan L, et al. The IPL gene on chromosome 11p15.5 is imprinted in humans and mice and is similar to TDAG51, implicated in Fas expression and apoptosis. *Hum Mol Genet* 1997; 6:2021–2029.

64. Lee MP, Hu RJ, Johnson LA, Feinberg AP. Human KVLQT1 gene shows tissue-specific imprinting and encompasses Beckwith-Wiedemann syndrome chromosomal rearrangements. *Nat Genet* 1997; 15:181–185.

65. Cooper PR, Smilinich NJ, Day CD, Nowak NJ, Reid LH, Pearsall RS, et al. Divergently transcribed overlapping genes expresed in liver and kidney and located in the 11p15.5 imprinted domain. *Genomics* 1998; 49:38–51.

66. De Pagter-Holthuizen P, Jansen M, Van der Kammen RA, Van Schaik FMA, Sussenbach JS. Differential expression of the human insulin-like growth factor II gene. Characterization of the IGF-II mRNAs and an mRNA encoding a putative IGF-II associated protein. *Biochim Biophys Acta* 1988; 950:282–295.

67. Leighton PA, Ingram RS, Eggenschwiler J, Efstratiadis A. Disruption of imprinting caused by deletion of the H19 gene. *Nature* 1995; 375:34–39.

68. Leighton PA, Saam JR, Ingram RS, Stewart CL, Tilghman SM. An enhancer deletion affects both H19 and IGF-2 expression. *Genes Dev* 1995; 9:2079–2089.

69. Hao Y, Crenshaw T, Moulton T, Newcomb E, Tycko B. Tumor-suppressor activity of H19 RNA. *Nature* 1993; 365:764–767.

70. Brunkow ME, Tilghman SM. Ectopic expression of the H19 gene in mice causes prenatal lethality. *Genes Dev* 1991; 5:1092–1101.

71. Matsuoka S, Edwards MC, Bai C, Parker S, Zhang PM, Baldini A, Harper JW, Elledge SJ. P57(KIP2), a structurally distinct member of the p21(CIP1) cdk inhibitor family, is a candidate tumor-suppressor gene. *Genes Dev* 1995; 9:650–662.

72. Steenman MJC, Rainier S, Dobry CJ, Grundy P, Horon IL, Feinberg AP. Loss of imprinting of IGF-2 is linked to reduced expression and abnormal methylation of H19 in Wilms' tumor. *Nat Genet* 1994; 7:433–439.

73. Ogawa O, Eccles MR, Szeto J, McNoe LA, Yun K, Maw MA, Smith PJ, Reeve AE. Relaxation of insulin-like growth factor-II imprinting implicated in Wilms' tumor. *Nature* 1993; 362:749–751.

74. Vu TH, Hoffman AR. Promoter-specific imprinting of the human insulin-like growth factor II gene. *Nature* 1994; 371:714–717.

75. Thompson JS, Reese KJ, De Baun MR, Perlman EJ, Feinberg AP. Reduced expression of the cyclin-dependent kinase gene p57(KIP2) in Wilms' tumor. *Cancer Res* 1996; 56:5723–5727.

76. Henry I, Bonaiti C, Puech A, Chenensse V, Beldjord C, Landrieu P, Junien C. Uniparental paternal disomy in sporadic Beckwith-Wiedemann syndrome. *Am J Hum Genet* 1991; 49:19.

77. Sun FL, Dean WL, Kelsey G, Allen ND, Reik W. Transactivation of Igf2 in a mouse model of Beckwith-Wiedemann syndrome. *Nature* 1997; 389:809–815.

78. Yan YM, Frisen J, Lee MH, Massague J, Barbacid M. Mice lacking p57(Kip2) display developmental defects in the gastrointestinal tract and in endochondral bone formation. *Eur J Cell Biol* 1997; 72:29.

79. Yan YM, Lee MH, Massague J, Barbacid M. Ablation of the CDK inhibitor p57(Kip2) results in increased apoptosis and delayed differentiation during mouse development. *Genes Dev* 1997; 11:973–983.

80. Reid LH, Davies C, Cooper PR, CriderMiller SJ, Sait SNJ, Nowak NJ, et al. 1-Mb physical map and PAC contig of the imprinted domain in 11p15.5 that contains TAPA1 and the BWSCR1/WT2 region. *Genomics* 1997; 43:366–375.

81. Karnik P, Chen P, Paris M, Yeger H, Williams BRG. Loss of heterozygosity at chromosome 11p15 in Wilms tumors: identification of two independent regions. *Oncogene* 1998; 17:237–240.

82. Hannigan GE, Bayani J, Weksberg R, Beatty B, Pita A, Dedhar S, Squire J. Mapping of the gene encoding the integrin-linked kinase, ILK, to human chromosome 11p15.5-p15.4. *Genomics* 1997; 42:177–179.

83. Cui HM, Hedborg F, He LM, Sandstedt B, Pfeifer-Ohlsson S, Ohlsson R. Inactivation of H19, an imprinted putative tumor repressor gene, is a preneoplastic event during Wilms' tumorigenesis. *Cancer Res* 1997; 57:4469–4473.

84. Besnard-Guerin C, Newsham I, Winqvist R, Cavenee WK. A common region of loss of heterozygosity in Wilms' tumor and embryonal rhabdomyosarcoma distal to the D11S988 locus on chromosome 11p15.5. *Hum Genet* 1996; 97:163–170.

85. Sait SNJ, Nowak NJ, Singkahlon P, Weksberg R, Squire J, Shows TB, Higgins MJ. Localization of Beckwith-Wiedemann and rhabdoid tumor chromosome rearrangements to a defined interval in chromosome band 11p15.5. *Genes Chromosomes Cancer* 1994; 11:97–105.

86. Winqvist R, Mannermaa A, Alavaikko M, Blanco G, Taskinen PJ, Kiviniemi H, Newsham I, Cavenee W. Refinement of regional loss of heterozygosity for chromosome 11p15.5 in human breast tumors. *Cancer Res* 1993; 53:4486–4488.

87. Viel A, Giannini F, Tumiotto L, Sopracordevolle F, Visentin MC. Chromosomal localization of 2 putative 11p oncosuppressor genes. *Br J Cancer* 1992; 66:1030–1036.

88. Baffa R, Negrini M, Mes B, Rugge M, Ranzani GN, Hirohashi S, Croce CM. Loss of heterozygosity for chromosome 11 in adenocarcinoma of the stomach. *Cancer Res* 1996; 56:268–272.

89. Li XR, Adam G, Cui HM, Sandstedt B, Ohlsson R, Ekstrom TJ. Expression, promoter usage and parental imprinting status of insulin-like growth factor II (IGF2) in human hepatoblastoma; uncoupling of IGF2 and H19 imprinting. *Oncogene* 1995; 11:221–229.

90. Uyeno S, Aoki Y, Nata M, Sagisaka K, Kayama T, Yoshimoto T, Ono T. IGF2 but not H19 shows loss of imprinting in human glioma. *Cancer Res* 1996; 56:5356–5359.

91. Steenman M, Redeker B, de Meulemeester M, Wiesmeijer K, Voute PA, Westerveld A, Slater R, Mannens M. Comparative genomic hybridization analysis of Wilms tumors. *Cytogenet Cell Genet* 1997; 77:296–303.

92. Klamt B, Schulze M, Thate C, Mares J, Goetz P, Kodet R, et al. Allele loss in Wilms tumors of chromosome arms 11q, 16q, 22q correlates with clinical pathological parameters. *Genes Chromosomes Cancer* 1998; 22:287–294.

93. Grundy RG, Pritchard J, Scambler P, Cowell JK. Loss of heterozygosity on chromosome 16 in sporadic Wilms' tumour. *Br J Cancer* 1998; 78:1181–1187.

94. Green DM, Coppes MJ. Future directions in clinical research in Wilms' tumor. *Hematol Oncol Clin North Am* 1995; 9:1329.

95. Mannens M, Alders M, Redeker B, Bliek J, Steenman M, Wiesmeyer C, et al. Positional cloning of genes involved in the Beckwith-Wiedemann syndrome, hemihypertrophy, and associated childhood tumors. *Med Pediatr Oncol* 1996; 27:490–494.

96. Wilmore HP, White GFJ, Howell RT, Brown KW. Germline and somatic abnormalities of chromosome 7 in Wilms' tumor. *Cancer Genet Cytogenet* 1994; 77:93–98.

97. Miozzo M, Perotti D, Minoletti F, Mondini P, Pilotti S, Luksch R, et al. Mapping of a putative tumor suppressor locus to proximal 7p in Wilms tumors. *Genomics* 1996; 37:310–315.

98. Bardeesy N, Falkof D, Petruzzi M-J, Nowak N, Zabel B, Adam M, et al. Anaplastic Wilms' tumour, a subtype displaying poor prognosis, harbours p53 gene mutations. *Nat Gene* 1995; 7:91–97.

99. Bardeesy N, Beckwith JB, Pelletier J. Clonal expansion and attenuated apoptosis in Wilms' Tumors are associated with p53 gene mutations. *Cancer Res* 1995; 55:215–219.

100. Green DM, Beckwith JB, Breslow NE, Faria P, Mokness J, Finklestein JZ, et al. Treatment of children with stage-II to stage-IV anaplastic Wilms' tumor: a report from the National Wilms' Tumor Study Group. *J Clin Oncol* 1994b; 12:2126–2131.

101. Matsunaga E. Genetics of Wilms' tumor. *Hum Genet* 1981; 57:231–246.

102. Altura RA, Valentine M, Li H, Boyett JM, Shearer P, Grundy P, Shapiro DN, Look AT. Identification of novel regions of deletion in familial Wilm's tumor by comparative genomic hybridization. *Cancer Res* 1996; 56:3837–3841.

103. Huff V, Compton DA, Chao LY, Strong LC, Geiser CF, Saunders GF. Lack of linkage of familial Wilms' tumor to chromosomal band 11p13. *Nature* 1988; 336:377–378.

104. Grundy P, Koufos A, Morgan K, Li FP, Meadows AT, Cavenee WK. Familial predisposition to Wilms' tumor does not map to the short arm of chromosome 11. *Nature* 1988; 336:374–376.

105. Rahman N, Arbour L, Tonin P, Renshaw J, Pelletier J, Baruchel S, et al. Evidence for a familial Wilms' tumour gene (FWT1) on chromosome 17q12–q21. *Nat Genet* 1996; 13:461–463.

106. Schwartz CE, Haber DA, Stanton VP, Strong LC, Skolnick MH, Housman DE. Familial predisposition to Wilms tumor does not segregate with the WT1 gene. *Genomics* 1991; 10:927–930.

107. Coppes MJ, Liefers GJ, Higuchi M, Zinn AB, Balfe JW, Williams BRG. Inherited WT1 mutations in Denys-Drash Syndrome. *Cancer Res* 1992; 52:6125–6128.

108. Pelletier J, Bruening W, Li FP, Haber DA, Glaser T, Housman DE. WT1 mutations contribute to abnormal genital system development and hereditary Wilms' tumour. *Nature* 1991; 353:431–434.

109. Kaplinsky C, Ghahremani M, Frishberg Y, Rechavi G, Pelletier J. Familial Wilms' tumor associated with a WT1 zinc finger mutation. *Genomics* 1996; 38:451–453.

110. Knudson AG, Strong LC. Mutation and cancer: a model for Wilms' tumor of kidney. *J Natl Cancer Inst* 1972; 48:313–324.

111. Tonin P, Ehrenborg E, Lenoir G, Feunteun J, Lynch H, Morgan K, et al. The human insulin-like growth factor binding protein-4 gene maps to chromosome region 17q12–q211 and is close to the gene for hereditary breast/ovarian cancer. *Genomics* 1993; 18:414–417.

112. McDonald JM, Douglass EC, Fisher R, Geiser CF, Krill CE, Strong LC, Virshup D, Huff V. Linkage of familial Wilms' tumor predisposition to chromosome 19: a two-locus model for the etiology of familial tumors. *Cancer Res* 1998; 58:1387–1390.

113. Miyashita T, Krajewski S, Krajewska M, Wang HG, Lin HK, Liebermann DA, Hoffman B, Reed JC. Tumor-suppressor P53 is a regulator of Bcl-2 Bax gene expression in-vitro and in-vivo *Oncogene* 1994; 9:1799–1805.

114. Matsuda Y, Kusano H, Tsujimoto Y. Chromosomal assignment of the Bcl2-related genes, Bcl21 Bax, in the mouse and rat. *Cytogenet Cell Genet* 1996; 74:107–110.

115. Van Duin M, Van den Tol J, Warmerdam P, Odijk H, Meijer D, Westerveld A, Bootsma D, Hoeijmakers JHJ. Evolution and mutagenesis of the mammalian excision repair gene ERCC-1. *Nucleic Acids Res* 1988; 16:5305–5322.

116. Lazo PS, Dorfman K, Noguchi T, Mattei MG, Bravo R. Structure and mapping of the FosB gene FosB downregulates the activity of the FosB promoter. *Nucleic Acids Res* 1992; 20:343–350.

117. Charles AK, Brown KW, Berry PJ. Microdissecting the genetic events in nephrogenic rests Wilms' tumor development. *Am J Pathol* 1998; 153:991–1000.

118. Kent J, Coriat A-M, Sharpe PT, Hastie ND, Van Heyningen V. The evolution of WT1 sequence and expression pattern in the vertebrates. *Oncogene* 1995; 11:1781–1792.

119. Miles C, Elgar G, Coles E, Kleinjan DJ, van Heyningen V, Hastie N. Complete sequencing of the Fugu WAGR region from WT1 to PAX6: Dramatic compaction and conservation of synteny with human chromosome 11p13. *Proc Natl Acad Sci USA* 1998; 95:13068–13072.

120. Pelletier J, Schalling M, Buckler A, Rogers A, Haber DA, Housman D. Expression of the Wilms' tumor gene WT1 in the murine urogenital system. *Genes Dev* 1991, 5:1345–1356.

121. Pabo CO, Sauer RT. Transcription factors: structural families and principles of DNA recognition. *Annu Rev Biochem* 1992; 61:1053–1095.

122. Mitchell PJ, Tjian R. Transcriptional regulation in mammalian cells by sequence specific DNA binding proteins. *Science* 1989; 245:371–378.

123. Haber DA, Sohn RL, Buckler AJ, Pelletier J, Call KM, Housman DE. Alternative splicing genomic structure of the Wilms tumor gene WT1. *Proc Natl Acad Sci USA* 1991; 88:9618–9622.

124. Morris JF, Madden SL, Tournay OE, Cook DM, Sukhatme VP, Rauscher FJ. Characterization of the zinc finger protein encoded by the WT1 Wilms' tumor locus. *Oncogene* 1991; 6:2339–2348.

125. Bruening W, Pelletier J. A non-AUG translation initiation event generates novel WT1 isoform. *J Biol Chem* 1996; 271:8646–8654.

126. Sharma PM, Bowman M, Madden SL, Rauscher FJ III, Sukumar S. RNA editing in the Wilms' tumor susceptibility gene, WT1. *Genes Dev* 1994b; 8:720–731.

127. Rauscher FJ, Morris JF, Tournay OE, Cook DM, Curran T. Binding of the Wilms' tumor locus zinc finger protein to the EGR-1 consensus sequence. *Science* 1990; 250:1259–1262.

128. Bickmore WA, Oghene K, Little MH, Seawright A, van Heyningen V, Hastie ND. Modulation of DNA binding specificity by alternative splicing of the Wilms tumor wt1 gene transcript. *Science* 1992; 257:235–237.

129. Nakagama H, Heinrich G, Pelletier J, Housman DE. Sequence and structural requirements for high-affinity DNA-binding by the WT1 gene product. *Mol Cell Biol* 1995; 15:1489–1498.

130. Hamilton TB, Barilla KC, Romaniuk PJ. High affinity binding sites for the Wilms' tumour suppressor protein WT1. *Nucleic Acids Res* 1995; 23:277–284.

131. Wang Z-Y, Qiu Q-Q, Enger KT, Deuel T. A second transcriptionally active DNA-binding site for Wilms tumor gene product, WT1. *Proc Natl Acad Sci USA* 1993; 90:8896–8900.

132. Hastie ND. The genetics of Wilms' tumor: a case of disrupted development. *Annu Rev Genet* 1994; 28:523–558.

133. Pavletich NP, Pabo CO. Zinc finger-DNA recognition: crystal structure of a ZIF268-DNA complex at 21A. *Science* 1991; 252:809–817.

134. Menke AL, van der Eb, AJ Jochemsen AG. The Wilms' tumor 1 gene: oncogene or tumor suppressor gene? *Int Rev Cytol* 1998a; 181:151–212.

135. Reddy JC, Hosono S, Licht JD. The transcriptional effect of WT1 is modulated by choice of expression vector. *J Biol Chem* 1995b, 270:29,976–29,982.

136. Wang Z, Qiu Q, Huang J, Gurrieri M, Deuel TF. Products of alternatively spliced transcripts of the Wilms' tumor suppressor gene, WT1, have altered DNA binding specificity and regulate transcription in different ways. *Oncogene* 1995; 10:415–422.

137. Madden SL, Cook DM, Morris JF, Gashler A, Sukhatme VP, Rauscher FJ III. Transcriptional repression mediated by the WT1 Wilms tumor gene product. *Science* 1991; 253:1550–1553.

138. Drummond IA, Rupprecht HD, Rohwer-Nutter P, Lopez-Guisa JM, Madden SL, Rauscher FJ III, Sukhatme VP. DNA recognition by splicing variants of the Wilms' tumor suppressor, WT1. *Mol Cell Biol* 1994; 14:3800–3809.

139. Rupprecht HD, Drummond IA, Madden SL, Rauscher FJ III, Sukhatme VP. The Wilms' tumor suppressor gene WT1 is negatively autoregulated. *J Biol Chem* 1994; 269:6198–6206.

140. Ryan G, Steele-Perkins V, Morris JF, Rauscher FJ III, Dressler GR. Repression of Pax-2 by WT1 during normal kidney development. *Development* 1995; 121:867–875.

141. Madden SL, Cook DM, Rauscher FJ III. A structure-function analysis of transcriptional repression mediated by the WT1, Wilms' tumor suppressor protein. *Oncogene* 1993; 8:1713–1720.

142. Nachtigal MW, Hirokawa Y, Enyeart-Van Houten DL, Flanagan JN, Hammer GD, Ingraham HA. Wilms' tumor 1 and DAX-1 modulate the orphan nuclear receptor SF-1 in sex-specific gene expression *Cell* 1998; 93:445–454.

143. Maheswaran S, Englert C, Zheng G, Lee SB, Wong J, Harkin DPJ, et al. Inhibition of cellular proliferation by the Wilms tumor suppressor WT1 requires association with the inducible chaperone Hsp70. *Genes Dev* 1998; 12:1108–1120.

144. Ladomery M. Multifunctional proteins suggest connections between transcriptional post- and transcriptional processes. *Bioessays* 1997; 119:903–909.

145. Clemens KR, Wolf V, McBryant SJ, Zhang PH, Liao XB, Wright PE, Gottesfeld JM. Molecular basis for specific recognition of both RNA and DNA by a zinc finger protein. *Science* 1993; 260:530–533.

146. Arranz V, Harper F, Florentin Y, Puvion E, Kress M, Ernoult-Lange M. Human and mouse MOK2 proteins are associated with nuclear ribonucleoprotein components and bind specifically to RNA and DNA through their zinc finger domains. *Mol Cel Biol* 1997; 17:2116–2126.

147. Ward A, Pooler JA, Miyagawa K, Duarte A, Hastie ND, Caricasole A. Repression of promoters for the mouse insulin-like growth-factor II-encoding gene (IGF-2) by products of the Wilms' tumour suppressor gene WT1. *Gene* 1995; 167:239–243.

148. Caricasole A, Duarte A, Larsson SH, Hastie ND, Little M, Holmes G, Todorov I, Ward A. RNA binding by the Wilms tumor suppressor zinc finger proteins. *Proc Natl Acad Sci USA* 1996; 93:7562–7566.

149. Larsson SH, Charlieu JP, Miyagawa K, Engelkamp D, Rassoulzadegan M, Ross A, et al. Subnuclear localization of WT1 in splicing or transcription factor domains is regulated by alternative splicing. *Cell* 1995; 81:391–401.

150. Nyman U, Hallman H, Hadlaczky G, Petterson I, Sharp G, Ringertz N. Intranuclear localization of SnRNP antigens. *J Cell Biol* 1986; 102:137–144.

151. Fu X-D, Maniatis T. Factor required for mammalian spliceosome assembly is localized to discrete regions in the nucleus. *Nature* 1990; 343:437–441.

152. Spector D, Fu X-D, Maniatis T. Associations between distinct pre-mRNA splicing components the cell nucleus. *EMBO J.* 1991; 10:3467–3481.

153. Spector D. Macromolecular domains within the cell nucleus. *Annu Rev Cell Biol* 1993; 9:265–315.

154. Menke AL, Riteco N, Van Ham RCA, De Bruyne C, Rauscher FJ III, Van der Eb AJ, Jochemsen AG. Wilms' tumor 1 splice variants have opposite effects on the tumorigenicity of adenovirus-transformed baby-rat kidney cells. *Oncogene* 1996; 12:537–546.

155. Menke AL, Shvarts A, Riteco N, Van Ham RCA, Van der Eb AJ, Jochemsen AG. Wilms' tumor 1: KTS isoforms induce p53-independent apoptosis that can be partially rescued by expression of the epidermal growth factor receptor or the insulin receptor. *Cancer Res* 1997; 57:1353–1363.

156. Englert C, Vidal M, Maheswaran S, Ge Y, Ezzell R, Isselbacher KJ, Haber DA. Truncated WT1 mutants alter the subnuclear localization of the wild-type protein. *Proc Natl Acad Sci USA* 1995; 92:11,960–11,964.

157. Kennedy D, Ramsdale T, Mattick J, Little M. RNA recognition motif in Wilms' tumour protein (WT1) revealed by structural modelling. *Nat Genet* 1996; 12:329–332.

158. Davies RC, Calvio C, Bratt E, Larsson SH, Lamond AI, Hastie ND. WT1 interacts with the splicing factor U2AF65 in an isoform-dependent manner and can be incorporated into spliceosomes. *Genes Dev* 1998; 12:3217–3225.

159. Bardeesy N, Pelletier J. Overlapping RNA and DNA binding domains of the wt1 tumor suppressor gene product. *Nucleic Acids Res* 1998; 26:1784–1792.

160. Armstrong JF, Pritchard-Jones K, Bickmore WA, Hastie ND, Bard JBL. Expression of the Wilms' tumour gene, WT1, in the developing mammalian embryo. *Mech Dev* 1992; 40:85–97.

161. Kreidberg JA, Sariola H, Loring JM, Maeda M, Pelletier J, Housman D, Jaenisch R. WT-1 is required for early kidney development. *Cell* 1993; 74:679–691.

162. Grubb GR, Yun K, Williams BRG, Eccles MR, Reeve AE. Expression of the WT1 protein in fetal kidneys and Wilms tumors. *Lab Invest* 1994; 71:472–479.

163. Lin LFH, Doherty DH, Lile JD, Bektesh S, Collins F. GDNF A glial-cell line derived neurotrophic factor for midbrain dopaminergic neurons. *Science* 1993; 260:1130–1132.

164. Sanchez MP, Silos-Santiago I, Frisen J, He B, Lira SA, Barbacid M. Renal agenesis and the absence of enteric neurons in mice lacking GDNF. *Nature* 1996; 382:70–73.

165. Schuchardt A, D'Agati V, Larsson-Blomberg L, Constantini F, Pachnis V. Defects in the kidney and enteric nervous system of mice lacking the tyrosine kinase receptor ret. *Nature* 1994; 367:380–383.

166. Herzlinger D, Qiao J, Cohen D, Ramakrishna N, Brown AMC. Induction of kidney epithelial morphogenesis by cells expressing Wnt-1. *Dev Biol* 1994; 166:815–818.

167. Stark K, Vanio S, Vassileva G, McMahon AP. Wnt-4 regulates epithelial transformation of metanephric mesenchyme in the developing kidney. *Nature* 1994; 372:679–683.

168. Dressler GR, Douglass EC. Pax-2 is a DNA-binding protein expressed in embryonic kidney and Wilms' tumor. *Proc Natl Acad Sci USA* 1992; 89:1179–1183.

169. Torres M, Gomez-Pardo E, Dressler GR, Gruss P. Pax-2 controls multiple steps of urogenital development. *Development* 1995; 121:4057–4065.

170. Dehbi M, Gharemani M, Lechner M, Dressler G, Pelletier J. The paired-box transcription factor, PAX2, positively modulates expression of the Wilms' tumor suppressor gene (WT1). *Oncogene* 1996; 13:447–453.

171. Moore AW, McInnes L, Kreidberg J, Hastie N, Schedl A. Yac complementation shows a requirement for WT1 in the development of epicardium, adrenal gland and throughout nephrogenesis. *Development* 1999; 126:1845–1857.

172. Mundel P, Kriz W. Structure and function of podocytes: an update. *Anat Embryol* 1995; 192:385–397.

173. Jadresic L, Leake J, Gordon I, Dillon MJ, Grant DB, Pritchard J, Risdon RA, Barrat TM. Clinicopathologic review of twelve children with nephropathy, Wilms' tumor, genital abnormalities (Drash syndrome). *J Pediatr* 1990; 117:717–725.

174. Kudoh T, Ishidate T, Moriyama M, Toyoshima K, Akiyama T. G1 phase arrest induced by Wilms tumor protein WT1 is abrogated by cyclin/cdk complexes. *Proc Natl Acad Sci USA* 1995; 92:4517–4521.

175. Hewitt SM, Saunders GF. Differentially spliced exon 5 of the Wilms' tumor gene WT1 modifies gene function. *Anticancer Res* 1996; 16:621–626.

176. Englert C, Maheswaran S, Garvin AJ, Kreidberg J, Haber DA. Induction of p21 by the Wilms' tumor suppressor gene WT1. *Cancer Res* 1997; 57:1429–1434.

177. Guan LS, Rauchman M, Wang ZY. Induction of Rb-associated protein (RbAp46) by Wilms' tumor suppressor WT1 mediates growth inhibition. *J Biol Chem* 1998; 273:27,047–27,050.

178. Yamagami T, Ogawa H, Tamaki H, Oji Y, Soma T, Oka Y, et al. Suppression of Wilms' tumor gene (WT1) expression induces G(2)/M arrest in leukemic cells. *Leukemia Res* 1998; 22:383–384.

179. Yamagami T, Sugiyama H, Inoue K, Ogawa H, Tatekawa T, Hirata M, et al. Growth inhibition of human leukemic cells by WT1 (Wilms tumor gene) antisense oligodeoxynucleotides: implications for the involvement of WT1 in leukemogenesis. *Blood* 1996; 87:2878–2884.

180. Brenner BM. Determinants of epithelial differentiation during early nephrogenesis. *J Am Soc Nephrol* 1990; 1:127–139.
181. Park S, Schalling M, Bernard A, Maheswaran S, Shipley GC, Roberts D. Wilms tumour gene WT1 is expressed in murine mesoderm-derived tissues and mutated in a human mesothelioma. *Nat Genet* 1993; 4:415–420.
182. Zhou J, Rauscher FJ III, Bondy C Wilms' tumor (WT1) gene expression in rat decidual differentiation. *Differentiation* 1993; 5:109–114.
183. Hosono S, Luo X, Hyink DP, Schnapp LM, Wilson PD, Burrow CR, et al. WT1 expression induces features of renal epithelial differentiation in mesenchymal fibroblasts. *Oncogene* 1999; 417–427.
184. Miyagawa K, Kent J, Moore A, Charlieu JP, Little MH, Williamson KA, et al. Loss of WT1 function leads to ectopic myogenesis in Wilms' tumour. *Nat Genet* 1998; 18:15–17.
185. Smith SI, Weil D, Johnson GR, Boyd AW, Li CL. Expression of the Wilms' tumor suppressor gene, WT1, is upregulated by leukemia inhibitory factor and induces monocytic differentiation in M1 leukemic cells. *Blood* 1998; 91:764–773.
186. Svedberg H, Chylicki K, Baldetorp B, Rauscher FJ, Gullberg U. Constitutive expression of the Wilms' tumor gene (WT1) in the leukemic cell line U937 blocks parts of the differentiation program. *Oncogene* 1998; 16:925–932.
187. Scharnhorst V, Kranenburg O, Van der Eb AJ, Jochemsen AG. Differential regulation of the Wilms' Tumor gene, WT1 during differentiation of embryonal carcinoma and embryonic stem cells. *Cell Growth Differ* 1997; 8:133–143.
188. Sekiya M, Adachi M, Hinoda Y, Imai K, Yachi A. Downregulation of Wilms' tumor 1 gene (WT1) during myelomonocytic differentiation in HL60 cells. *Blood* 1994; 83:1876–1882.
189. Phelan SA, Linberg C, Call KM. Wilms' tumor gene WT1, mRNA is downregulated during induction of erythroid and megakaryocytic differentiation of K562 cells. *Cell Growth Differ* 1994; 5:677–686.
190. Algar EM, Khromykh T, Smith SI, Blackburn DM, Bryson GJ, Smith PJ. A WT1 antisense oligonucleotide inhibits proliferation and induces apoptosis in myeloid leukaemia cell lines. *Oncogene* 1996; 12:1005–1014.
191. Coles HSR, Burne JF, Raff MC. Large scale normal cell death in the developing rat kidney and its reduction by epidermal growth factor. *Development* 1993; 118:777–784.
192. Maheswaran S, Englert C, Bennett P, Heinrich G, Haber DA. The WT1 gene product stabilizes p53 and inhibits p53 mediated apoptosis. *Genes Dev* 1995; 9:2143–2156.
193. Englert C, Hou X, Maheswaran S, Bennett P, Ngwu C, Re GG, et al. WT1 suppresses synthesis of the epidermal growth factor receptor and induces apoptosis. *EMBO J* 1995b; 14:4662–4675.
194. Bellamy COC, Malcomson RDG, Harrison DJ, Wyllie A. Cell death in health and disease: the biology regulation of apoptosis. *Cancer Biol* 1995; 6:3–16.
195. Kerr JFR, Winterford CM, Harmon BV. Apoptosis: its significance in cancer and cancer therapy. *Cancer* 1994; 73:2013–2026.
196. Sen S. Programmed cell death: concept, mechanism and control. *Biol Rev* 1992; 67:287–319.
197. Hoffman B, Liebermann DA. Molecular controls of apoptosis: differentiation/growth arrest primary response genes, proto-oncogenes and tumor suppressor genes as positive and negative modulators. *Oncogene* 1994; 9:1807–1812.
198. Miyashita T, Reed JC. Bcl-2 gene transfer increases relative resistance of S491 and WEHI72 lymphoid cells to cell death and DNA fragmentation induced by glucocorticoids and multiple chemotherapeutic drugs. *Cancer Res* 1992; 52:5407–5411.
199. Miyashita T, Reed JC. Bcl-2 oncoprotein blocks chemotherapy-induced apoptosis in a human leukemia cell line. *Blood* 1993; 81:151–157.
200. Vaux DL, Cory S, Adams JM. Bcl-2 gene promotes haemopoietic cell survival and cooperates with c-myc to immortalize pre-B cells. *Nature* 1988; 355:440–442.
201. Selvakumaran M, Lin HK, Sjin RT, Reed JC, Liebermann DA, Hoffman B. The novel primary response gene MyD118 and the proto-oncogenes myb, myc, and bcl-2 modulate transforming growth factor beta 1-induced apoptosis of myeloid leukemia cells. *Mol Cell Biol* 1994; 14:2352–2360.
202. Zong L, Sarafian T, Kane DJ, Charles AC, Mah SP, Edwards RH, Bredesen, DE. Bcl-2 inhibits death of central neural cells induced by multiple agents. *Proc Natl Acad Sci USA* 1993; 90:4533–4537.
203. Evans GI, Wyllie AH, Gilber CS, Littlewood TDH, Brooks M, Waters CM, Penn LZ, Hancock DC. Induction of apoptosis in fibroblasts by c-myc protein. *Cell* 1992; 69:119–128.

204. Sakamuro D, Eviner V, Elliot KJ, Showe L, White E, Prendergast GC. c-Myc induces apoptosis in epithelial cells by both p53-dependent and p53-independent mechanisms. *Oncogene* 11:2411–2418, 1995.

205. Hoffman-Liebermann B, Liebermann DA. Interleukin-6 leukemia inhibitory factor-induced terminal differentiation of myeloid leukemic cells is blocked at an intermediate stage by constitutive c-myc expression. *Mol Cell Biol* 1991; 11:2375–2381.

206. Clarke MF, Kukowska-Latallo JF, Westin E, Smith M, Prochovnik EV. Constitutive expression of a c-myb cDNA blocks erythroleukemia cell differentiation. *Mol Cell Biol* 1988; 8:884–892.

207. Lin J-K, Chou C-K. In vitro apoptosis in the human hepatoma cell line induced by transforming growth factor b1. *Cancer Res* 1992; 52:385–388.

208. Ponchel F, Puisieux A, Tabone E, Michot JP, Froschl G, Morel AP, et al. Hepatocarcinoma-specific mutant p53-249ser induces mitotic activity but has no effect on transforming growth factor b1-mediated apoptosis. *Cancer Res* 1994; 54:2064–2068.

209. Oberhammer FA, Pavelka M, Sharma S, Tiefenbacher R, Purchio AF, Bursch W, Schulte-Hermann R. Induction of apoptosis in cultured hepatocytes and in regressing liver by transforming growth factor b1. *Proc Natl Acad Sci USA* 1992; 89:5408–5412.

210. Resnicoff M, Burgaud J-L, Rotman HL, Abraham D, Baserga R. Correlation between apoptosis, tumorigenesis, and levels of insuline growth factor I receptors. *Cancer Res* 1995; 55:3739–3741.

211. Frisch SM, Francis H. Disruption of epithelial cell-matrix interactions induces apoptosis. *J Cell Biol* 1994; 124:619–626.

212. Bates RC, Buret A, Van Helden DF, Horton MA, Burns GF. Apoptosis induced by inhibition of intercellular contact. *J Cell Biol* 1994; 125:403–415.

213. Kudoh T, Ishidate T, Nakamura T, Toyoshima K, Akiyama T. Constitutive expression of the Wilms tumor suppressor gene WT1 in F9 embryonal carcinoma cells induces apoptotic cell death in response to retinoic acid. *Oncogene* 1996; 13:1431–1439.

214. Liu J-P, Baker J, Perkins AS, Robertson EJ, Efstradiadis A. Mice encoding null mutations of the genes encoding insulin-like growth factor I (Igf-1) and type 1 IGF receptor (Igf1r). *Cell* 1993a; 75:59–72.

215. Liu ZZ, Wada J, Alvares K, Kumar A, Wallner EI, Kanwar YS. Distribution and relevance of insulin-like growth factor-I receptor in metanephric development. *Kidney Int* 1993b; 44:1242–1250.

216. Moxham CM, Hod Y, Malbon CC. Gi alpha 2 mediates the inhibitory regulation of adenylcyclase in vivo: analysis in transgenic mice with Gi alpha suppressed by inducible antisense RNA. *Dev Genet* 1993; 14:266–273.

217. Keller SA, Jones JM, Boyle A, Barrow LL, Killen PD, Green DG, et al. Kidney and retinal defects (krd), a transgene-induced mutation with a deletion of mouse chromosome 19 that includes the Pax2 locus. *Genomics* 1994; 23:309–320.

218. Veis DJ, Sorenson CM, Shutter JR, Korsmeyer SJ. Bcl-2-deficient mice demonstrate fulminant lymphoid apoptosis, polycystic kidneys, and hypopigmented hair. *Cell* 1993; 75:229–240.

219. Gerald WL, Gramling TS, Sens DA, Garvin AJ. Expression of the 11p13 Wilms' tumor gene, WT1, correlates with histologic category of Wilms' tumor. *Am J Pathol* 1992; 140:1031–1037.

220. Yeger H, Cullinane C, Flenniken A, Chilton-MacNeil S, Campbell C, Huang A, et al. Coordinate expression of Wilms' tumor genes correlates with Wilms' tumor phenotypes. *Cell Growth Differ* 1992; 3:855–864.

221. Pritchard-Jones K, Flemming S. Cell types expressing the Wilms' tumour gene (WT1) in Wilms' tumours: implications for tumour histogenesis. *Oncogene* 1991; 6:2211–2220.

222. Schumacher V, Schneider S, Figge A, Wildhardt G, Harms D, Schmidt D, et al. Correlation of germline mutations and two-hit inactivation of the WT1 gene with Wilms tumors of stromal-predominant histology. *Proc Natl Acad Sci USA* 1997; 94:3972–3977.

223. Royer-Pokora B, Ragg S, Heckl-Ostreicher B, Held M, Loos U, Call K, et al. Direct pulsed field gel electrophoresis of Wilms tumors shows that DNA deletions in 11p13 are rare. *Genes, Chromosomes Cancer* 1991; 3:89–100.

224. Mannens M, Hoovers JM, Bleeker-Wagemakers EM, Redeker E, Bliek J, Overbeeke-Melkert M, et al. The distal region of 11p13 and associated genetic-diseases. *Genomics* 1991; 11:284–293.

225. Cowell JK, Wadey RB, Haber DA, Call KM, Housman DE, Pritchard J. Structural rearrangements of the WT1 gene in Wilms' tumour cells. *Oncogene* 1991; 6:595–599.

226. Tadokoro K, Fujii H, Oshima A, Kakizawa Y, Shimizu K, Sakai A, et al. Intragenic homozygous deletion of the WT1 gene in Wilms' tumor. *Oncogene* 1992; 7:1215–1221.

227. Kikuchi H, Akasaka Y, Nagai T, Umezawa A, Iri H, Kato S, Hata J. Genomics changes in the WT1-gene (WT1) in Wilms' tumors and their correlation with histology. *Am J Pathol* 1992; 140:781–786.

228. Radice P, Pilotti S, De Benedetti V, Mondini P, Miozzo M, Luksch R, et al. Homozygous intragenic loss of the WT1 locus in a sporadic intralobular Wilms' tumor. *Int J Cancer* 1993; 55:174–176.

229. Waber PG, Chen J, Nisen PD. Infrequency of ras, p53, WT1, or Rb gene alterations in Wilms' Tumor. *Cancer* 1993; 72:3732–3738.

230. Little MH, Dunn R, Byrne JA, Seawright A, Smith PJ, Pritchard-Jones K. Equivalent expression of paternally and maternally inherited WT1 alleles in normal fetal tissue and Wilms' tumours. *Oncogene* 1992a; 7:635–641.

231. Benedict WF, Weissman BE, Mark C, Stanbridge EJ. Tumorigenicity of human HT1080 fibroblasts x normal fibroblast hybrids is chromosome dosage dependent. *Cancer Res* 1984; 44:3471–3479.

232. Denys P, Malvaux P, Van den Berghe H, Tanghe W, Proesmans W. Association d'un syndrome anato-mopathologique de pseudohermaphrodisme masculin, d'une tumeur de Wilms d'une nephropathie parenchymateuse et d'un mosaicisme XX/XY. *Arch Fr Pediatr* 1967; 24:729–739.

233. Drash A, Sherman F, Hartmann WH, Blizzard RM. A syndrome of pseudohermaphroditism, Wilms' tumor, hypertension degenerative renal disease. *J Pediatr* 1970; 76:585–593.

234. Reddy JC, Morris JC, Wang J, English MA, Haber DA, Shi Y, Licht JD. WT1-mediated transcriptional activation is inhibited by dominant negative mutant proteins. *J Biol Chem* 1995a; 270:10,878–10,884.

235. Guan LS, Liu JJ, Xu YH, Wang ZY. A point mutation within exon 5 of the WT1 gene of a sporadic unilateral Wilms' tumor alters gene function. *Cancer Res* 1998; 58:4180–4184.

236. Park S, Tomlinson G, Nisen P, Haber DA. Altered trans-activational properties of a mutated WT1 gene product in a WAGR-associated Wilms' tumor. *Cancer Res* 1993; 53:4757–4760.

237. Williamson KA, Van Heyningen V. Towards an understanding of Wilms' tumour. *Int J Exp Pathol* 1994; 75:147–155.

238. Coppes MJ, Liefers GJ, Paul P, Yeger H, Williams BRG. Homozygous somatic WT1 point mutations in sporadic unilateral Wilms' tumor. *Proc Natl Acad Sci USA* 1993; 90:1416–1419.

239. Van Heyningen V, Hastie ND. Wilms' tumour: reconciling genetics and biology. *Trends in Genetics* 1992; 8:16–21.

240. Reddy JC, Licht JD. The WT1 Wilms' tumor suppressor gene: How much do we really know? *Biochim Biophys Acta* 1996; 1287:1–28.

241. Gessler M, Konig A, Arden K, Grundy P, Orkin S, Sallan S, et al. Infrequent mutation of the WT1 gene in 77 Wilms' tumors. *Hum Mutat* 1994; 3:212–222.

242. Gessler M, Konig A, Bruns GA. The genomic organization expression of the WT1 gene. *Genomics* 1992; 12:807–813.

243. Brenner B, Wildhardt G, Schneider S, Royer-Pokora B. RNA polymerase chain reaction detects different levels of four alternatively spliced WT1 transcripts in Wilms' tumors. *Oncogene* 1992; 7:1431–1433.

244. Little MH, Prosser J, Condie A, Smith PJ, Van Heyningen V, Hastie ND. Zinc finger point mutations within the WT1 gene in Wilms tumor patients. *Proc Natl Acad Sci USA* 1992b; 89:4791–4795.

245. Rackley RR, Flenniken AM, Kuriyan NP, Kessler PM, Stoler MH, Williams BRG. Expression of the Wilms' tumor suppressor gene WT1 during mouse embryogenesis. *Cell Growth Differ* 1993; 4:1023–1031.

246. Hewitt SM, Fraizer GC, Wu YJ, Rauscher FJ III, Saunders GF. Differential function of the Wilms' tumor gene WT1 splice isoforms in transcriptional regulation. *J Biol Chem* 1996a; 271:8588–8592.

247. Hofmann W, Royer H-D, Drechsler M, Schneider S, Royer-Pokora B. Characterization of the transcriptional regulatory region of the human WT1 gene. *Oncogene* 1993; 8:3123–3132.

248. Cohen HT, Bossone SA, Zhu G, McDonald GA, Sukhatme VP. Spl is a critical regulator of the Wilm'tumor 1 gene. *J Biol Chem* 1997; 272:2901–2913.

249. Dehbi M, Hiscott J, Pelletier J. Activation of the wt1 Wilms' tumor suppressor gene by NF-kappa b. *Oncogene* 1998; 16:2033–2039.

250. Dehbi M, Pelletier J. Pax-8-mediated activation of the WT1 tumor suppressor gene. *EMBO J* 1996b; 16:4297–4306.

251. Hewitt SM, Fraizer GC, Saunders GF. Transcriptional silencer of the Wilms' tumor gene WT1 contains an alu repeat. *J Biol Chem* 1995b; 270:17,908–17,912.

252. Fraizer GC, Wu Y-J, Hewitt SM, Maity T, Ton CCT, Huff V, Saunders GF. Transcriptional regulation of the human wilms tumor gene (WT1). *J Biol Chem* 1994; 269:8892–8900.

253. Wu Y-J, Fraizer GC, Saunders GF. GATA-1 transactivates the WT1 hematopoietic specific enhancer. *J Biol Chem* 1995; 270:5944–5949.

254. Grubb GR, Yun K, Reeve AE, Eccles MR. Exclusion of the Wilms tumour gene (WT1) promoter as a site of frequent mutation in Wilms tumour. *Oncogene* 1995; 10:1677–1681.

255. Eccles MR, Grubb G, Ogawa O, Szeto J, Reeve AE. Cloning of novel Wilms tumor gene (WT1) cDNAs; evidence for antisense transcription of WT1. *Oncogene* 1994; 9:2059–2063.

256. Malik KTA, Wallace JI, Ivins SM, Brown KW. Identification of an antisense WT1 promoter in intron 1: implications for WT1 gene regulation. *Oncogene* 1995; 11:1589–1595.

257. Menke AL, Van Ham RCA, Sonneveld E, Shvarts A, Stanbridge EJ, Miyagawa K, Van der Eb AJ, Jochemsen AG. Human chromosome 11 suppresses the tumorigenicity of adenovirus transformed baby rat kidney cells: Involvement of the Wilms' tumor 1 gene. *Int J Cancer* 1995; 63:76–85.

258. McMaster ML, Gessler M, Stanbridge EJ, Weissman BE. WT1 expression alters tumorigenicity of the G401 kidney derived cell line. *Cell Growth Differ* 1995; 6:1609–1617.

259. Simms LA, Algar EM, Smith PJ. Splicing of exon 5 in the WT1 gene is disrupted in Wilms' tumour. *Eur J Cancer* 1995; 31:2270–2276.

260. Liu J-J, Wang Z-Y, Deuel TF, Xu Y-H. Imbalanced expression of functionally different WT1 isoforms may contribute to sporadic unilateral Wilms' tumor. *Biochem Biophy Res Commun* 1999; 254:197–199.

261. Paik S, Rosen N, Jung W, You JM, Lippman ME, Perdue JF, Yee D. Expression of insulin-like growth factor-II mRNA in fetal kidney and Wilms' tumor. *Lab Invest* 1989; 61:522–526, 1989.

262. Reeve AE, Eccles MR, Wilkins RJ, Bell GI, Millow LJ. Expression of insulin-like growth factor-II transcripts in Wilms' tumour. *Nature* 1985; 317:258–260.

263. Haselbacher GK, Irminger J-C, Zapf J, Ziegler WH, Humbel RE. Insulin-like growth factor-II in human adrenal pheochromocytomas and Wilms' tumors: expression at the messenger-RNA and protein level. *Proc Natl Acad Sci USA* 1987; 84:1104–1106.

264. Scott J, Cowell J, Robertson ME, Priestley LM, Wadey R, Hopkins B, et al. Insulin-like growth factor-II gene expression in Wilms' tumour embryonic tissues. *Nature* 1985; 317:260–262.

265. Nichols KE, Re GG, Yan YX, Garvin AJ, Haber DA. WT1 induces expression of insulin-like growth factor 2 in Wilms' tumor cells. *Cancer Res* 1995; 55:4540–4543.

266. Caceres JF, Stamm S, Helman D, Krainer A. Regulation of alternative splicing in vivo by overexpression of antagonistic splicing factors. *Science* 1994; 256:1706–1709.

267. Li L, Zhou J, James G, Heller-Harrison R, Cxech MP, Olson EN. FGF activates myogenic helix-loop-helix proteins through phosphorylation of a conserved protein kinase C site in their DNA binding domains. *Cell* 1992; 71:1181–1194.

268. Mosialos G, Hamer P, Capobianco AJ, Laursen RA, Gilmore TD. A protein kinase-A recognition sequence is structurally linked to transformation by 59v-rel and cytoplasmic retention of p68c-rel. *Mol Cell Biol* 1991; 11:5867–5877.

269. Rihs HP, Jans DA, Fan H, Peters R. The rate of nuclear cytoplasmic protein transport is determined by the casein kinase II site flanking the nuclear localization sequence of the SV40 T-antigen. *EMBO J* 1991; 10:633–639.

270. Li CCH, Dai RM, Chen E, Longo DL. Phosphorylation of NF-kappa-B activation and stable DNA-binding. *J Cell Biochem* 1995; S19A:39.

271. Ye Y, Raychaudhuri B, Gurney A, Campbell CE, Williams BRG. Regulation of WT1 by phosphorylation: inhibition of DNA-binding, alteration of transcriptional activity and cellular translocation. *EMBO J* 1996; 15:5606–5615.

272. Sakamoto Y, Yoshida M, Semba K, Hunter T. Inhibition of the DNA-binding and transcriptional repression activity of the Wilms' tumor gene product, WT1, by cAMP-dependent protein kinase-mediated phosphorylation of Ser-365 and Ser-393 in the zinc finger domain. *Oncogene* 1997; 15:2001–2012.

273. Wang Z, Qiu Q, Gurrieri M, Huang J, Deuel TF. WT1, the Wilms' tumor suppressor gene product, represses transcription through an interactive nuclear protein. *Oncogene* 1995b; 10:1243–1247.

274. Maheswaran S, Park S, Bernard A, Morris JF, Rauscher FJ, Hill DE, Haber DA. Physical functional interaction between WT1 and p53 proteins. *Proc Natl Acad Sci USA* 1993; 90:5100–5104.

275. Johnstone RW, See RH, Sells SF, Wang J, Muthukkumar S, Englert C, et al. A novel reppressor, par-4, modulates transcription and growth suppression functions of the Wilms' tumor suppressor WT1. *Mol Cell Biol* 1996; 16:6945–6956.

276. Wang Z-Y, Qiu Q-Q, Seufert W, Taguchi T, Testa JR, Whitmore SA, et al. Molecular cloning of the cDNA and chromosome localization of the gene for human ubiquitin-conjugating enzyme 9. *J Biol Chem* 1996; 271:24811–24816.

277. Johnstone RW, Wang J, Tommerup N, Vissing H, Roberts T, Shi Y. Ciao 1 is a novel WD40 protein that interacts with the tumor suppressor protein WT1. *J Biol Chem* 1998; 273:10,880–10,887.

278. Seufert W, Futcher B, Jentsch S. Role of a ubiquitin-conjugating enzyme in degradation of S- M-phase cyclins. *Nature* 1995; 373:78–81.

279. Gashler A, Sukhatme VP. Early growth response protein 1 (Egr-1): prototype of a zinc-finger family of transcription factors. *Prog Nucleic Acid Res Mol Biol* 1995; 50:191–224.

280. Rauscher FJ III. WT1 Wilms tumor gene product: a developmentally regulated transcription factor in the kidney that functions as a tumor suppressor. *FASEB J* 1993; 7:896–903.

281. Wang ZY, Madden SL, Deuel TF, Rauscher FJ III. The Wilms' tumor gene product, WT1, represses transcription of the platelet-derived growth factor A-chain gene. *J Biol Chem* 1992; 267:21,999–22,002.

282. Kinane TB, Finder JD, Kawashima A, Brown D, Abbate M, Shang C, et al. Growth of LLC-PK1 renal cells is mediated by Egr-1 up-regulation of G-protein alpha-I-2 protooncogene transcription. *J Biol Chem* 1994; 269:27,503–27,509.

283. Kinane TB, Finder JD, Kawashima A, Brown D, Abbate M, Fredericks WJ, et al. LLC-PK1 cell growth is repressed by WT1 inhibition of G-protein a_{i-2} protooncogene transcription. *J Biol Chem* 1996; 271:30760–30764.

284. Moshier JA, Skunca M, Wu W, Boppana SM, Rauscher FJ III, Dosescu J. Regulation of ornithine decarboxylase gene expression by the Wilms' tumor suppressor WT1. *Nucleic Acids Res* 1996; 24:1149–1157.

285. Khachigian LM, Williams AJ, Collins T. Interplay of Sp1 and Egr-1 in the proximal platelet-derived-growth-factor-A-chain promoter in cultured vascular endothelial-cells. *J Biol Chem* 1995; 270:27,679–27,686.

286. McCann S, Sullivan J, Guerra J, Arcinas M, Boxer LM. Repression of the c-myb gene by WT1 protein in T and B cell lines. *J Biol Chem* 1995; 270:23,785–23,789.

287. Werner H, Shen-Orr Z, Rauscher FJ II, Morris JF, Roberts CT Jr, LeRoith D. Inhibition of cellular proliferation by the Wilms tumr suppressor WT1 is associated with suppression of insulin-like growth factor I receptor gene expression. *Mol Cell Biol* 1995; 15:3516–3522.

288. Werner H, Hernez-Sanchez C, Karnieli E, Leroith D. The regulation of the IGF-I receptor expression. *Int J Biochem Cell Biol* 1995; 27:987–994.

289. Deb SP, Munoz RM, Brown DR, Subler MA, Deb S. Wild-type human p53 activates the human epidermal growth factor receptor promoter. *Oncogene* 1994; 9:1341–1349.

290. Liebermann DA, Hoffman B. Differentiation primary response genes and proto-oncogenes as positive and negative regulators of terminal hematopoietic cell differentiation. *Stem Cells* 1994; 12:352–369.

291. Kistner A, Gossen M, Zimmermann F, Jerecic J, Ullmer C, Lubbert H, Bujard H. Doxycycline-mediated quantitative tissue-specific control of gene expresion in transgenic mice. *Proc Natl Acad Sci USA* 1996; 93:10,933–10,938.

292. Menke A, McInnes L, Hastie ND, Schedl A. The Wilms' tumor suppressor WT1: approaches to gene function. *Kidney Int* 1998b; 53:1512–1518.

293. Inoue K, Sugiyama H, Ogawa H, Nakagawa M, Yamagami T, Miwa H, et al. WT1 as a new prognostic factor and a new marker for the detection of minimal residual disease in acute leukemia. *Blood* 1994; 84:3071–3079.

294. Lemoine NR, Hughes CM, Cowell JK. Abberant expression of the tumor suppressor gene P53 is very frequent in Wilms' tumors. *J Pathol* 1992; 168:237–242.

295. Kikuchi H, Takata A, Akasaka Y, Fukuzawa R, Yoneyama H, Kurosawa Y, et al. Do intronic mutations affecting splicing of WT1 exon 9 cause Frasier syndrome? *J Med Genet* 1998; 35:45–48.

296. Menssen HD, Renkl HJ, Rodeck U, Maurer J, Notter M, Schwartz S, Reinhardt R, Thiel E. Presence of Wilms' tumor gene (WT1) transcripts and the WT1 nuclear protein in the majority of human acute leukemias. *Leukemia* 1995; 9:1060–1067.

297. Menssen HD, Renkl HJ, Rodeck U, Kari C, Schwartz S, Thiel E. Detection by monclonal antibodies of the Wilms' Tumor (WT1) nuclear protein in patients with acute leukemia. *Int J Cancer* 1997; 70:518–523.

298. Brieger J, Weidmann E, Fenchel K, Mitrou PS, Hoelzer D, Bergmann L. The expression of the Wilms' tumor gene in acute myelocytic leukemias as a possible marker for leukemic blast cells. *Leukemia* 1994; 8:2138–2143.

299. Coppes MJ, Ye Y, Rackley R, Zhao X-I, Liefers GJ, Casey G, Williams BRG. Analysis of WT1 in granulosa cell and other sex cord-stromal tumors. *Cancer Res* 1993b; 53:2712–2714.
300. Bruening W, Gros P, Sato T, Stanimir J, Nakamura Y, Housman D, Pelletier J. Analysis of the 11p13 Wilms' tumor suppressor gene (WT1) in ovarian tumors. *Cancer Invest* 1993; 11:393–399.
301. Miwa H, Beran M, Saunders GF. Expression of the Wilms tumor gene (WT1) in human leukemias. *Leukemia* 1992; 6:405–409.
302. Rodeck U, Bossler A, Kari C, Humphreys CW, Gyorfi T, Maurer J, Thiel E, Menssen HD. Expression of the WT1 Wilms' tumor gene by normal and malignant melanocytes. *Int J Cancer* 1994; 59:78–82.
303. Amin KM, Litzky LA, Smythe WR, Mooney AM, Morris JM, Mews DJY, et al. Wilms' Tumor 1 susceptibility (WT1) gene products are selectively expressed in malignant mesothelioma. *Am J Pathol* 1995; 146:344–356.
304. Silberstein GB, Van Horn K, Strickl P, Roberts CT, Daniel CW. Altered expression of the WT1 Wilms tumor suppressor gene in human breast cancer. *Proc Natl Acad Sci USA* 1997; 94:8132–8137.
305. Gerald WL, Rosai J, Ladanyi M. Characterization of the genomic breakpoint and chimeric transcripts in the EWS-WT1 gene fusion of desmoplastic small round cell tumor. *Proc Natl Acad Sci USA* 1995; 92:1028–1032.
306. Huff V, Reeve AE, Leppert M, Strong LC, Douglass EC, Geiser CF, et al. Nonlinkage of 16q markers to familial predisposition to Wilms' tumor. *Cancer Res* 1992; 52:6117–6120.
307. Malik KTA, Poirier V, Ivins SM, Brown KW. Autoregulation of the human WT1 gene promoter *FEBS Lett* 1994; 349:75–78.
308. Drummond IA, Madden SL, Rohwer-Nutter P, Bell GI, Sukhatme VP, Rauscher III FJ. Repression of the insulin-like growth factor II gene by the wilms tumor suppressor WT1. *Science* 1992; 257:674–678.
309. Werner H, Re GG, Drummond IA, Sukhatme VP, Rauscher FJ III, Sens DA, et al. Increased expression of the insulin-like growth factor I receptor gene, IGF1R, in Wilms tumor is correlated with modulation of IGF1R promoter activity by the WT1 Wilms tumor gene product. *Proc Natl Acad Sci USA* 1993; 90:5828–5832.
310. Werner H, Rauscher FJ III, Sukhatme VP, Drummond IA, Roberts CT Jr, LeRoith D. Transcriptional represion of the insulin-like growth factor 1 receptor (IGF-I-R) gene by the tumor supressor WT1 involves binding to sequences both upstream and downstream of the IGF-I-R gene transcription start site. *J Biol Chem* 1994; 269:12,577–12,582.
311. Wang Z-Y, Qiu Q-Q, Deuel TF. The Wilms' tumor gene product WT1 activates or suppresses transcription through separate functional domains. *J Biol Chem* 1993; 268:9172–9175.
312. Gashler AL, Bonthron DT, Madden SL, Rauscher FJ III, Collins T, Sukhatme VP. Human plateletderived growth factor A chain is transcriptionally repressed by the Wilms' Tumor suppressor WT1 *Proc Natl Acad Sci USA* 1992; 89:10,984–10,988.
313. Harrington MA, Konicek B, Song A, Xia X-l, Fredericks WJ, Rauscher FJ III. Inhibition of colonystimulating factor-1 promoter activity by the product of the Wilms' tumor locus. *J Biol Chem* 1993; 268:21,271–21,275.
314. Dey BR, Sukhatme VP, Roberts AB, Sporn MB, Rauscher FJ III, Kim S-J. Repression of the transforming growth factor-b1 gene by the Wilms' tumor suppressor WT1 gene product. *Mol Endoscrinol* 1994; 8:595–602.
315. Martinerie C, Chevalier G, Rauscher FJ III, Perbal B. Regulation of nov by WT1: a potential role for nov in nephrogenesis. *Oncogene* 1996, 12:1479–1492.
316. Goodyer P, Dehbi M, Torban E, Bruening W, Pelletier J. Repression of the retinoic acid receptor-a gene by the Wilms' tumor suppressor gene product, WT1. *Oncogene* 1995; 10:1125–1129.
317. Hsu SY, Kubo M, Chun S-Y, Haluska FG, Housman DE, Hsueh AJW. Wilms' tumor protein WT1 as an ovarian transcription factor: decreases in expression during follicle development repression of inhibin-a gene promoter. *Mol Endocrinol* 1995; 9:1356–1366.
318. Hewitt SM, Hamada S, McDonnell TJ, Rauscher FJ III, Saunders GF. Regulation of the proto-oncogenes bcl-2 and c-myc by the Wilms' tumor suppressor gene WT1. *Cancer Res* 1995; 55:5386–5389.
319. Webster NJG, Kong Y, Sharma P, Haas M, Sukumar S, Seely BL. Differential effects of Wilms tumor WT1 splice variants on the insulin receptor promoter. *Biochem Mol Med* 1997; 62:139–150.
320. Cook DM, Hinkes MT, Bernfield M, Rauscher FJ III. Transcriptional activation of the syndecan-1 promoter by the Wilms' tumor protein WT1. *Oncogene* 1996; 13:1789–1799.
321. Adachi Y, Matsubara S, Pedraza C, Ozawa M, Tsutsui J-I, Takamatsu H, et al. Midkine as a novel target gene for the Wilms' tumor suppressor gene (WT1). *Oncogene* 1996; 13:2197–2203.

322. Poukka H, Kallio PJ, Janne OA, Palvimo JJ. Regulation of the rat p75 neurotrophin receptor promoter by GC element binding proteins. *Biochem Biophys Res* 1996; 229:565–570.

323. Shimamura R, Fraizer GC, Trapman J, Lau YFC, Saunders GF. The Wilms' tumor gene WT1 can regulate genes involved in sex determination and differentiation: SRY, mullerian-inhibiting substance, and the androgen receptor. *Clin Cancer Res* 1997; 3:2571–2580.

324. Kim J, Prawitt D, Bardeesy N, Torban E, Vicaner C, Goodyer P, Zabel B, Pelletier J. The Wilms' tumor suppressor gene (wt1) product regulates Dax-1 gene expression during gonadal differentiation. *Mol Cell Biol* 1999; 19:2289–2299.

325. Miyamoto N, Yoshida M, Kuratani S, Matsuo I, Aizawa S. Defects of urogenital development in mice lacking Emx2. *Development* 1997; 124:1653–1664.

326. Moore MW, Klein RD, Farinas I, Sauer H, Armanini M, Philips H, et al. Renal and neural abnormalities in, mice lacking GDNF. *Nature* 1996; 382:76–79.

327. Pichel JG, Shen L, Sheng HZ, Granholm A, Drago J, Grinberg A, et al. Defects in enteric innervation and kidney development in mice lacking GDNF. *Nature* 1996; 382:73–76.

328. Shawlot W, Behringer RR. Requirement for Lim1 in head-organiser function. *Nature* 1995; 374:425–430.

329. Muller U, Wang D, Denda S, Meneses JJ, Pedersen RA, Reichardt LF. Integrin a8b1 is critically important for epithelial mesenchymal interactions during kidney morphogenesis. *Cell* 1997; 88:603–613.

330. Quaggin SE, Vanden Heuvel GB, Igarashi P. Pod-1, a mesoderm-specific basic-helix-loop-helix protein expressed in mesenchymal and glomerular epithelial cells in the developing kidney. *Mech Dev* 1998; 71:37–48.

331. Luo G, Hofmann C, Bronckers ALJJ, Sohocki A, Bradley A, Klarsenty G. BMP-7 is an inducer of nephrogenesis, and is also required for eye development and skeletal patterning. *Genes Dev* 1995; 9:2808–2820.

332. Dudley AT, Lyons KM, Robertson EJ. A requirement for bone morphogenic protein-7 during development of mammalian kidney and eye. *Genes Dev* 1995; 9:2795–2807.

333. Leveen P, Pekney M, Gebre-Medhin S, Swolin B, Larsson E, Betsholtz C. Mice deficient for PDGF B show renal, cardiovascular, and hematological abnormalities. *Genes Dev* 1994; 8:1875–1887.

334. Soriano P. Abnormal kidney development and hematological disorders in PDGF b-receptor mutant mice. *Genes Dev* 1994; 8:1888–1896.

335. Kreidberg JA, Donovan MJ, Goldstein SL, Rennke H, Shepherd K, Jones RC, Jaenisch R. Alpha 3 beta 1 integrin has a crucial role in kidney and lung organogenesis. *Development* 1996; 122:3537–3547.

14 Fanconi Anemia Pathway and Cancer Susceptibility

Yanan Kuang, PhD, Irene Garcia-Higuera, PhD, Eric Nisbet-Brown, MD, PhD, Anna Savoia, MD, and Alan D. D'Andrea, MD

CONTENTS

1. INTRODUCTION

Fanconi anemia (FA) is a rare autosomal recessive disease characterized by multiple congenital abnormalities, bone marrow failure (BMF), and cancer susceptibility. The mean age of onset of anemia is 8 years, and mean survival is 16 years. Death in FA usually results from complications of BMF. Considerable progress in the field of FA research has resulted from the recent identification and cloning of three FA genes. The purpose of this chapter is to describe the clinical and diagnostic features of FA, review the cellular phenotype of FA, review the structure and putative function of the cloned FA genes, and discuss the possible function of the FA genes as tumor suppressors.

From: *Tumor Suppressor Genes in Human Cancer*
Edited by: D. E. Fisher © Humana Press Inc., Totowa, NJ

2. CLINICAL COURSE OF FA

The congenital abnormalities and clinical course of FA have been extensively reviewed *(1,2)*. Patients with FA have growth retardation and abnormalities of the skin, upper extremities (frequently with defects in the thumbs or forearms), kidneys, and gastrointestinal system. The large range of organ systems affected in FA implicates the FA genes in a general developmental process required during normal human embryogenesis.

The hematological complications of FA have also been extensively reviewed *(3,4)*. FA patients develop macrocytosis and pancytopenia, typically during the first decade of life. Deficiencies in platelets or red cells usually precede white blood cell abnormalities. The patients have fetal-like erythropoiesis, with increased i antigen and hemoglobin F, and generally have high serum erythropoietin levels. The progression to pancytopenia is highly varied among FA patients, and some patients develop hematological malignancies even before anemia is clinically evident.

At least 20% of patients with FA develop cancers *(3)*. Of the FA patients with cancer, approx one-half develop acute myeloblastic leukemia; however, cancers of several organ systems, including skin, gastrointestinal, and gynecological systems, have been described. There appears to be an increased incidence of head and neck carcinomas and esophageal carcinomas in FA patients. Cancer tends to be a disease of older FA patients, with an average age of 15 yr for leukemia, 16 yr for liver tumors, and 23 yr for other tumors *(5)*.

3. DIAGNOSIS OF FA

The diagnosis of FA exploits the sensitivity of FA cells to bifunctional alkylating agents, diepoxybutane (DEB) or mitomycin C (MMC). FA cells have increased spontaneous chromosome (chr) breakage that is amplified by the addition of these crosslinking agents. Similar, spontaneous, but not DEB-induced, chr changes are observed in other inherited chr instability syndromes, such as Bloom syndrome (BS) and ataxia telangiectasia (AT). The DEB test is a highly sensitive and specific test for FA *(6)*, and has been used successfully in prenatal diagnosis of FA.

Other diagnostic tests for FA are also available. In one flow cytometric technique, cells that have been fluorescently prelabeled are treated with alkylating agents. The percentage of cells arrested in the G2–M phase of the cell cycle is measured. In this way, FA cells are indistinguishable from normal cells by their increased accumulation in the G2–M phase of the cell cycle. In some cases, FA carrier cells can also be distinguished from normal cells *(7)*.

New diagnostic approaches have resulted from the cloning of the FA genes. For instance, specific antisera for the FANCA, FANCC, and FANCG proteins have been described, which are useful for diagnostic immunoblotting of FA cell lines. Also, retroviruses, which transduce the FANCA, FANCC, and FANCG cDNAs, have been generated. Transduction of FA cell lines with these retroviral supernatants allows efficient subtyping of FA patients *(8)*. Finally, FA can be diagnosed on the basis of direct mutational analysis of the three cloned FA genes.

The diagnosis of FA has recently been confounded by the observation that some patients may have significant mosaicism. Patients with mosaicism have two populations of cells exhibiting either a normal or an FA phenotype. Such mosaicism may generate a false-negative DEB chr breakage assay, if the percentage of cells with a normal phenotype is sufficiently high. Approximately 25% of patients with FA have spontaneously occurring mosaicism.

The molecular basis of mosaicism in FA has not yet been completely determined. In some cases *(9)*, a somatic mutation in a mutant FA gene allele results in correction of the mutant allele and subsequent expression of a functional FA protein. For instance, the somatic mutation of a mutant (frameshifted) FANCC allele could theoretically result in the correction of the mutant allele and the expression of a functional protein product. A hematopoietic progenitor (FA) cell could theoretically be corrected by such a somatic reversion, giving rise to a clone of hematopoietic cells with a normal (non-FA) phenotype. Such a normal cell may have a selective advantage, and may be more likely to proliferate faster than other FA cells of the host. The comparatively-high frequency of hematopoietic cell mosaicism in FA may result from a combination of the increased mutational (reversion) frequency in FA, and the selective advantage of functionally corrected cells. Mosaicism has already been described in other diseases, such as Bloom Sydrome (BS) and adenosine deaminase deficiency *(10,11)*.

4. TREATMENT OF FA

The treatment of FA is similar, but not identical, to the treatment of other forms of acquired aplastic anemia. Patients are treated with supportive care (i.e., blood transfusions) for their BMF. The use of the cytokines, granuloate colony-stimulating factor, erythropoietic, and granuloate-macrophage colony-stimulating factor has also been efficacious for some FA patients *(12–14)*. The treatment of choice for FA is allogenic BM transplant with a histocompatible sibling donor. Umbilical cord blood offers a potential source of hematopoietic stem cells for FA patients without sibling matches. Gene therapy (GT) also offers an alternative strategy for FA treatment, but this approach requires prior knowledge of the patient's complementation group *(15–17)*. A more detailed discussion of GT for FA is provided in Sections.

5. CELLULAR PHENOTYPE OF FA CELLS

Because of the cellular sensitivity to crosslinking agents, FA is often compared to other syndromes of drug sensitivity and genomic instability, including AT, xeroderma pigmentosum (XP), Cockayne syndrome, BS, and hereditary nonpolyposis colorectal cancer *(2)*. The chr instability of these syndromes may result from a cellular defect in any one of several processes, including DNA repair, cell cycle regulation, or DNA replication.

In addition to crosslinking agent sensitivity, FA cells have several other phenotypic abnormalities. FA cells have cell cycle abnormalities, as demonstrated by the prolongation of the G2 phase of the cell cycle *(7,18)*. FA cells also have pronounced hypersensitivity to oxygen and reactive oxygen species *(19–22)*, and a G2 phase specific hypersensitivity to ionizing radiation *(23)*. Extracts of FA cells have been shown to have direct defects in DNA repair *(24–28)*. FA cells have been shown to have increased apoptosis *(29,30)* and abnormal p53 responsiveness *(30,31)*. FA patients have also been shown to have an intrinsic hematopoietic *(32,33)* and gonadal *(34)* stem cell defect. More recently, FA cells were shown to be hypersensitive to the effects of interferon (IFN-γ) and tumor necrosis factors (TNF-α) (see below), accounting, at least in part, for the aplastic anemia in FA patients *(34,35)*. Many of these abnormal phenotypic characteristics are also evident in primary cells derived from mice homozygous for a disrupted *FANCC* gene. How the absence of the FA proteins leads to these cellular abnormalities remains unknown.

6. MOLECULAR BIOLOGY OF FA GENES

6.1. Cloning of FA Genes

The complementation analysis of FA cells, using somatic cell fusion studies, has allowed the identification of at least eight complementation groups *(36–40;* Table 1*)*. *At least four of the eight groups (A, C, D, and G) map to discrete chr loci* (41–44). FA is therefore a genetically heterogenous disorder, unlike the syndromes, AT and BS, which arise from mutations in single genes. The three cloned FA genes (for A,C, and G) encode orphan proteins, with no sequence similarity to each other or to other proteins in GenBank.

6.2. Molecular Biology of FANCC

FANCC was cloned by functional complementation of an Epstein-Barr virus (EBV)-immortalized type C FA cell line *(45)*. As predicted by the complementation test, the FANCC cDNA corrects the MMC sensitivity and DEB sensitivity of FA-C cell lines, but does not correct the MMC sensitivity of FA cells derived from other FA groups. Cells derived from FA-C patients have mutations in both alleles of the *FANCC* gene, consistent with the autosomal recessive inheritance pattern of the FA syndrome. The human *FANCC* gene is composed of 14 exons *(46)*, spans approx 120 kb, maps to human chr 9q22.3 *(46)*, and encodes a 558-amino-acid (aa) polypeptide (63 kDa) (Table 2). The murine, rat, and bovine FANCC cDNA homologs have also been cloned *(47,48)*. The murine FANCC protein is 78% identical to the human FANCC protein. When murine FANCC is expressed in human FA-C cells, functional complementation is established *(47)*.

Mutational analysis of the *FANCC* gene has revealed a relatively small number of characteristic mutations, represented in specific ancestral backgrounds (Fig. 1). The IVS4+4 A > T mutation is found in patients of Ashkenazi-Jewish ancestry, and accounts for >80% of FA in this population *(49–50)*. Patients homozygous for this mutation have severe FA, with multiple congenital abnormalities and early onset of hematological disease *(51)*. Whether this mutant *FANCC* allele corresponds to a severe FA phenotype in other (non-Ashkenazi-Jewish) genetic backgrounds remains unknown. The 322delG mutation is found in patients of Northern European ancestry, particularly from Holland. Patients homozygous for this mutation have comparatively mild FA, with fewer congenital abnormalities and later onset of hematological disease *(51)*. Based on the relative prevalence of mutations in exon 14 and the high conservation in this exon across species, the carboxy-terminal region of FANCC is likely to contain a critical functional domain. Consistent with this hypothesis, recent studies suggest that the C-terminal region of FANCC may be required for interaction with FANCA *(52)*.

Analysis of the FANCC mRNA and protein have provided some insight into the cellular function of the *FANCC* gene. The FANCC mRNA is expressed in multiple cell types and organ systems, consistent with a general function of FANCC in organism

Table 1
Complementation Groups of FA

Subtype	Estimated percentage of FA patients	Chromosome location	Protein product
A	66	16q24.3	163 kDa
B	4	?	?
C	12	9q22.3	63 kDa
D	4	3p22–26	?
E	12	?	?
F	Rare	?	?
G (XRCC9)	Rare	9p13	68 kDa
H	Rare	?	?

Table 2
Structural Features of FA Genes

Gene	Cloned cDNA (kb)	Full length gene (kb)	Structure (no. exons)	mRNA transcripts (kb)	Expression	Comments
FANCC	4.6	120	14	2.3 3.2 4.6	Ubiquitous	5'UTR has two alternatively spliced exons
FANCA	5.5	80	43	4.7 (major) 7.5 3.0 2.0	Ubiquitous	–
FANCG	2.5	6	14	2.2 2.5	Ubiquitous	–

development. Increased expression of the FANCC mRNA has been observed in the skeletal system, suggesting a more specialized function of FANCC in bone development *(53)*.

The FANCC protein is primarily a soluble cytoplasmic protein *(54,55)*, although a nuclear complex of FANCA, FANCC, and FANCG proteins has also been detected *(52,56); (see* below). The function of the FANCC protein remains mostly unknown, but it has recently been shown to physically interact with other cellular proteins, including GRP94 *(57)*, Cdc2 *(58)*, and NADPH reductase *(59)*.

6.3. Molecular Biology of FANCA

FANCA was cloned by two independent strategies. One group cloned FANCA by functional complementation of an EBV-immortalized FA-A cell line *(60)*. The second group cloned the *FANCA* gene by a positional strategy *(61)*. As predicted by the complementation test, the FANCA cDNA corrects the MMC and DEB sensitivity of FA-A

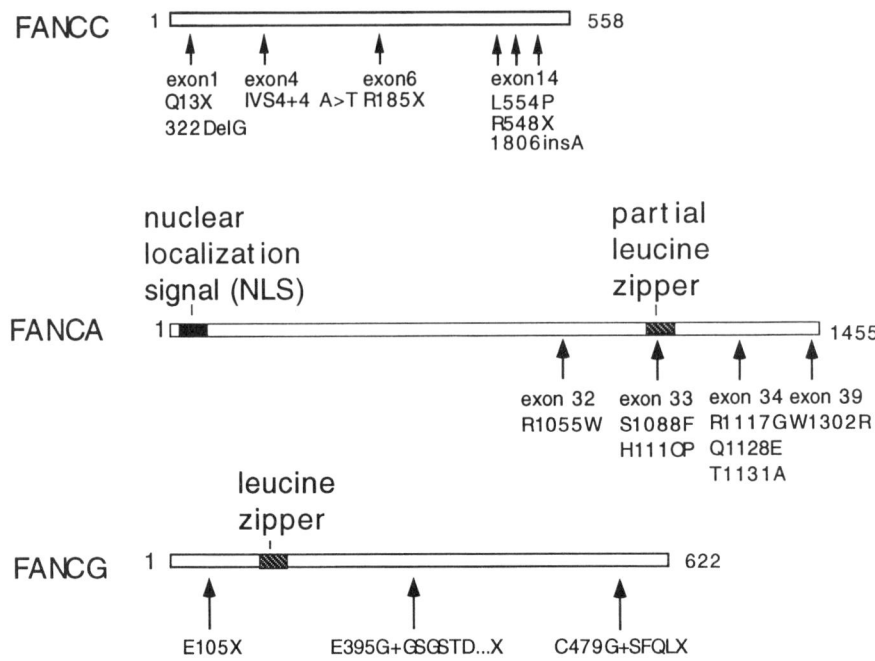

Fig. 1. Patient-derived mutations of the FANCC, FANCA, and FANCG proteins. The wt FANCC protein (558 aa) is shown schematically at the top. The IVS4+4 A > T mutant allele encodes a protein with an in-frame deletion of 37 aa *(49,55)* The 322delG has a frameshift mutation in exon 1 *(45)*. The IVS4 + 4 A > T mutation and 322delG account for approx 75% of FANCC mutations. Less common mutant alleles of FANCC are also shown. The relative frequency of these various FANCC alleles has recently been described *(51)*. The wt FANCA protein (1455 aa) is also shown. The protein contains a bipartite nuclear localization signal at its N-terminus and a partial Leu zipper. Mutation analysis has demonstrated that the *FANCA* gene is highly polymorphic. The region of the FANCA protein from aa 1046 to 1320, encoded by exons 32–39, appears to be critical for FANCA function, based on the prevalence of missense mutations in this region. The wt FANCG protein (622 aa) is also shown. The protein contains a Leu zipper between aa 135 and 163, but contains no other recognizable protein motifs. Based on published analysis of FA-G patients, only a few mutations are known. The mutant *FANCG* allele, 313G > T, is predicted to encode a truncated FANCG protein, E105X. The mutant *FANCG* allele, 1184–1194 del, encodes FANCG (E395G+GSGSTDPGRQSPRCCDSMX). The mutant allele IVS11+1G>C encodes FANCG (C479G+SFQLX). Whether these mutant *FANCG* alleles encode mutant FANCG proteins remains unknown.

cells, but does not correct the drug sensitivity of FA cell derived from FA groups B–H. The *FANCA* gene is composed of 43 exons, spans approx 80 kb, maps to human chr 16q23.4, and encodes an orphan protein of 1455 amino acids (163 kDa) (*see* Table 2).

The *FANCA* gene generates multiple FANCA mRNA transcripts, including a 4.7-kb transcript (major) and transcripts of 7.5, 3.0, or 2.0 kb. The mRNAs for FANCA and FANCC are ubiquitously expressed, although considerable variation is observed in specific tissues. In human adult tissues, the highest level of expression of both genes was observed in testis, thymus, and BM. In general, transcription levels were elevated in fetal tissues and in actively proliferating cells. Discordant expression patterns were observed for FANCA and FANCC. The FANCC transcripts were found in similar abundance in all tissues; the FANCA transcripts demonstrated more tissue-specific variation, with high levels in liver and kidney. Taken together, these data suggest alternative transcriptional regulatory mechanisms for the two genes. The 5′ *cis*-acting regulatory (promoter) regions of the *FANCC* and *FANCA* genes contain highly GC-rich regulatory sequences, a feature of several known housekeeping genes *(62,63)*.

The FANCA protein contains a nuclear localization signal (NLS) at its N-terminus *(64)*, and a partial leucine (Leu) zipper motif between aa 1069 and 1090. The importance of this Leu zipper region remains unclear. Some, but not all, mutations in this region of FANCA result in loss of FANCA function *(65)*. Although the primary aa sequence of FANCA provides little insight into its biochemical function, missense mutations derived from FA-A patients may help identify critical functional domains of the FANCA protein.

Based on mutational screens published to data *(66–68)* the region of the FANCA protein from aa 1046 to 1320, encoded by exons 32–39 *(62)*, appears to be critical for FANCA function. Multiple patient-derived missense mutations have been identified in this region (Fig. 2; *66,67*). The FANCA1263delF (exon 34), encoded by the mutant *FANCA* allele, 3788–3790del, was found in 30 FA-A patients, and appears to account for 5% of known *FANCA* mutations *(66)*. In many cases, it is difficult to distinguish between pathogenic point mutations and simple polymorphisms. As for the FANCC protein, the cellular function of the FANCA protein is unknown. Recent studies demonstrate that the FANCA protein forms a complex with FANCC and FANCG (*see* model in Subheading 6.4). FANCA has also been reported to form a complex with the kinase, IKK-2 *(69)*. FANCA is phosphorylated on serine residues *(70)*, and requires nuclear localization for its function *(64)*.

6.4. Molecular Biology of FANCG

The demonstration that the human *FANCG* gene *(71)* is identical to the previously cloned human *XRCC9* gene is recent *(72)*, and, accordingly, little is known about the encoded FANCG protein. The human *XRCC9* cDNA was originally expression-cloned by its ability to partially complement the MMC sensitivity of a mutant CHO cell line *(72)*. de Winter et al. *(71)* subsequently cloned the same cDNA by expression, based on its ability to complement an FA-G cell line. The FANCG/XRCC-9 protein is yet another orphan protein that contains an internal Leu zipper as its only recognizable motif. The *FANCG* gene is composed of 14 exons, spans approx 6 kb, maps to human chr 9p13 *(44,72)*, and encodes a 68 kDsa protein (Table 2). For FANCG, major mRNA bands of 2.2 kb and 2.5 kb were detected in all human tissues tested *(72)*. As for FANCA and FANCC, the expression was 10–100-fold higher in testis, compared to other tissues.

Only one study has been published that identifies patient-derived mutations in the *FANCG* gene *(71)*. One mutation, found in three patients with German ancestry, was a G-to-T transition at nucleotide 313, which changes codon 105 from glutamic acid to a stop (E105X). A second mutation was an 11-bp deletion in exon 10 and a splice-site mutation (IVS11 + 1 G to C). A third mutation was a splice-site mutation, IVS13 – 1 G to C, found in two affected children (Fig. 1). Whether any of these mutant *FANCG* alleles direct the synthesis of a truncated, nonfunctional FANCG protein remains unknown (Fig. 2). Further identification of relevant functional domains of the FANCG protein

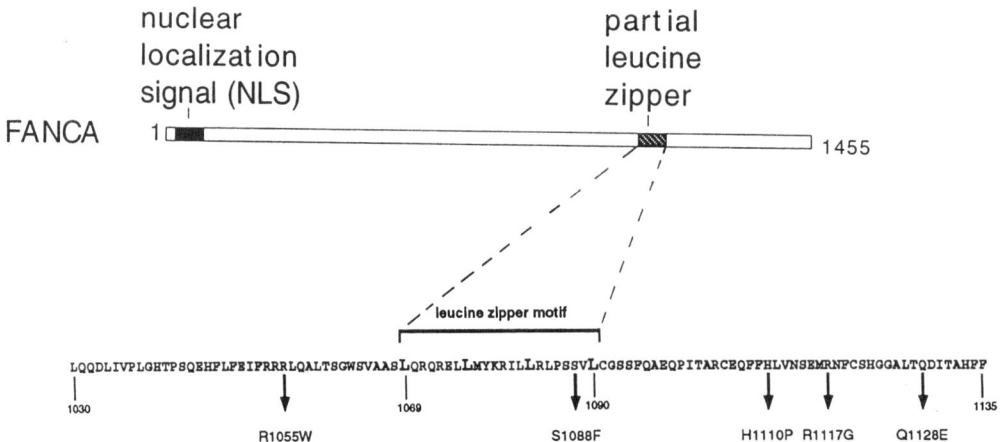

Fig. 2. The C-Terminal region of the FANCA protein contains a domain required for functional activity. A schematic representation of the FANCA protein is shown. The region of the protein, from aa 1030 to 1135 is enlarged. This region contains a high incidence of mutations in the FA-A patient population, suggesting that it contains a critical functional domain. Recent data supports the notion that this domain is required for FANCC binding, nuclear accumulation, and functional activity *(117)*.

may be gained by a comparison of the human and murine FANCG proteins (Fig. 3). The human and murine proteins have 72.2% identity. One Leu residue of the putative Leu zipper is not conserved in the mouse protein. Whether the murine FANCG cDNA corrects the MMC sensitivity of FA-G cells remains unknown.

7. MOUSE MODELS OF FA

Two murine models of FA-C have been developed, using targeted recombination in embryonic stem cells *(34,73)*. The phenotype of homozygous FANCC mutant animals was similar in both models. FANCC mutants are viable and show no obvious birth defects of the skeletal system or urinary system. Cells derived from these animals show classic hypersensitivity to bifunctional crosslinking agents. Nonetheless, spontaneous pancytopenia did not develop during the first year of life, and no leukemia or increased cancer susceptibility was observed.

The FANCC–/– mice have a germ cell defect. The germ cell defect in FANCC–/– male mice, resulting in subnormal fertility, appears to be secondary to decreased sperm viability, and not to a defect in meiosis. The cells from the FANCC–/– mice also are hypersensitivity to the apoptotic effects of IFN-γ and TNF-α *(34)*. This TNF-α and IFN-γ hypersensitivity is also observed in primary cells and cell lines derived from human FA patients.

FANCC–/– mice do not show any of the baseline hematologic defects found in FA patients, but the mice were found to be highly sensitive to DNA crosslinking agents in vivo *(74)*. Specifically, acute MMC exposure caused extreme BM aplasia. Sequential, nonlethal MMC exposure caused a progressive decrease in peripheral blood counts. BMF in these MMC-treated FANCC–/– mice was shown to result from a reduction in the number of early and committed hematopoietic progenitor cells. Taken together,

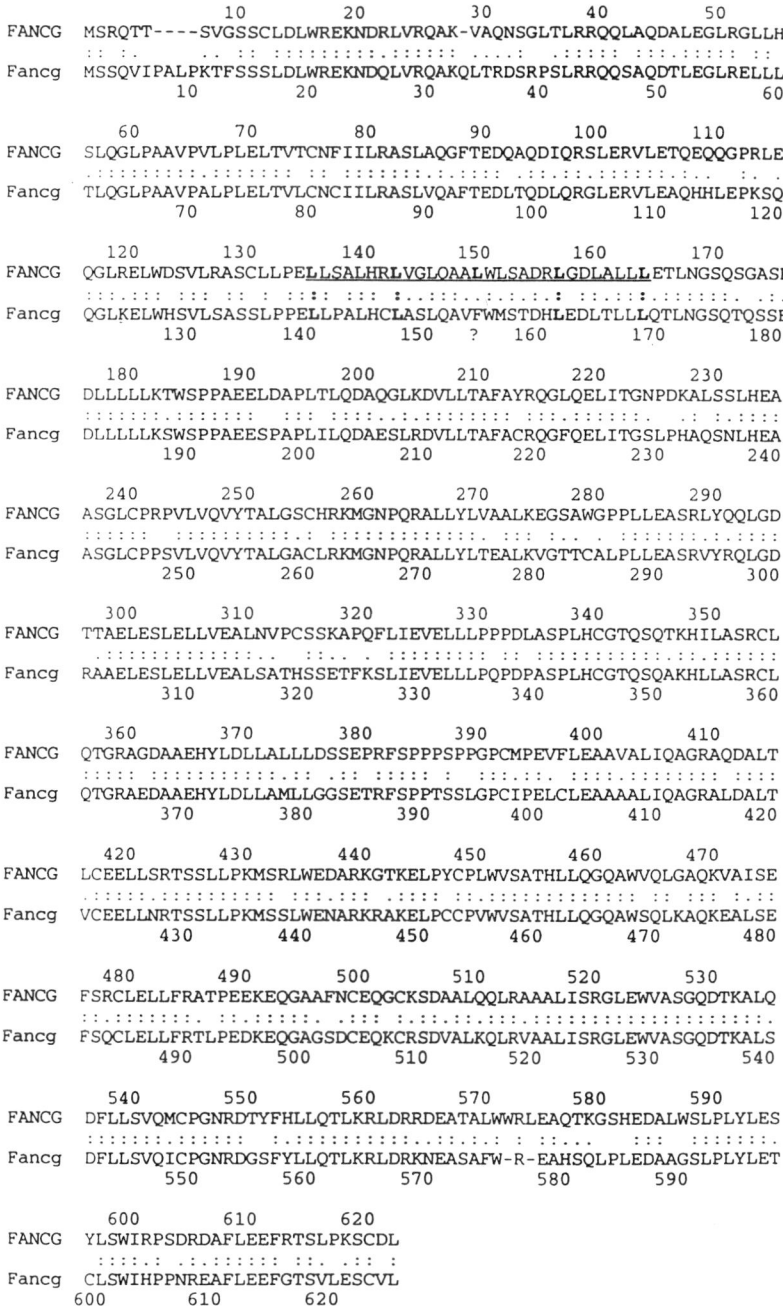

```
              10        20        30        40        50
FANCG  MSRQTT----SVGSSCLDLWREKNDRLVRQAK-VAQNSGLTLRRQQLAQDALEGLRGLLH
       ::  :.      ..  :: :::::::::.:::::::  .....:  .::::: :::.::::: ::
Fancg  MSSQVIPALPKTFSSSLDLWREKNDQLVRQAKQLTRDSRPSLRRQQSAQDTLEGLRELLL
              10        20        30        40        50        60

              60        70        80        90       100       110
FANCG  SLQGLPAAVPVLPLELTVTCNFIILRASLAQGFTEDQAQDIQRSLERVLETQEQQGPRLE
       .:::::::::.::::::: :: :::::::.:.:::: .::.:::.:::::  :. .
Fancg  TLQGLPAAVPALPLELTVLCNCIILRASLVQAFTEDLTQDLQRGLERVLEAQHHLEPKSQ
              70        80        90       100       110       120

             120       130       140       150       160       170
FANCG  QGLRELWDSVLRASCLLPELLSALHRLVGLQAALWLSADRLGDLALLLETLNGSQSGASK
       ::::.::: :: :  ::: ::   :::*::: ::.:  :: ::.:::.:.::::::.:::
Fancg  QGLKELWHSVLSASSLPPELLPALHCLASLQAVFWMSTDHLEDLTLLLQTLNGSQTQSSE
             130       140       150       ?       160       170       180

             180       190       200       210       220       230
FANCG  DLLLLLKTWSPPAEELDAPLTLQDAQGLKDVLLTAFAYRQGLQELITGNPDKALSSLHEA
       ::::::::.:::::::  ::: ::::..:.::::::::: :::.::::::.  .: :.::::
Fancg  DLLLLLKSWSPPAEESPAPLILQDAESLRDVLLTAFACRQGFQELITGSLPHAQSNLHEA
             190       200       210       220       230       240

             240       250       260       270       280       290
FANCG  ASGLCPRPVLVQVYTALGSCHRKMGNPQRALLYLVAALKEGSAWGPPLLEASRLYQQLGD
       ::::::  :::::::::::::.:::::::::::::.::::: ::: ..  : :::::::::::
Fancg  ASGLCPPSVLVQVYTALGACLRKMGNPQRALLYLTEALKVGTTCALPLLEASRVYRQLGD
             250       260       270       280       290       300

             300       310       320       330       340       350
FANCG  TTAELESLELLVEALNVPCSSKAPQFLIEVELLLPPPDLASPLHCGTQSQTKHILASRCL
       .::::::::::::::::.  ::.. :: ::::::::: :::::::::::::::::.:::::::
Fancg  RAAELESLELLVEALSATHSSETFKSLIEVELLLPQPDPASPLHCGTQSQAKHLLASRCL
             310       320       330       340       350       360

             360       370       380       390       400       410
FANCG  QTGRAGDAAEHYLDLLALLLDSSEPRFSPPPSPPGPCMPEVFLEAAVALIQAGRAQDALT
       :::::  :::::::::::::.:::. :: ::: ::: ::.:::.:  :::::::::::::.:::
Fancg  QTGRAEDAAEHYLDLLAMLLGGSETRFSPPTSSLGPCIPELCLEAAAALIQAGRALDALT
             370       380       390       400       410       420

             420       430       440       450       460       470
FANCG  LCEELLSRTSSLLPKMSRLWEDARKGTKELPYCPLWVSATHLLQGQAWVQLGAQKVAISE
       .:::::.:::::::::::.:::: :..::.. .::::: ::.:::::::::::  ::: :.::
Fancg  VCEELLNRTSSLLPKMSSLWENARKRAKELPCCPVWVSATHLLQGQAWSQLKAQKEALSE
             430       440       450       460       470       480

             480       490       500       510       520       530
FANCG  FSRCLELLFRATPEEKEQGAAFNCEQGCKSDAALQQLRAAALISRGLEWVASGQDTKALQ
       ::.:::::::.:: :.::::.:::  .::: :.::::.:::::::::::::::::::::::::::.
Fancg  FSQCLELLFRTLPEDKEQGAGSDCEQKCRSDVALKQLRVAALISRGLEWVASGQDTKALS
             490       500       510       520       530       540

             540       550       560       570       580       590
FANCG  DFLLSVQMCPGNRDTYFHLLQTLKRLDRRDEATALWWRLEAQTKGSHEDALWSLPLYLES
       :::::::.:::::: :.:::::::::::::::::.:::.::::    .: :::::::::::::.
Fancg  DFLLSVQICPGNRDGSFYLLQTLKRLDRKNEASAFW-R-EAHSQLPLEDAAGSLPLYLET
             550       560       570       580       590

             600       610       620
FANCG  YLSWIRPSDRDAFLEEFRTSLPKSCDL
       :::::.: ..::::::: ::.. .:: :
Fancg  CLSWIHPPNREAFLEEFGTSVLESCVL
             600       610       620
```

Fig. 3. Sequence comparison of human and murine FANCG proteins. The human FANCG/XRCC9 protein sequence is shown, compared to the predicted sequence of the murine homolog (fancg), obtained from Genbank. Overall, the two proteins have 72.2% identity. The Leu zipper region is underlined. One Leu (L150) is not conserved in the murine sequence.

these results are consistent with a function of the FANCC protein product in the viability or differentiation of early hematopoietic progenitor cells or stem cells.

8. MODELS OF FA GENE PATHWAY

Little is known about the cellular function of the FA proteins. Multiple studies have been performed in order to identify binding proteins of FANCC and FANCA. To date, it is known that FANCC interacts with several cytoplasmic binding proteins (75), as well as cdc2 (58), GRP94 (76), and NADPH reductase (59). More recently, FANCA has been shown to interact with the cytoplasmic kinase, IKK2 (69). The physiological relevance of these protein–protein interactions has not yet been determined. A more unifying hypothesis has recently emerged, based on the recognition that the three FA proteins (FANCA, FANCG, and FANCC) interact in a functional nuclear complex (56).

8.1. Interaction of FA Proteins in a Common Pathway

Mutation of any of the cloned FA genes leads to a conserved cellular and clinical phenotype, which suggests that the FA genes encode proteins that function in a coordinate manner (Fig. 4). For instance, one can envision the FA proteins functioning in a common, multisubunit complex (protein complex hypothesis) or in a common biochemical pathway (pathway hypothesis). Considerable experimental data now exist to support the model of a multisubunit complex (52,56).

The authors have recently determined that FANCA, FANCG, and FANCC form a complex that is localized in both the cytoplasm and the nucleus of normal cells (56). Initial studies provided evidence for a FANCA–FANCC interaction. First, for lymphoblasts, primary fibroblasts, and primary BM cells, expressing normal (endogenous) levels of the FANCA and FANCC proteins, this laboratory detected a physical complex of FANCA and FANCC (52). The FANCA–FANCC protein complex was detected by reciprocal immunoprecipitation–Western blotting protocols, with either anti-FANCA or anti-FANCC antisera. Second, the FANCA–FANCC complex was detected in protein fractions from the cytoplasm and the nucleus of primary cells. The co-immunoprecipitation was more efficient from nuclear extracts (64), suggesting a relative enrichment of the complex in the nucleus. Other studies have also used confocal microscopy to localize FANCC to the nucleus (77). Third, the interaction of FANCA and FANCC in a complex is critical to the function of the proteins. For lymphoblast lines derived from FA patients, mutant FANCC proteins fail to bind to FANCA (52), and mutant FANCA proteins fail to bind to FANCC (70). Functional complementation of these cells rescued FANCA–FANCC binding. Fourth, additional evidence, independent of the use of anti-FANCA antisera, demonstrated that cellular FANCA and FANCC bind in a complex (I. Garcia-Higuera, unpublished observation). For these studies, the authors generated an amino-terminal Flag-tagged FANCA protein, and expressed this protein in an FA-A cell line, GM6914. The Flag-tagged FANCA protein corrected the MMC sensitivity of the transfected cells, and co-fractionated with FANCC from an anti-Flag column. Fifth, the authors have shown that the FANCA protein is a phosphoprotein, and that its phosphorylation correlates with FANCC binding (70). FANCA is not phosphorylated, and the FANCA–FANCC complex is not detected in FA cells derived from other FA complementation groups (groups B, E, F, G, and H), suggesting that products of other FA genes regulate the assembly of the nuclear complex (70).

Protein Complex Hypothesis   **Pathway Hypothesis**

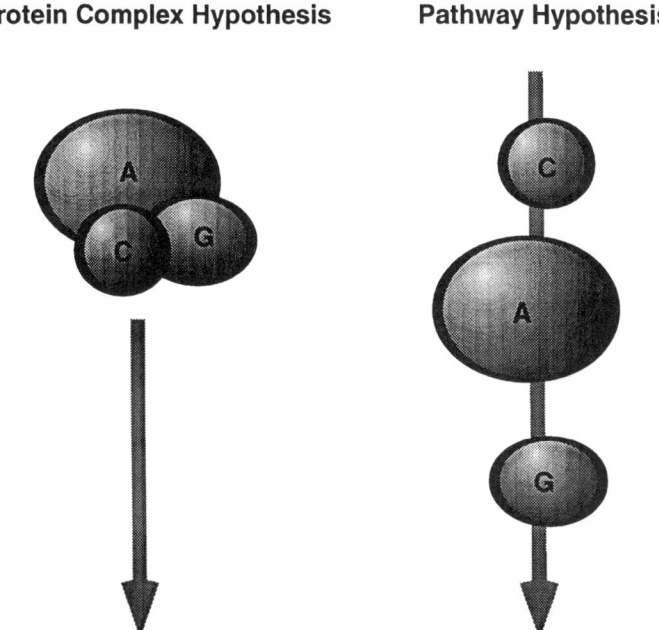

Fig. 4. Functional interaction of the FA proteins. Shown are two possible models for the interaction of the FA proteins, FANCA, FANCG, and FANCC. Because loss or mutation of these proteins yields the same specific FA clinical and cellular phenotype, it is assumed that the proteins interact, either directly (protein complex hypothesis) or indirectly (pathway hypothesis). Recent data discussed in the text supports the pathway hypothesis.

More recent data demonstrate that the FANCG protein is also part of the FANCA–FANCC complex (56). Expression of the FANCG protein in a FA-G lymphoblast line corrects the MMC sensitivity of the cell line, and restores the binding interaction between FANCA and FANCC. In these FANCG-corrected FA-G cells, an antibody to FANCG co-immunoprecipitates a complex of FANCA, FANCG, and FANCC proteins. The binding of FANCA and FANCG appears to be a constitutive, direct interaction. The two proteins bind to each other when the proteins are translated in vitro and mixed. Also, the two proteins bind to each other in extracts of FA cells from multiple complementation groups. The binding between FANCA and FANCG is therefore different from the binding interaction between FANCA and FANCC. The FANCA–FANCC interaction appears to be a weaker or regulated interaction. The FANCA and FANCC proteins do not bind in vitro, suggesting that their binding requires additional adaptor proteins or posttranslational modifications. For instance, the binding of FANCA to FANCC requires the phosphorylation of FANCA. Finally, the binding of FANCA and FANCC is not observed in cell lines derived from several FA complementation groups, including groups B, E, F, G, and H, suggesting that the products of other FA genes may be required for the binding interaction to occur. A schematic representation of the molecular interactions among the three proteins is shown (Fig. 5). The trimolecular complex of FANCA, FANCC, and FANCG is found in both the cytoplasm and the

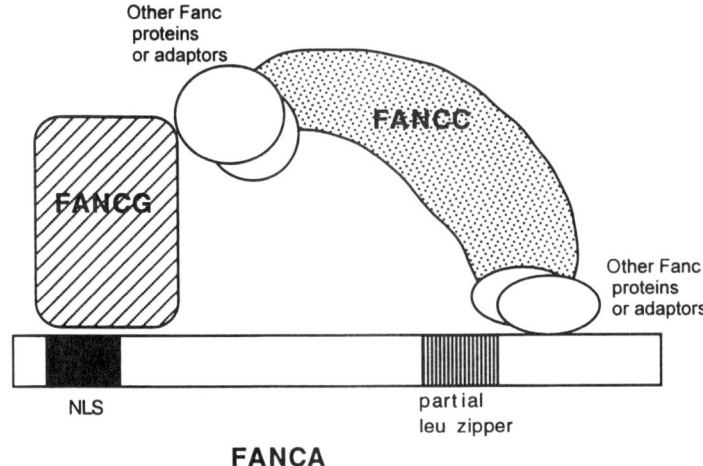

FANCA

Fig. 5. Schematic model of molecular interactions of the FANCA, FANCG, and FANCC proteins. Based on available data, at least two regions of the FANCA polypeptide are required for the FANCA–FANCG–FANCC interaction. First, the amino-terminal region of the FANCA protein, which includes the bipartite NLS region, is required for constitute interaction with the FANCG protein. Second, a region of the C-terminus of FANCA is required for the FANCC interaction. Patient-derived point mutations in this region disrupt the FANCC interaction. The FANCC interaction is a weaker interaction or a regulated interaction, requiring FANCA–FANCG binding and the products of other FA genes *(70)*. The region of FANCG required for FANCA binding is not known. The region of FANCC required for FANCA binding may be the C-terminus of FANCC. The C-terminus of FANCC is highly conserved among human, murine, and bovine FANCC proteins. Also, an FA-C patient-derived point mutation (FANCC-L554P) ablates FANCA binding.

nucleus, suggesting the three proteins act in a coordinate manner to execute a common function, such as DNA repair, transcription, or chr segregation.

Based on this recent evidence of a regulated interaction of FANCA, FANCG, and FANCC, a model of the FA pathway can be devised (Fig. 6). According to this model, monomeric FANCA and FANCG protein bind constitutively in the cytoplasm of the cell. This interaction appears to be stoichiometric: Complexes of FANCA and FANCG are efficiently immunoprecipitated with antisera to either FANCA or FANCG. Upon some cellular stimulus, FANCA phosphorylation occurs and FANCC binds to the FANCA–FANCG protein complex. Products of other FA genes may mediate the phosphorylation of FANCA and the formation of the FA protein complex. For instance, other FA genes may encode the FANCA kinase, or may encode adaptor proteins in the FA protein complex. The trimolecular complex of FANCA–FANCG–FANCC then translocates to the cell nucleus. Defects in the FA pathway result in a failure of the complex to translocate to the nucleus, and thereby result in chr instability, the hallmark of FA. Consistent with this model, the FANCA protein is not phosphorylated, and FANCA–FANCG–FANCC interaction is not detected in cells from other FA subtypes, including groups A B C E F G, and H. FA-D cells are distinct. These cells remain sensitive to MMC, despite FANCA phosphorylation, FANCA–FANCG–FANCC binding, and nuclear accumulation of the complex. Taken together, these results suggest that the FANCD protein may act independently (either upstream or downstream) of the complex.

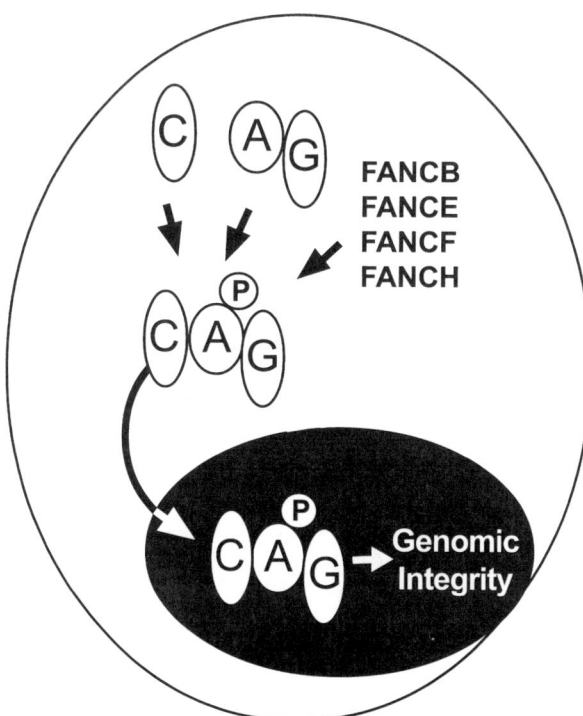

Fig. 6. Model for the regulated molecular interactions in the FA pathway. Based on available data, the FANCA and FANCG proteins bind constitutively in the cytoplasm of the normal cell. FANCC protein binds to the FANCA–FANCG complex, but this binding requires FANCA phosphorylation and requires the products of other FA genes. The trimolecular complex subsequently translocates to the cell nucleus. The FANCA–FANCG complex does not accumulate in the nucleus in the absence of FANCC protein binding. Whether FANCC is required for nuclear transport or stabilization of the complex in the nucleus is not known. After nuclear translocation, the complex executes a nuclear function, such as the repair of DNA or the orderly segregation of sister chromatids. Recent evidence suggests that the FA proteins comlex acts upstream of the FANCD protein (Higuera I and D'Andrea AD, unpublished observation).

Although considerable evidence exists for the interaction of the FA proteins in a complex, little is known regarding the biochemical and cellular function of this complex. Recent lines of evidence suggests that the FA protein complex may play a cellular role in the suppression of IFN-α— and TNF-α—induced apoptosis, the modulation of cell cycle events, or the repair of crosslinked DNA. Evidence supporting these models is provided below.

8.2. Suppression of IFN-— and TNF-—Induced Apoptosis by FA Pathway

Recent evidence suggests a direct relationship between the FA pathway and the IFN-γ signaling pathway. First, multiple studies *(34,35)* have demonstrated that FA cells have decreased cellular survival and an increased IFN-α—mediated apoptotic response. FA-C overexpression has been shown to protect Ba/F3 cells from apoptosis, following interleukin-3 deprivation *(78)*. Also, primary BM cells from an FANCC–/–

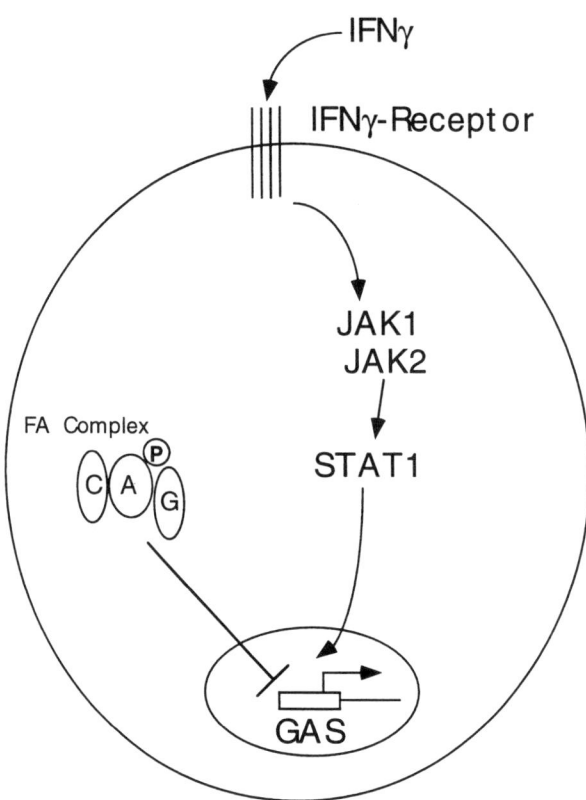

Fig. 7. Possible role of the FA pathway in the suppression of IFN-α—mediated apoptotic responses. Stimulation of the IFN-γ receptor results in activation of JAK1 and JAK2 kinases and phosphorylation of the STAT kinase, STAT1. STAT1 phosphorylation leads to the induction of specific genes, some of which mediate cell cycle arrest and apoptosis. According to this model, the FA protein pathway may normally function to downregulate the IFN-γ response. Absence of the FA protein complex would lead to increased IFN-γ signaling and increased apoptosis in response to IFN-γ.

mouse have increased IFN-α—mediated cell death *(34)*. Second, FA cells have been shown to express increased levels of IFN-α—inducible genes. IFN-γ is known to activate the (JAK1)/JAK2/STAT1) pathway, resulting in the induction of immediate-early genes that regulate apoptosis. STAT1-deficient cells are defective in the IFN-α—mediated cell death response *(79)*. FA cells also have increased expression of the IFN-α—inducible genes, *IRF1, fas,* and *MxA (80)*. This may be a selective or partial IFN-γ response, since not all IFN-α—inducible genes are upregulated in FA cells. For instance, HLA class I antigen, IP10, and STAT2 are not upregulated in FA cells. Taken together, these results suggest that the FA pathway may normally function to suppress the apoptotic response to IFN-γ. According to this model, absence of a functional FA pathway, resulting from mutations of an FA gene, may lead to an increased IFN-γ response and increased apoptosis. A model depicting this possible suppressive mechanism of the FA pathway is shown in Fig. 7.

FA cells are also hypersensitive to the apoptotic responses of the cytokine, TNFα *(35)*. This hypersensitivity may result from increased expression of nuclear factor-κB

(NF-κB)-inducible genes. Taken together, these results also suggest that the FA pathway may normally function to suppress the TNF-α—induced NF-κB response. A recent report demonstrates that the TNF-α—activated kinase, IKK2, binds to FANCA *(69)*. Whether or not IKK2 phosphorylates or affects the activity of the FA protein complex remains unknown.

8.3. Cell cycle and FA pathway

Several lines of evidence suggest that FA cells have an underlying molecular defect in cell cycle regulation. First, FA cells have a prolonged G2 transit time, which is enhanced by treatment with chemical crosslinking agents *(7,18)*. Second, the G2 arrest and reduced proliferation of FA cells can be partially corrected by overexpression of a protein, SPHAR, which is a member of the cyclin family of proteins *(81)*. Third, caffeine abrogates the G2 arrest of FA cells *(30)*. Consistent with these results, caffeine is known to constitutively activate cdc2 and to override a normal G2 cell cycle checkpoint that may be defective in FA cells. Fourth, other diseases characterized by genomic instability, including Li-Fraumeni syndrome and AT, result from abnormalities in cell cycle regulation, suggesting a possible relationship with FA. Finally, a specific interaction between the FANCC protein and the cyclin dependent kinase, cdc2 *(58)*, has been demonstrated, suggesting that FA proteins may be substrates or modifiers of the cyclin B/cdc2 complex. How these cell cycle abnormalities in FA cells result from the absence of a functional FA pathway remains unknown.

8.4. DNA repair and FA pathway

Recent studies suggest that the FA protein complex may have DNA repair activity and/or DNA-binding activity. As stated previously, FA cells have a high level of chromosome breakage, both spontaneous and induced by crosslinking agents. This increased chr instability is reflected by an overproduction of intragenic deletions. For instance, in FA cells, there is a high proportion of deletions at the endogenous HPRT locus *(82)*. Recent studies *(83)* have also suggested that FA cells are defective in error-free processing of blunt-end double-strand breaks (DSB). Studies with transfected DNA templates further demonstrate that FA cells are defective in the fidelity of specific blunt-end DNA end joining, *(84)*.

Lambert et al. *(25,85)* have isolated a nuclear protein complex with increased binding affinity for psorolen-crosslinked DNA. The complex contains a 230-kDa protein (p230), along with the FANCA and FANCC proteins *(86)*. These provocative recent studies suggest that a nuclear complex containing the FA proteins may function to bind and repair DNA that is selectively crosslinked by MMC. To this extent, a nuclear FA complex of proteins is reminiscent of the excision repair complex, composed of XP proteins, and directly involved in the repair of DNA templates containing ultraviolet light-activated thymine dimers *(87)*. Further studies are needed to explore the relative DNA-binding and possible endonuclease activities of the nuclear FA protein complex.

Given the selective sensitivity of FA cells to DNA-crosslinking agents, it is possible that FA cells have a specific defect in DNA-crosslink repair. DNA crosslinks represent a unique class of DNA lesions, capable of blocking both DNA replication and RNA transcription. Little is known about the mechanism of crosslink repair in mammalian cells. Crosslink repair has been studied in *Saccharomyces cerevisiae,* and has been shown to require both nucleotide excision repair (NER) and double-strand break repair

(DSBR) by the homologous recombination function *(88)*. For the repair of psoralen-crosslinked DNA, the NER pathway initially generates DSBs near the interstrand crosslink. Then Rad51-dependent homologous recombination function subsequently repairs the DSBs in yeast *(89)*. In *Escherichia coli,* crosslink repair appears to occur through homologous recombination, without the generation of DSB intermediates *(90)*.

Based on the requirement for DSBR in the repair of crosslinks in yeast, the FA protein complex may directly or indirectly interact with the protein complexes involved in this function. In general, DSBR can be executed by two separate, nonoverlapping pathways, the nonhomologous end-joining pathway (NHEJ) and the homologous recombination pathway (HR) (reviewed in refs. [*91* and *92*]. Accordingly, the FA protein complex may interact with the proteins in the NHEJ pathway (Ku, DNA-PK, XRCC4, and DNA ligase IV), or with proteins in the HR pathway (Rad51 family members) *(93,94)*. Future studies are needed to determine whether the FA protein complex interacts with proteins in these other repair pathways.

9. RELEVANCE OF FA PATHWAY TO OTHER HUMAN DNA REPAIR SYNDROMES AND TUMOR SUPPRESSOR PATHWAYS

The FA pathway is analogous to genetic pathways for other human chr instability syndromes, such as XP and AT. XP is an autosomal recessive disease with seven complementation groups, and mutations in XP genes result in cellular sensitivity to UV light. The protein products of at least five of the XP genes assemble in an excision repair complex that functions to repair UV-induced thymine dimers *(95–97)*. The XP protein products have variable binding affinity within the excision repair complex *(98)* and the related transcription complex, TFIIB *(99)*. Some proteins are tightly bound, and some are loosely bound to the complex. AT, and the related disorder, Nijmegens breakage syndrome (NBS), are genetic cancer susceptibility syndromes characterized by cellular sensitivity to ionizing radiation. The ATM protein is a protein kinase that regulates a pathway involved in the repair of double-strand DNA breaks, induced by ionizing irradiation. The ATM protein directly phosphorylates the tumor suppressor protein, p53 *(100)*. The NBS protein, in contrast, interacts in a complex with at least two other cellular protein (RAD50 and Mre11) *(101–103)*, and thereby presumably regulates (DSBR or chr segregation. How the ATM protein interacts with the NBS–RAD50–Mre11 complex remains unknown. In addition to these human syndromes, other gene products are known to play a direct or indirect role in DNA repair. These gene products were originally isolated, in large part, through the characterization of CHO cell mutants (XRCC gene products) *(93,104)*.

Based on these other model systems, several themes have emerged regarding these genetic chr instability syndromes. First, DNA repair results from the concerted effort of several gene products. Second, these protein products form protein complexes dedicated to the repair of specific classes of DNA adducts. Third, inducible changes in phosphorylation, assembly, or translocation of these complexes are often observed. The authors hypothesize that the FA proteins may comprise a parallel pathway, dedicated to the repair of DNA crosslinks. How mutations in any of these pathways lead to specific kinds of cancers, such as lymphomas in AT, remains mostly unexplained.

In some cases, proteins involved in DNA repair processes have been shown to co-localize, suggesting that at least some of these proteins participate in coordinated signaling pathways. For example, immunofluorescence studies have demonstrated that human Rad51, XP type G protein, and proliferating cell nuclear antigen relocalize in response to DNA damage, implicating these proteins in the repair of specific classes of DNA damage. More recently, it was shown *(105,106)* that Rad51 co-localizes with BRCA1 and BRCA2 following cellular treatment during S phase, thereby directly implicating BRCA1 in a DNA repair response. Whether the FA proteins co-localize with any of these other proteins, known to be involved in regulating chr integrity, remains to be determined. Future analysis of the FA protein complex, in the absence or presence of cellular MMC treatment, may help to resolve this question.

10. GT FOR FA

FA is a particularly attractive candidate disease for GT approaches. The genes associated with three of the major complementation groups have been cloned and sequenced, and expression of these genes in FA mutant cells has been shown to complement the mutant phenotype, and restore resistance to clastogenic agents. The FA proteins need only be expressed at low levels in cells to restore normal phenotype, but can also be expressed at supraphysiological levels without interfering with normal cellular functions *(60,71)*. This is similar to adenosine deaminase deficiency *(107–109)*, and contrasts with β-globin disorders, such as sickle cell disease and thalassemia, in which expression of the introduced gene must be closely regulated. Finally, corrected FA cells have a selective growth advantage over the corresponding parental cells in vitro and in vivo *(17,110)*, and successful correction of even a small proportion of hematopoietic stem cells may lead to substantial reconstitution of host hematopoiesis in vivo.

For successful GT of primary hematopoietic disorders, it is necessary to stably express the introduced gene in hematopoietic stem cells having extensive self-replicative potential *(111–113)*. Correction of more mature cells would lead only to transient correction of the defect. The phenotypic identity of these stem cells remains controversial, but increasing evidence based on repopulation of the *nod/scid* mouse-strain with human cells suggests that the stem cells reside in the CD34$^-$ lin$^-$ subpopulation of BM and cord blood *(114,115)*.

Retroviral vectors suitable for stably expressing introduced genes in a variety of cell types were initially described nearly 15 years ago. The majority of vectors used are based on the Moloney murine leukemia virus, into which the gene(s) of interest have been introduced by standard recombinant DNA techniques. For the production of infectious virus, the proviral genome of interest is transfected into packaging cell lines, which produce retroviral vectors, without the concomitant production of replication-competent virus. These packaging lines express a helper genome, which encodes viral structural proteins; the provirus contains the gene(s) to be transferred and the *cis*-active elements required for viral encapsidation, replication, and integration. A number of genetic modifications have been made in packaging cell lines, to minimize the likelihood of production of replication-competent virus by recombination between the proviral and helper genomes. Typical packaging cell lines produce approx 10^6 infectious viral particles per milliliter of supernatant. Infection of human hematopoietic precursors is thought to require minimum titers of 10^7, and possibly as much as 10^9 particles/mL, and supernatants must therefore be concentrated by ultrafiltration and/or ultracentrifugation *(116)*. Retroviral vectors are able to infect only actively replicating cells, and G_0 cells are therefore refractory to proviral integration.

Vectors incorporating the *FANCA, FANCC,* or *FANCG* cDNAs have been constructed, and shown to successfully correct the sensitivity of FA lymphoblast lines and fibroblasts to clastogenic agents in vitro. These vectors thus provide a simple alternative to conventional complementation assays for determining the complementation group of individual FA patients *(8,16)*. Similar vectors have been used to transduce purified human CD34+ BM cells from FA-A and FA-C patients in vitro. In-vitro-derived colonies from these cells were shown to contain the corresponding proviral DNA, using polymerase chain reaction techniques. Limited studies, in which FA-C human marrow cells infected with a FA-C retroviral vector were reintroduced into patients, failed to show engraftment of the transduced cells *(17)*. However, this area remains one of active research interest, and a number of groups are currently pursuing preclinical studies of GT in FA.

11. FUTURE DIRECTIONS

The cloning of three FA genes *(FANCC, FANCA,* and *FANCG)* has provided an unprecedented opportunity to understand the molecular basis of FA. With these genes, new approaches can be taken to explore the cellular role of the FA proteins. First, the mammalian genes may be used to identify FA gene homologs in yeast or Drosophila. A genetic dissection of the FA gene pathway in one of these organisms may implicate the pathway in a basic cellular function, such as cell growth, cell cycle, or genomic stabilization. Second, the FA genes may be used for the generation of additional mouse models of FA. Cells derived from mice deficient in the *FANCC* gene product exhibit genomic instability, similar to that observed in cell lines derived from FA patients. Similarly, knockouts of the *FANCA* and *FANCG* genes are now possible. Interactions of FA genes could be further analyzed by crossing different knockout strains. Third, the recognition that the three cloned FA proteins interact in a nuclear complex will provide the basis of future biochemical assays. For instance, it will be important to examine the nuclear complex for its ability to bind to chrs or to regulate in vitro DNA repair processes.

The availability of the FA genes will directly impact the diagnosis and therapy of FA. Mutations in FA genes can be directly screened as an adjunct diagnostic procedure to the DEB test. Retroviruses that transduce the FA genes can be used for rapid diagnosis and complementation analysis of primary cells from suspected FA patients. GT studies with autologous peripheral blood CD34 positive stem cells can be initiated for FA patients who do not have sibling-matched histocompatible donors. The observation of somatic reversion in FA provides some reason for optimism regarding the efficacy of GT. Finally, once the molecular basis of the disease is established, more rational therapies or preventive measures may be devised for patients and families with FA.

ACKNOWLEDGMENTS

Supported by National Institutes of Health grants R01-HL5725 and PO1-CA39542. I.G.H. is supported by a fellowship from the Cancer Research Institute. A.D.D. is a Scholar of the Leukemia Society of America.

REFERENCES

1. Auerbach A, Buchwald M, Joenje H. Fanconi anemia. In: Vogelstein B, Kinzler KW, eds., *Genetics of Cancer* New York: McGraw Hill. 1997:

2. D'Andrea AD, Grompe M. Molecular biology of Fanconi anemia: implications for diagnosis and therapy. *Blood* 1997; 90:1725–1736.

3. Alter BP, Young NS. The Bone marrow failure syndromes. In: Nathan DG, Oski FA, eds., *Hematology of Infancy and Childhood,* 4th ed., vol 1. Philadelphia: WB Saunders. 1993:216–316.

4. Liu J, Buchwald M, Walsh CE, Young NS. Fanconi anemia and novel strategies for therapy. *Blood* 1994; 84:3995–4007.

5. Alter BP. Fanconi's anemia and malignancies. *Am J Hematol* 1996; 53:99–110.

6. Auerbach AD. Fanconi anemia diagnosis and the diepoxybutane (DEB) test. *Exp Hematol* 1993; 21:731–733.

7. Kaiser TN, Lojewski A, Dougherty C, Juergens L, Sahar E, Latt SA. Flow cytometric characterization of the response of Fanconi's anemia cells to mitomycin C treatment. *Cytometry* 1982; 2:291–297.

8. Pulsipher M, Kupfer GM, Naf D, Suliman A, Lee J-S, Jakobs P, et al. Subtyping analysis of Fanconi anemia by immunoblotting and retroviral gene transfer. *Mol Med* 1998; 4:468–479.

9. Lo Ten Foe J, et al. Somatic mosaicism in Fanconi anemia: molecular basis and clinical significance. *Eur J Hum Genet* 1997; 5:137–148.

10. Ellis NA, Lennon DJ, Proytcheva M, Alhadeff B, Henderson EE, German J. Somatic intragenic recombination within the mutated locus BLM cn correct the ihgih sister-chromatid exchange phenotype of Bloom syndrome cells. *Am J Hum Genet* 1995; 57:1019–1027.

11. Hirschhorn R, Yang DR, Puck JM, Huie ML, Jiang CK, Kurlandsky LE. Spontaneous in vivo reversion to normal of an inherited mutation in a patient with adenosine deaminase deficiency. *Nat Genet* 1996; 13:290–295.

12. Rackoff WR, Orazi A, Robinson CA, Cooper RJ, Alter BP, Freedman MH, Harris RE, Williams DA. Prolonged administration of granulocyte colony-stimulating factor (Filgrastim) to patients with Fanconi anemia: a pilot study. *Blood* 1996; 88:1588–1593.

13. Guinan EC, Gillio AP. Cytokine therapy of Fanconi anemia (FA). Workshop on molecular, cellular, and clinical aspects of Fanconi anemia. Bethesda, MD 1992; abstract.

14. Guinan EC, Lopez KD, Huhn RD, Felser JM, Nathan DG. Evaluation of granulocyte-macrophage colony-stimulating factor for treatment of pancytopenia in children with Fanconi anemia. *J Pediatr* 1994; 124:144.

15. Fu KL, Foe JR, Joenje H, Rao KW, Liu JM, Walsh CE. Functional correction of Fanconi anemia group A hematopoietic cells by retroviral gene transfer. *Blood* 1997; 90:3296–3303.

16. Fu KL, Thuss PC, Fujino T, Digweed M, Liu JM, Walsh CE. Retroviral gene transfer for the assignment of Fanconi anemia (FA) patients to a FA complementation group. *Hum Genet* 1998; 102:166–169.

17. Liu JM, Young NS, Walsh CE, Cottler-Fox M, Carter C, Dunbar C, Barrett AJ, Emmons R. Retroviral mediated gene transfer of the Fanconi anemia complementation group C gene to hematopoietic progenitors of group C patients. *Hum Gene Ther* 1997; 8:1715–1730.

18. Kubbies M, Schindler D, Hoehn H, Schinzel A, Rabinovich PS. Endogenous blockage and delay of the chromosome cycle despite normal recruitment and growth phase explain poor proliferation and frequent edomitosis in Fanconi anemia cells. *Am J Hum Genet* 1985; 37:1022–1030.

19. Schlindler D, Hoehn H. Fanconi anemia mutation causes cellular susceptibility to ambient oxygen. *Am J Hum Genet* 1988; 43:429–435.

20. Korkina LG, Samochatova EV, A.A. M, Suslova TB, Cheremisina ZP, Afanaslev IB. Release of active oxygen radicals by leukocytes of Fanconi anemia patients. *J Leukoc Biol* 1992; 52:357.

21. Gille JJP, Wortelboer HM, Joenje H. Antioxidant status of Fanconi anemia fibroblasts. *Hum Genet* 1987; 77:28–31.

22. Joenje H, Eriksson AW, Frants RR, Arwert F, Houween B. Erythrocyte superoxide dismutase deficiency in Fanconi's anaemia. *Lancet* 1978; 1:204.

23. Bigelow SB, Rary JM, Bender MA. G2 chromosomal radiosenitivity in Fanconi's anamia. *Mutat Res* 1979; 63:189.

24. Lambert MW, Tsongalis GJ, Lambert WC, Hang B, Parrish DD. Defective DNA endonuclease activities in Fanconi's anemia cells, complementation groups A and B. *Mutat Res* 1992; 273:57.

25. Kumaresan KR, Hang B, Lambert MW. Human endonucleolytic incision of DNA 3' and 5' to a site-directed psoralen monoadduct and interstrand crosslinu. *J Biol Chem* 1995; 270:30709–30716..

26. Lambert M, Tsongalis G, Lambert W, Parrish D. Correction of the DNA repair defect in Fanconi anemia complementation groups A ·and D cells. *Biochem Biophys Res Cummun* 1997; 230:587–591.

27. Fujiwara Y, Tatsumi M, Sasaki MS. Cross-link repair in human cells and its possible defect in Fanconi's anaemia cells. *Biochim Biophys Acta* 1977; 113:635–649.

28. Fujiwara Y. Defective repair of mitomycin C crosslinks in Fanconi's anemia and loss in confluent normal human and Xerodermia pigmentosum cells. *Biochim Biophys Acta* 1982; 699:217–225.

29. Wang J, Youssoufian H, Lo Ten Foe JR, Devetten M, Cumming RC, Buchwald M, Liu JM. Overexpression of the Fanconi anemia group C gene (FAC) protects hematopoietic progenitors from FAS-mediated apoptosis. *Blood* 1996; 88:548a.

30. Kupfer GM, D'Andrea AD. The effect of the Fanconi anemia polypeptide, FAC, upon p53 induction and G2 checkpoint regulation. *Blood* 1996; 88:1019–1025.

31. Rosselli F, Ridet A, Soussi T, Duchaud E, Alapetite C, Moustacchi E. p53-dependent pathway of radio-induced apoptosis is altered in Fanconi anemia. *Oncogene* 1995; 10:9–17.

32. Daneshbod-Skibba G, Martin J, Shahidi NT. Myeloid and erythroid colony growth in non-anaemic patients with Fanconi's anaemia. *Br J Haematol* 1980; 44:33.

33. Alter BP, Knobloch ME, Weinberg RS. Erythropoiesis in Fanconi's Anemia. *Blood* 1991; 78:602–608.

34. Whitney MA, Royle G, Low MJ, Kelly MA, Axthelm MK, Reifsteck C, et al. Germ cell defects and hematopoietic hypersensitivity to gamma-interferon in mice with a targeted disruption of the Fanconi anemia C gene. *Blood* 1996; 88:49–58.

35. Rathbun R, Faulkner G, Ostroski M, Christianson T, Hughes G, Jones G, et al. Inactivation of the Fanconi Anemia group C gene augments interferon-gamma induced apoptotic responses in hematopoietic cells. *Blood* 1997; 90:974.

36. Buchwald M. Complementation groups: one or more per gene. *Nature Genet* 1995; 11:228–230.

37. Duckworth-Rysiecki G, Cornish K, Clarke CA, Buchwald M. Identification of two complementation groups in Fanconi anemia. *Somatic Cell Mol Gen* 1985; 11:35–41.

38. Joenje H. Fanconi anemia complementation groups in Germany and The Netherlands. *Hum Genet* 1996; 97:280–282.

39. Joenje H, Ten FL, Oostra A, Berkel CV, Rooimans M, Schroeder S, et al. Classification of Fanconi anemia patients by complementation analysis: evidence for a fifth genetic subtype. *Blood* 1995; 86:2156.

40. Strathdee CA, Duncan AMV, Buchwald M. Evidence for at least four Fanconi anemia genes including FACC on chromosome 9. *Nat Genet* 1992; 1:196–198.

41. Strathdee CA, Duncan AM, Buchwald M. *Nat Genet* 1992; 1:196–198.

42. Pronk JC, et al. Localization of the Fanconi anemia complementation group A gene to chromosome 16q24.3. *Nat Genet* 1995; 11:338–340.

43. Whitney M, et al. Microcell mediated chromosome transfer maps the Fanconi anemia group D gene to chromosome 3p. *Nat Genet* 1995; 11:341–343.

44. Saar K, Schindler D, Wegner R-D, Reis A, Wienker TF, Hoehn H, et al. Localisation of a Fanconi anaemia gene to chromosome 9p. *Eur J Hum Genet* 1998; 6:501–508.

45. Strathdee CA, Gavish H, Shannon WR, Buchwald M. Cloning of cDNAs for Fanconi's anaemia by functional complementation. *Nature* 1992; 356:763–767.

46. Gibson RA, Buchwald M, Roberts RG, Mathew CG. Characterisation of the exon structure of the Fanconi anaemia group C gene by vectorette PCR. *Hum Mol Genet* 1993; 2:35–38.

47. Wevrick R, Clarke CA, Buchwald M. Cloning and analysis of the murine Fanconi anemia group C cDNA. *Hum Mol Genet* 1993; 2:655–662.

48. Ching-Ying-Wong J, Alon N, Buchwald M. Cloning of the bovine and rat Fanconi anemia group C cDNA. *Mamm Genome* 1997; 8:522–525.

49. Whitney MA, Saito H, Jakobs PM, Gibson RA, Moses RE, Grompe M. A common mutation in the FACC gene causes Fanconi anemia in Ashkenazi Jews. *Nat Genet* 1993; 4:202–205.

50. Gillio AP, Verlander PC, Batish SD, Giampietro PF, Auerbach AD. Phenotypic consequences of mutations in the Fanconi anemia FAC gene: an international Fanconi anemia registry study. *Blood* 1997; 90:105–110.

51. Verlander PC, Lin JD, Udono MU, Q Zhang, Gibson RA, Mathew CG, Auerbach AD. Mutation analysis of the Fanconi anemia gene FACC. *Am J Hum Genet* 1994; 54:595.

52. Kupfer GM, Naf D, Suliman A, Pulsipher M, D'Andrea AD. The Fanconi anemia proteins, FAA and FAC, interact to form a nuclear complex. *Nat Genet* 1997; 17:487–490.

53. Krasnoshtein F, Buchwald M. Developmental expression of the Fac gene correlates with congenital defects in Fanconi anemia patients. *Hum Mol Genet* 1996; 5:85–93.

54. Youssoufian H. Localization of Fanconi anemia C protein to the cytoplasm of mammalian cells. *Proc Natl Acad Sci USA* 1994; 91:7975–7979.

55. Yamashita T, Barber DL, Zhu Y, Wu N, D'Andrea AD. The Fanconi anemia polypeptide FACC is localized to the cytoplasm. *Proc Natl Acad Sci USA* 1994; 91:6712–6716.

56. Garcia-Higuera I, Kuang Y, Naf D, Wasik J, D'Andrea AD. The Fanconi anemia proteins, FANCA, FANCC, and FANCG/XRCC9 interact in a functional nuclear complex. *Mol Cell Biol* 1999; 19:4866–4873

57. Hoshino T, Youssoufian H, Wang J, Devetten MP, Iwata N, Kajigaya S, Liu JM. The molecular chaperone GRP94 binds to a central domain within the group C Fanconi anemia protein. *Blood* 1996; 88:1734a.

58. Kupfer G, Yamashita T, Naf D, Suliman A, Asano S, D'Andrea AD. Fanconi anemia protein, FAC, binds to the cyclin-dependent kinase, cdc2. *Blood* 1997; 90:1047–1054.

59. Kruyt FAE, Hoshino T, Liu JM, Joseph P, Jaiswal AK, Youssoufian H. Abnormal microsomal detoxification implicated in Fanconi anemia group C by interaction of the FAC protein with NADPH cytochrome P450 reductase. *Blood* 1998; 92:3050–3056.

60. Lo Ten Foe JR, Rooimans MA, Bosnoyan-Collins L, et al. Expression cloning of a cDNA for the major Fanconi anemia gene, *FAA*. *Nat Genet* 1996; 14:320–323.

61. The Fanconi Anemia/Breast Cancer Consortium. Positional cloning of the Fanconi anaemia group A gene. *Nat Genet* 1996; 14:324–328.

62. Ianzano L, D'Apolito M, Centra M, Savino M, Levran O, Auerbach AD, et al. The genomic organization of the Fanconi anemia group A (FAA) gene. *Genomics* 1997; 41:309–314.

63. D'Apolito M, Zelante L, Savoia A. Molecular basis of Fanconi anemia. *Haematologica* 1998; 83:533–542.

64. Naf D, Kupfer GM, Suliman A, Lambert K, D'Andrea AD. Functional activity of the Fanconi anemia protein, FAA, requires FAC binding and nuclear localization. *Mol Cell Biol* 1998; 18:5952–5960.

65. Kruyt FAE, Waisfisz Q, Dijkmans LM, Hermsen MAJA, Youssoufian H, Arwert F, Joenje H. Cytoplasmic localization of a functionally active Fanconi Anemia group A-green fluorescent protein chimera in human 293 cells. *Blood* 1997; 90:3288–3295.

66. Levran O, Erlich T, Magdalena N, Gregory JJ, Batish SD, Verlander PC, Auerbach AD. Sequence variation in the Fanconi anemia gene FAA. *Proc Natl Acad Sci USA* 1997; 94:13,051–13,056.

67. Wijker M, Morgan NV, Herterich S, van Berkel CGM, Tipping AJ, Gross HJ, et al. Heterogeneous spectrum of mutations in the Fanconi anaemia group A gene. *Eur J Hum Genet* 1998; 7(1):52–59.

68. Savino M, Ianzano L, Strippoli P, Ramenghi U, Arslanian A, Bagnara GP, et al. Mutations of the Fanconi anemia group A (FAA) gene in Italian patients. *Am J Hum Genet* 1997; 61:1246–1253.

69. Otsuki T, Mercurio F, Liu JM. The Fanconi Anemia Group A Product is a component of the IκB kinase signalsome complex involved in NF-κB activation. *Blood* 1998;92, abst. 2847:

70. Yamashita T, Kupfer GM, Naf D, Suliman A, Joenje H, Asano S, D'Andrea AD. Fanconi anemia pathway requires FAA phosphorylation and FAA/FAC nuclear accumulation. *Proc Natl Acad Sci USA* 1998; 95:13,085–13,090.

71. de Winter JP, Waisfisz Q, Rooimans MA, van Berkel CGM, Bosnoyan-Collins L, Alon N. Fanconi anaemia group G gene is identical with human XRCC9. *Nat Genet* 1998; 20:281–283.

72. Liu N, Lamerdin JE, Tucker JD, Zhou Z-Q, Walter CA, Albala JS, Busch DB, Thompson LH. The human *XRCC9* gene corrects chromosomal instability and mutagen sensitivities in CHO UV40 cells. *Proc Natl Acad Sci USA* 1997; 94:9232–9237.

73. Chen M, Tomkins DJ, Auerbach W, McKerlie C, Youssoufian H, Liu L, et al. Inactivation of Fac in mice produces inducible chromosomal instability and reduced fertility reminiscent of Fanconi anaemia. *Nat Genet* 1996; 12:448–451.

74. Carreau M, Gan OI, Liu L, Doedens M, McKerlie C, Dick JE, Buchwald M. Bone marrow failure in the Fanconi anemia group C mouse model after DNA damage. *Blood* 1998; 91:2737–2744.

75. Youssoufian H, Auerbach AD, Verlander PC, Steimle V, Mach B. Identification of cytosolic proteins that bind to the Fanconi anemia complementation group C polypeptide *in vitro*. *J Biol Chem* 1995; 270:9876–9882.

76. Hoshino T, Wang J, Devetten MP, Iwata N, Kajigaya S, Wise RJ, Liu JM, Youssoufian H. Molecular chaperone GRP94 binds to the Fanconi anemia group C protein and regulates its intracellular expression. *Blood* 1998; 91:4379–4386.

77. Hoatlin ME, Christianson TA, Keeble WW, Hammond AT, Zhi Y, Heinrich MC, Tower PA, Bagby GC, Jr. The Fanconi anemia group C gene product is located in both the nucleus and cytoplasm of human cells. *Blood* 1998; 91:1418–1425.

78. Cumming RC, Liu JM, Youssoufian H, Buchwald M. Suppression of apoptosis in hematopoietic factor-dependent progenitor cell lines by expression of the *FAC* gene. *Blood* 1996; 88:4558–4567.

79. Kumar A, Commane M, Flickinger TW, Horvath CM, Stark GR. Defective TNF-—induced apoptosis in STAT1-Null cells due to low constitutive levels of caspases. *Science* 1997; 278:1630–1635.

80. Li Y, Youssoufian H. MxA overexpression reveals a common genetic link in four Fanconi anemia complementation groups. *J Clin Invest* 1997; 100:2873–2880.

81. Digweed M, Gunthert U, Schneider R, Seyschab H, Friedl R, Sperling K. Irreversible repression of DNA synthesis in Fanconi anemia cells is alleviated by the product of a novel cyclin-related gene. *Mol Cell Biol* 1995; 15:305–314.

82. Papadopoulo D, Guillouf C, Mohnrenweiser H, Moustacchi E. Hypomutability in Fanconi anemia cells is associated with increased deletion frequency at the HPRT locus. *Proc Natl Acad Sci USA* 1990; 87:8383–8387.

83. Escarceller M, Rousset S, Moustacchi E, Papadopoulo D. The fidelity of double strand breaks processing is impaired in complementation groups B and D of Fanconi anemia, a genetic instability syndrome. *Somatic Cell Mol Genet* 1997; 23:401–411.

84. Escarceller M, Buchwald M, Singleton BK, Jeggo PA, Jackson SP, Moustacchi E, Papadopoulo D. Fanconi anemia C gene product plays a role in the fidelity of blunt DNA end-joining. *J Mol Biol* 1998; 279:375–385.

85. Hang B, Yeung AT, Lambert MW. A damage-recognition protein which binds to DNA containing interstrand cross-links is absent or defective in Fanconi anemia, complementation group A, cells. *Nucleic-Acids Res* 1993; 21:4187.

86. Lambert MW, Walsh CE, D'Andrea AD, McMahon LW. A 230 kD DNA repair protein which is deficient in Fanconi Anemia complementation group A and C cells forms a complex with the FAA and FAC proteins in the nucleus. *Blood* 1998; 92.

87. van Vuuren AJ, Appeldoorn E, Odijk H, Yasui A, Jaspers NGJ, Bootsma D, Hoeijmakers JHJ. Evidence for a repair enzyme complex involving ERCC1 and complementing activities of ERCC4, ERCC11 and xeroderma pigmentosum group F. *EMBO J* 1993; 12:3693–3701.

88. Jachymczyk WJ, von Borstel RC, Mowat MR, Hastings PJ. Repair of interstrand cross-links in DNA of Saccharomyces cerevisiae requires two systems for DNA repair: the RAD3 system and the RAD51 system. *Mol Gen Genet* 1981; 182:196–205.

89. Magana-Schwencke N, Henriques JA, Chanet R, Moustacchi E. The fate of 8-methoxypsoralen photoinduced crosslinks in nuclear and mitochondrial yeast DNA: comparison of wild-type and repair-deficient strains. *Proc Natl Acad Sci USA* 1982; 79:1722–1726.

90. Friedberg E. Relationships between DNA repair and transcription. *Ann Rev Biochem* 1996; 65:15–42.

91. Chu G. Double strand break repair. *J Biol Chem* 1997; 272:24,097–24,100.

92. Weaver DT. What to do at an end: DNA double-strand-break repair. *Trends Genet* 1995; 11:388–392.

93. Liu N, Lamerdin JE, Tebbs RS, Schild D, Tucker JD, Shen MR, et al. XRCC2 and XRCC3, new human Rad51-family members, promote chromosome stability and protect against DNA cross-links and other damages. *Mol Cell* 1998; 1:783–793.

94. Baumann P, West SC. Role of the human RAD51 protein in homologous recombination and double-stranded break repair. *TIBS* 1998; 23:247–251.

95. Arrand JE, Bone NM, Johnson RT. Molecular cloning and characterization of a mamalian excision repair gene that partially restores UV resistance to xeroderma pigmentosum complementation group D cells. *Proc Natl Acad Sci USA* 1989; 86:6997–7001.

96. Legerski R, Peterson C. Expression cloning of a human DNA repair gene involved in xeroderma pigmentosum group C. *Nature* 1992; 359:70–73.

97. Park MS, Knauf JA, Pendergrass SH, Coulon CH, Strniste GF, Marrone BL, MacInnes MA. Ultraviolet-induced movement of the human DNA repair protein, *Xeroderma pigmentosum* type G, in the nucleus. *Proc Natl Acad Sci USA* 1996; 93:8368–8373.

98. Wood RD. DNA repair in eukaryotes. *Annu Rev Biochem* 1996; 65:135–167.

99. Schaeffer L, Moncollin V, Roy R, Staub A, Mezzina M, Sarasin A, et al. The ERCC2/DNA repair protein is associated with the class II BTF2/TFIIH trascription factor. *EMBO J* 1994; 13:2388–2392.

100. Banin S, Moyal L, Shieh S, Taya Y, Anderson C, Chessa L, et al. Enhance phosphorylation of p53 by ATM in response to DNA damage. *Science* 1998; 281:1674–1677.

101. Carney JP, Maser RS, Olivares H, Davis EM, Le Beau M, Yates JR III, et al. The hMre11/hRad50 protein complex and nijmegen breakage syndrome: linkage of double-strand break repair to the cellular DNA damage response. *Cell* 1998; 93:477–486.
102. Dolganov GM, Maser RS, Novikov A, Tosto L, Chong S, Bressan DA, Petrini JHJ. Human Rad50 is physically associated with hMre11: identification of a conserved multiprotein complex implicated in recombinational DNA repair. *Mol Cell Biol* 1996; 16:4832–4841.
103. Maser RS, Monsen KJ, Nelms BE, Petrini JHJ. hMre11 and hRad50 nuclear foci are induced during the normal cellular response to DNA double strand breaks. *Mol Cell Biol* 1997; 17:6087–6096.
104. Zdzienicka MZ. Mammalian X ray sensitive mutants: a tool for the elucidation of the cellular response to ionizing radiation. *Cancer Surv* 1996; 28:281–293.
105. Scully R, Chen J, Plug A, Xiao Y, Weaver D, Feunteun J, Ashley T, Livingston DM. Association of BRCA1 with Rad51 in mitotic and meiotic cells. *Cell* 1997; 88:265–275.
106. Scully R, Chen J, Ochs RL, Keegan K, Hoekstra M, Feunteun J, Livingston DM. Dynamic changes of BRCA1 subnuclear location and phosphorylation state are initiated by DNA damage. *Cell* 1997; 90:425–435.
107. Blaese RM, Culver KW, Chang L, Anderson WF, Mullen C, Nienhuis A, et al. Treatment of severe combined immunodeficiency disease (SCID) due to adenosine deaminase deficiency with CD34+ selected autologous peripheral blood cells transduced with a human ADA gene. Amendment to clinical research project, Project 90-C-195, January 10, 1992. *Hum Gene Ther* 1993; 4:521–527.
108. Mullen CA, Snitzer K, Culver KW, Morgan RA, Anderson WF, Blaese RM. Molecular analysis of T lymphocyte-directed gene therapy for adenosine deaminase deficiency: long-term expression in vivo of genes introduced with a retroviral vector. *Hum Gene Ther* 1996; 7:1123–1129.
109. Riviere I, Brose K, Mulligan RC. Effects of retroviral vectors design on expression of human adenosine deaminase in murine bone marrow transplant recipients engrafted with genetically modified cells. *Proc Natl Acad Sci USA* 1995; 92:6733–6737.
110. Fu K-L, Lo Ten Foe JR, Joenje H, Rao KW, Liu JM, Walsh CE. Functional correction of Fanconi anemia group A hematopoietic cells by retroviral gene transfer. *Blood* 1997; 90:3296–3303.
111. Anderson WF. Human gene therapy. *Nature* 1998; 392:25–30.
112. Kohn DB, Parkman R. Gene therapy for newborns. *FASEB J* 1997; 11:635–639.
113. Mulligan RC. The basic science of gene therapy. *Science* 1993; 260:926–932.
114. Bhatia M, Bonnet D, Murdoch B, Gan OI, Dick JE. A newly discovered class of human hematopoietic cells with SCID-repopulating activity. *Nat Med* 1998; 4:1038–1045.
115. Goodell MA, Brose K, Paradis G, Conner AS, Mulligan RC. Isolation and functional properties of murine hematopoietic stem cells that are replicating in vivo. *J Exp Med* 1996; 183:1797–1806.
116. Ory D, Neugeboren B, Mulligan R. A stable human-derived packaging cell line for production of high-titer retrovirus/vesicular stomatitis virus G pseudotypes. *Proc Natl Acad Sci USA* 1996; 93:11,400–11,406.
117. Kupfer G, Naf D, Garcia-Higuera I, Wasik J, Cheng A, Yamashita T, Tipping A. A patient-derived mutant form of the Fanconi anemia protein, FANCA, is defective in nuclear accumulation. *Exp Hem* 1999; 27:587–593.

Index